FM 3-05.212
MCRP 3-11.3A

Special Forces Waterborne Operations

September 2009

DISTRIBUTION RESTRICTION: Distribution authorized to U.S. Government agencies and their contractors only to protect technical or operational information from automatic dissemination under the International Exchange Program or by other means. This determination was made on 29 May 2009. Other requests for this document must be referred to Commander, United States Army John F. Kennedy Special Warfare Center and School, ATTN: AOJK-DTD-SF, Fort Bragg, NC 28310-9610.

DESTRUCTION NOTICE: Destroy by any method that will prevent disclosure of contents or reconstruction of the document.

FOREIGN DISCLOSURE RESTRICTION (FD 6): This publication has been reviewed by the product developers in coordination with the United States Army John F. Kennedy Special Warfare Center and School foreign disclosure authority. This product is releasable to students from foreign countries on a case-by-case basis only.

Headquarters, Department of the Army

This publication is available at
Army Knowledge Online (www.us.army.mil) and
General Dennis J. Reimer Training and Doctrine
Digital Library at (www.train.army.mil).

*FM 3-05.212
MCRP 3-11.3A

Field Manual
No. 3-05.212
Marine Corps Reference
Publication No. 3-11.3A

Headquarters
Department of the Army
Washington, DC, 30 September 2009

Special Forces Waterborne Operations

Contents

		Page
	PREFACE	x
Chapter 1	MISSION PLANNING	1-1
	Perspective	1-1
	METT-TC	1-2
	Phases of Waterborne Operations	1-3
	Beach Landing Site Selection Criteria	1-6
	Beach Selection	1-8
	Planning Considerations for Waterborne Operations	1-8
	Reverse Planning Considerations	1-8
	Waterborne Employment Considerations	1-10
	Environmental Constraints	1-11
	Exfiltration	1-11
	Unit Training and Capabilities	1-13
Chapter 2	ENVIRONMENTAL FACTORS	2-1
	Wind	2-1
	Clouds	2-3
	Air Masses and Fronts	2-8

DISTRIBUTION RESTRICTION: Distribution authorized to U.S. Government agencies and their contractors only to protect technical or operational information from automatic dissemination under the International Exchange Program or by other means. This determination was made on 29 May 2009. Other requests for this document must be referred to Commander, United States Army John F. Kennedy Special Warfare Center and School, ATTN: AOJK-DTD-SF, Fort Bragg, NC 28310-9610.

DESTRUCTION NOTICE: Destroy by any method that will prevent disclosure of contents or reconstruction of the document.

FOREIGN DISCLOSURE RESTRICTION (FD 6): This publication has been reviewed by the product developers in coordination with the United States Army John F. Kennedy Special Warfare Center and School foreign disclosure authority. This product is releasable to students from foreign countries on a case-by-case basis only.

*This publication supersedes FM 3-05.212, 31 August 2004.

Contents

	Surface Weather Analysis Charts	2-10
	Forecasting	2-12
	Waves	2-14
	Ice	2-17
Chapter 3	**TIDES AND TIDAL CURRENTS**	**3-1**
	Tides	3-1
	Tide Tables	3-3
	Currents	3-6
	Offset Navigation	3-13
Chapter 4	**NAUTICAL CHARTS AND PUBLICATIONS**	**4-1**
	Nautical Charts	4-1
	Accuracy of Charts	4-16
	Publications	4-16
	Aids to Navigation	4-19
	Nautical Compass	4-23
Chapter 5	**SMALL BOAT NAVIGATION**	**5-1**
	Navigation	5-1
	Basic Navigation Equipment	5-4
	Navigation Planning	5-9
	Dead Reckoning	5-13
	Piloting	5-24
Chapter 6	**COMBAT RUBBER RAIDING CRAFT OPERATIONS**	**6-1**
	Maritime Proponency	6-1
	Responsibilities	6-1
	Training	6-2
	Operations and Planning	6-5
	Environmental Personnel Limitations	6-10
Chapter 7	**INFLATABLE BOATS**	**7-1**
	Basic Boat-Handling Skills	7-1
	Zodiac F-470 Inflatable Boat	7-11
Chapter 8	**KAYAKS**	**8-1**
	Military Background Data	8-1
	Terminology	8-4
	Kayak Operational Techniques	8-7
	Kayak Considerations	8-14
Chapter 9	**OVER-THE-BEACH OPERATIONS**	**9-1**
	Beach Landing Site Procedures	9-1
	Scout Swimmers	9-4
	Signals	9-7
	Detachment Infiltration Swimming	9-9
	Waterproofing	9-11
	Bundle Transporting	9-12
Chapter 10	**COMBAT DIVING AND SWIMMING OPERATIONS**	**10-1**
	Organization and Duties	10-1
	Operations	10-7

Contents

	Safety	10-9
	Equipment	10-12
	Specific Manning Requirements	10-14
Chapter 11	**OPEN-CIRCUIT DIVING**	**11-1**
	Open-Circuit Dive Operations	11-1
	Diving Injuries and Medicine	11-3
	Dive Tables	11-7
	Equipment and Basic Techniques	11-13
	Diving Considerations	11-17
	Contaminated Water Diving	11-20
	Altitude Diving	11-22
	Cold Weather Diving	11-23
	Dive Operations Planning	11-28
Chapter 12	**CLOSED-CIRCUIT DIVING**	**12-1**
	Safety Considerations	12-1
	Description	12-3
	Function	12-4
	Components	12-4
	Characteristics	12-11
	Operational Duration of the MK 25 UBA	12-12
	Predive Procedures	12-12
	Postdive Procedures	12-16
	Malfunction Procedures	12-17
	Closed-Circuit Oxygen Exposure Limits	12-18
	Water Entry and Descent	12-22
	Transport and Storage of Prepared UBA	12-24
Chapter 13	**SURFACE INFILTRATION**	**13-1**
	Mother Ship Operations	13-1
	CRRC Mission Planning Factors	13-3
	Planning for a Rendezvous at Sea	13-7
	Raiding Craft Detectability and Classification Countermeasures	13-7
Chapter 14	**SUBMARINE OPERATIONS**	**14-1**
	Submarines	14-1
	Debarkation	14-3
	Withdrawal by Submarine	14-11
	The Future	14-13
Chapter 15	**RIVERINE OPERATIONS**	**15-1**
	Environment	15-1
	Approach to Planning	15-2
	Organization and Command	15-2
	Security Responsibilities	15-3
	Concept of Riverine Operations	15-3
	Riverine Movement	15-6
	River Piloting	15-7
	Emergencies	15-11

Contents

Chapter 16	**AIRCRAFT IN SUPPORT OF MARITIME OPERATIONS**	**16-1**
	Delivery Methods	16-1
	Recovery Operations	16-6
	Over-the-Horizon Computations	16-7
Chapter 17	**SEARCH DIVES**	**17-1**
	Searches	17-1
	Search Methods	17-6
	Ship Bottom Search	17-15
	River Searches	17-15
Appendix A	**WEIGHTS, MEASURES, AND CONVERSION TABLES**	**A-1**
Appendix B	**DESCRIPTION OF SUBJECT AREA AND CRITICAL TASK LIST**	**B-1**
Appendix C	**EQUIPMENT MAINTENANCE**	**C-1**
Appendix D	**DIVING SUPERVISOR PERSONNEL INSPECTION**	**D-1**
Appendix E	**MK 25 MOD 2 PREDIVE AND POSTDIVE CHECKLISTS**	**E-1**
Appendix F	**DIVER PROPULSION DEVICE**	**F-1**
	GLOSSARY	**Glossary-1**
	REFERENCES	**References-1**
	INDEX	**Index-1**

Figures

Figure 1-1. Waterborne operations planning phases 1-3
Figure 1-2. Typical beach profile 1-9
Figure 1-3. Nominal planning ranges 1-10
Figure 1-4. Required capabilities 1-14
Figure 2-1. Synoptic weather chart 2-2
Figure 2-2. High- and low-pressure wind circulation 2-3
Figure 2-3. Cirrus clouds 2-4
Figure 2-4. Cirrus and cirrostratus clouds 2-4
Figure 2-5. Altocumulus clouds 2-5
Figure 2-6. Altostratus clouds 2-5
Figure 2-7. Stratocumulus clouds 2-6
Figure 2-8. Cumulus clouds 2-6
Figure 2-9. Cumulonimbus clouds 2-7
Figure 2-10. Cross-section of a cold front 2-9
Figure 2-11. Cross-section of a warm front 2-10
Figure 2-12. Cross-section of an occluded front 2-10
Figure 2-13. Sample weather chart 2-11
Figure 2-14. International Weather Code 2-12
Figure 2-15. Generalizations for weather forecasting 2-13

Contents

Figure 2-16. Wave characteristics 2-15
Figure 2-17. Types of breaking waves 2-16
Figure 3-1. High and low tides 3-2
Figure 3-2. Neap and spring tides 3-2
Figure 3-3. Table 1 excerpt, Daily Tide Predictions (Key West, Florida) 3-4
Figure 3-4. Table 2 excerpt, Tidal Differences (Florida Subordinate Stations) 3-5
Figure 3-5. Plotting a tide curve 3-7
Figure 3-6. Table 1 excerpt, Tidal Currents (Key West, Florida) 3-10
Figure 3-7. Table 2 excerpt, Tidal Currents (Florida Subordinate Stations) 3-11
Figure 3-8. Plotting an approximate tidal current curve 3-14
Figure 3-9. Offset navigation techniques 3-16
Figure 4-1. Example of nautical chart information 4-3
Figure 4-2. Navigation guides 4-4
Figure 4-3. Example of a logarithmic speed scale 4-5
Figure 4-4. Example of a time–speed–distance logarithmic scale 4-5
Figure 4-5. The nautical slide rule 4-6
Figure 4-6. Global latitude and longitude lines 4-8
Figure 4-7. Conversion factors 4-9
Figure 4-8. The compass rose 4-10
Figure 4-9. Determining distance on charts 4-11
Figure 4-10. Symbols for lighthouses and other fixed lights 4-12
Figure 4-11. Symbols for ranges and day beacons 4-13
Figure 4-12. Range markers 4-14
Figure 4-13. Prominent landmarks 4-14
Figure 4-14. Symbols for wrecks, rocks, and reefs 4-15
Figure 4-15. Bottom characteristics 4-15
Figure 4-16. Structures 4-15
Figure 4-17. Coastlines 4-16
Figure 4-18. Guides within the "Sailing Directions" 4-18
Figure 4-19. Buoys 4-20
Figure 4-20. Light characteristics 4-21
Figure 4-21. Additional light characteristics 4-22
Figure 4-22. Types of ranges 4-22
Figure 4-23. Nominal and luminous ranges 4-23
Figure 4-24. Compass direction 4-24
Figure 4-25. Guide to applying variation and deviation 4-26
Figure 4-26. Determining variation and deviation 4-27
Figure 5-1. Navigation terms 5-2
Figure 5-2. Plotting board 5-7
Figure 5-3. Navigation tools 5-8
Figure 5-4. Standardized plotting symbols 5-10

Contents

Figure 5-5. Example of labeling a dead-reckoning plot 5-11
Figure 5-6. Basic equations 5-15
Figure 5-7. D/ST formula examples 5-15
Figure 5-8. Terms used in current sailing 5-17
Figure 5-9. Steps required to prepare a vector diagram 5-18
Figure 5-10. Example of actual current triangle with labeling 5-19
Figure 5-11. Opposing currents 5-20
Figure 5-12. Currents in different directions 5-21
Figure 5-13. Determination of actual course and speed made good 5-22
Figure 5-14. Determination of course to steer and speed to run 5-23
Figure 5-15. Determination of course to steer and actual speed 5-23
Figure 5-16. Determining a fix 5-26
Figure 6-1. CRRC surf passage table 6-9
Figure 7-1. Expedient harness 7-4
Figure 7-2. Capsize line 7-5
Figure 7-3. Lifting sling 7-6
Figure 7-4. Boat navigation console 7-7
Figure 7-5. Zodiac F-470 inflatable boat 7-12
Figure 7-6. Outboard motor 7-12
Figure 7-7. Maritime operations gear 7-18
Figure 8-1. Typical expedition-grade folding kayak 8-3
Figure 8-2. Characteristics of the folding kayak 8-3
Figure 8-3. Selected kayaking terms 8-4
Figure 8-4. Basic ocean items 8-5
Figure 8-5. Drogue and sea chute 8-6
Figure 8-6. War chest 8-6
Figure 8-7. Folding kayak components (Long-Haul Commando shown) 8-7
Figure 8-8. Bow and stern completed assemblies 8-8
Figure 8-9. Launching and landing methods 8-10
Figure 8-10. Basic kayaking strokes 8-11
Figure 8-11. Capsize drills 8-12
Figure 8-12. Pump and reentry method 8-13
Figure 8-13. Manuevering in surf conditions 8-13
Figure 8-14. Maintenance procedures for folding kayaks 8-16
Figure 9-1. Swim line formations 9-9
Figure 11-1. Residual nitrogen timetable 11-11
Figure 11-2. Repetitive dive flowchart 11-12
Figure 11-3. Repetitive dive profile 11-12
Figure 11-4. Hyperthermia symptoms 11-21
Figure 11-5. Water temperature protection chart 11-24
Figure 11-6. Wet and dry suits 11-25

Figure 12-1. Safety guidelines 12-1
Figure 12-2. MK 25 UBA component locations (as worn) 12-6
Figure 12-3. Gas flow path 12-11
Figure 12-4. Oxygen consumption rates 12-13
Figure 12-5. Filling canister with absorbent 12-14
Figure 12-6. Inhalation and exhalation hoses 12-15
Figure 12-7. Breathing bag attached to scrubber canister 12-15
Figure 12-8. Breathing bag attached to demand valve 12-16
Figure 12-9. Transit-with-excursion limits 12-19
Figure 12-10. Predive purge procedures 12-23
Figure 13-1. Military shipping considerations 13-2
Figure 13-2. Civilian shipping considerations 13-2
Figure 13-3. Desirable characteristics for choosing a craft 13-3
Figure 13-4. CRRC mission planning checklist 13-4
Figure 13-5. Sample rendezvous plan 13-8
Figure 14-1. Sequence of lockout procedures 14-5
Figure 14-2. Submarine lockout rigging 14-6
Figure 14-3. Dry-deck shelter (without fairing) 14-9
Figure 14-4. Dry-deck shelter (underway) 14-9
Figure 14-5. Fulton recovery 14-11
Figure 15-1. Determining intelligence requirements 15-6
Figure 15-2. Indicators for river piloting 15-9
Figure 16-1. Team deployment using air assets 16-2
Figure 16-2. Flight in relation to radar line of sight 16-9
Figure 17-1. Standard troop-leading procedure 17-1
Figure 17-2. Estimating circle of probable error 17-4
Figure 17-3. Probable river search sites 17-4
Figure 17-4. Characteristics of tended-line search 17-6
Figure 17-5. Tended-line search 17-7
Figure 17-6. Open water tended-line search 17-7
Figure 17-7. Characteristics of circle line search 17-8
Figure 17-8. Circle line search 17-8
Figure 17-9. Characteristics of running jackstay search 17-9
Figure 17-10. Running jackstay search 17-9
Figure 17-11. Modified running jackstay (open water) search 17-10
Figure 17-12. Characteristics of checkerboard jackstay search 17-10
Figure 17-13. Checkerboard search 17-11
Figure 17-14. Modified checkerboard search 17-12
Figure 17-15. Characteristics of in-line search 17-12
Figure 17-16. In-line search 17-12
Figure 17-17. Characteristics of expanding-box and reciprocal-pattern searches 17-13

Contents

Figure 17-18. Expanding-box search 17-13
Figure 17-19. Reciprocal-pattern search 17-14
Figure 17-20. Characteristics of towed search 17-14
Figure 17-21. Characteristics of ship bottom search 17-16
Figure 17-22. Search zone setup 17-18
Figure 17-23. River search procedures using a rope bridge 17-19
Figure 17-24. River search procedures using a boat 17-20
Figure D-1. Table-top inspection D-4
Figure F-1. Diver propulsion device characteristics F-1
Figure F-2. DPD collapsed for launch or storage F-2
Figure F-3. Hull subsections F-2
Figure F-4. DPD view port F-3
Figure F-5. DPD interior view F-4
Figure F-6. Cargo area F-4
Figure F-7. Deck features F-5
Figure F-8. T-bar F-6
Figure F-9. Thruster shroud F-6
Figure F-10. Lithium-ion battery F-7
Figure F-11. Battery box F-8
Figure F-12. Thruster F-8
Figure F-13. Throttle controls F-9
Figure F-14. Battery status indicator F-9
Figure F-15. Rear deck mounting screws F-10
Figure F-16. Stern plane and bow plane F-11
Figure F-17. Depth and steering control yoke F-12
Figure F-18. Gauge console F-12
Figure F-19. Universal mounting brackets F-13
Figure F-20. Performance specifications F-13
Figure F-21. Freshwater foam wedge F-15
Figure F-22. Extended range battery components F-16
Figure F-23. Installation of dummy plug to thruster port connector F-16
Figure F-24. Cable plugs into the E-link port and routes through bulkhead F-17
Figure F-25. Foam cradle sits in the hull F-17
Figure F-26. Battery is secured F-18
Figure F-27. Preoperational checklist F-19
Figure F-28. Neutral buoyancy unit F-21
Figure F-29. Postoperational checklist F-23
Figure F-30. Troubleshooting guide F-24

Tables

Table 1-1. METT-TC analysis 1-2
Table 2-1. The Beaufort Wind Scale 2-2
Table 3-1. Constructing a tide table matrix 3-6
Table 3-2. Constructing a tidal current matrix 3-13
Table 5-1. Example of a navigation plan 5-12
Table 5-2. Current types, effects, and corrections for navigational errors 5-17
Table 7-1. Zodiac F-470 specifications 7-13
Table 7-2. CRRC characteristics (inflatables) 7-13
Table 7-3. Sample fuel usage for CRRCs 7-17
Table 11-1. Minimum standards of compressed air 11-3
Table 11-2. Injuries during descent 11-4
Table 11-3. Injuries during ascent 11-5
Table 11-4. Injuries caused by indirect effects of pressure 11-6
Table 11-5. No-decompression limits and repetitive group designators for
no-decompression air dives 11-9
Table 12-1. MK 25 UBA components list 12-5
Table 12-2. Transit-with-excursion limits table 12-19
Table 12-3. Single-depth oxygen exposure limits 12-20
Table 12-4. Adjusted oxygen exposure limits for successive oxygen dives 12-21
Table 16-1. Distance to horizon 16-8
Table 16-2. Distance to point of tangency 16-8
Table A-1. Linear measure A-1
Table A-2. Liquid measure A-1
Table A-3. Weight A-1
Table A-4. Square measure A-2
Table A-5. Cubic measure A-2
Table A-6. Temperature A-2
Table A-7. Approximate conversion factors A-2
Table A-8. Area A-3
Table A-9. Volume A-3
Table A-10. Capacity A-3
Table A-11. Statute miles to kilometers and nautical miles A-3
Table A-12. Nautical miles to kilometers and statute miles A-4
Table A-13. Kilometers to statute and nautical miles A-4
Table A-14. Yards to meters A-5
Table A-15. Meters to yards A-5

Preface

Field Manual (FM) 3-05.212/Marine Corps Reference Publication (MCRP) 3-11.3A presents a series of concise, proven techniques and guidelines that are essential to safe and successful waterborne operations. It provides a consolidated reference for training and employing Special Forces (SF) and Marines in all types of waterborne operations.

PURPOSE

FM 3-05.212 with United States Army Special Operations Command (USASOC) Regulation 350-20, *USASOC Dive Program*, and Army Regulation (AR) 611-75, *Management of Army Divers*, will assist commanders and staffs in selecting and preparing personnel to execute waterborne operations. This manual with AR 611-75 will also assist commanders in selecting and preparing personnel to attend the SF underwater operations course.

SCOPE

Waterborne operations involve the employment of forces from waterborne platforms to meet objectives ashore; they may be in support of or independent from amphibious operations. This FM covers the entire spectrum of waterborne operations as they apply to United States (U.S.) Army SF and U.S. Marines. It provides detailed operational planning considerations for the following: small boat operations, surface swimming operations, dive operations, submarine operations (SUBOPS), riverine operations, and air operations (AIROPS). It also provides information on a variety of associated subject areas. The most common measurements used in waterborne operations are expressed throughout the text and in many cases are U.S. standard terms rather than metric. Appendix A consists of conversion tables that may be used when mission requirements or environments change.

APPLICABILITY

This publication applies to the Active Army, Marine Corps, Army National Guard (ARNG)/Army National Guard of the United States, and the United States Army Reserve (USAR) unless otherwise stated.

ADMINISTRATIVE INFORMATION

Unless this publication states otherwise, masculine nouns and pronouns do not refer exclusively to men. The proponent of this manual is the United States Army John F. Kennedy Special Warfare Center and School (USAJFKSWCS). Submit comments and recommended changes on DA Form 2028 (Recommended Changes to Publications and Blank Forms) directly to Commander, USAJFKSWCS, ATTN: AOJK-DTD-SF, Fort Bragg, NC 28310-9610.

Chapter 1

Mission Planning

SF units must conduct a detailed mission analysis to determine an appropriate method of infiltration. SF maritime operations (MAROPS) are one of many options available to a commander to infiltrate and exfiltrate a detachment into (or out of) a designated area of operations (AO) for the purpose of executing any of the SF missions. A thorough understanding of all factors impacting waterborne-related missions is essential due to the inherently higher levels of risk associated with even the most routine waterborne operations. The objective of this chapter is to outline mission, enemy, terrain and weather, troops and support available—time available and civil considerations (METT-TC) and planning considerations needed to successfully execute all types of waterborne operations.

PERSPECTIVE

1-1. Over five-eighths of the earth's surface is covered by water. SF units conduct waterborne operations to infiltrate a designated target area from these water-covered areas. Regardless of whether an AO has exposed coastlines, coastal river junctions, or harbors, many areas will have large rivers, lakes, canals, or other inland waterways located within their boundaries. These maritime or riverine features represent exploitable characteristics that special operations forces (SOF) can use to their advantage.

1-2. Throughout the world, military equipment sales programs by various countries have caused a proliferation of advanced radar technologies and coastal air defense systems. As governments seek to exercise greater control over their indigenous territories, these facilities and the associated risks to SOF units' normal means of infiltration and exfiltration will continue to increase. Waterborne infiltrations and exfiltrations keep the high-value air, surface, or subsurface infiltration assets offshore and out of the detection and threat ranges of coastal defense installations.

1-3. Waterborne operations are a means to an end. Despite the increased use of sophisticated coastal surveillance systems and active surface and air interdiction efforts on the part of regional governments, local inhabitants continue to engage in various illicit activities; for example, smuggling and illegal fishing just as their forebears have for centuries. In many parts of the world, long coastlines, extensive waterways, and small undermanned and underpaid navies exacerbate these problems. The clandestine nature and high probability of success for these illicit operations mirror SF units' requirements for successful infiltrations into, or exfiltrations out of, potentially hostile areas of responsibility (AORs).

1-4. Personnel involved in waterborne operations require extensive knowledge of hydrography, meteorology, navigation, and MAROPS. They must be able to conduct realistic premission training, gather information, plan, rehearse, and use the appropriate waterborne operations technique to accomplish their assigned mission.

1-5. Commanders use sophisticated techniques and equipment to conduct waterborne operations. When used correctly, these factors give commanders another means to move teams and influence the battlefield. The skills and techniques used in waterborne operations are equally applicable to all SF core tasks, especially direct action (DA), special reconnaissance (SR), unconventional warfare (UW), and foreign internal defense (FID). A wide variety of peacetime and combat missions may require waterborne operations capability.

METT-TC

1-6. The successful execution of any operation requires thorough and detailed planning. Mission planning begins with a detailed analysis of METT-TC (Table 1-1).

Table 1-1. METT-TC analysis

Factors	Questions
Mission	<ul><li>What is the specified mission?</li><li>What are the implied missions?</li><li>Is the objective within range of a coastline or waterway that a detachment can use as a route to infiltrate the AOR?</li></ul>
Enemy	<ul><li>What are the ranges of enemy detection capabilities?</li><li>Does the enemy have the ability to detect or interdict conventional infiltration techniques; for example, static-line parachute, high-altitude low-opening (HALO) parachute technique, or air-mobile insertion?</li><li>Does the enemy have an extensive coastline or internal waterways that are vulnerable to a maritime or waterborne infiltration?</li><li>Is the enemy capable of detecting and interdicting U.S. maritime assets outside the range of the detachment's intermediate (or final) transport method?</li></ul>
Terrain and Weather	<ul><li>Is the weather conducive to a maritime operation?</li><li>Are sea states within the operational limits of the detachment's chosen infiltration method?</li><li>Are any weather patterns expected in the AO that might cause unacceptable sea conditions?</li><li>What is the percentage of illumination? Can the detachment infiltrate during periods of limited visibility and illumination?</li><li>Are the tides and currents favorable during the desired infiltration window?</li><li>Are there suitable primary and alternate beach landing sites (BLSs) available?</li><li>What are the surf conditions at the tentative BLS?</li></ul>
Troops and Support Available	<ul><li>Does the detachment have the training and experience to successfully execute the selected infiltration method? Is additional training required?</li><li>Does the detachment have time to do the required training and rehearsals?</li><li>What equipment is required to execute the primary mission?</li><li>Does the detachment have the means to transport the required equipment?</li><li>Does the equipment require special waterproofing? Can it be submerged?</li><li>Does the equipment have special handling and storage requirements during the transit portion of the mission?</li></ul>
Time Available	<ul><li>Is the mission time-critical?</li><li>Given the complexity of MAROPS and the extended transit times required, is time available for the detachment to plan, rehearse, and execute a long-range infiltration?</li><li>Is the mission flexible enough to allow time for a MAROPS infiltration window that is dependent on meteorological and oceanographic conditions?</li><li>How far is it from the BLS and enemy detection capabilities to an over the horizon (OTH) debarkation point?<ul><li>Can the detachment make it from the debarkation point to the BLS and complete actions during the hours of darkness?</li><li>Will the detachment have the time and fuel to move to an alternate BLS if the primary site is unsuitable or compromised?</li></ul></li></ul>
Civil Considerations	<ul><li>Can the operation be executed clandestinely so that the civilian populace is unaware of it?</li><li>If the operation is compromised, what will be the repercussions to the local populace?</li><li>If the detachment is receiving local support, is there a risk of reprisals?</li></ul>

Mission Planning

1-7. A thorough METT-TC analysis concentrating on the above questions will determine if water infiltration is appropriate. The detachment must then complete the remainder of the mission planning process.

PHASES OF WATERBORNE OPERATIONS

1-8. To aid the Special Forces operational detachment (SFOD) in planning and executing a maritime operation, waterborne operations are divided into seven phases. Figure 1-1 shows each phase and the following paragraphs provide the details for each.

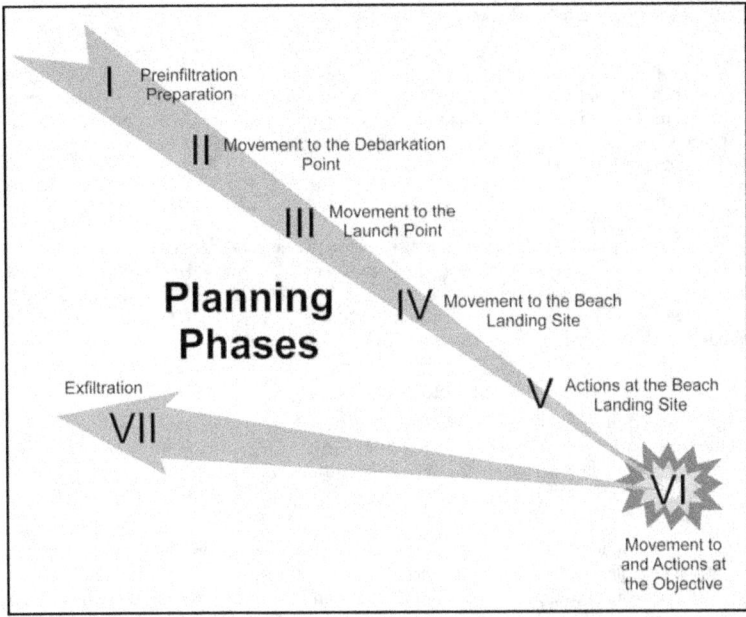

Figure 1-1. Waterborne operations planning phases

PHASE I—PREINFILTRATION PREPARATION

1-9. Preinfiltration preparation starts with preparing an estimate of the situation. The SFOD uses the military decision making process (MDMP) to identify critical nodes in the mission and to develop courses of action (COAs) to address them. During this phase, the detachment will plan the mission, prepare orders, conduct briefbacks and training, prepare equipment, and conduct inspections and rehearsals.

1-10. There are three methods of waterborne infiltration, which may be used individually or combined. They are aircraft, surface craft, or submarine. The key to the premission process is determining the infiltration method that the detachment will use from its home station to the debarkation point. The type of infiltration will determine the planning and preparations required during Phase I. The method selected depends upon the mission, time, and distance.

Aircraft

1-11. There is a wide variety of fixed- and rotary-wing aircraft that can deliver infiltrating detachments. Aircraft provide the most practical and rapid means of transporting infiltrating detachments to a debarkation point. They may be delivered by—

- Conventional static-line parachute techniques.
- Military free-fall parachute techniques.

Chapter 1

- Water landing by amphibious aircraft.
- Helocast or free-drop from helicopter.

Surface Craft

1-12. Combat swimmers, divers, and boat teams may also use surface craft to reach the debarkation point. This method is generally considered the most efficient means of transporting infiltrating detachments. Surface craft can transport large quantities of supplies and are relatively unaffected by weather up to the point of debarkation. This method also allows the detachment to conduct operational planning and preparations en route to the debarkation point.

Submarine

1-13. The submarine is an excellent insertion vehicle. However, it requires extensive training and coordination to use effectively. SF commanders may prefer this method to insert combat divers, swimmers, or boat teams because it is a very effective means for a clandestine insertion or recovery. (Chapter 14 further explains SUBOPS.)

Combinations of Transport

1-14. In addition to aircraft, surface craft, and submarines, commanders may use various combinations of methods to deliver detachments. These combinations can enable commanders to create deception, increase the range of the mission, or decrease the time required for transport. To increase the range of the mission, operational elements may fly to a staging area and transfer to an aircraft carrier for transport to the debarkation point. Once there, aircraft can transport the detachment to the next leg of the infiltration.

PHASE II—MOVEMENT TO THE DEBARKATION POINT

1-15. In Phase II of infiltration, the primary transport craft moves the swimmers to the selected debarkation point where they begin the initial transit to the BLS. All personnel should understand the following procedures while en route in the aircraft and surface craft.

Aircraft

1-16. During flight to the debarkation point, the pilot informs the troop commander of the aircraft's location and any changes in the infiltration plan. All personnel must know their relative position along the route in the event of a bailout, an emergency abort, or enemy action. As the aircraft nears the debarkation point, the pilot provides preplanned time warnings for final personnel and equipment preparations. The infiltrating detachments arrive at the debarkation point by parachute operations, helocasting, or water landing.

1-17. In most situations, the delivery of swimmers by parachute during periods of limited or reduced visibility onto unmarked water drop zones (DZs) requires the use of the computed air release point (CARP). Parachutists exit the aircraft on the pilot's command at a point designated by the navigator, and attempt to group in the air as close as possible. Once in the water, they sink their air items and sterilize the DZ by sinking everything to ensure that nothing floats ashore. The detachment then begins movement to the BLS or launch point (LP).

1-18. Enemy activity will usually prevent detachments parachuting within swimming range of a BLS. Normally a detachment will jump with inflatable boats, assemble on the boats, and motor the extended distances involved. When using inflatable assault boats, personnel should rig them for airdrop before loading the aircraft. At the command to exit the aircraft, personnel drop the rigged boatloads before they exit. Upon landing, predesignated personnel swim to the boats and begin derigging; they dispose of air items and prepare the boats for movement to the BLS. Remaining personnel link up with their buddy team members and rendezvous with the boats. (Chapters 6 and 7 discuss small-boat information.)

Surface Craft

1-19. Designated personnel continue mission planning and preparations en route to the debarkation point. As they receive new or revised intelligence, they update the infiltration plan and inform all personnel of changes in the situation or mission. While en route, all personnel conduct rehearsals for each phase of the infiltration. The detachment should thoroughly rehearse debarkation procedures with those members of the crew assigned specific duties for the operation.

1-20. Upon arrival at the debarkation point, all personnel man their respective debarkation stations. The troop commander will be oriented in relation to the BLS and briefed on sea and surf conditions. Debarkation begins on orders from the vessel commander.

1-21. One of the simplest methods of debarkation is for swimmers to slip over the side of the vessel into the water; debarkation from large surface craft may require the use of landing nets. Once in the water, the swimmers link up with their swim buddies and begin movement to the BLS. Swimmers should only use this method when the hydrography and enemy situation permit debarkation within swimming range of the beach.

1-22. Swimmers can use a variation of this method when the situation requires debarkation beyond the swimmers' range. Here, the swimmers debark over the side of the surface craft and move by small watercraft to an intermediate transfer point. From there, the detachment uses inflatable assault boats to transport the swimmers to a LP within swimming range of the beach.

Phase III—Movement to the Launch Point

1-23. After the detachment reaches its debarkation point, it must conduct a transit to the LP where it begins BLS procedures. The extended distances involved in any OTH operation require an intermediate transport system. The detachment normally uses inflatable assault boats and kayaks for this purpose. This equipment gives the detachment a planning range far greater than any combat swimmer operation and allows a larger mission equipment load. Inflatable assault boats or kayaks are especially useful when—

- Enemy air or coastal defense systems, hydrographic characteristics, or navigational errors prevent aircraft, surface craft, or submarines from delivering swimmers within range of the BLS.
- Tide, current, and wind conditions could cause swimmer fatigue.

1-24. Using boats requires detailed planning, extensive rehearsals, and consistent training that should take place in Phase I. Personnel must have a thorough knowledge of small-boat handling, dead reckoning or offset navigation techniques, and tide and current computations. They must maintain strict noise and light discipline throughout the operation, and adhere to the principles of patrolling as adapted from land operations, to include modifications of the movement formations.

1-25. The transit portion of the mission ends at the LP. A LP is the location where scout swimmers are released or combat divers and swimmers enter the water to begin the detachment's infiltration swim. The enemy situation, hydrography, the type of equipment used, and the detachment's ability to swim the required distance determine how far the LP should be from the BLS. The LP can also be synonymous with the boat holding pool. Although well-conditioned personnel are capable of swimming extended distances, the maximum planning range for surface swimmers should not exceed 3,000 meters. Subsurface infiltrations should not exceed 1,500 meters (open circuit [O/C]) or 2,000 meters (closed circuit [C/C]) unless diver propulsion vehicles (DPVs) are used.

1-26. Once at the LP, the detachment sinks the inflatable boat, a designated member returns the boat to the primary surface transport vessel, or the detachment caches the boat at the BLS.

Phase IV—Movement to the Beach Landing Site

1-27. After the infiltrating detachment enters the water at the launch point, its first action is to link up on a swim line and then swim along the predesignated azimuth toward the primary BLS. The equipment, enemy situation, and level of unit training determine whether they swim on the surface or subsurface. In some cases, a combination of surface and subsurface swimming techniques, commonly called turtle backing, can greatly extend the swimmers' range. However, it requires a closed-circuit underwater breathing apparatus (UBA), a special compass, and increased unit training.

1-28. As the detachment approaches the BLS, the commander signals swimmers to halt outside the surf and small arms fire zones. At this holding area a predesignated security team (scout swimmers) moves on the surface (or subsurface) to the BLS to determine the enemy situation. Once the scout swimmers determine that the site is clear of the enemy, they signal the remaining swimmers to come ashore. While waiting for the detachment, the scout swimmers establish left, right, and farside security (as appropriate) at the limits of visibility from the landing point. When the remaining swimmers reach the BLS, they sterilize it to obscure tracks, and remove equipment and debris to conceal evidence of the detachment's presence. The detachment then moves immediately to the assembly area or cache point.

1-29. In some situations, a reception committee may be present to assist with the infiltration by marking the BLS, providing guides, and transporting accompanying supplies. The detachment must coordinate the initial contact with the reception committee before infiltration. The scout swimmers conducting the security check of the BLS make contact with the reception committee. Whenever the mission includes contact with unknown agencies, the detachment should make sure the communications (COMM) plan includes a "duress" signal from the scout swimmers to the detachment and an abort contingency for the detachment.

PHASE V—ACTIONS AT THE BEACH LANDING SITE

1-30. At the BLS, the commander immediately accounts for his personnel and equipment. Infiltrating detachments are especially vulnerable to enemy action during this phase. To minimize the chances of detection the detachment must complete landing operations, clear the beach as rapidly as possible, and move directly inshore to the preselected assembly area. This area must provide cover and concealment and facilitate subsequent movement to the objective area. The swimmers cache equipment not required for the inland operation. If a reception committee is present, its leader coordinates personnel movement and provides current intelligence on the enemy situation. Finally, the detachment sterilizes the assembly area and begins moving to the objective area.

PHASE VI—MOVEMENT TO AND ACTIONS AT THE OBJECTIVE

1-31. Infiltration of the target area is considered complete in Phase V. All actions pertinent to infiltration, such as cache of infiltration equipment, or movement away from infiltration site, are completed before Phase VI. This phase is conducted tactically regardless of the infiltration method. This phase consists of movement to the objective, actions on the objective, and withdrawal from the objective. At times, the infiltration method facilitates the actions on the objective itself, such as waterborne missions. In the cases where the infiltration method facilitates the mission, this phase begins in the objective rally point.

PHASE VII—EXFILTRATION

1-32. Exfiltration planning considerations for waterborne operations require the same preparations, tactics, and techniques as for infiltration. However, in exfiltration, the planners are primarily concerned with recovery methods. Distances involved in exfiltration usually require additional means of transport. Aircraft, surface craft, submarines, or various combinations of these three methods, can be used to recover infiltration swimmers. In addition, inflatable assault boats may be needed for seriously wounded personnel or for equipment.

BEACH LANDING SITE SELECTION CRITERIA

1-33. Before selecting a specific waterborne infiltration method, the SFOD examines the objective, the BLS, and the shipping and air assets available.

1-34. The BLS is of primary importance because it must facilitate and support the inland objective. The factors that determine the feasibility of a proposed BLS include hydrography, enemy situation, navigational aids (navaids), distance from the debarkation point to the BLS, beach vegetation and conditions, and routes of egress from the objective.

1-35. Hydrography deals with measuring and studying oceans and rivers along with their marginal land areas. Hydrographic conditions of interest to waterborne operations are ocean depth, beach depth, beach gradient, tide and surf conditions, and beach composition. Detailed hydrographic information can be obtained from a wide variety of sources, to include—
- Surf observation reports.
- BLS reports.

Mission Planning

- Nautical charts.
- Tide and current data.
- Aerial photoreconnaissance.
- Hydrographic surveys.

1-36. Hydrographic surveys are the most accurate means of obtaining detailed and specific information concerning the BLS. These are normally prepared for large-scale amphibious landings by United States Marine Corps Special Operations Forces (MARSOF) units or U.S. Navy sea-air-land (SEAL) units. Army terrain teams may also be able to supply applicable hydrographic information. Up-to-date nautical charts, tide and current data, and photo reconnaissance provide the minimum data needed.

1-37. The enemy's situation and capabilities have a direct impact on the location of a proposed BLS. Ideally, a BLS would be located away from enemy observation and fields of fire. If this is not possible, the SFOD must consider the enemy's ability to locate and interdict the infiltration route based on—

- Location of coastal patrol boats.
- Coastal fortifications.
- Security outposts.
- Defensive obstacles.
- Artillery, mortar, and missile positions.
- Armor and mechanized units.
- Major communication and command posts.

1-38. Regardless of the hydrography or enemy situation, any SFOD must be skilled in basic navigation techniques when operating in oceans, large lakes, and rivers. Without basic skills, the infiltration detachment is severely limited and will not be able to apply the full range of techniques associated with waterborne operations. These skills include—

- Using dead-reckoning navigation.
- Reading nautical charts.
- Interpreting tidal data.
- Understanding international marine traffic buoys.
- Computing tidal current data for offset navigation.

1-39. The detachment determines what infiltration technique to use by the distance from the debarkation point to the BLS. For example, if the debarkation point from the primary vessel is 20 nm offshore, the detachment uses a combat rubber raiding craft (CRRC) to bring them within range of the beach. Airdrops or helocasts can also deliver the detachment to a debarkation point. If the enemy situation permits, the same delivery systems could transport the detachment within swimming range of the BLS. Regardless of the combinations of delivery craft used, the distance should be no more than what the operational detachment can swim or travel by boat within the time available. The detachment should always use the simplest method possible to reduce potential equipment failures.

1-40. Hydrographic surveys, charts, photoreconnaissance, and tidal data outline the beach and vegetation conditions but may not match conditions at the BLS. The detachment should immediately determine the available cover and concealment once they arrive at the BLS. The BLS is the first danger area where combat swimmers are exposed to enemy observation and fire. The faster the detachment clears the beach of men and equipment, the less the chance of enemy detection. Therefore, the distance to the first covered and concealed location is extremely important. Personnel should also keep combat loads to an absolute minimum to make rapid movement off the beach easier.

1-41. Once the detachment completes infiltration, all members will need to move with the same stealth inland to the target area. Planners must evaluate the suitability of a BLS by carefully analyzing the approach and withdrawal routes to the objective. Since the detachment transported the equipment ashore during infiltration, they can use this same equipment in exfiltration. Waterborne exfiltrations involve the same planning, preparation, and techniques required for infiltration. Offshore recoveries can be made by sea, air, or a combination of both.

BEACH SELECTION

1-42. In addition to the factors that form a suitable BLS already discussed, the SFOD also needs the latest detailed information to accurately plan for securing the BLS and for troop movement. These additional factors include—

- Offshore navigation conditions to include weather, tides, currents, surf, winds, and underwater obstacles or mines.
- Security precautions.
- Distance from the launch point to the BLS.
- Access to the hinterland.

The detachment must also plan for an alternate BLS in case, upon arrival, the primary BLS is unsuitable or compromised. This adjustment may be the result of a delay in the operation, unexpected sea conditions, or a change in the enemy situation. The selection of an alternate BLS must meet the same mission requirements as the primary BLS. However, the alternate BLS should have a different azimuth orientation to counter any adverse weather effects on the primary BLS.

PLANNING CONSIDERATIONS FOR WATERBORNE OPERATIONS

1-43. After approving a COA and determining waterborne operations to be the primary means of infiltration, the SFOD commander divides the waterborne operation into four major phases. These are—

- Preinfiltration preparation.
- Movement to the debarkation point.
- Movement to the launch point.
- Movement to the beach landing site.
- Actions at the beach landing site.
- Movement to and actions at the objective.
- Exfiltration.

1-44. The detachment should start at a common focal point. The easiest way to divide the operation is the transition point where the water meets the land. The focal point for both the waterborne infiltration and the land movement to the objective is at the BLS. Therefore, planning tasks for the waterborne infiltration end once the SFOD reaches the onshore assembly area. Consequently, the land movement to the objective starts once the detachment occupies the assembly area.

1-45. There are a few areas of waterborne planning that are vitally important for mission success. First, the infiltration should take place at night to provide the stealth and secrecy that the detachment needs. Second, the environmental factors produced by tides and currents must be suitable for successful infiltration. Therefore, the detachment must accurately plan the mission time to satisfy these two requirements.

REVERSE PLANNING CONSIDERATIONS

1-46. Personnel should use the reverse planning sequence to allow enough time for the most important phases of the operation to take place during the most favorable conditions. This sequence also enables the detachment to place emphasis on areas that are most critical to mission success. The following paragraphs are areas where additional time may be required and should be taken into consideration when planning.

LAUNCH POINT

1-47. The launch point is the designated point where swimmers exit the boat within swimming range of the beach. In small-boat operations where the SFOD is taking watercraft all the way onto the beach, the launch point is where designated scout swimmers exit the boat to survey and secure the BLS before the main element arrives. The launch point should be no farther than 2 nm offshore to prevent unduly fatiguing the swimmers, and no closer than 500 yards (small arms range). Two nm is the visible horizon for an observer six feet tall, standing on the beach. At this range, a CRRC with seated passengers is below the observer's

Mission Planning

line of sight. Additional factors limiting visibility, such as percent of illumination, sea state, and weather conditions, may allow the launch point to be moved much closer to the BLS. Personnel waiting offshore must also consider their ability to see features on shore so they can maintain position relative to the BLS in case the swimmers must return to the boats. Based on the tactical situation, swimmers can also designate the launch point as a noise abatement area in which they maintain strict noise and light discipline.

TRANSFER POINT

1-48. The transfer point is where the detachment transitions from the initial infiltration platform to an intermediate platform. Commanders use a transfer point when the primary support vessel cannot deliver the SFOD within range of the debarkation point due to hydrographic conditions or the danger of enemy detection. Depending upon the complexity of the infiltration plan, there may be several transfer points.

DEBARKATION POINT

1-49. The debarkation point is the location where the SFOD departs the mother ship and begins movement to the BLS. The detachment's debarkation location is extremely important. It must start from a known location to execute dead-reckoning navigation. This method applies to surface craft as well as aircraft debarkation. During mission planning, detachment members are given specific tasks to conduct at the debarkation point and rehearse debarkation procedures. At the debarkation point, the vessel commander orients the detachment to the BLS with a final update on weather and sea conditions.

1-50. Equally important as the reverse planning sequence is the ability to gather detailed information necessary to put together a highly accurate infiltration profile. The "typical beach profile" depicted in Figure 1-2 identifies portions of the ocean and marginal land areas of interest to waterborne operations. Any information that the detachment is lacking about one of these areas can be requested by name from the intelligence unit providing terrain analysis support. Once acquired, detachment personnel can use this information to update nautical charts. They can also use the identified changes in beach profile as checkpoints or phase lines when planning and coordinating troop movements. The terms used in identifying these areas of interest are standardized throughout military and civilian agencies, allowing for coordination between joint and sister Service commands.

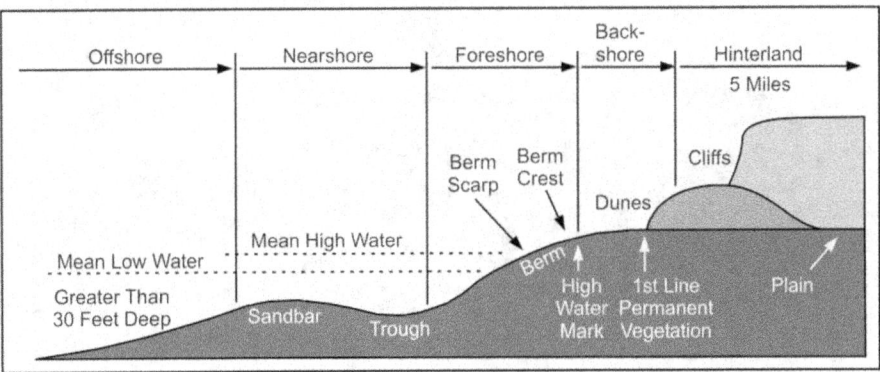

Figure 1-2. Typical beach profile

ROUTE SELECTION

1-51. The route selected by the SFOD depends primarily upon three elements: enemy situation, vessel capabilities, and vessel limitations. The detachment plans the route from the debarkation point to the BLS, taking advantage of all known navaids and checkpoints en route. After selecting the route, the detachment will prepare a navigation plan detailing time, distance, speed, course to steer, intended track (ITR), and navaids used to verify position. The SFOD should exercise caution when using navaids. During periods of conflict, they may be unavailable or even inaccurate and tampered with.

WATERBORNE EMPLOYMENT CONSIDERATIONS

1-52. If the detachment is fully trained, they can successfully execute the three methods of waterborne infiltration. Personnel must be able to calculate the time required to transit from release/launch/transfer point to the BLS for all types of infiltration methods. Figure 1-3 shows the nominal planning ranges. The time required can be calculated using the D = SxT formula, where D equals the distance in nautical miles (nm), S equals the speed in knots (kts), and T equals the time in hours. This formula calculates distance covered from average speed x time elapsed.

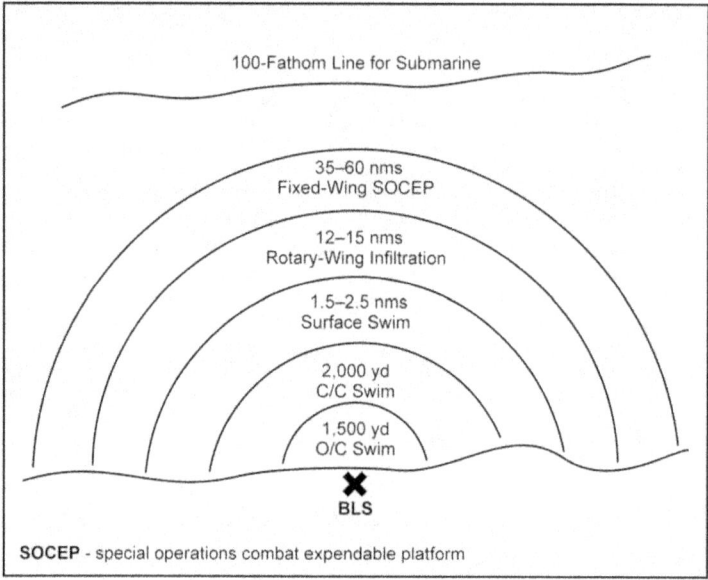

Figure 1-3. Nominal planning ranges

1-53. For infiltration planning purposes, the average swim speed for both surface and subsurface swimming is about 1 kt, equal to 1 nm per hour on the surface of the water. Ideally, swims—especially underwater swims—are conducted at 3 minutes per 100 yards. This equals 2,000 yards per hour or about 1 nm per hour (1 nm = 2024.25 yards). This measure is true for both open- and closed-circuit swimming. This speed must not be exceeded when swimming closed-circuit due to the high probability of oxygen toxicity. Units should practice attaining this swim speed by using the 100-yard pace line. By using 1 kt as a planning speed, swimmers can easily apply calculations to nautical charts and other related maritime data.

> *Note.* The speed and direction of the current must be taken into consideration when planning the average time to swim to the BLS. If a swimmer is swimming into a .5 nm-per-hour current his swim speed is only .5 nm per hour. However, if the swimmer is swimming with a .5 nm-per-hour current, his speed would be 1.5 nm per hour.

1-54. In addition to planning the amount of time required for a combat swimmer or diver to travel from the launch point to the BLS, the detachment also plans for the amount of time required to travel from the debarkation point to the launch point. Personnel use the same formula (D = ST) to determine the travel time, regardless of the type of inflatable small boat and engine size. They can best determine this time during training, by running the boat a measured nautical mile course over the ocean and recording the time. The small boat should be fully combat-loaded whenever checking for speed. Combat loads also include the total amount of fuel needed for the planned infiltration distance.

1-55. The detachment can establish a realistic mission capability once they determine the total travel time from the debarkation point to the BLS. The detachment can then execute an accurate, time-phased, and coordinated waterborne operation as long as they maintain a reasonable training sustainment program.

ENVIRONMENTAL CONSTRAINTS

1-56. Environmental elements pose an immediate threat to the infiltration plan. The detachment must compensate for these to ensure successful completion of the operation. Two of the environmental elements that can impede or halt the conduct of waterborne operations are the height of the tide and the speed and direction of the tidal current. The detachment can predict both the tide height and tidal current with great accuracy by using Tide Tables and Tidal Current Tables that are published annually by the National Oceanic and Atmospheric Administration (NOAA). Chapter 3 describes techniques for calculating tides, tidal currents, and the steps to compensate for their effects.

1-57. Other factors that may negatively influence a detachment's ability to conduct a maritime infiltration include wind, waves, and weather. The prudent MAROPS detachment will study these factors as they pertain to basic seamanship. Chapter 2 explains in detail additional meteorological and oceanographic data that will influence a detachment's decision-making process.

EXFILTRATION

1-58. The two main exfiltration methods used for waterborne exfiltration are small boats and surface swimming. Commanders generally use these methods to deliver the exfiltrating detachment from the beach departure site to a rendezvous point (RP) at sea. The most important part of exfiltration is the at-sea rendezvous (Chapter 13). The detachment is then either transferred to, or picked up by, one of the following secondary exfiltration methods:
- Submarine.
- Larger vessel.
- Rotary-wing aircraft.

1-59. In a submarine linkup, the boat teams and swimmers will do a wet deck or dry deck recovery. Divers will do a subsurface lock-in, wet deck, or dry deck recovery, depending on tactical considerations. Although a wet or dry deck recovery is simplest, a subsurface recovery is more secure. Boat teams or surface swimmers rendezvousing with larger vessels generally come aboard the vessels using cargo nets. Small boats can be recovered with a ship's hoist system. When recovering swimmers or boat teams by helicopter, they are recovered by ladder or the boat may drive right up into the helicopter (depending on helicopter type).

CONSIDERATIONS

1-60. The detachment must thoroughly analyze tidal current data, offset navigation, and time distance criteria in infiltration methods as well as in exfiltration planning and execution. Personnel must direct detailed attention toward prearranged signals between the exfiltration detachment and the recovery systems.

1-61. MAROPS are very equipment intensive. Detachments conducting waterborne exfiltrations must have assets available that are sufficient to exfiltrate the detachment, any casualties, and mission-essential equipment or personnel. Depending on the situation, the detachment may be able to use the same equipment it infiltrated with for its exfiltration. If using this COA, the detachment must ensure initial infiltration goes undetected because they will return to the same location to recover their cached equipment. In particular, personnel must ensure the BLS area is totally sanitized. Nothing must indicate the presence of infiltration swimmers or an equipment-cached area. Personnel can cache equipment either inland (using normal caching procedures) or subsurface. Other methods of supplying the detachment with the requisite equipment include pre-positioning and resupply.

CACHE

1-62. If a decision is made to emplace an underwater cache, the detachment should attempt to identify suitable areas of sea floor during premission planning. If adequate charts or hydrographic surveys of the AOR exist, it may be possible to identify ideal bottom conditions for emplacing underwater caches. Ideal conditions presuppose sufficient structures; for example, coral, rocky reefs, debris, or artificial constructs (pier or bridge pilings, docks) for the detachment to secure or anchor the cache so that it will not be disturbed. Flat, sandy bottoms that are exposed to the full effects of wave action make successful long-term caches almost impossible.

1-63. Underwater caches can be quite difficult to find and are normally not marked. Therefore, the detachment must use and record very accurate reference points and distances. In planning for the cache, personnel must consider how much weight will be required to make the cache negative-buoyant and ensure that it remains subsurface. To do so will require a tremendous amount of weight—perhaps more than the detachment can carry. The weights attached to the cache must be easy to jettison during recovery. The amount of weight must not only keep the cache negative-buoyant, it must also be unaffected by currents. The detachment should attach lifting devices to the outside of the cache so that upon recovery the weights are jettisoned and the lifting devices inflated. This method will allow for rapid recovery and ease of movement out of the water.

1-64. In most cases, the detachment will have to plan for some type of anchor system. Caches placed in water shallow enough to recover by breath-hold diving are susceptible to wave action. If the cache is not anchored with all of the bundles interconnected, it is susceptible to being dispersed by wave action. This is especially true if severe weather causes an increased sea state with its attendant surge. Detachments must investigate available technologies to determine what will be most effective in their particular circumstances. Small craft anchors that can be wedged into bottom structures may present the most efficient solution. Personnel may also consider using mountaineering anchors (for example, "friends" or chocks) if suitable crevasses exist. Detachments must also determine how to protect the anchor ropes from chafing. Underwater structure becomes extremely abrasive over time and a frayed or broken line endangers the integrity of the cache.

1-65. Certain waterborne-related items are extremely sensitive to the environment. Personnel should completely seal buried items in airtight waterproof containers. Normally, heavy-duty plastic bags are adequate for small items. Large items, such as rubber boats, generally have their own heavy-duty rubber or canvas containers. Subsurface caching is much more time-consuming and difficult. Personnel must totally waterproof all sensitive items. The cache must be at a depth that can be reached by surface swimmers or divers. The detachment must also consider tide height and tidal current data when weighing where and at what depth to put the cache. The rise and fall of the tide will affect the depth. In any case, personnel should position the cache as close to the shore or beach as possible.

1-66. Because the detachment will return to its infiltration site, it should place the cache site under observation for a period before recovery execution to ensure the cache area is secure. To ensure the best possible execution of the operation, the detachment should always rehearse the recovery operation.

1-67. Another type of cache is pre-positioning. If the support mechanism in the AO allows it, personnel should pre-position the equipment for the exfiltration. This task will normally occur or be possible in waterborne scenarios with active auxiliary or underground forces. The caching element emplaces the cache and submits a cache report that is forwarded to the detachment. The detachment then establishes surveillance, emplaces security, and recovers the cache in accordance with (IAW) the recovery plan.

RESUPPLY

1-68. A detachment may need to be resupplied with the equipment required to conduct a waterborne exfiltration. This additional support can be either a preplanned, on-call, or emergency resupply and is usually a contingency in case the primary exfiltration plan fails. It is very hard to conduct an airborne resupply since the equipment must, in most cases, be airdropped very close to the beach departure point or in the water near the shore. If the area to be exfiltrated has a coastal air defense system, this resupply method becomes extremely dangerous for the mission aircraft. However, if possible, one of the most effective means is to drop a "combat rubber raiding craft" (or two). The aircraft can fly parallel to and just off the coast and put out a CRRC packed with the needed equipment. The exfiltrating detachment then simply swims to the equipment, unpacks it, and exfiltrates.

1-69. If the equipment is not dropped in the water, it must be dropped very close to water because it may be hard to transport equipment overland. Because of the difficulties associated with airdropping resupplies, they will normally be delivered by sea or overland. Resupply by sea can be done by simply infiltrating the required equipment and caching it if the resupply takes place a long time before exfiltration. The personnel bringing in the resupply can secure the equipment and exfiltrate with the detachment. If possible, and if a support mechanism within the AO will allow it, resupply personnel can deliver the exfiltration equipment overland (by vehicle) to the beach departure point. Personnel can cache the equipment for later removal, or the detachment can begin the exfiltration as soon as the equipment arrives.

UNIT TRAINING AND CAPABILITIES

1-70. The ability of any unit to conduct a specified mission ultimately depends on its level of training and capabilities. The level of training is the responsibility of the unit commander. He ensures that his troops are prepared to carry out their wartime missions. A properly balanced training program must focus on the detachment as well as each individual to produce a reasonably proficient detachment.

1-71. The unit's capability to execute the mission is directly related to the amount and type of equipment available. Regardless of the amount or type of equipment, the unit should train to the utmost with the available assets to maintain a viable waterborne operations capability. Surface swimming, self-contained underwater breathing apparatus (scuba) techniques, SUBOPS, small boat operations, diver propulsion device (DPD) techniques, and waterborne insertion or extraction techniques require special training programs to attain and maintain proficiency.

SUSTAINMENT TRAINING

1-72. The complexity of waterborne operations demands additional training, both for proficiency and for safety. At least one block of instruction in the detachment's weekly training schedule should focus on some aspect of MAROPS. Classes should range the entire spectrum of MAROPS to include infiltration and exfiltration tactics, means of delivery, equipment maintenance, and medical treatment of diving injuries. The detachment's senior diving supervisor should coordinate these classes. Appendix B outlines waterborne operations sustainment training requirements.

1-73. The following paragraphs provide the SFOD commanders an overview of the minimum training required to sustain their MAROPS detachments in a mission-ready status. It also provides the detachment a list of training requirements for developing their long- and short-range training plans. There are many techniques available to a detachment for use in a maritime environment to successfully infiltrate an operational area or to reach a specific target. Considerable training time is required to maintain the specialized MAROPS skills. To be fully mission-capable, the detachment must have the commander's support to allocate the required training time and resources. The required capabilities show the minimum skills and topical areas to be covered during training. For each skill or area, there is a determination of how often training must be conducted. A description of each skill or area and details on what must be accomplished during the training. Finally, there is a listing of some of the critical tasks associated with waterborne operations. They may be used as aids for both training and evaluation.

REQUIRED CAPABILITIES

1-74. This required capabilities list (Figure 1-4, page 1-14) identifies what a paid combat diver must be able to perform IAW AR 611-75 and USASOC Reg 350-20.

Chapter 1

- Inspect and maintain basic diving equipment to include open- and closed-circuit scuba
- Conduct open-circuit dives up to 130 FSW within a 1-year period.
- Dive while using closed-circuit equipment within operational limits.
- Take appropriate action in underwater emergencies.
- Apply concepts of physics to diving.
- Recognize and apply first aid to divers with underwater injuries or illnesses.
- Assist in planning diving operations, to include use of United States Navy diving tables.
- Apply techniques of infiltration and exfiltration using scuba and surface swimming.
- Rig bundles and combat equipment for underwater or surface swimming operations.
- Navigate 1,500 meters underwater with a compass to required time and accuracy standards as prescribed by USAJFKSWCS.
- Conduct underwater searches to include ship bottom searches.
- Perform a surface swim (3,000 meters) to required time standard as prescribed by USAJFKSWCS.
- Apply techniques used to waterproof and transport demolitions.
- Perform free ascents from 25 and from 50 feet, respectively.
- Swim 50 meters underwater without breaking the surface as prescribed by USAJFKSWCS.
- Complete a 1,500-meter, closed-circuit combat infiltration swim as a team member of at least six men with load-carrying equipment, rucksack, and weapon.

Figure 1-4. Required capabilities

Pay Dives

1-75. Divers are required to perform diving duties IAW AR 611-75 to maintain proficiency and draw special duty pay. As a minimum requirement, a combat diver must perform six qualifying dives within 6 months, one deep dive (70 to 130 feet of seawater [FSW]) within 12 months and be in a qualified status. The criteria for qualifying dives and status are listed in AR 611-75 and USASOC Reg 350-20. (Requirements for diving duty pay for Marine Corps personnel are set forth in Marine Corps Order (MCO) 3150.4, *Marine Corps Diving Policy and Program Administration*.) To maintain proficiency at infiltration swimming, detachments should conduct underwater compass swims monthly using open- or closed-circuit breathing apparatus. Swims should be done with properly waterproofed and neutrally buoyant rucksacks. Once each quarter, the team members waterproof and pack (according to the standing operating procedure [SOP]) the team's equipment into their rucksacks. This operation should use the team swim concept—the detachment is linked together by a buddy line, if need be, and moves through the water and onto the shore as one unit.

Requalification

1-76. To maintain currency, all combat divers are required to perform certain diving tasks at least once annually. These tasks are outlined in AR 611-75. Divers who maintain their currency do not have a formal requirement to "requalify" annually. Divers who have allowed their qualifications to lapse or who have returned to diving duty after a period of inactivity must requalify for diving duty IAW AR 611-75.

1-77. Diving supervisors and diving medical technicians (DMTs) are required to perform duties as a combat dive supervisor (CDS) or DMT at least once every 6 months and the dive medical technician must attend a training seminar every 2 years. DMTs supporting special operations (SO) diving must also maintain all of their other medical qualifications IAW the applicable policies and procedures developed by the USASOC commander. The minimum required subjects for the CDS and DMT seminars are listed in USASOC Reg 350-20.

1-78. In addition to the above-stated annual requalification requirements, each combat diver must undergo a Type B medical examination every 3 years with a minimum of a Type B update annually.

Mission Planning

Operational Exercise

1-79. The goal of all sustainment training is for the detachment to be able to execute a full mission profile. To that end, the detachment must conduct a semiannual operational mission exercise that puts as many of their mission-ready skills to use at one time as can be realistically coordinated. This exercise can be conducted in conjunction with other training requirements mentioned above. Multiple delivery methods should be used, coupled with a surface swim, an underwater navigation team swim, the DPD, or a combination of all infiltration and exfiltration techniques. An example would be an airborne or airmobile OTH insertion with CRRCs with an offshore navigation to a drop-off point. This method would be followed by a turtle-back swim to a point 1,500 meters offshore, a closed-circuit underwater team compass swim with equipment to the BLS, and an over-the-beach (OTB) infiltration. Following a UW or DA mission, the team would execute an OTB exfiltration and some form of marine extraction. It may of course be impossible to include all of these phases in one tactical exercise. However, multiple phases must be conducted in each exercise. Realistic training challenges the detachment to excel and gives the commander an effective tool to assess mission capabilities.

1-80. Meeting these requirements does not guarantee an individual combat diver or combat diver detachment to be mission-ready. These are the minimum requirements which, when met, allow combat divers to engage in the training necessary to achieve a combat, mission-ready status.

This page intentionally left blank.

Chapter 2

Environmental Factors

Weather and its effects on the friendly and enemy situations is a critical factor in mission planning and the safe execution of any operation. The maritime environment in which waterborne operations take place is always changing. Those changes have immediate and urgent effects on the types of small vessels available to an infiltrating or exfiltrating detachment. This chapter focuses on the effects weather has on the water, and the potential problems detachments face while operating small boats or conducting an infiltration swim.

Timely weather forecasts coupled with an understanding of the basic principles of weather patterns and their effects are a key element in the planning of waterborne operations. The environmental factors that have the greatest impact on waterborne operations include wind, storms, waves, surf, tides, and currents. This chapter examines the basic elements of weather, types of storm systems, storm propagation, and how to forecast weather from local observations. It also briefly explains how to read a weather map.

WIND

2-1. High winds are a genuine concern for personnel conducting waterborne operations. High winds can greatly impact on almost every type of waterborne operation. High seas are directly related to wind speed. The Beaufort Wind Scale is the internationally recognized guide to expected wave height and sea states under varying wind conditions (Table 2-1, page 2-2). When planning waterborne operations, planners should use this scale to define a particular state of wind and wave.

2-2. Without wind, weather would remain virtually unchanged. Wind is a physical manifestation of the movement of air masses. It is the result of horizontal differences in air pressure. Air flows from a high-pressure area to a low-pressure area producing winds. Solar radiation and the resultant uneven heating of the earth's surface is the driving force creating the pressure differentials that cause wind. The pressure gradient is shown on weather maps as a series of isobars (pressure contours) connecting places of equal barometric pressure. The closer together the isobars appear, the steeper the pressure gradient and the higher the wind speed. A synoptic weather chart is a comprehensive view of the AO (Figure 2-1, page 2-2).

2-3. Wind direction is determined based on where it is coming from. If a person is looking north and the wind is in his face, it is a North wind. Wind speed is reported in kts and direction is reported in degrees true.

2-4. Major air masses move on a global scale. One of the modifiers for this movement is the Coriolis effect. The earth's rotation exerts an apparent force, which diverts air from a direct path between the high- and low-pressure areas. This diversion of air is toward the east in the Northern Hemisphere and toward the west in the Southern Hemisphere.

Chapter 2

Table 2-1. The Beaufort Wind Scale

Beaufort Wind Force	Wind Range (Knots)	Sea Indications	Wave Height (Feet)
0	Less Than 1	Mirror like.	0
1	1–3	Ripples with appearance of scales.	0.25
2	4–6	Small wavelets; glassy appearance; no breaking.	0.5–1
3	7–10	Large wavelets; some crests begin to break; scattered whitecaps.	2–3
4	11–16	Small waves becoming longer; fairly frequent whitecaps.	3.5–5
5	17–21	Moderate waves; pronounced long form; many whitecaps.	6–8
6	22–27	Large waves begin to form; white foam crests are more extensive; some spray.	9.5–13
7	28–33	Sea heaps up; white foam from breaking waves begins to blow in streaks along the direction of the wind.	13.5–19
8	34–40	Moderately high waves of greater length; edges of crests break down into spindrift foam blown in well-marked streaks in the directions of the wind.	18–25
9	41–47	High waves; dense streaks of foam; sea begins to roll; visibility affected.	23–32
10	48–55	Very high waves with overhanging crests; foam in great patches blown in dense white streaks; whole surface of sea takes a white appearance; visibility affected.	29–41

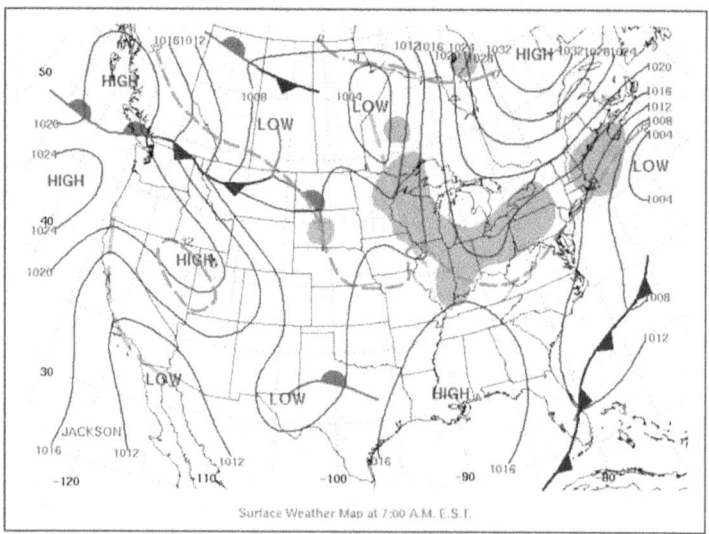

Figure 2-1. Synoptic weather chart

2-5. The uneven heating of the earth's surface results in differentially warmed air masses. Warm air expands and rises creating areas of lower air density and pressure. Cool denser air, with its greater pressure, flows in as wind to replace the rising warm air. These cells of high and low pressure have their own internal rotation influenced by the global modifiers (Figure 2-2). High-pressure cells have clockwise rotating winds and are called anticyclones. Low-pressure cells have counterclockwise rotating winds and are called cyclones.

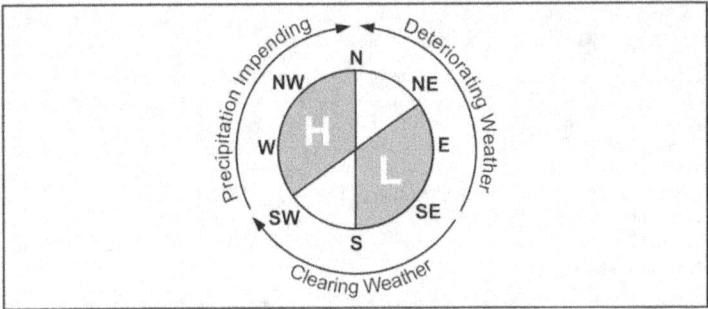

Figure 2-2. High- and low-pressure wind circulation

2-6. Local weather patterns are strongly affected by terrain and daily heating and cooling trends. Desert regions heat and cool rapidly; wooded or wet areas change temperature more slowly. Mountainous areas experience updrafts and downdrafts in direct proportion to the daily cycles. Coastal areas will experience onshore and offshore winds because of the differential solar heating and cooling of coastal land and adjoining water.

2-7. The advent of a high-pressure cell usually denotes fair weather. A high-pressure cell is evident to the observer when skies are relatively clear and winds are blowing from the southwest, west, northwest, and north. This fact is the same for the Northern and Southern Hemisphere.

2-8. Low-pressure cells are the harbingers of unsettled weather. They are evident when clouds gather and winds blow from the northeast, east, southeast, and south. Winds that shift from the north toward the east or south signal deteriorating weather. Impending precipitation is signaled by winds (especially a north wind) shifting to the west and then to the south. Wind shifts from the east through south to west are an indicator of clearing weather. The approach of high- and low-pressure systems can also be tracked and anticipated with a barometer. Some wristwatches have an altimeter and barometer function. Barometers are most useful when monitoring trends. Normal barometric pressure is 29 inches of mercury (Hg). Planners should take readings at regular intervals and record changes. A drop in barometric pressure signals deteriorating weather (an approaching low-pressure system). A rise in barometric pressure indicates clearing weather (the approach of a high-pressure system). This fact is the same for the Northern and Southern Hemisphere.

CLOUDS

2-9. Clouds are the most visible manifestation of weather. Cloud formations are valuable in determining weather conditions and trends. Clouds form when the moisture in rising warm air cools and condenses. They may or may not be accompanied by precipitation.

2-10. Naming conventions for cloud types are intended to convey crucial information about the altitude and type of cloud. Different cloud types have descriptive names that depend mainly on appearance, but also on the process of formation as seen by an observer. Cloud nomenclature that describes the cloud type is usually combined with prefixes or suffixes that describe the altitude of that cloud. Key descriptive terms are as follows:

- *Cumulo-Cumulus*. Prefix and suffix that describes a vertical heaping of clouds. Castellated refers to a "turreted" cumulo-type cloud. These clouds grow vertically on summer afternoons to produce showers and thunderstorms.

- *Fracto-Fractus.* Prefix and suffix that describes broken-up clouds.
- *Nimbo-Nimbus.* Prefix and suffix that describes clouds that are full of rain or already have rain falling from them.

2-11. Despite an almost infinite variety of shapes and forms, it is still possible to define ten basic types. These types are grouped by altitude and are further divided into three levels: high, middle, and low.

2-12. High-altitude clouds have their bases at or above 18,000 feet and consist of ice crystals. *Cirro* is a prefix denoting a high-altitude cloud form. *Cirrus* is the name of a particular, very high, wispy cloud comprised of ice crystals (Figures 2-3 and 2-4). It appears as delicate curls or feathers miles above the earth. When these clouds form feathery curls, they denote the beginning of fair weather. When they strand out into "Mare's Tails," caused by strong winds aloft, they indicate the direction of a low-pressure cell. Cirrocumulus is a layer of cloud without shading, comprised of grains or ripples, and more or less regularly arranged. Cirrostratus is a transparent, whitish veil of cloud (Figure 2-4). It partially or totally covers the sky and produces the characteristic "halo" around the sun or moon. These clouds thickening or lowering indicate an approaching front with precipitation.

Figure 2-3. Cirrus clouds

Figure 2-4. Cirrus and cirrostratus clouds

2-13. Middle clouds have bases located between 7,000 and 18,000 feet. They consist of water droplets at lower levels of altitude and ice crystals at higher levels. *Alto* is the prefix describing middle zone clouds. Altocumulus is a layer of white, gray, or mixed white and gray clouds, generally with shading, and made up of round masses. Several layers of altocumulus clouds (Figure 2-5) indicate confused patterns of air currents, rising dew points, and impending rain and thunderstorms. Altocumulus clouds massing and thickening is a sign of coming thunderstorms. Altostratus is a grayish or bluish layer of cloud with a uniform appearance that partially or totally covers the sky. It may or may not be thick enough to obscure the sun. Altostratus clouds (Figure 2-6) indicate an oncoming warm front. Rain from thickening altostratus clouds may evaporate before reaching the ground.

Figure 2-5. Altocumulus clouds

Figure 2-6. Altostratus clouds

2-14. Low clouds have bases below 7,000 feet. Low clouds consist mostly of water droplets; however, in colder climates, ice crystals may predominate. These clouds can develop into multilevel clouds and go through various phases of cloud formation. *Strato* is the prefix referring to the low part of the sky. When used as a suffix, stratus describes a spread-out cloud that looks sheetlike or layered. When used by itself, stratus is a complete name, describing a low-level cloud that is usually gray and covers a great portion of the sky. Other low-level clouds are nimbostratus, stratocumulus (Figure 2-7, page 2-6), cumulus (Figure 2-8, page 2-6), and cumulonimbus.

Chapter 2

Figure 2-7. Stratocumulus clouds

Figure 2-8. Cumulus clouds

2-15. The cloud formation that signals conditions with the greatest risks to the detachment (or any maritime operator) is the cumulonimbus (Figure 2-9, page 2-7), which is the classic thundercloud. Thunderstorms are particularly dangerous not only because of lightning, but also because of the strong winds and the rough, confused seas that accompany them. On the ocean, a thunderstorm may manifest itself as a localized phenomenon or as a squall line of considerable length. Before a storm, the winds are generally from the south and west in the middle latitudes of the Northern Hemisphere, and the air is warm and humid. Sharp, intermittent static on the amplitude modulation (AM) radio is also an indicator of approaching thunderstorms. If the sky is not obscured by other clouds, an observer may determine the direction a thunderstorm will move; it will follow the direction the incus (the anvil-shaped top of the thunderhead) is pointing. The incus is the streamer blown forward from the top of the cumulonimbus cloud by high-altitude winds. The observer should not be deceived by surface winds. The strong updrafts inside the cloud that contribute to its vertical development will cause surface winds to blow in toward the base of the cloud. As the storm approaches, it may be preceded by the "calm before the storm," an apparent slackening of surface winds. Detachment personnel should always be alert to conditions that favor the development of thunderstorms. If possible, they should seek shelter; if not, they should batten down the hatches; it can be a rough ride.

Environmental Factors

Figure 2-9. Cumulonimbus clouds

LIGHTNING

2-16. Lightning is associated with some storms and is a potentially life-threatening event. It is caused by a buildup of dissimilar electrical charges within a vertically developing cumulonimbus cloud. Lightning occurs most often as a cloud-to-cloud strike. The lightning most likely to threaten mariners is the cloud-to-ground (water) strike. Thunder is caused by the explosive heating of the air as the lightning strike occurs. There are no guaranteed safeguards against lightning. It is very unpredictable and has immense power. Staying in port (assuming there are higher objects about) during thunderstorms can minimize the danger of having a boat struck. Personnel can also lessen the danger by installing a grounding system like those on buildings and other land structures. The grounding system provides lightning with a path to reach the ground without causing damage or injury.

2-17. A detachment member can judge his distance from a thunderstorm by knowing that light travels at about 186,000 miles per second and sound at about 1,100 feet per second (or about 1 mile in 5 seconds). If he times how long the sound of the thunder takes to reach him after he sees the lightning flash, he can roughly estimate the distance to the storm. (Counting ONE THOUSAND ONE, ONE THOUSAND TWO, ONE THOUSAND THREE will aid him in counting seconds.) Detachment personnel should reduce their exposure or risk of being struck by lightning by getting out of or off of the water and seeking shelter. If a detachment member is caught in a lightning strike area in a relatively unprotected boat (as in a CRRC), he should—

- Stay inside the boat, keep the crew centrally positioned and low down in the vessel, and stay dry.
- Avoid touching metal objects such as weapons, equipment, outboard motors, shift and throttle levers, and metal steering wheels.
- Avoid contact with the radio, lower and disconnect antennas, and unplug the antenna to save the radio in case lightning strikes.
- Quickly remove people from the water.
- If a lightning strike occurs, expect the compass to be inaccurate and to have extensive damage to onboard electronics.

FOG

2-18. Fog is a cloud in contact with the ground and water. It is composed of a multitude of minute water droplets suspended in the atmosphere. These are sufficiently numerous to scatter the light rays and thus reduce visibility.

2-19. The most troublesome type of fog to mariners is advection fog. Advection means horizontal movement. It is also the name given to fog produced by air in motion. This type of fog is formed when warm air is transported over colder land or water surfaces. The greater the difference between the air temperature and the ocean temperature, the deeper and denser the fog. Unlike radiation fog, advection fog is little affected by sunlight. It can and does occur during either the day or night. Advection fog is best dispersed by an increase in wind velocity or change in wind direction.

2-20. Fog awareness is a local knowledge item. Personnel should ask the local people questions such as, "Where does it usually occur? What time of day? What time of year?" If available, a dedicated staff weather officer (SWO) may be able to provide the desired information.

2-21. Unless small-boat operators have a compelling reason to go out in dense fog, they should not do so. If fog seems to be developing, they should try to run in ahead of it. The small boat operating on larger bodies of water or oceans should always maintain a running dead reckoning (DR) plot. If, for some reason, no plot has been maintained and fog rolls in, the boat operator should attempt to get a position-fix immediately. With an accurate heading to port and an accurate knowledge of speed over the bottom, it is possible to plot a course back home.

2-22. When in fog, the boat operator should slow down so he will have time to maneuver or stop if another vessel approaches. If the boat's size and configuration permit, an observer should stand lookout well forward and away from the engine sounds and lights to listen and look for other signals. The observer should also listen for surf in case the DR is incorrect. He should use all electronic aids available, but not depend upon them without reserve. Boat personnel should even consider anchoring to await better visibility, especially if their return to port includes transiting congested areas or narrow channels.

AIR MASSES AND FRONTS

2-23. Reading and understanding weather forecasts requires a basic knowledge of air masses and fronts. An air mass is another name for a high-pressure cell, which is a buildup of air descending from high-altitude global circulation. There is no corresponding term for the low-pressure equivalent. However, "air mass" refers to the volume of air and its physical mass, rather than its pressure characteristics. An air mass is capable of covering several hundred thousand square miles with conditions in which temperature and humidity are essentially the same in all directions horizontally.

2-24. Air masses derive their principal characteristics from the surface beneath them. They are characterized as continental or maritime, tropical or polar, and warmer or colder than the surface over which they are then moving. Maritime and continental air masses differ greatly. Maritime air masses tend to change less with the seasons. Oceans have less variation in their heating and cooling cycles, so maritime air masses moving over land tend to moderate conditions of excess heat and cold. Continental air masses are subject to large changes in humidity and temperature, which in turn, bring varying weather.

2-25. Cold air masses consist of unstable internal conditions as the air at the earth's surface attempts to warm and rise through the cooler air. Warm air masses are more stable as the air cooled by contact with the ground sinks and warm air above tends to stay there or rise. These changes cause strong, gusty winds in cold air masses, and weaker, steadier winds in warm air masses. Cumulus clouds usually mean cold air masses. As a cold air mass moves a warm air mass, these clouds may change to cumulonimbus clouds and produce thunderstorms. Warm air masses usually mean stratus clouds and extended drizzle.

2-26. Weather fronts are the boundaries between two different air masses. These bodies do not tend to mix but instead will move under, over, or around each other. The passage of a front brings a change in the weather.

COLD FRONT

2-27. An oncoming cold air mass will push under a warm air mass, forcing it upward (Figure 2-10). In the Northern Hemisphere, cold fronts are normally oriented along a northeast to southwest line and they advance toward the southeast at a rate of 400 to 500 miles per day—faster in winter, slower in summer. A strong, fast cold front will bring intense weather conditions that do not last long. A slow-moving front may pass over an area before precipitation starts. A weak cold front may not have any precipitation. An approaching cold front may have a squall line form in front of it. Winds will shift towards the south, then southwest, and barometric pressure will fall. Clouds will lower and build up; rain will start slowly but increase rapidly. As a cold front passes, skies will clear quickly and temperatures will drop. Barometric pressure will climb quickly. Winds will continue to veer toward the north and northeast. For a few days, at least, the weather will have the characteristics of a high-pressure cell.

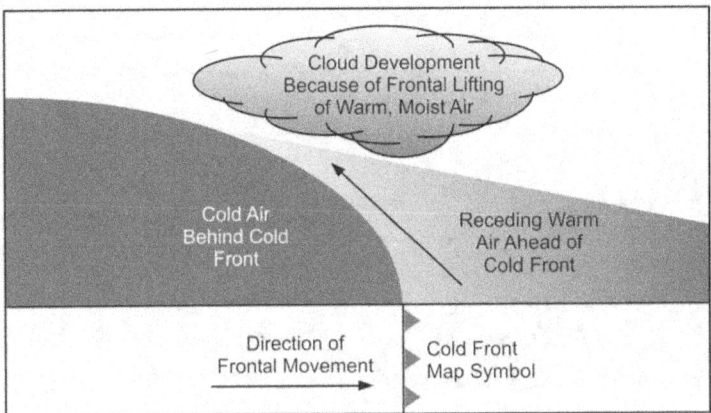

Figure 2-10. Cross-section of a cold front

WARM FRONT

2-28. A warm front is an advancing warm air mass that meets a colder mass and rides up over it (Figure 2-11, page 2-10). Warm fronts are generally oriented north to south, northwest to southeast, or east to west, and change their direction more often than cold fronts do. Warm fronts advance between 150 to 200 miles per day. Because of their slower speed, they are often overtaken by a following cold front. Approaching warm fronts signal milder weather than do cold fronts. Warm fronts are preceded by low stratus cloud formations and moderate but extended rains. Another indicator of a warm front is a slowly falling barometer. As a warm front passes, cumulus clouds replace stratus clouds, and the temperature and barometer both rise.

2-29. A stationary front is a front that has slowed to a point where there is almost no forward movement. The neighboring air masses are holding their positions or moving parallel to one another. The result is clouds and rain similar to a warm front.

OCCLUDED FRONT

2-30. An occluded front is a more complicated situation where there is warm, cold, and very cold air. An occluded front occurs after a cold front (with its greater speed) has overtaken a warm front and lifted it off the ground (Figure 2-12, page 2-10). The appearance on a weather map is that of a curled tail extending outward from the junction of the cold and warm fronts. This front is a low-pressure area with counterclockwise winds in the Northern Hemisphere and clockwise winds in the Southern Hemisphere.

Chapter 2

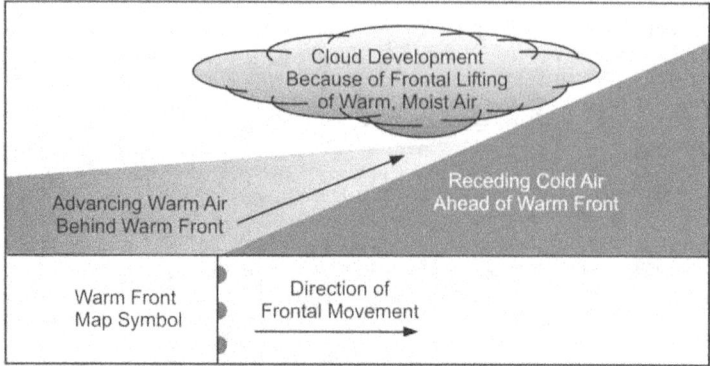

Figure 2-11. Cross-section of a warm front

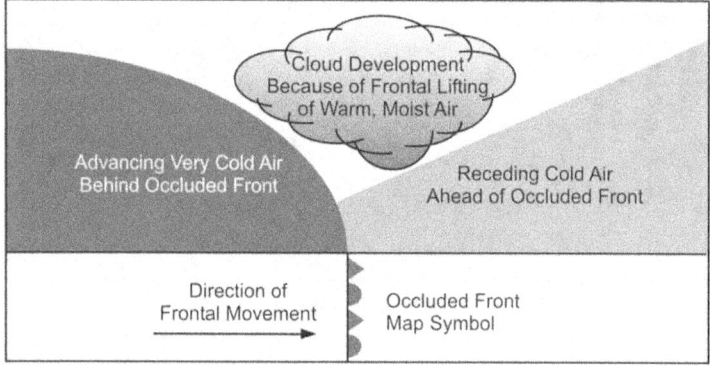

Figure 2-12. Cross-section of an occluded front

SURFACE WEATHER ANALYSIS CHARTS

2-31. Surface weather maps are one of the most useful charts for ascertaining current weather conditions just above the surface of the earth for large geographic regions. These maps are called surface analysis charts if they contain fronts and analyzed pressure fields, with the solid lines representing isobars (Figure 2-13, page 2-11). By international agreement, all meteorological observations are taken at the same time according to Zulu time. Most charts list somewhere in the title the Zulu time when the observations were made.

2-32. Synoptic weather analysis requires the simultaneous observation of the weather at many widely located sites using standardized instruments and techniques. A display of this information would be difficult to make and interpret unless a uniform system of plotting was adopted. Weather data from each reporting station is plotted around the reporting station's symbol (a small circle on the base map) using a standardized methodology known as a station model. The symbology used comes from the International Weather Code (Figure 2-14, page 2-12). Information that may appear on a synoptic chart include—
- Temperature and dew point temperature.
- Wind speed and direction.

Environmental Factors

- Barometric pressure.
- Weather (rain, snow).
- Precipitation intensities.
- Sky cover.

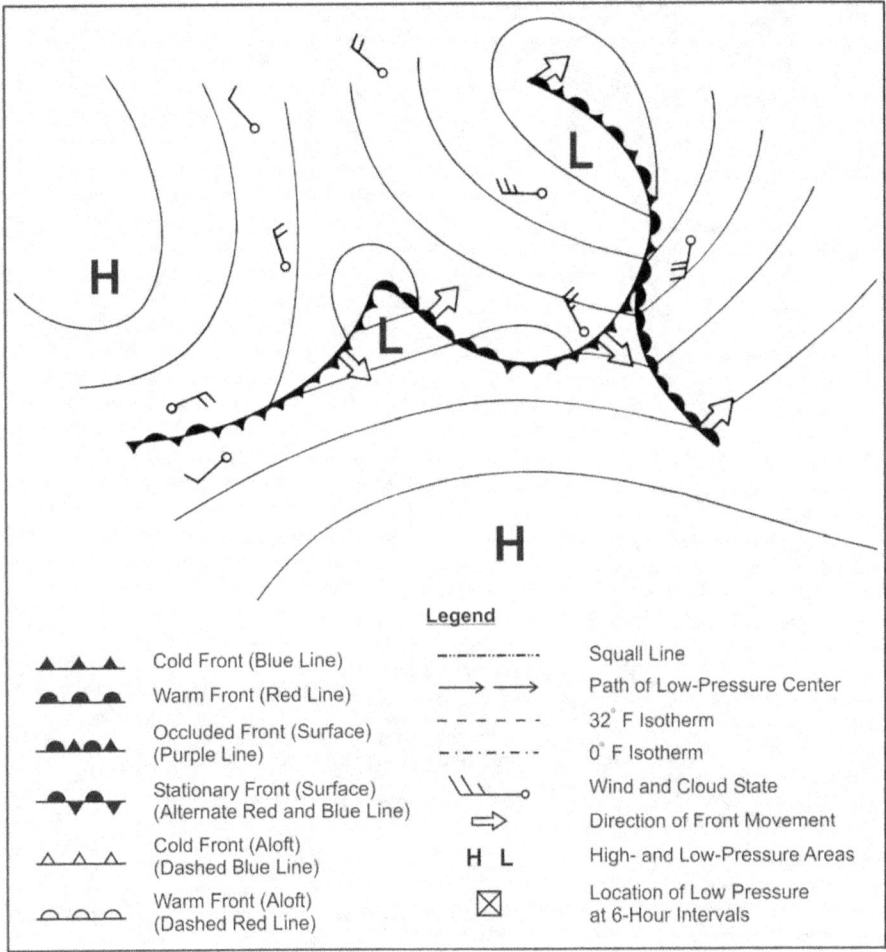

Figure 2-13. Sample weather chart

2-33. The surface analysis charts permit a person to identify and locate the large-scale features of the sea level pressure field and the surface fronts. Isobars with the lowest value will encircle the region with the lowest pressure, while closed isobars with the highest value will encircle the area with the highest pressure. How tightly the isobars are packed reveals the steepness of the pressure gradient between any two points. Wind flows tend to parallel the isobars, with low pressure to the left of the wind flow in the Northern Hemisphere. A slight cross-deflection of the winds toward the area of lower pressure is often seen. Thus (in the Northern Hemisphere), winds appear to spiral in toward a surface low-pressure cell in a

Chapter 2

counterclockwise direction and spiral out from a high-pressure cell in a clockwise direction. The more tightly packed the isobars, the higher the wind speeds. If surface charts are available for the previous reporting periods, a person may be able to distinguish trends and make a reasonable short-range weather forecast based on the movement of high- or low-pressure cells. Radar overlays can also be helpful for showing the extent and intensity of precipitation in the area of interest.

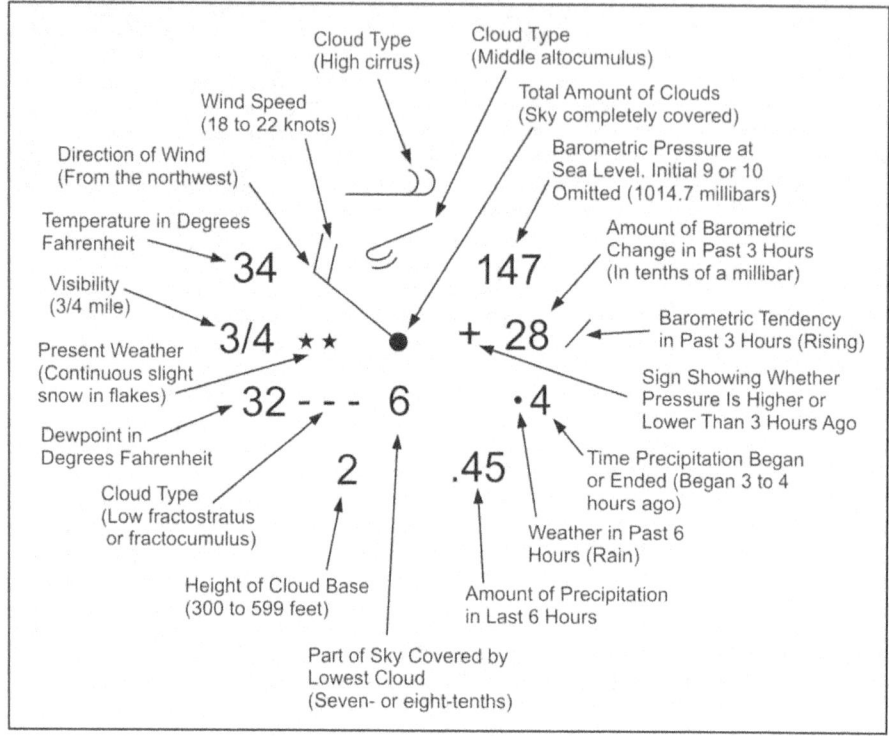

Figure 2-14. International Weather Code

FORECASTING

2-34. Fifty years ago, U.S. World War II fleet admirals had some of the best weather experts in this country on their staffs. Today the military can receive a better weather briefing than Admiral Halsey ever had. In the United States, the satellite receiver-equipped television weatherman and the National Weather Service's continuous very high frequency (VHF) weather briefings, coupled with local knowledge, should make personnel effective and safe weather-wise. But even experts are far from 100 percent correct concerning the weather. Figure 2-15, pages 2-13 and 2-14, provides some generalizations that can indicate a change in the weather.

Indicators of Deteriorating Weather

- Clouds lower and thicken; ceiling lowers.
- Puffy clouds begin to develop vertically and darken.
- Sky darkens and looks threatening to the west.
- Clouds increase in numbers and move rapidly across the sky.
- Clouds move in different directions at different heights.
- Clouds move from east or northeast toward the south.
- Heavy rains occur at night.
- Barometer falls steadily or rapidly.
- Smoke from stacks lowers.
- Static develops on AM radio.
- North wind shifts east and possibly through east to south.
- A ring (halo) shows around the moon.
- If on land, leaves that grow according to prevailing winds turn over and show their backs.
- Strong wind or a red sky in the morning.
- Temperatures far above or below normal for the time of the year.

Indicators of Impending Strong Winds

- Light, scattered clouds alone in a clear sky.
- Sharp, clearly defined edges to clouds.
- A yellow sunset.
- Unusually bright stars.
- Major changes in temperature.

Indicators of Impending Precipitation

- Distant objects seem to stand above the horizon.
- Sounds are very clear and can be heard for great distances.
- Transparent, veil-like cirrus clouds thicken; ceiling lowers.
- Hazy and sticky air. (Rain can occur in 18 to 38 hours.)
- Halo shows around the sun or the moon.
- South wind increases with clouds moving from the west.
- North wind shifts to west and then to south.
- Barometer falls steadily.
- A pale sunset.
- A red sky to the west at dawn.

Indicators of Clearing Weather

- Cloud bases rise.
- Smoke from stacks rises.
- Wind shifts to west, especially from east through south.
- Barometer rises quickly.
- A cold front passes in the last 4 to 7 hours.
- A gray early morning sky that shows signs of clearing.
- A morning fog or dew.
- Rain stops and clouds break away at sunset.

Figure 2-15. Generalizations for weather forecasting

Indicators of Continuing Fair Weather
• An early morning fog that clears. • A gentle wind from west to northwest. • Barometer steady or rising slightly. • A red sky to the east with a clear sky to the west at sunset. • A bright moon and a light breeze. • A heavy dew or frost. • A clear blue morning sky to the west. • Clouds that dot the afternoon summer sky. • Sounds do not carry; dull hearing. • Clouds do not increase or actually decrease. • Altitude of cloud bases near mountains increases.

Figure 2-15. Generalizations for weather forecasting (continued)

Note. The observations discussed about weather phenomena and observational forecasting are regional in nature. The generalizations mentioned in this manual are based on the global circulation patterns in the temperate zones of the Northern Hemisphere. Although the general observations are accurate, they may require modification when used outside the temperate regions and in the Southern Hemisphere.

WAVES

2-35. Wave formations and wave activity are extremely important in planning the successful execution of any waterborne operation. Waves impact on all surface-related activities, including boating or swimming. Waves can likewise affect the subsurface activities of the combat diver; therefore, divers must be totally familiar with the effects of waves.

2-36. Waves are a transfer of energy through water particles. The actual water particles in a wave trace the outline of a circular orbit and return very nearly to their exact starting point at the end of a wave cycle. Like a cork bobbing freely, water particles in the open ocean move up and down in circular motion. There is very little lateral or horizontal movement.

2-37. When planners study waves, they break them into deep water waves and shallow water waves. In deep water waves, the orbit is intact or circular. In shallow water waves, the orbit reflects off the bottom and becomes elliptical. As the orbit becomes elliptical, the energy transfers and the breaking wave becomes surf. The following paragraphs discuss wave terms, the formation of waves, and types of waves.

TERMINOLOGY

2-38. Planners and operators must understand the basic wave activity and wave formation terminology (Figure 2-16, page 2-15). The common definitions that can assist personnel in understanding waves are as follows:

- *Crest* is the very top of the wave or the highest point in the wave.
- *Trough* is the lowest portion of the wave and is that point between two crests.
- *Wavelength* is the horizontal distance between a wave crest and the crest of the preceding wave. Wavelength can also be measured from trough to trough.
- *Wave height* is the total vertical distance from the crest of a wave to its trough.
- *Amplitude* is the height of a wave above or below sea level. Amplitude is equal to one-half of the wave height.
- *Wave period* is the time it takes, in seconds, for two consecutive wave crests to pass a fixed observation point.

Environmental Factors

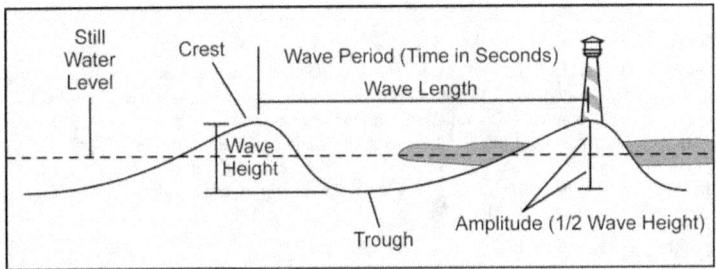

Figure 2-16. Wave characteristics

WAVE FORMATION

2-39. Planners and operators must also know how waves are formed. With this knowledge, the combat swimmer can anticipate what type of wave activity and wave action will most likely be prevalent. He must also know how waves react under certain conditions so that he can anticipate and react to sea conditions. Wave formation is primarily a wind function. As the wind blows across the waters of the ocean, it imparts energy onto the surface of the water, causing it to oscillate. Wave height depends on three factors: force, duration, and fetch. Force is the speed of the wind. Duration is how long the wind blows; it takes roughly 12 hours for fully developed waves to build. Fetch is the open distance over which the wind blows uninterrupted by land masses such as islands or reefs. As a rule of thumb, the maximum height of a wave will be equal to one-half of the wind velocity, providing the fetch is great and the duration is sustained.

2-40. Secondary causes of wave formation include geological disturbances such as earthquakes, landslides, volcanic action, or nuclear explosions. Any of these events can cause tsunamis (commonly referred to as tidal waves). Tsunamis may only be 3 feet high on the open ocean but can routinely reach heights of 60 feet as they approach shallow water. Other types of waves are storm surges, tidal bores, seiches, and internal waves.

Storm Surges

2-41. These surges always occur during bad weather. They result from the combination of tides and rising sea level. The low atmospheric pressure, coupled with the high winds and rising tides, forces large amounts of water inland, causing extensive flooding that can last through several tidal cycles. The worst storm surges normally occur with hurricanes, due to the extreme low pressure and high winds.

Tidal Bores

2-42. Tidal bores occur when land masses serve to restrict the flow of water. Normally, as in the Amazon Basin and the Bay of Fundi in Nova Scotia, the area is fed by a freshwater river. As the large amount of water is rapidly channelized, its speed increases dramatically. Large amounts of water are rapidly carried upstream due to the incoming tide, creating a wall of water. These occur throughout the tidal cycle, moving upstream during flood tides and downstream during ebb tides. Asia has the most tidal bores.

Seiche

2-43. This type of standing wave is normally found in lakes or semiclosed or confined bodies of water. It is a phenomenon where the entire body of water oscillates between fixed points without progression. Depending on the natural frequency of the body of water, these oscillations may be monomodal or multimodal. A seiche can be caused by changes in barometric pressure or strong winds of longer duration pushing the water in a lake to the opposite shore. When the wind diminishes, this buildup of water seeks to return to its normal level and the lake level will oscillate between the shores until it stabilizes. It is a long wave, usually having its crest on one shore and its trough on the other. Its period may be anything from a few minutes to an hour or more. Strong currents can accompany this movement of water, especially if it passes through a restriction.

Chapter 2

Internal Waves

2-44. Internal waves, or boundary waves, form below the surface at the boundaries between water strata of different densities. The density differences between adjacent water strata in the sea are considerably less than that between sea and air. Consequently, internal waves are much more easily formed than surface waves, and they are often much larger. The maximum height of wind waves on the surface is about 60 feet, but internal wave heights as great as 300 feet have been encountered. The full significance of internal waves has not yet been determined, but it is known that they may cause submarines to rise and fall like a ship at the surface, and they may also affect sound transmission in the sea.

BREAKING WAVES

2-45. Breaking waves (breakers) are another area of concern and interest. These waves form the different types of surf. There are three types of breakers: spilling, plunging, or surging (Figure 2-17). The actual type of breaker is normally dependent upon the bottom gradient.

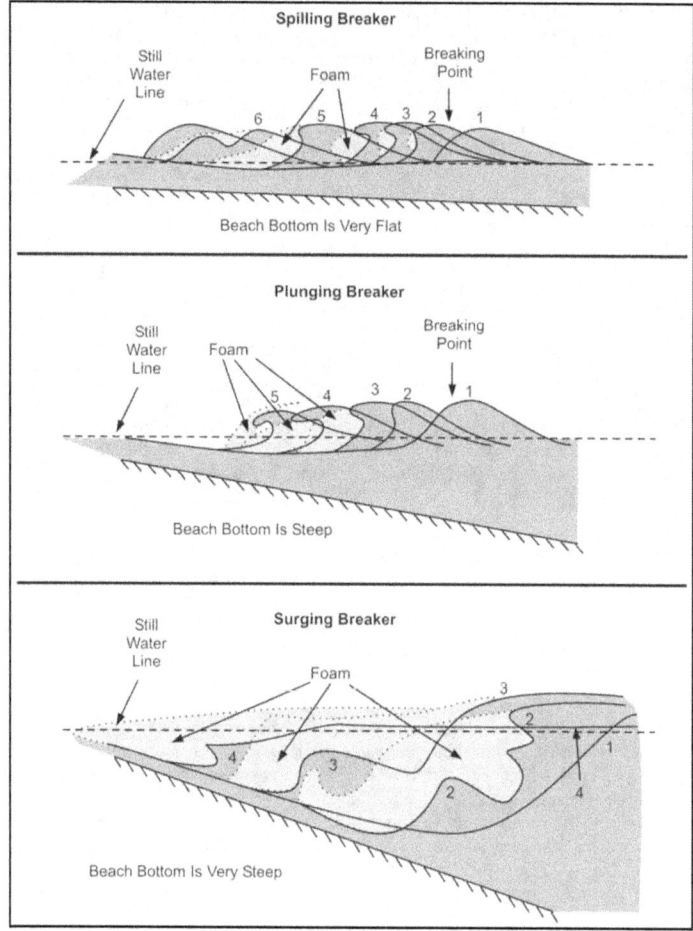

Figure 2-17. Types of breaking waves

2-16　　　　　　　　　　　　　FM 3-05.212/MCRP 3-11.3A　　　　　　　　　　30 September 2009

2-46. If the slope of the bottom is very gentle or gradual, the breakers' force will be very gentle. Thus, the waves will create a spilling action or what is normally the white water at the crest of the wave. These breaking waves are called spillers.

2-47. If the slope of the bottom is steep, the breakers' force will be more pronounced. Thus the wave's crest, as it is unsupported, causes a plunging effect. These waves are called plungers and literally pound the beach.

2-48. With an extremely steep or near vertical slope, the wave literally surges onto shore all at once. These are surging waves and are very violent.

2-49. Knowing the types of breaking waves and what causes them is important for conducting any boat or swimmer operation in a surf area. Also, knowing the type of bottom (from a chart or survey) allows the swimmer to determine wave activity in his AO. From another standpoint, the swimmer can determine bottom slope by observing wave action in a surf zone. This information, when reported, could prove useful for future operations.

IMPACT OF WAVE ACTIVITY

2-50. Obviously, wave activity can have a positive or negative impact on any waterborne operation. Careful planning and consideration of all possible wave activity will greatly enhance the operational success of any mission. The operational planner should remember the following points:
- The height of the waves is about one-half the speed of the wind.
- The depth of the water is four-thirds the height of a breaking wave. (Example: Height of wave is equal to 6 feet, 1/3 of 6 = 2, 2 x 4 = 8; depth of water is about 8 feet.)
- The likely existence of a sandbar or reef just under the water when waves are observed breaking offshore and again onshore.
- The wind must blow across the water about 12 hours to generate maximum wave activity.

ICE

2-51. Operational personnel must consider ice and its effects on men and equipment when planning mission requirements. The freezing of a body of water is governed primarily by temperature, salinity, and water depth. However, winds, currents, and tides may retard the formation of ice. When strong gusty winds are present, the mixing of the water brings heat from lower depths and raises the temperature enough to prevent the forming of ice, even if the air is at subzero temperatures. Freshwater freezes at 0 degrees centigrade (C) (32 degrees Fahrenheit [F]), but the freezing point of seawater decreases about 0.28 degrees C (1.91 degrees F) per 5 percent increase in salinity. Shallow bodies of low-salinity water freeze more rapidly than deeper basins because a lesser volume must be cooled. Once the initial cover of ice has formed on the surface, no more mixing can take place from wind or wave action, and the ice will thicken. The first ice of autumn usually appears in the mouths of rivers that empty over a shallow continental shelf. During the increasingly longer and colder nights of autumn, ice forms along the shorelines as a semipermanent feature. It then widens by spreading into more exposed waters. When islands are close together, ice blankets the sea surface and bridges the waters between the land areas.

2-52. If personnel must enter an ice field, they should proceed cautiously. One-inch thick ice will stop most recreational boats and can do serious damage to the hull. Boat operators should take into account the time of ebb and flood tides; ice is generally more compact during the flood and is more likely to break up on the ebb. They should move at idle speed, but keep moving. It is important to be patient. Personnel will not be able to tell how thick ice is just by looking at the field in front of them. They should look at the broken ice at the stern of the boat. The boat should make no sharp turns. Operators should watch engine temperatures carefully because ice slush causes problems with water intakes; it rapidly clogs up filters and strainers. Personnel should also keep a good watch on the propellers, especially if encountering large chunks of ice. When backing down, operators should keep the rudder amidships to minimize damage.

2-53. One of the most serious effects of cold weather is that of topside icing, caused by wind-driven spray, particularly if the ice continues to accumulate. Ice grows considerably thicker as a result of splashing, spraying, and flooding. It causes an increased weight load on decks and masts (radar and radio). It

Chapter 2

introduces complications with the handling and operation of equipment. It also creates slippery deck conditions. Ice accumulation (known as ice accretion) causes the boat to become less stable and can lead to a capsizing.

2-54. Crew members should break ice away by chipping it off with mallets, clubs, scrapers, and even stiff brooms. However, crew members must be very careful to avoid damage to electrical wiring and finished surfaces.

Chapter 3
Tides and Tidal Currents

Successful amphibious landings are based on careful planning and a comprehensive knowledge of the environmental conditions that influence the landings. Weather, with its immediate effects on wind and waves, and the hydrography and topography of the BLS, are two of these environmental conditions. The third element is tide and tidal current data. Mistiming tides and tidal currents will have an immediate and obvious effect on the potential success or failure of a waterborne operation. History is full of invasion forces and raiding teams trapped and wiped out while crossing tidal flats at low tide. There were also many reconnaissance elements lost at sea or compromised by daylight because they could not make headway against contrary currents. Many of these operational disasters could have been avoided with proper prior planning.

Environmental conditions can affect every operation in a positive or negative manner. The height of the tide and the speed and direction of the tidal current can impede or halt a waterborne operation. These two elements require the detachment to properly conduct mission planning to ensure a positive impact. For operational teams to be successful, the height, direction, and speed of the tide must be compatible with the chosen infiltration method and must coincide with the hours of darkness. Adverse environmental elements can pose immediate threats to the detachment's infiltration plan. Therefore, compensating for these elements helps ensure successful completion of the mission.

After the detachment examines the environmental conditions and selects an infiltration time, it must develop viable contingency plans. To assist in developing contingency plans, the detachment should calculate the tide and tidal current data for at least 3 days before and 3 days after the desired time on target. For some missions, the environmental data may need to be computed for several weeks.

TIDES

3-1. Tide is the periodic rise and fall of the water accompanying the tidal phenomenon. A rising or incoming tide is called a **flood tide**, and a falling or outgoing tide is called an **ebb tide**. This variation in the ocean level is caused by the interaction of gravitational forces between the earth and the moon and, to a lesser extent, between the earth and the sun. Because the **lunar day** or **tidal day** is slightly longer than 24 hours (it averages 24 hours and 50 minutes), the time between successive high and low tides is normally a little more than 12 hours. When a high or low tide occurs just before midnight, the next high or low tide occurs approximately at noon on the following day; the next, just after the ensuing midnight, and so on.

3-2. The highest level reached by an ascending tide is called **high water**; the minimum level of a descending tide is called **low water**. The rate of rise and fall is not uniform. From low water, the tide begins to rise slowly at first but at an increasing rate until it is about halfway to high water. The rate of rise then decreases until high water is reached and the rise ceases. The detachment can then graphically plot the rate of rise and fall, as well as the speeds of the accompanying tidal currents, to determine optimal conditions for MAROPS. At high and low water, there is a brief period during which there is no change in the water level. This period is called **stand**.

Chapter 3

3-3. The total rise or fall from low water to high, or vice versa, is called the **range** of the tide. The actual height of the water level at high and low water varies with phases of the moon, variations of wind force and direction, atmospheric pressure, and other local causes. The average height of high water, measured over an extended period, is called **mean high water**. The average height of low water, measured in the same way, is called **mean low water**. The plane midway between mean high and mean low water is called **mean sea level** (at sea) and called **mean tide level** near the coast and in inshore waters. Figure 3-1 shows each of these levels.

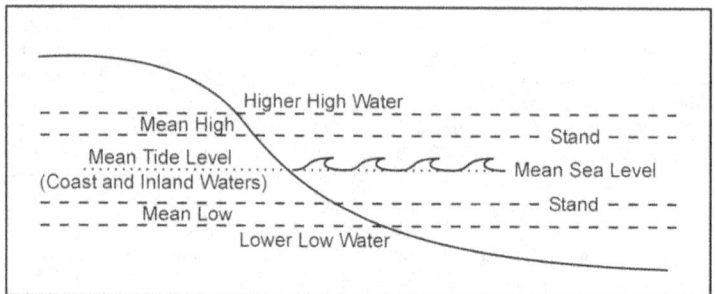

Figure 3-1. High and low tides

3-4. **Spring tides** occur near the time of full moon and new moon, when the sun and moon act together to produce tides higher and lower than average. When the moon is in its first or last quarter, it and the sun are opposed to each other, and **neap tides** of less than average range occur (Figure 3-2).

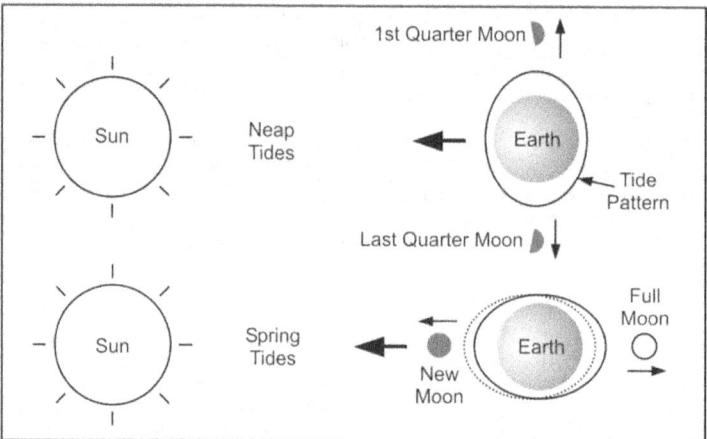

Figure 3-2. Neap and spring tides

3-5. Tides at a particular location are classified as one of three types: semidiurnal, diurnal, or mixed. In the semidiurnal tide, there are two high and two low waters each tidal day, with relatively small inequality in the high and low water heights. Tides on the Atlantic Coast of the United States are representative of the semidiurnal type.

3-6. In the diurnal tide, only one high and low tide (diurnal tide) occur each tidal day. These tides occur along the north coast of the Gulf of Mexico, in the Java Sea, in the Gulf of Tonkin (off the Vietnam-China coast), and in a few other localities.

3-7. In the mixed type of tide, the diurnal and semidiurnal oscillations are both important factors, and the tide is characterized by a large inequality in the high water heights, low water heights, or in both. There are usually two high and two low waters each day (semidiurnal), but occasionally the tide may become diurnal. Such tides are prevalent along the Pacific coast of the United States and in many other parts of the world. Where an inequality in the heights of high or low tides exists, the higher (or lower) of the two tides will be referred to as **higher high water** (or **lower low water**). Nautical charts in areas affected by mixed tides normally use this information to determine the depth or overhead obstruction datum.

3-8. A **tidal datum** is a level from which heights and depths are measured. Because water depths (soundings) measured during the initial charting of an area vary with tidal conditions, all soundings are converted to a common **chart sounding datum**. Many of these levels are important to the mariner.

3-9. Detachment personnel should already be familiar with the tiny figures that indicate depth of water on a nautical chart. For example, a 6. Under the title on the chart, it says "Soundings in feet at mean low water." It is important to remember that mean low water is only an average of the various depths actually sounded in one particular area at low water during the survey. When working in shallow water areas, the navigator should know the minimum depth of water the vessel will pass through. Depth varies with stages of the tide. The actual water level at low water may be above or below mean low water at different times because of the height of tide. The charted depth, shown by one of the small figures on the chart, is an average. It does not indicate the lowest depth to be found at all times at that particular point.

3-10. The charted depth is the vertical distance from the reference plane, called datum, on which soundings are based (usually, but not always, mean low water), to the ocean bottom. As discussed earlier, the actual depth of water can be less than the charted depth or below the reference plane. This number is shown by a minus (-) sign placed before the height of tide in the tide tables. The depth of water is equal to the algebraic sum of the charted depth and the height of tide.

3-11. Frequently, operations take place near reefs, rocks, shallows, flats, sandbars, or shoals. Navigators should use the Tide Tables to determine the actual depth of the water at a particular time and place. As a further safety measure, a lead line is a valuable adjunct.

3-12. Currently, charts are being changed to use **mean lower low water** as datum. Mean lower low water is the average of the lowest of the low waters each day and can differ significantly from mean low water.

TIDE TABLES

3-13. Navigators can predict both the tide height and tidal current with great accuracy by using Tide Tables and Tidal Current Tables published annually by the NOAA.

3-14. The NOAA Tide Tables are divided into four volumes that provide worldwide coverage. Within each volume are six tables that provide specific information about tide calculations. The tables are as follows:
- Table 1—Daily Tide Predictions.
- Table 2—Tidal Differences.
- Table 3—Height of the Tide at Any Time.
- Table 4—Local Mean Time of Sunrise and Sunset.
- Table 5—Reduction of Local Mean Time to Standard Time.
- Table 6—Moonrise and Moonset.

3-15. Navigators need specific data from Tables 1 and 2. Table 3 may be useful in certain areas. Tables 4, 5, and 6 provide meteorological data useful to celestial navigators. The G-2 or S-2 that supports the operational team can supply the latter information.

3-16. Explanations are given for particular cases and particular tides. Tides are predicted specifically for a number of principal ports. These ports are referred to, for prediction purposes, as reference stations. The water level for each high and low tide and the time of each tide is listed for every day of the year for each reference station. Volume 1, *East Coast of North and South America*, Table 1, lists only 48 reference stations. Figure 3-3, page 3-4, is an excerpt from Table 1 that shows daily tide predictions for Key West, Florida.

Chapter 3

Key West, Florida, 2001
Times and Heights of High and Low Waters

	April							May							June					
	Time	Height		Time	Height			Time	Height		Time	Height			Time	Height		Time	Height	
	h m	ft	cm	h m	ft	cm		h m	ft	cm	h m	ft	cm		h m	ft	cm	h m	ft	cm
1 Su ◐	0420 0758 1445 2227	0.8 0.6 1.6 -0.2	24 18 49 -6	16 M	0539 0843 1604 2334	0.9 0.7 1.4 0.2	27 21 43 6	1 Tu	0515 0932 1610 2307	1.0 0.6 1.6 0.0	30 18 49 0	16 W	0519 1035 1633 2323	1.1 0.7 1.3 0.3	34 21 40 9	1 F	0612 1214 1852	1.6 0.1 1.3	49 3 40	16 Sa	0532 1202 1818 2333	1.5 0.3 1.1 0.4	46 9 34 12
2 M	0547 0928 1615 2338	0.8 0.6 1.0 -0.2	24 18 49 -6	17 Tu	0632 1107 1728	1.0 0.6 1.4	30 18 43	2 W	0610 1101 1741	1.2 0.5 1.5	37 15 46	17 Th	0601 1143 1750	1.2 0.5 1.3	37 15 40	2 Sa	0003 0654 1313 1957	0.3 1.8 -0.1 1.3	9 55 -3 40	17 Su	0612 1253 1925	1.6 0.1 1.1	49 3 34
3 Tu	0647 1057 1747	0.9 0.5 1.7	27 15 52	18 W	0024 0707 1212 1836	0.2 1.1 0.5 1.4	6 34 15 43	3 Th	0000 0653 1215 1858	0.0 1.4 0.2 1.5	0 43 6 46	18 F ●	0002 0836 1236 1854	0.3 1.4 0.3 1.3	9 43 9 40	3 Su	0046 0733 1405 2052	0.3 1.9 -0.2 1.2	9 58 -6 37	18 M	0013 0652 1339 2022	0.5 1.8 -0.1 1.1	15 55 -3 34
4 W	0036 0731 1213 1903	-0.2 1.1 0.3 1.7	-6 34 9 52	19 Th	0103 0735 1303 1929	0.1 1.2 0.3 1.4	3 37 9 43	4 F	0046 0732 1317 2001	0.1 1.6 0.0 1.5	3 49 0 46	19 Sa	0037 0708 1322 1949	0.3 1.5 0.1 1.3	9 46 3 40	4 M	0127 0810 1451 2140	0.4 2.0 -0.3 1.2	12 61 -9 37	19 Tu	0053 0732 1424 2113	0.5 2.0 -0.3 1.2	15 61 -9 37
5 Th	0124 0809 1317 2006	-0.2 1.3 0.1 1.7	-6 40 3 52	20 F	0136 0802 1346 2015	0.1 1.3 0.2 1.5	3 40 6 46	5 Sa	0126 0807 1411 2056	0.1 1.7 -0.2 1.5	3 52 -6 46	20 Su	0109 0740 1403 2039	0.3 1.7 -0.1 1.3	9 52 -3 40	5 Tu ○	0207 0848 1535 2222	0.4 2.1 -0.4 1.2	12 64 -12 37	20 W	0134 0814 1509 2201	0.5 2.1 -0.4 1.2	15 64 -12 37
6 F	0205 0843 1414 2101	-0.2 1.5 -0.1 1.7	-6 46 -3 52	21 Sa	0204 0829 1425 2058	0.2 1.5 0.0 1.5	6 46 0 46	6 Su	0205 0841 1500 2145	0.2 1.9 -0.3 1.4	6 58 -9 43	21 M	0140 0812 1443 2126	0.4 1.8 -0.2 1.3	12 55 -6 40	6 W	0245 0922 1617 2302	0.4 2.1 -0.4 1.1	12 64 -12 34	21 Th ●	0216 0859 1556 2247	0.5 2.2 -0.5 1.2	15 67 -15 37
7 Sa ○	0243 0917 1506 2151	-0.1 1.6 1.6	-3 49 -9 49	22 Su	0231 0856 1503 2140	0.3 1.6 -0.1 1.5	9 49 -3 46	7 M ○	0241 0915 1546 2231	0.2 2.0 -0.4 1.3	6 61 -12 40	22 Tu	0213 0845 1524 2212	0.4 1.9 -0.4 1.3	12 59 -12 40	7 Th	0324 0958 1658 2339	0.5 2.0 -0.3 1.1	15 61 -9 34	22 F	0301 0945 1643 2334	0.4 2.3 -0.5 1.1	12 70 -15 34
8 Su	0320 0949 1556 2239	0.0 1.8 -0.4 1.5	0 55 -12 46	23 M ●	0259 0924 1541 2222	0.2 1.7 -0.3 1.4	6 52 -9 43	8 Tu	0318 0948 1630 2314	0.3 2.0 -0.4 1.2	9 61 -12 37	23 W	0248 0920 1608 2259	0.4 2.1 -0.5 1.2	12 64 -15 37	8 F	0401 1035 1740	0.5 2.0 -0.2	15 61 -6	23 Sa	0349 1034 1732	0.4 2.3 -0.5	12 70 -15
9 M	0355 1022 1644 2325	0.1 1.9 -0.5 1.4	3 58 -15 43	24 Tu	0327 0953 1621 2306	0.3 1.8 -0.4 1.3	9 55 -12 40	9 W	0354 1022 1714 2356	2.0 -0.4 1.2	61 -12 37	24 Th	0325 0959 1654 2347	0.4 2.1 -0.5 1.2	12 64 -15 37	9 Sa	0017 0440 1114 1824	1.1 0.5 1.9 -0.1	34 15 58 -3	24 Su	0021 0441 1126 1821	1.1 0.4 2.2 -0.4	34 12 67 -12
10 Tu	0430 1055 1731	0.2 1.9 -0.4	6 58 -12	25 W	0358 1024 1704 2353	0.3 1.9 -0.4 1.2	9 58 -12 37	10 Th	0430 1057 1800	0.5 2.0 -0.3	15 61 -9	25 F	0404 1041 1743	0.5 2.1 -0.5	15 64 -15	10 Su	0056 0521 1156 1908	1.0 0.6 1.8 0.0	30 18 55 0	25 M	0109 0540 1227 1911	1.2 0.4 2.0 -0.2	37 12 61 -6
11 W	0010 0505 1129 1821	1.2 0.3 1.8 -0.3	37 9 58 -9	26 Th	0431 1058 1751	0.4 1.9 -0.4	12 58 -12	11 F	0039 0506 1135 1848	1.1 0.5 1.8 -0.2	34 15 58 -6	26 Sa	0039 0449 1128 1836	1.1 0.5 2.1 -0.4	34 15 64 -12	11 M	0139 0609 1241 1955	1.0 0.6 1.6 0.1	30 18 49 3	26 Tu	0159 0648 1323 2002	1.2 0.4 1.8 -0.1	37 12 55 -3
12 Th	0057 0541 1206 1913	1.1 0.4 1.7 -0.2	34 12 55 -6	27 F	0043 0507 1138 1844	1.1 0.5 1.9 -0.3	34 15 58 -9	12 Sa	0124 0546 1217 1941	1.0 0.6 1.8 0.0	30 18 55 0	27 Su	0133 0540 1222 1933	1.1 0.5 2.0 -0.3	34 15 61 -9	12 Tu	0226 0709 1332 2042	1.1 0.7 1.5 0.2	34 21 46 6	27 W ○	0251 0808 1433	1.3 0.4 1.5 0.1	40 12 46 3
13 F	0148 0619 1248 2012	1.0 0.5 1.7 -0.1	30 15 52 -3	28 Sa	0141 0549 1225 1941	1.0 0.5 1.9 -0.3	30 15 58 -9	13 Su	0216 0632 1306 2039	1.0 0.7 1.6 0.1	30 21 49 3	28 M	0232 0644 1324 2033	1.1 0.6 1.8 -0.2	34 18 55 -6	13 W	0316 0827 1431 2128	1.1 0.7 1.4 0.3	34 21 43 9	28 Th	0344 0934 1554 2143	1.5 0.3 1.3 0.4	46 9 40 12
14 Sa	0250 0705 1338 2120	0.9 0.6 1.6 0.0	27 18 49 3	29 Su	0249 0643 1324 2054	0.9 0.6 1.8 -0.2	27 18 55 -6	14 M	0318 0737 1404 2140	0.9 0.8 1.5 0.2	27 24 46 6	29 Tu ◐	0333 0806 1437 2132	1.1 0.6 1.6 -0.1	34 18 49 -3	14 Th	0405 0950 1542 2212	1.2 0.6 1.2 0.3	37 18 37 9	29 F	0438 1055 1724 2253	1.6 0.2 1.2 0.3	49 6 37 9
15 Su ◐	0413 0811 1442 2231	0.8 0.7 1.4 0.1	24 21 43 3	30 M ○	0405 0757 1439 2204	0.9 0.7 1.7 -0.1	27 21 52 -3	15 Tu	0424 0907 1514 2236	1.0 1.0 1.4 0.2	30 30 43 6	30 W	0432 1104 1604 2227	1.2 0.5 1.5 0.1	37 15 46 3	15 F	0450 1103 1701 2253	1.3 0.4 1.2 0.4	40 12 37 12	30 Sa	0529 1205 1846 2323	1.7 0.1 1.1 0.4	52 3 34 12
											31 Th	0525 1103 1734 2318	1.4 0.3 1.3 0.2	43 9 40 6								

Figure 3-3. Table 1 excerpt, Daily Tide Predictions (Key West, Florida)

3-17. Navigators may want to predict tide levels and times at a greater number of locations than could possibly be listed in Table 1. Therefore, Table 2 (Figure 3-4, page 3-5) was devised to show the subordinate stations, or secondary points between the reference points listed in Table 1. It lists the time and height of the tide for the subordinate station as a correction to the times and heights of the tide at one of the reference stations (identified on each page of Table 2). Depending upon local conditions, the height of the tide at a subordinate station is found in several ways. If a difference for the height of high or low water is given, this is applied as the tables explain. The index and other explanations are provided to assist in finding the proper page of the table to use and to clarify any other unusual situations.

TABLE 2 – TIDAL DIFFERENCES AND OTHER CONSTANTS

No.	PLACE	POSITION		DIFFERENCES				RANGES		Mean Tide Level
		Latitude	Longitude	Time		Height		Mean	Spring	
				High Water	Low Water	High Water	Low Water			
		North	West	h m	h m	ft	ft	ft	ft	ft
	Florida Keys cont. Time meridian, 75° W			on Key West, p.164						
4187	Sugarloaf Key, Pirates Cove	24° 39.2'	81° 30.9'	-0 48	+1 41	*0.59	*0.75	0.74	0.92	0.55
4189	Cudjoe Key, Cudjoe Bay	24° 39.6'	81° 29.5'	-0 38	+0 41	*0.87	*0.71	1.18	1.48	0.76
4191	Summerland Key, southwest side, Kemp Channel	24° 39.0'	81° 26.8'	-0 26	+0 50	*0.81	*0.54	1.12	1.40	0.69
4193	Cudjoe Key, Kemp Channel Bridge	24° 39.7'	81° 28.1'	---	---	*0.59	*0.50	0.79	0.99	0.52
4195	Cudjoe Key, northeast side, Kemp Channel	24° 41.2'	81° 29.0'	+3 45	---	---	---	---	---	---
4197	Cudjoe Key, north end, Kemp Channel	24° 42.0'	81° 30.6'	+3 32	+4 40	*1.63	*1.46	2.17	2.71	1.43
4199	Sugarloaf Key, northeast side, Bow Channel	24° 40.3'	81° 32.0'	+3 47	+3 24	*1.01	*0.71	1.40	1.75	0.87
4201	Cudjoe Key, Pirates Cove	24° 39.8'	81° 30.8'	+3 50	-2 55	*0.77	*0.79	1.01	1.26	0.69
4203	Sugarloaf Key, north end, Bow Channel	24° 41.6'	81° 33.3'	+3 37	-5 20	*1.29	*0.75	1.82	2.28	1.09
4205	Pumpkin Key, Bow Channel	24° 43.0'	81° 33.7'	+3 17	+4 39	*1.56	*1.17	2.14	2.68	1.35
4207	Sawyer Key, outside, Cudjoe Channel	24° 45.5'	81° 33.7'	+2 45	+5 24	*1.57	*0.50	2.32	2.90	1.28
4209	Sawyer Key, inside, Cudjoe Channel	24° 45.5'	81° 33.7'	+2 37	+5 19	*1.43	*0.70	2.10	2.62	1.17
4211	Johnston Key, southwest end, Turkey Basin	24° 42.6'	81° 35.6'	+3 26	-5 38	*1.10	*0.50	1.59	1.99	0.92
	Upper Sugarloaf Sound									
4213	Perky	24° 38.9'	81° 34.2'	+5 37	+8 25	*0.28	-0.08	0.42	0.52	0.23
4215	Park Channel Bridge	24° 39.3'	81° 32.4'	+5 47	+8 33	*0.26	*0.29	0.34	0.42	0.24
4217	North Harris Channel	24° 39.0'	81° 33.2'	+5 32	+8 04	*0.25	*0.25	0.33	0.41	0.22
4219	Sugarloaf Shores East <26>	24° 38.6'	81° 33.6'	---	---	---	---	---	---	---
4221	Tarpon Creek	24° 37.8'	81° 31.0'	-0 29	-0 17	*0.35	*0.38	0.46	0.58	0.32
	Lower Sugarloaf Sound <27>									
4223	Sugarloaf Shores <27>	24° 38.0'	81° 33.1'	---	---	---	---	---	---	---
4225	Sugarloaf Beach <27>	24° 36.4'	81° 34.0'	---	---	---	---	---	---	---
4227	Sugarloaf Shores North <27>	24° 38.4'	81° 34.2'	---	---	---	---	---	---	---
4229	Saddlebunch Keys, south end <27>	24° 36.1'	81° 34.9'	---	---	---	---	---	---	---
4231	Lower Sugarloaf Channel Bridge <27>	24° 38.0'	81° 35.2'	---	---	---	---	---	---	---
4233	Saddlebunch Keys, Channel No. 2 <27>	24° 37.6'	81° 35.9'	---	---	---	---	---	---	---
4235	Saddlebunch Keys <27>	24° 37.1'	81° 36.1'	---	---	---	---	---	---	---
4237	Snipe Keys, southeast end, Inner Narrows	24° 39.5'	81° 36.5'	+3 25	+5 39	*1.28	*0.83	1.79	2.24	1.10
4239	Snipe Keys, Middle Narrows	24° 40.0'	81° 37.8'	+3 45	+5 54	*1.02	*0.67	1.42	1.78	0.87
4241	Snipe Keys, Snipe Point	24° 41.5'	81° 40.4'	+2 15	+3 33	*1.69	*1.29	2.31	2.89	1.47
4243	Waltz Key, Waltz Key Basin	24° 38.8'	81° 39.2'	+3 53	+5 33	*1.03	*0.96	1.36	1.70	0.91
4245	Duck Key Point, Duck Key, Waltz Key Basin	24° 37.4'	81° 41.1'	+3 27	+4 57	*1.19	*0.96	1.61	2.01	1.03
4247	O'Hara Key, north end, Waltz Key Basin	24° 37.0'	81° 38.7'	+3 53	+5 39	*1.03	*0.83	1.40	1.75	0.90
4249	Saddlebunch Keys, Channel No. 5	24° 36.7'	81° 37.5'	+4 32	+6 58	*0.66	*1.12	0.76	0.95	0.65
4251	Saddlebunch Keys, Channel No. 4	24° 36.9'	81° 37.0'	+4 35	+5 36	*0.54	*0.29	0.76	0.95	0.45
4253	Saddlebunch Keys, Channel No. 3	24° 37.4'	81° 36.2'	+1 44	-0 10	*0.43	*0.21	0.62	0.78	0.36
4255	Bird Key, Similar Sound	24° 35.3'	81° 36.2'	-0 21	+1 03	*0.59	*0.42	0.82	1.02	0.51
4257	Shark Key, southeast end, Similar Sound	24° 36.2'	81° 38.7'	-0 18	+1 51	*0.52	*0.46	0.70	0.88	0.46
4259	Saddlebunch Keys, Similar Sound	24° 36.0'	81° 37.3'	+0 39	+2 41	*0.37	*0.21	0.52	0.65	0.31
4261	Geiger Key, inside <26>	24° 35.0'	81° 39.3'	---	---	---	---	---	---	---
4263	Big Coppitt Key, northeast side, Waltz Key Basin	24° 36.1'	81° 39.3'	+4 21	+6 54	*0.84	*0.33	1.22	1.52	0.69
4265	Rockland Key, Rockland Channel Bridge	24° 34.5'	81° 40.1'	+5 02	+6 06	*0.76	*0.88	0.97	1.21	0.69
4267	Boca Chica Key, Long Point	24° 36.2'	81° 41.9'	+3 54	+5 22	*0.94	*0.71	1.28	1.60	0.81
4269	Channel Key, west side	24° 36.2'	81° 43.5'	+3 09	+3 07	*0.70	*0.71	0.91	1.14	0.62
4271	Boca Chica Channel Bridge	24° 34.6'	81° 43.2'	+1 23	+1 29	*0.57	*0.67	0.72	0.90	0.52
4273	Key Haven – Stock Island Channel	24° 34.8'	81° 44.3'	+2 25	+5 27	*0.73	*0.79	0.94	1.18	0.66
4275	Sigsbee Park, Garrison Bight Channel	24° 35.1'	81° 46.5'	+1 59	+2 06	*0.81	*0.88	1.04	1.30	0.73
4277	Key West, south side, Hawk Channel	24° 32.7'	81° 47.0'	-0 52	-0 30	*1.07	*0.92	1.44	1.80	0.94
4279	KEY WEST	24° 33.2'	81° 48.5'	Daily predictions				1.31	1.64	0.90
4281	Sand Key Lighthouse, Sand Key Channel	24° 27.2'	81° 52.6'	-1 03	-0 39	*0.94	*0.79	1.26	1.58	0.82
4283	Garden Key, Dry Tortugas	24° 37.6'	82° 52.3'	+0 29	+0 33	*0.94	*1.33	1.14	1.42	0.89
	Gulf Coast							Mean Diurnal		
4285	Cape Sable, East Cape	25° 07'	81° 05'	+3 56	+4 43	*2.26	*2.20	2.9	3.8	2.0
4287	Shark River entrance	25° 21'	81° 08'	+3 20	+4 38	*2.71	*2.50	3.8	4.5	2.4
4289	Whitewater Bay	25° 19'	81° 02'	---	---	*0.38	*0.38	0.5	0.8	0.4
4291	Lostmans River entrance	25° 33'	81° 13'	+3 22	+4 42	*2.31	*2.31	3.0	3.9	2.1
4293	Onion Key, Lostmans River	25° 37'	81° 08'	+5 32	+7 46	*0.46	*0.46	0.6	0.9	0.4
4295	Chatham River entrance	25° 41'	81° 17'	+3 22	+4 46	*2.82	*2.50	3.3	4.2	2.1
				on St. Marks River Ent., p.172						
4297	Pavilion Key	25° 42'	81° 21'	-0 57	-0 43	*1.23	*0.71	3.5	4.3	2.2
4299	Chokoloskee	25° 49'	81° 22'	+4 11	+1 07	*0.92	*0.55	2.8	3.2	1.6
4301	Everglades City, Barron River	25° 52'	81° 23'	+0 23	+1 18	*0.82	*0.47	2.3	2.9	1.4
4303	Indian Key	25° 48'	81° 28'	-1 05	-0 48	*1.23	*0.93	3.4	4.3	2.3
4305	Round Key	25° 50'	81° 32'	-1 06	-0 55	*1.23	*0.93	3.4	4.3	2.3
4307	Pumpkin Bay	25° 55'	81° 33'	+0 39	+1 00	*0.77	*0.77	2.1	2.7	1.3
4309	Coon Key	25° 54'	81° 38'	-0 45	-0 35	*0.99	*0.86	2.6	3.5	1.9
4311	Cape Romano	25° 51'	81° 41'	-1 17	-1 03	*0.99	*1.02	2.8	3.5	1.9
4313	Marco, Big Marco River	25° 58'	81° 44'	-1 04	-1 08	*0.68	*0.68	1.7	2.6	1.2
4315	Naples, Naples Bay, north end	26° 08.2'	81° 47.3'	-1 17	-1 11	*0.80	*0.80	2.06	2.85	1.58
4317	Naples (outer coast)	26° 08'	81° 48'	-1 59	-2 04	*0.85	*0.85	2.1	2.8	1.6
				on St. Petersburg, p.168						
4319	Estero Bay Little Hickory Island	26° 21'	81° 51'	-0 58	-1 05	*1.09	*1.09	---	2.5	1.3
4321	Coconut Point	26° 24'	81° 50'	-0 47	-0 40	*1.17	*1.17	---	2.7	1.3
4323	Carlos Point	26° 24'	81° 53'	-1 08	-1 28	*1.17	*1.17	---	2.7	1.4
4325	Matanzas Pass (fixed bridge) Estero Island	26° 27'	81° 57'	-1 10	-1 34	*1.22	*1.22	---	2.8	1.4
4327	Point Ybel, San Carlos Bay entrance	26° 27'	82° 01'	-1 50	-1 12	*1.21	*1.21	---	2.6	1.4
4329	Punta Rassa, San Carlos Bay	26° 29'	82° 01'	-1 01	-1 19	*1.04	*1.04	---	2.4	1.2

Figure 3-4. Table 2 excerpt, Tidal Differences (Florida Subordinate Stations)

Chapter 3

3-18. Detachment planners can better determine the optimum conditions if they can visualize the tide (and tidal current) during their proposed infiltration window. They can determine tide height at any point during the tidal cycle using the matrix (Table 3-1) or graphical methods (Figure 3-5, pages 3-7 and 3-8). How-to instructions for each method are explained in the charts. Tide and current predictions are affected by meteorological conditions, and therefore may not be 100-percent accurate. Local knowledge is also important.

Table 3-1. Constructing a tide table matrix

	High Tide		Low Tide	
	Time	Height (Ft)	Time	Height (Ft)
Reference Station	0148	1.0	0619	0.5
Subordinate Station 4275	+01:59	0.8	+02:06	0.88
Correction	0347	1.81	0825	1.38
	Time	Height (Ft)	Time	Height (Ft)
Reference Station	1248	1.7	2012	-0.1
Subordinate Station 4275	+01:59	0.81	+02:06	0.88
Correction	1447	2.51	2218	0.78

INSTRUCTIONS: Refer to Tables 1 and 2, NOAA Tide Tables. Data computed from Key West, Florida (Reference Station) and Sigsbee Park Florida (Subordinate Station 4275) on 13 April 2001.

Table 1 contains the predicted times and heights of the high and low waters for each day of the year at designated reference stations.

Table 2 lists tide predictions for many other places called subordinate stations. By applying the differences or ratios listed under the subordinate station to predictions listed at the reference station, the times and height of the tide can be corrected (Correction) nearest the desired location.

The height of the tide, referred to as the datum of the charts, is obtained by applying the correction of height differences or ratios to the charted depth.

Note. Non applicable low- or high-tide datum is ignored.

CURRENTS

3-19. As discussed earlier, tide is the vertical rise and fall of the ocean's water level caused by the attraction of the sun and moon. A **tidal current** is the result of a tide. Tidal current is the **horizontal motion of water** resulting from the vertical motion caused by a tide, distinguished from ocean or river currents or from those created by the wind. Tidal currents are of particular concern in small-boat operations.

3-20. The horizontal motion of water toward the land caused by a rising tide is called **flood current**. The horizontal motion away from the land caused by a falling tide is known as **ebb current**. Between these two, while the current is changing direction, is a brief period when no horizontal motion is visible. This time is called **slack water**.

3-21. An outgoing or ebb current running across a bar builds up a more intense sea than the incoming or flood current. This sea results from the rush of water out against the incoming ground swell that slows the wave speed and steepens the wave prematurely.

Tides and Tidal Currents

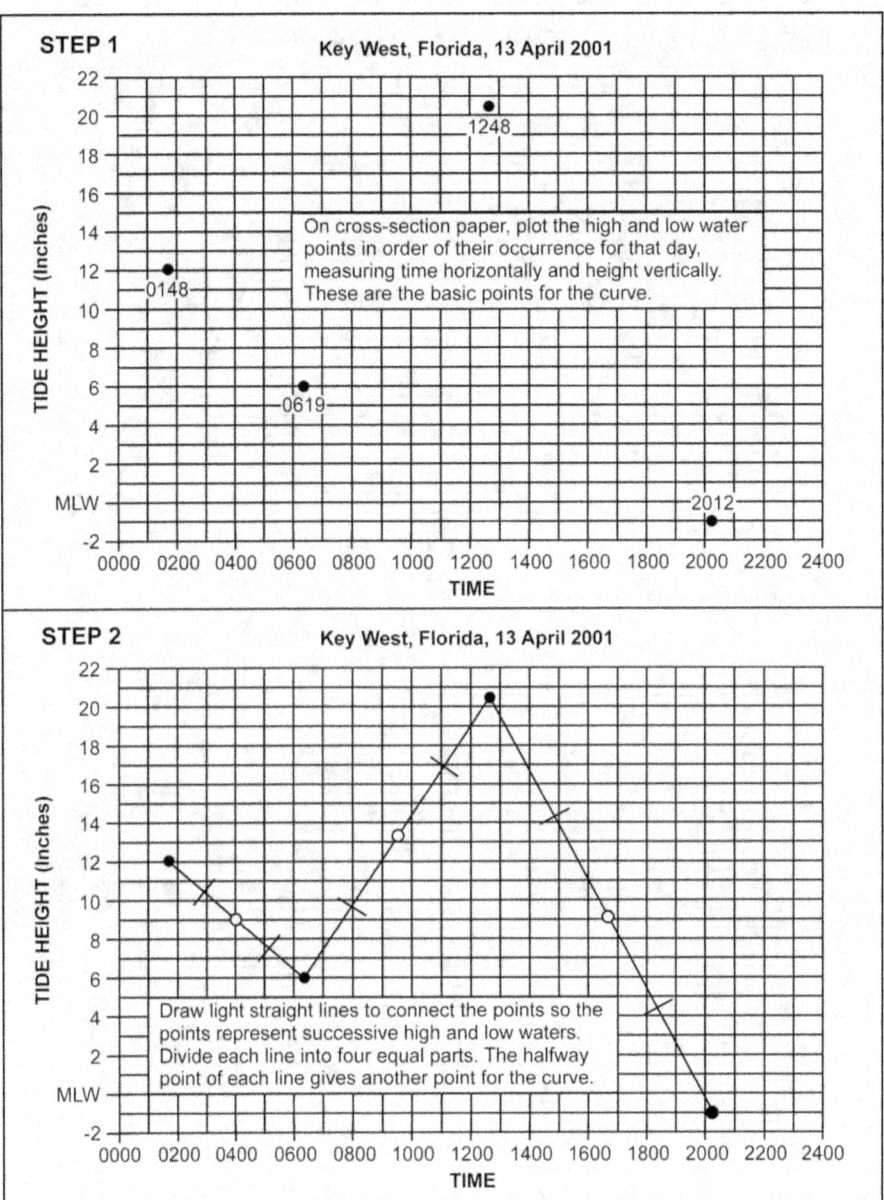

Figure 3-5. Plotting a tide curve

Chapter 3

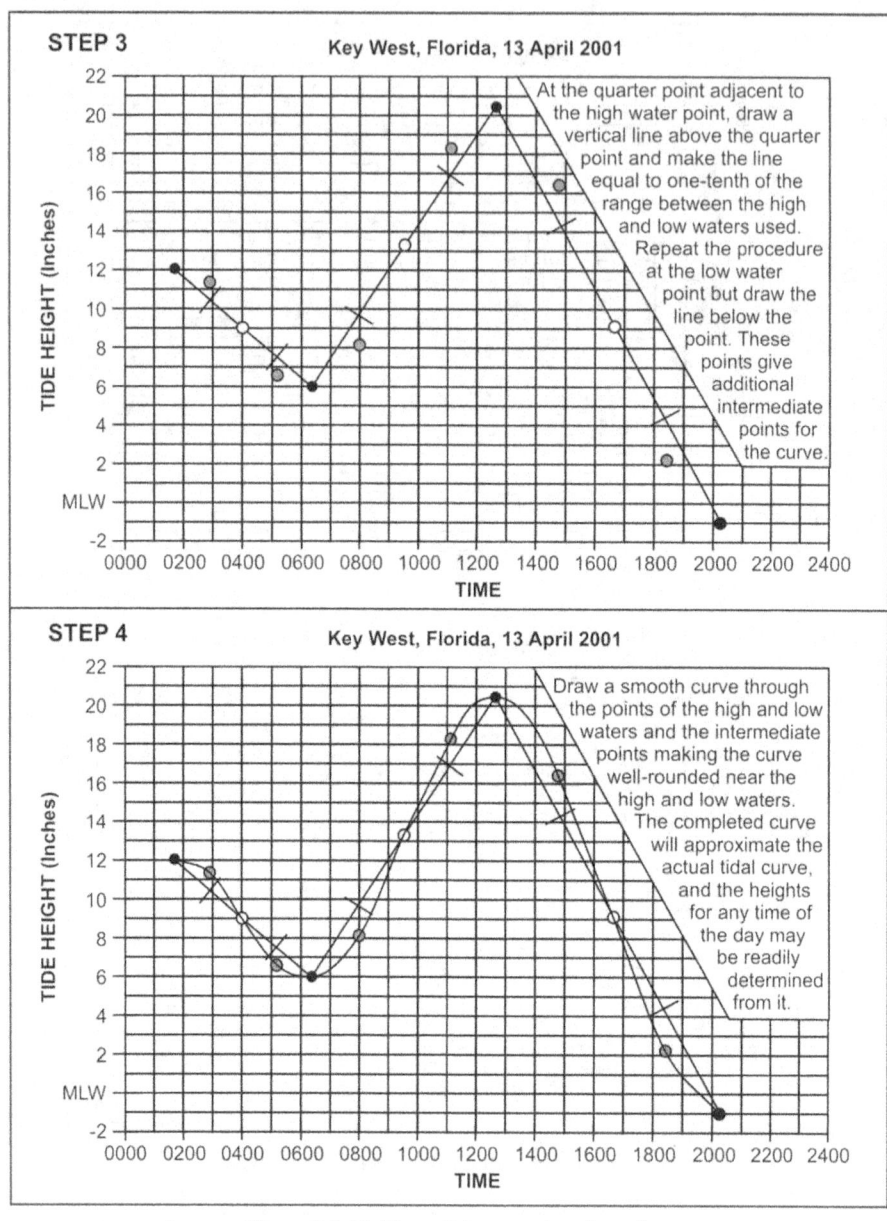

Figure 3-5. Plotting a tide curve (continued)

3-22. Some currents run parallel to the shore and inside the breakers. The water the waves carry to the beach causes these currents. They are called **longshore currents**. A navigator should pay close attention to this type of current because it can cause his boat to broach (capsize), or cause an object that he is searching for to move farther than he would expect.

3-23. Currents affect boat speed. When going with the current, the boat's speed over the ground is faster than the speed or revolutions per minute (rpm) indication. The effect is the same as that experienced by an aircraft affected by head or tail winds. When going against the current, the boat's speed over the ground is slower than the speed or rpm indication.

3-24. Currents affect boat maneuverability. When working in current, the navigator must remember that the boat's maneuverability depends on its speed through the water. The boat may have significant speed in relation to fixed objects (a pier, for example), but because the current is carrying it, the boat may lack maneuverability if too little water is flowing past its rudder.

3-25. When crossing the current to compensate for the set, the navigator may have to put the boat in a "crab." That is, he must turn the bow slightly into the current or wind. As a result of this maneuver, the boat's heading and the actual course made good will be different. Therefore, the navigator must "play" the current or wind by either sighting on a fixed object, such as a range, or by marking bearing drift on some object as nearly in line with his destination as possible.

3-26. Eddy currents (eddies) are swirling currents, sometimes quite powerful, that occur downstream (down current) of obstructions (for example, islands, rocks, or piers) at channel bends, near points of land, and at places where the bottom is uneven. Eddies can be dangerous to small boats. Navigators should watch for and avoid them.

3-27. Wind affects current speed. A sustained wind in the same direction as the current increases current speed by a small amount. A wind in the opposite direction slows it down and may create a chop. A very strong wind blowing directly into the mouth of an inlet or bay can produce an unusually high tide by piling up the water. (Similarly, a very strong wind blowing out of a bay can cause an unusually low tide and change the time of the high or low tide.)

3-28. The time of a tidal current's change of direction does not coincide with the time of high or low tide. The current's change of direction always lags behind the tide's turning. This time interval varies according to the physical characteristics of the land around the body of tidewater. For instance, there is usually little difference between the times of high or low tide and the time of slack water along a relatively straight coast with only shallow indentions. But where a large body of water connects with the ocean through a narrow channel, the tide and the current may be out of phase by as much as several hours. In this case, the current in the channel may be running at its greatest speed during high or low tide.

3-29. Each navigator operating in tidal waters must know the set (direction toward) and drift (velocity or speed expressed in kts) of the tidal currents in the area. He can use the NOAA Tidal Current Tables to predict the force and direction of tidal currents in most oceans. The NOAA publishes these tables annually and organizes them in a manner similar to the Tide Tables. The Tidal Current Tables are also divided into four volumes. Each volume consists of five tables that provide specific information on tidal current characteristics. They are—

- Table 1—Daily Current Predictions.
- Table 2—Current Differences and Other Constants.
- Table 3—Velocity of the Current at Any Time.
- Table 4—Duration of Slack.
- Table 5—Rotary Tidal Currents.

3-30. Navigators need the specific data contained in Tables 1 and 2. Like the Tide Tables, they are divided into a table for reference stations and a table for subordinate stations. Table 1 lists predicted times of slack water and predicted times and speed of maximum flood and ebb at the reference stations for each day of the year.

Chapter 3

3-31. Table 2 includes the latitude and longitude of each subordinate station (and reference stations), time and differences for slack water and maximum current, speed ratios for maximum flood and ebb, and direction and average speed for maximum flood and ebb currents. Figure 3-6 below, and Figure 3-7, page 3-11, show examples of each table.

Figure 3-6. Table 1 excerpt, Tidal Currents (Key West, Florida)

3-10 FM 3-05.212/MCRP 3-11.3A 30 September 2009

TABLE 2 – CURRENT DIFFERENCES AND OTHER CONSTANTS

No.	PLACE	Meter Depth	POSITION		TIME DIFFERENCES				SPEED RATIOS		AVERAGE SPEEDS AND DIRECTIONS							
			Latitude	Longitude	Min. before Flood	Flood	Min. before Ebb	Ebb	Flood	Ebb	Minimum before Flood		Maximum Flood		Minimum before Ebb		Maximum Ebb	
		ft	North	West	h m	h m	h m	h m			knots	Dir.	knots	Dir.	knots	Dir.	knots	Dir.
	MIAMI HARBOR–cont. Time meridian, 75° W				on Miami Harbor Entrance, p.72													
	Fishermans Channel																	
8136	Fisher Is., 0.2 mi. NW of	15d	25° 45.87'	80° 09.08'	+0 14	+0 38	+0 17	+0 39	0.6	0.7	--	--	1.0	280°	0.1	014°	1.1	090°
8141	Lummus Is., 0.15 mi. off SW corner	15d	25° 45.89'	80° 09.69'	+0 20	–0 20	+0 10	+0 22	0.3	0.6	--	--	0.6	271°	--	--	0.9	095°
8146	West end, SW of Dodge Island	11d	25° 46.36'	80° 10.74'	–0 05	–0 32	–0 15	–0 21	0.1	0.2	0.1	002°	0.2	277°	--	--	0.3	089°
8151	Miami River entrance	10d	25° 46.21'	80° 11.23'	+0 15	–0 02	–0 01	+0 46	0.1	0.4	--	--	0.2	261°	--	--	0.6	071°
8156	Fowey Rocks Light, 1.5 miles SW of		25° 35'	80° 07'	See table 5.													
	FLORIDA REEFS to MIDNIGHT PASS				on Key West, p.76													
8161	Caesar Creek, Biscayne Bay		25° 23.2'	80° 13.6'	+0 07	–0 08	–0 14	–0 05	1.2	1.0	0.0	--	1.2	316°	0.0	--	1.8	123°
8166	Long Key, drawbridge east of		24° 50.4'	80° 46.2'	+0 58	+1 27	+2 21	+1 33	1.1	0.7	0.0	--	1.1	000°	0.0	--	1.2	202°
8171	Long Key Viaduct		24° 48.1'	80° 51.9'	+1 34	+1 28	+2 02	+1 57	0.9	0.7	0.0	--	0.9	349°	0.0	--	1.2	170°
8176	Moser Channel, swingbridge		24° 42.0'	81° 10.2'	+1 07	+1 30	+1 50	+1 47	1.4	1.0	0.0	--	1.4	339°	0.0	--	1.8	166°
8181	Bahia Honda Harbor, bridge		24° 39.4'	81° 17.3'	+1 01	+0 39	+1 53	+1 05	1.4	1.2	0.0	--	1.4	004°	0.0	--	2.1	182°
8186	No Name Key, northeast of		24° 42.3'	81° 18.8'	+0 55	+1 24	+1 20	+0 53	0.7	0.5	0.0	--	0.7	312°	0.0	--	0.9	142°
	Key West																	
8191	Main Ship Channel entrance		24° 28.4'	81° 48.1'	–0 44	–0 12	+0 10	+0 10	0.2	0.3	0.0	--	0.2	040°	0.0	--	0.4	178°
8196	Main Ship Channel <42>		24° 30.5'	81° 48.3'	---	---	---	+0 30	--	0.2	0.0	--	--	064°	0.0	--	0.4	133°
8201	KEY WEST, 0.3 mi. W of Ft. Taylor		24° 32.9'	81° 49.0'	*Daily predictions*						0.0	--	1.0	022°	0.0	--	1.7	194°
8206	Ft. Taylor, 0.6 mile N of		24° 33.5'	81° 48.6'	+0 20	+0 13	–0 11	+0 13	0.6	0.7	0.0	--	0.6	042°	0.0	--	1.2	202°
8211	Turning Basin		24° 34.0'	81° 48.25'	+0 43	+0 44	+0 29	+1 06	0.8	0.6	0.0	--	0.8	048°	0.0	--	1.1	216°
8216	Northwest Channel		24° 35.0'	81° 50.9'	–0 08	–0 03	–0 09	–0 07	1.2	0.8	0.0	--	1.2	353°	0.0	--	1.4	162°
8221	Northwest Channel		24° 37.3'	81° 52.8'	–0 28	–0 19	–0 20	–0 20	0.6	0.4	0.0	--	0.6	346°	0.0	--	0.6	168°
8226	Boca Grande Channel		24° 34'	82° 04'	–0 40	–0 45	–0 01	–0 06	1.1	0.8	0.0	--	1.1	353°	0.0	--	1.2	194°
8231	New Ground <43>		24° 39.0'	82° 25.0'	+1 36	+1 55	+1 28	+1 18	0.7	0.4	0.2	356°	0.7	068°	0.2	142°	0.7	244°
8236	Isaac Shoal		24° 33.5'	82° 32.2'	+1 00	+0 54	+1 52	+1 55	1.0	0.5	0.0	--	1.0	002°	0.0	--	0.8	181°
8241	Southeast Channel		24° 37.62'	82° 51.07'	–0 27	–0 06	+0 37	+0 36	0.6	0.4	0.0	--	0.6	004°	0.0	--	0.6	172°
8246	Southwest Channel		24° 36.92'	82° 54.70'	+0 45	+0 59	+1 25	+2 04	0.4	0.4	0.0	--	0.4	001°	0.0	--	0.6	209°
					on Tampa Bay Entrance, p.80													
8251	Point Ybel, 0.4 mile northwest of		26° 27.40'	82° 01.12'	–0 25	–0 52	+0 17	+0 35	0.8	0.7	0.0	--	1.0	255°	0.0	--	0.9	080°
8256	Captiva Pass <44>		26° 36.56'	82° 13.34'	–0 53	–1 29	–1 14	–0 23	1.4	0.9	0.0	--	1.8	067°	0.0	--	1.9	251°
8261	Boca Grande Pass, Charlotte Harbor		26° 42.86'	82° 15.40'	–0 15	–0 37	–0 15	+0 05	1.7	1.3	0.0	--	2.2	057°	0.0	--	1.8	251°
8266	Pine Island Sound		26° 40.90'	82° 11.87'	---	---	---	---	--	--	0.0	--	0.5	011°	0.0	--	0.5	191°
8271	Little Pine I. bridge, Matlacha Pass		26° 37.9'	82° 04.1'	---	–0 19	---	---	0.4	--	0.0	--	0.6	132°	0.0	--	--	--
8276	Cape Haze, 2.3 mi. S of, Charlotte Hbr		26° 44.7'	82° 09.1'	+0 30	+0 41	–0 20	+1 18	0.4	0.4	0.0	--	0.5	080°	0.0	--	0.5	268°
8281	Punta Gorda, Peace River Bridge		26° 56.7'	82° 03.4'	---	---	---	---	0.3	0.2	0.0	--	0.4	047°	0.0	--	0.3	230°
8286	Myakka River bridge <53>		26° 57.5'	82° 12.8'	+1 48	+1 18	+1 47	---	0.4	--	0.0	--	0.5	304°	0.0	--	--	--
8291	Gasparilla Pass		26° 48.74'	82° 16.86'	–1 15	–1 13	–0 35	–0 41	0.8	0.8	0.0	--	1.0	066°	0.0	--	1.1	236°
8296	Venice Inlet		27° 06.8'	82° 26.0'	–2 05	–2 08	–1 57	–1 59	0.8	0.7	0.0	--	1.1	087°	0.0	--	0.9	262°
8301	Blackburn Bay, south end, bridge		27° 07.4'	82° 28.2'	–0 55	–1 20	–1 28	–0 10	0.7	0.5	0.0	--	0.9	357°	0.0	--	0.7	180°
8306	Little Sarasota Bay, south end, bridge		27° 10.8'	82° 29.7'	–1 19	–0 56	–0 57	–0 06	1.1	0.5	0.0	--	1.4	167°	0.0	--	0.7	357°
8311	Midnight Pass entrance		27° 12.4'	82° 30.6'	–1 43	–1 59	–1 49	–1 13	1.4	1.1	0.0	--	1.8	061°	0.0	--	1.4	242°
	SARASOTA BAY																	
8316	Big Sarasota Pass		27° 18.0'	82° 33.8'	–1 54	–1 49	–1 34	–2 03	1.2	0.8	0.0	--	1.5	006°	0.0	--	1.0	183°
8321	Sarasota Bay, south end, bridge		27° 18.1'	82° 32.8'	–1 25	–1 39	–1 13	–0 32	0.2	0.2	0.0	--	0.3	196°	0.0	--	0.3	013°
8326	New Pass		27° 19.9'	82° 34.9'	–2 06	–2 48	–1 18	–1 25	1.2	0.8	0.0	--	1.6	046°	0.0	--	1.0	231°
8331	Golden Gate Point, off		27° 19.7'	82° 33.4'	–1 38	–1 57	–1 25	–1 19	0.3	0.2	0.0	--	0.4	344°	0.0	--	0.3	159°
8336	Longboat Pass		27° 26.5'	82° 41.4'	–2 32	–2 42	–1 51	–1 56	1.4	1.2	0.0	--	1.8	088°	0.0	--	1.6	267°
8341	Cortez, north of bridge		27° 28.2'	82° 41.6'	–1 47	–1 10	–0 25	–1 11	0.5	0.1	0.0	--	0.6	346°	0.0	--	0.1	162°

Endnotes can be found at the end of table 2.

159

Figure 3-7. Table 2 excerpt, Tidal Currents (Florida Subordinate Stations)

3-32. Winds, variations in stream discharges produced by heavy rain or snow and ice melt, and other weather factors frequently affect current direction and speed. When any of these occur, actual current conditions vary from those predicted. The ability to estimate the amount by which they vary can be acquired only through experience in a particular area.

3-33. Like the tidal difference in time, the time differences are applied to the slack and maximum current times at the reference station to obtain the corresponding times at the subordinate station. Maximum speed at the subordinate station is found by multiplying the maximum speed at the reference station by the appropriate flood or ebb ratio.

3-34. Flood direction is the approximate true direction toward which the flooding current flows. Ebb direction is generally close to the reciprocal of the flood direction. Average flood and ebb speeds are averages of all the flood and ebb currents. Tidal Current Table 3 is similar to Table 3 of the Tide Tables. It is used to find current speed at a specific time.

3-35. Actual conditions often vary considerably from those predicted in the Tide Tables and the Tidal Current Tables. Changes in wind force and direction or in atmospheric pressure produce changes in ocean water level, especially the high-water height. For instance, the hurricane that struck the New England coast in September 1938 piled up a huge wall of water in Narragansett Bay. This wall of water increased to such a point that it became a huge storm wave when it struck the city of Providence. Generally, with an onshore wind or a low barometer, the high-water and low-water heights are higher than the predicted heights. With a high barometer or offshore wind, those heights are usually lower than predicted.

3-36. When working with the tidal current tables, the navigator should always remember that the actual times of slack or strength of current may sometimes differ from the predicted times by as much as 1/2 hour. On rare occasions, the difference may be as much as 1 hour. However, comparison between predicted and observed slack times shows that more than 90 percent of slack water predictions are accurate to within 1/2 hour. Thus, to fully take advantage of a favorable current or slack water, the navigator should plan to reach an entrance or strait at least 1/2 hour before the predicted time.

3-37. Tidal current calculations are by far the most critical factor pertaining to environmental conditions. The entire waterborne operation can be jeopardized without precise knowledge of the speed and direction of the tidal current. For example, a 1-kt ebb tidal current will halt any forward movement of a combat swimmer in the water. Even a 0.5-kt current will cause excessive fatigue in a combat swimmer.

CAUTION
Precise knowledge of speed and direction of a tidal current is essential to the success of a waterborne operation. Any current against the infiltration direction will slow down or even stop the forward progress of a combat swimmer.

3-38. The navigator can also make tidal current calculations using the matrix (Table 3-2, page 3-13) or the graphical method (Figure 3-8, pages 3-14 and 3-15). The results provide the information needed to determine the time period that is most suitable to conduct the waterborne operation. Once the navigator has the tide height and tidal current data, he can compare the data and execute a "backwards planning process" to determine the best start time for the infiltration.

3-39. The predicted slacks and strengths given in tidal current tables refer to the horizontal motion of water, not to the vertical height of the tide. Therefore, it is important to compute both tide height and tidal current to gain a complete picture of the tidal forces in the chosen AO.

Tides and Tidal Currents

Table 3-2. Constructing a tidal current matrix

	Slack	Maximum	Speed (kn)
Reference Station	0144	0443	1.3 ebb
Subordinate Station 8211	+00:29	+1:06	0.6
Correction	**0213**	**0549**	**0.78 ebb**
Reference Station	0843	1030	0.4 flood
Subordinate Station 8211	+00:43	+00:44	0.8
Correction	**0926**	**1114**	**0.32 flood**
Reference Station	1249	1649	1.6 ebb
Subordinate Station 8211	+00:29	+1:06	0.6
Correction	**1318**	**1755**	**0.96 ebb**
Reference Station	2058	2329	0.7 flood
Subordinate Station 8211	+00:43	+00:44	0.8
Correction	**2141**	**0013**	**0.56 flood**

INSTRUCTIONS: Refer to NOAA Tidal Current Tables 1 and 2. This data was computed from Key West, Florida (Reference Station) on 13 April 2001, and Key West, Turning Basin (Subordinate Station 8211).

Table 1 contains the list of reference stations for the predicted times of slack water and the predicted times and velocities of maximum current—flood and ebb—for each day of the year.

Table 2 enables the navigator to determine the approximate times of minimum currents (slack water) and the times and speeds of maximum currents at numerous subordinate stations.

By applying specific data given in Table 2 to the times and speeds of the current at the reference station, the navigator can compile reasonable approximations (Correction) of the current of the subordinate station.

3-40. The graphical method of depicting the tide and current predictions is an excellent tool for extended calculations. It provides a complete visual picture of the tidal forces during the operational time period.

OFFSET NAVIGATION

3-41. The navigator must apply the final result of the tide and tidal current calculations to the BLS. The environmental factors do not always coincide with the orientation of the launch point and the BLS. Therefore, the direction of the tidal current may not be perpendicular to the landing point. Combat divers and swimmers are very vulnerable to the effects of tidal currents. These currents will cause them to arrive downstream of their intended BLS. Navigators use offset navigation to compensate for the effects of currents not perpendicular to the shore. For combat swimmers to arrive at the intended BLS, they must compensate for two types of current: longshore currents moving parallel to the shore and flood currents that are other than 90 degrees to the BLS.

3-42. The determination to use offset navigation is based on the criticality of the currents. Criticality is determined based on the current's projected effect on swimmers. Because the swimmers need the most time to traverse any given distance, the farther away from shore the launch point is, and the longer the period of time that the swimmer is exposed to it, the more effect the current will have. For launches within 460 meters of the beach, currents of 0.5 kt or greater are considered critical. For launches in excess of 460 meters, a 0.2 kt current is considered critical.

Chapter 3

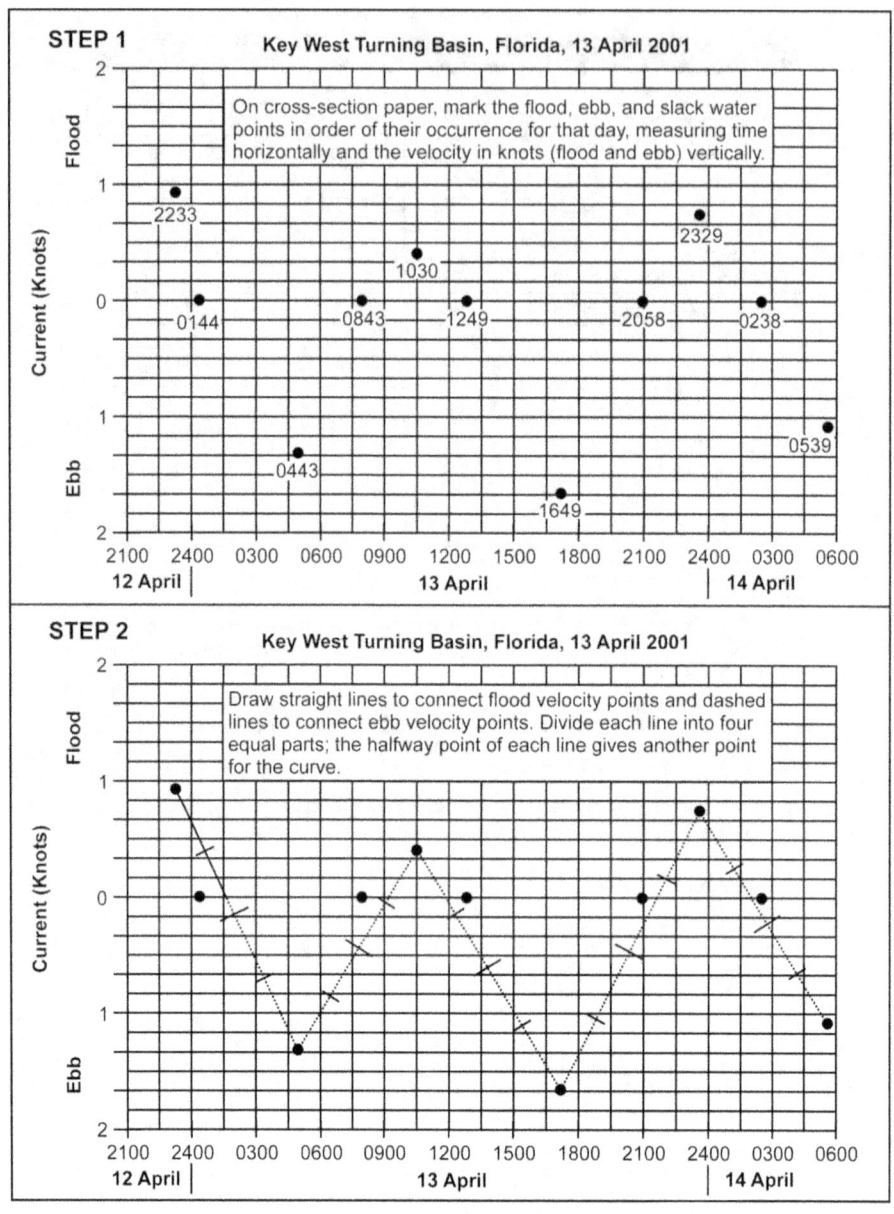

Figure 3-8. Plotting an approximate tidal current curve

Tides and Tidal Currents

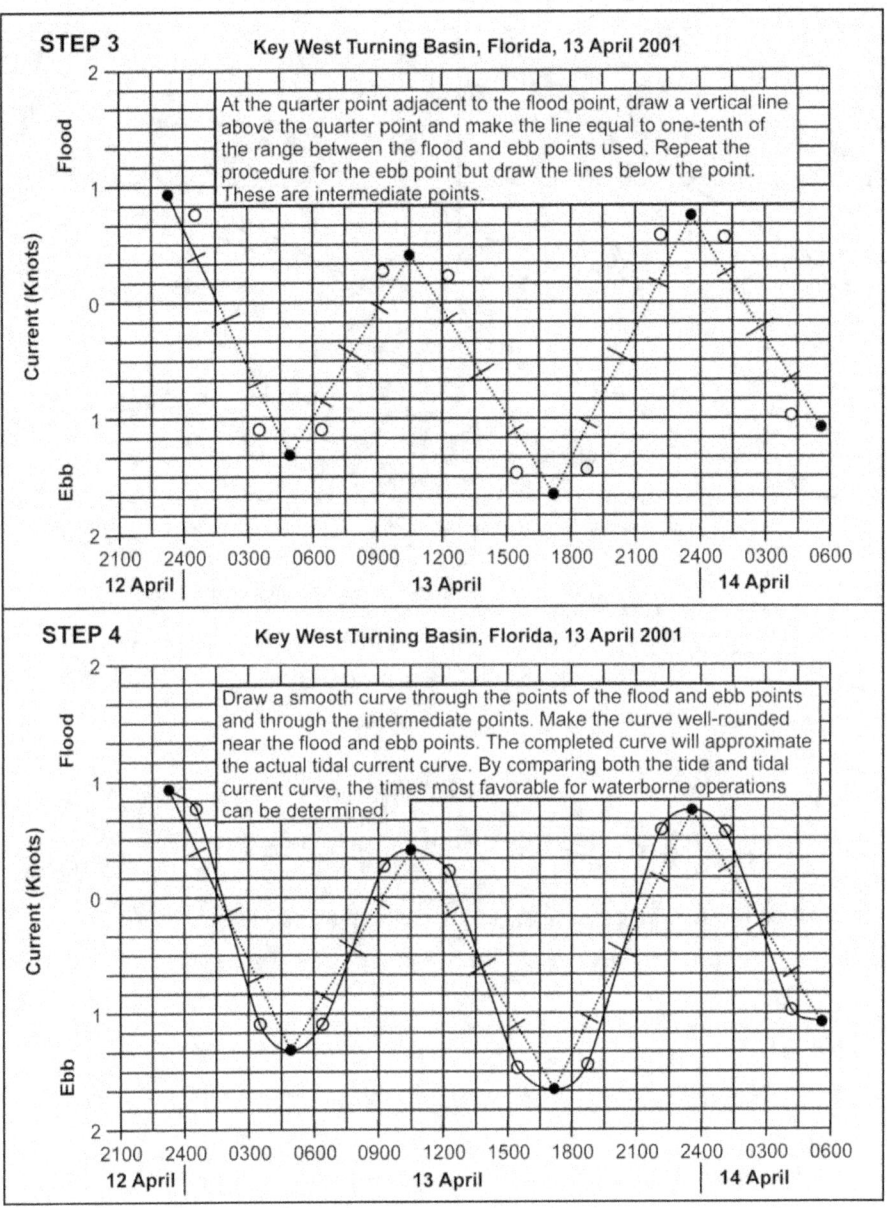

Figure 3-8. Plotting an approximate tidal current curve (continued)

3-43. Figure 3-9, page 3-16, illustrates the construction of a tidal current offset used to compensate for longshore and flood currents that may or may not be perpendicular to the shore.

30 September 2009   FM 3-05.212/MCRP 3-11.3A   3-15

Chapter 3

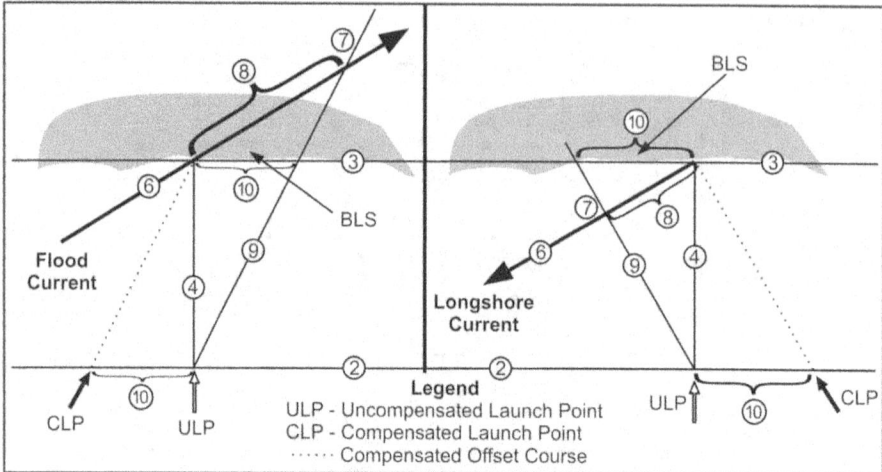

INSTRUCTIONS: From Tables 1, 2, and 3 of the NOAA current tables, compute the set and drift of the tidal current for the planned launch time at the subordinate station nearest to the launch point.

1) On the chart or map that includes the BLS, project a line parallel to the coastline. This line represents the track of the transporting vessel.
   A) The track is normally 2 miles offshore (the limit of horizontal visibility for an observer 3 feet above the surface of the water).
   B) The distance from the shoreline must be measured to scale.

2) Project a second line (BLS line) parallel along the coastline through the BLS.

3) Extend a perpendicular line from the BLS to the track. This line represents the course of the boat or swimmer unaffected by a current. The intersection of this line and the track is called the ULP.

4) Calculate the time required for passage from the ULP to the landing point: $T \text{ (time)} = \dfrac{D \text{ (distance)}}{S \text{ (speed)}}$

5) From the BLS, protract a line (azimuth) representing the set of the current. The direction of set of the current is listed as degrees **true** as listed in Table 2 of the current tables.

6) Compute the effect of current: duration of passage (Step 4) multiplied by the drift (speed) of the current.

7) Measure this value (effect on the current) along the set line (Step 5) using the same scale used in Step 1.

8) Draw a line connecting the ULP through the set of the current value on the set line to the BLS parallel line. This represents the course determined by the exposure to the current.

9) Compensate for the effect of the current on the BLS line by offsetting an equal value on the **upcurrent** side of the track. This produces a **minimum offset** from the derived CLP.

Figure 3-9. Offset navigation techniques

Chapter 4

Nautical Charts and Publications

Charts and publications are the navigator's library. This chapter introduces the detachment to some of the reference materials available for planning and executing MAROPS. The mission planner must learn as much as possible about his designated AO to improve the chances of mission success.

Without accurate, updated charts, mission planning and safe navigation are virtually impossible. These resources contain a wealth of invaluable information, and reflect channels, water depths, buoys, lights, lighthouses, prominent landmarks, rocks, reefs, sandbars, and much more information to aid in navigation. A thorough and complete understanding of the nautical chart is absolutely essential in ensuring the safe and successful navigation of a vessel.

Publications are the supplemental tools of the navigator's trade. They are the supporting reference materials that explain, amplify, update, or correct charts. They also provide important information, such as tides and current data that cannot be depicted on a chart.

This chapter will provide the SF maritime operator with a basic understanding of nautical charts and publications. To do mission planning and safe navigation, operators must be able to identify the required chart, order it, update (correct) it, interpret the information contained on it, and use it to plan and execute maritime movements.

NAUTICAL CHARTS

4-1. Charts, like maps, provide a graphic representation of features on the earth's surface. Unlike most maps, charts are primarily concerned with hydrography: the measurement and description of the physical features of the oceans, seas, lakes, rivers, and their adjoining coastal areas, with particular reference to their use for navigation.

4-2. Nautical charts are the mariner's most useful and widely used navigation aid. They are maps of waterways specifically designed for nautical navigation. They are a graphic depiction of a portion of the earth's surface, emphasizing natural and man-made features of particular interest to a navigator. Nautical charts cover an area that is primarily water, and include such information as the depth of water, bottom contours and composition, dangers and obstructions, the location and type of aids to navigation, coastline features, currents, magnetic variation, and prominent landmarks.

4-3. Charts are essential for plotting and determining mission position whether operating in familiar or unfamiliar waters. Navigators can order charts from the National Imagery and Mapping Agency (NIMA) Catalog of Hydrographic Products, 10th Edition, April 2000. Waterborne units can order all charts from the National Geospatial-Intelligence Agency (NGA). The other publications are available from the National Ocean Survey (a part of the NOAA), the Defense Mapping Agency Hydrographic and Topographic Center (DMAHTC), and the United States Coast Guard (USCG).

Chapter 4

4-4. NGA Catalog, Part 2, contains 12 volumes. It is used to locate and order charts. Volumes 1 through 9 list the charts for the nine regions of the world, Volume 10 contains the numbers for special purpose charts, Volume 11 lists classified charts, and Volume 12 details the ordering procedures. Commercial telephone access is available at 1-800-826-0342. Defense Switched Network (DSN) access is available at DSN 695-6500; callers should press the number corresponding to information needed after connection. Maritime operators should never get underway without the appropriate charts. They should also always ensure that their charts are corrected, up-to-date, and adequately prepared.

4-5. Basic chart information is contained in a number of places on a nautical chart. The general information block (Figure 4-1, page 4-3) contains the following items:
- The chart title, which is usually the name of the prominent navigable body of water within the area covered in the chart.
- A statement of the type of projection and the scale.
- The depth measurement unit (feet, meters, or fathoms).

4-6. Nautical charts contain a great deal of information. Throughout the next few paragraphs, specific information and parts of the charts will be explained.

CHART CLASSIFICATION AND SCALE

4-7. The scale of a nautical chart is the ratio between the distance (measurement unit) on the chart and the actual distance on the surface of the earth. Since this is a ratio, such as 1:2,500,000, it does not matter what size the unit is or in what system it was measured (inch, foot, meter). For example, the scale of 1:2,500 (1 inch on the chart equals 2,500 inches on the earth's surface) is a much larger number and is referred to as a large-scale chart. This concept is confusing at first since a large-scale chart actually represents an area smaller than one-half that of a small-scale chart. There is no firm definition for the terms large scale and small scale; the two terms are only relative to each other. It is easier just to remember "large scale = small area, and small scale = large area."

4-8. The primary charts used in small-boat navigation are classified into "series" according to their scale. Charts are made in many different scales, ranging from about 1:2,500 to 1:14,000,000 (and even smaller for some world maps). Navigators should always use the largest scale chart available for navigation. Navigators commonly use the types of nautical charts listed in Figure 4-2, page 4-4.

4-9. Charts used for tactical CRRC navigation should be tailored and prepared for a specific target area. A larger-scale chart (1:50,000) is recommended. A navigator should plot his key navigation aids; he can then cut or fold the chart to a usable size. After cutting down the chart, he can move the compass rose and latitude scale near the ITR. After all work has been completed and inspected, he ensures the chart is laminated front and back. A navigator must always ensure operations security (OPSEC) requirements have been considered when marking charts for infiltration routes.

MARGINAL INFORMATION

4-10. The **title block** shows the official name of the chart, the type of projection, scale, datum plane used, and unit of measurement for water depths. It may show other navigational information about the area that is useful to the navigator.

4-11. The **edition number and date** appear in the margin at the lower left-hand corner. Immediately following these figures will be a date of the latest revised edition, if any. The latest revision date shows all essential corrections concerning lights, beacons, buoys, and dangers that have been received to the date of issue. Corrections occurring after the date of issue are published in the Notice to Mariners. These must be entered by hand on the local area chart upon receipt of the notice.

Nautical Charts and Publications

Figure 4-1. Example of nautical chart information

NAUTICAL CHARTS	
Types	**Uses**
Sailing	Produced at scales of 1:6,000,000 and smaller. They are the smallest-scale charts used for planning a long voyage. Mariner uses them to fix his position, to approach the coast from the open ocean, or to sail between distant coastal ports. On such charts, the shoreline and topography are generalized. Show only offshore soundings, the principal lights, outer buoys, and landmarks visible at considerable distances. Also useful for plotting the track of a major tropical storm.
General	Produced at scales between 1:150,000 and 1:600,000. Used for coastal navigation outside of outlying reefs and shoals, when a vessel is generally within sight of land or aids to navigation, and its course can be directed by piloting techniques.
Coast	Produced at scales between 1:50,000 and 1:150,000. Intended for inshore coastal navigation where the course may lie inside outlying reefs and shoals. Also used for entering large bays and harbors or navigating intracoastal waterways.
Harbor	Usually used for navigation and anchorage in harbors and small waterways. The scale is generally larger than 1:50,000
Small-Craft	Produced at scales of 1:40,000 and larger. They are special charts of inland waters, including the intracoastal waterways. Special editions of conventional charts that are printed on lighter-weight paper and folded. Contain additional information of interest to small-craft operators, such as data on facilities, tide predictions, and weather broadcast information.
Approach	NIMA charts that incorporate the features of coast and harbor charts in one. Smaller than about 1:150,000 and especially useful for planning reconnaissance operations.
Combat	Special topographic and hydrographic products that contain all of the information of both charts and maps. Prepared to support combined and joint operations.

Figure 4-2. Navigation guides

4-12. Charts will often have printed **notes** with information of considerable importance, such as navigation regulations and hazardous conditions. All notes should be read attentively because they contain information that cannot be presented graphically, such as—

- The meaning of abbreviations used on the chart.
- Special notes of caution regarding danger.
- Tidal information.
- Reference to anchorage areas.

4-13. Most charts will have sets of graphic bar **scales**. Each set consists of a scale of nautical miles and one of yards. Charts of intracoastal waterways will also include a set of statute miles. With increasing conversion to the metric system, the planner may also find scales of kilometers on future charts.

4-14. The logarithmic speed–time–distance scale provides a very useful graphic calculation of any one of the three variables if the other two are known. Speed is expressed in how far a ship will travel in sixty minutes and the distance can be expressed in yards or nautical miles. For example, a ship is going 10 kts. How far will it go in thirty minutes? To find distance, the planner must first place the right point of dividers on 60 and the left point on the ship's speed of 10 kts. He then, without changing the spread of the dividers, must place the right point on minutes run; the left point will then indicate distance. Or, place the left point on distance; the right point will indicate time (Figure 4-3). To find speed, reverse the process.

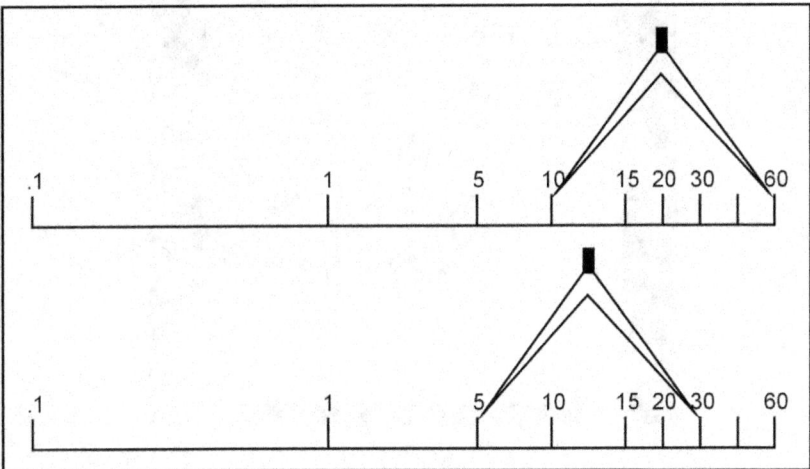

Figure 4-3. Example of a logarithmic speed scale

4-15. Similar logarithmic scales can be found on maneuvering board speed–time–distance nomograms where there are separate time, speed, and distance lines. To use the nomogram, the navigator draws a line between two known factors. He then can find the third factor at the intersection of the extended line at the third scale (Figure 4-4). For example, to use the nomogram, place pencil marks on the two given quantities, and the third quantity is ready by placing a straight edge over the two marks and observing the point of intersection on the third scale.

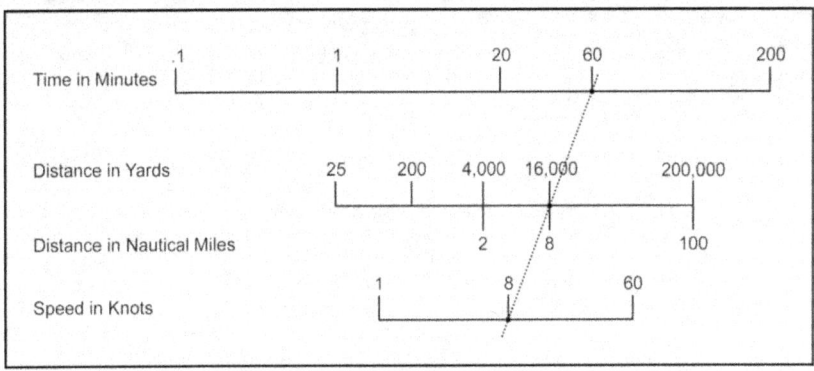

Figure 4-4. Example of a time–speed–distance logarithmic scale

THE NAUTICAL SLIDE RULE

4-16. The nautical slide ruler is similar to the maneuvering board, time-speed-distance nomogram, except the three scales, have been bent into circular form on a plastic base and covered by a plastic faceplate. To use the instrument, the known values are set by rotating the slide rule to the appropriate positions, and the third factor appears by the appropriate arrow (Figure 4-5).

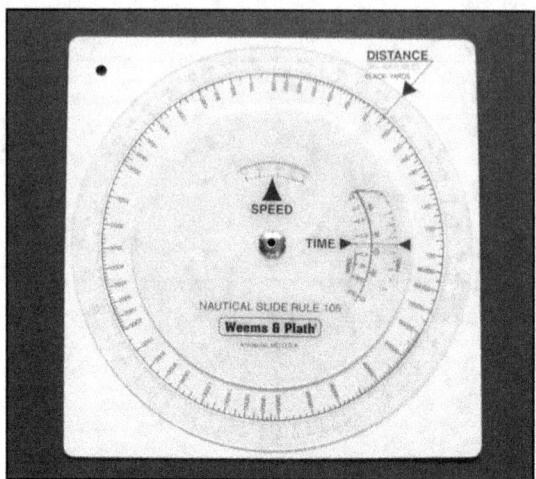

Figure 4-5. The nautical slide rule

4-17. The nautical slide rule consists of two circular dials mounted on a plastic base. The planner must always set the distance first when it is one of the known factors. Due to the fact that the speed scale is read through both dials, this setting should always be made last when speed is one of the known factors.

4-18. The time scale gives hours in green figures and minutes and seconds in black figures. Seconds are listed only to 120 and cannot be used as minutes and seconds, but only as total seconds for time runs of less than two minutes. Likewise, the separate hour scale and minute scale are not combined as hours and minutes but used only as hours and fractions of an hour, or as total minutes. To illustrate, the planner sets the time line on 1.5 minutes and notes that it is also on 90 seconds. Either unit may be used. Now he sets the time line on 150 minutes and notes that it also reads 2.5 hours.

4-19. The speed scale reads from 1 to 100 kts or miles per hour (mph). The distance scales are given in nautical miles (green figures) or yards (black figures). For example, the planner sets 3 nm on the distance line and notes that this can also be called 6,000 yards depending on which unit is being used for distance.

4-20. The green distance figures may also be used as statute miles, but in this case the speed will be in statute mph instead of kts. When using statute miles, the yard scales must not be used. Some examples follow:

- *Example 1:* To find speed when time and distance are known. The planner assumes that he has run a distance of 12 miles as measured on his chart between buoys and it took him 80 minutes to cover this distance. First, he turns the outer dial until 12 miles is on the DISTANCE line, then turns the inner dial until 80 minutes is under the TIME line. He should now read his speed as 9 kts.
- *Example 2:* In this type of problem, the planner knows that for a given run his boat will make a good speed of 15 kts and he wants to know how far he has traveled in 2 1/2 hours (150 minutes). The planner turns the inner dial until the time line is over 2.5 hours, then he turns the two dials together until 15 is opposite the speed marker. He should now read 37 1/2 miles at the distance line.

- *Example 3:* To find the length of time it will take to go a given distance at a known speed. For this type of problem, the planner wants to make a trip of 30 miles and the speed of his boat is 19 kts. First, he sets the outer dial so that the distance line is on 30 miles. He holds this setting in place with his thumb and turns the inner dial to a speed of 19 knots and reads off his time of 95 minutes for the trip.

4-21. The number of **colors** used on the chart varies with the agency that published the chart. The National Ocean Survey uses several basic colors and shades of them on its regular charts. They are as follows:

- *Buff or Yellow*—represents land areas, except on DMAHTC charts, which will show land in a gray tint.
- *White and Blue*—represents water. White is used for deep water and blue for shallow water.
- *Green*—signifies areas that may be covered during different tidal ranges and uncovered during others, such as mud flats, sandbars, marshes, and coral reefs. A greenish halftone will also be used for places that have been wire-dragged with the depth of drag indicated alongside.
- *Nautical Purple*—appears extensively throughout the chart. This color is used because it is easy to read under a red light. For example, red buoys, red day beacons, lighted buoys of any color, caution and danger areas, compass roses, and recommended courses.
- *Black*—represents most symbols and printed information.
- *Gray*—shows land masses on some charts.

4-22. **Vertical lettering** (nonslanted) is used for features that are dry at high water and are not affected by the movement of water. Slanted lettering is used for water, under water, and floating features with the exception of depth figures.

4-23. **Bottom composition** can be abbreviated and be in slanted lettering. For example, GRS, HRD, and S. Figure 4-15, page 4-15, provides a complete list of bottom characteristics with approved abbreviations.

4-24. **Fathom curves** are connecting lines of equal depths usually shown as 1, 2, 3, 5, 10, and multiples of ten (1 fathom = 6 feet).

4-25. **Depths** are indicated in feet or fathoms. Some charts use feet for shallow water and fathoms for deep water, but this practice is usually rare. Some newer charts use meters and decimeters.

GEOGRAPHIC COORDINATES

4-26. Nautical charts are oriented with north at the top. The frame of reference for all chart construction is the system of parallels of latitude and longitude. Any location on a chart can be expressed in terms of latitude and longitude. The latitude scale runs along both sides of the chart, while the longitude scale runs across the top and bottom of the chart.

4-27. Navigators use the latitude and longitude scales to pinpoint objects on the chart just as six- and eight-digit grid coordinates are used on a military map. There are scales on the top and bottom for longitude and on each side for latitude. Each scale is broken down in degrees, minutes, and seconds; the size of the scale will vary according to the scale of the chart.

Lines of Longitude (Meridians)

4-28. Meridians of longitude run in a north-and-south direction and intersect at the poles (Figure 4-6, page 4-8). The meridian that passes through Greenwich, England, is the reference for measurements of longitude. It is designated as the Prime Meridian or 0 degrees longitude. The longitude of any position is related to degrees East (E) or degrees West (W) from Greenwich to the maximum of 180 degrees (International Date Line) either way. East and West are an essential part of any statement using longitude. Degrees of longitude are always written as three-digit numbers and include the Easting or Westing, for example, 034° E or 123° W. The scales used to determine the longitude of any particular point are located at the top and bottom of the chart.

Chapter 4

> *Note.* A degree of longitude is equal to 60 nm only at the equator. As the meridians approach the poles, the distance between them becomes proportionally less. For this reason, longitudinal lines are never used to measure distance in navigational problems.

Lines of Latitude (Parallels)

4-29. Parallels of latitude run in an east-and-west direction. They are measured from the equator starting at 0 degrees and increasing to 90 degrees at each pole (Figure 4-6). Navigators should put N for North or S for South when using parallels of latitude. Lines running from side-to-side indicate the latitude parallels on a nautical chart; the latitude scales are indicated along the side margins by divisions along the black and white border. One degree of latitude (arc) is equal to 60 nm on the surface of the earth.

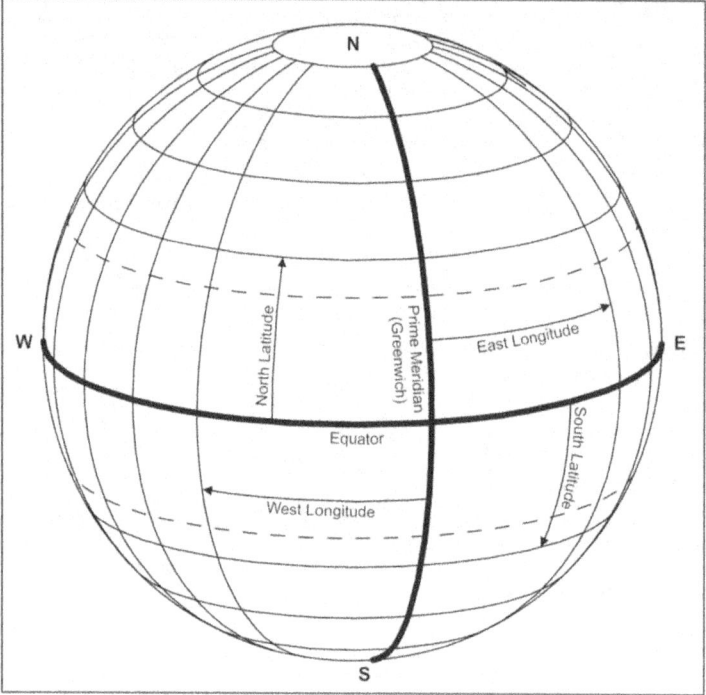

Figure 4-6. Global latitude and longitude lines

4-30. When navigators use latitude and longitude to locate necessary objects, they should keep the following points in mind:
- For greater precision in position, degrees are subdivided into minutes (60 minutes equals 1 degree) and seconds (60 seconds equals 1 minute).
- When moving closer to the poles, the meridians of longitude get closer together due to the elliptical shape of the earth; therefore, there is no set distance between them.
- For all practical purposes, parallels of latitude are essentially equally spaced and the distance is basically the same. One degree of latitude equals 60 nm, 1 minute of latitude equals 1 nautical mile, and 1 second of latitude equals 100 feet.

Conversion Factors

4-31. An example of a proper geographic coordinate should read as follows: 32° 40' N 117° 14' W. The following conversion factors will help navigators interpret data and apply it to nautical charts (Figure 4-7).

---

1 second (") = 100 feet (ft)
60 seconds = 1 minute of latitude
1 minute (') = 1 nm
60 minutes = 1 degree (°)
6,076.10 feet = 1 nm
2,024.5 yards = 1 nm
1.85 km = 1 nm
1 km = 0.539 nm
5,280 feet = 1 statute mile
1,760 yards = 1 statute mile
0.868 nm = 1 statute mile

Note. The ground distance covered by 1 degree of longitude at the equator is also about 30 meters (100 feet) but decreases as one moves north or south, until it becomes zero at the poles. For example: 1 second of longitude at the equator is 100 feet, but at the latitude of Washington, DC, 1 second of longitude is approximately 78 feet.

Figure 4-7. Conversion factors

Plotting a Geographic Coordinate

4-32. Plotting a position when latitude and longitude are known is a simple procedure. Navigators should—
- Mark the given latitude on a convenient latitude scale along the meridian bar.
- Place a straightedge at this point running parallel to the latitude line.
- Hold the straightedge in place, and set a pair of dividers to the desired longitude using a convenient longitude scale.
- Place one point on the meridian at the edge of the straightedge moving in the direction of the given longitude, making sure not to change the spread of the dividers.
- Lightly mark the chart at this point.

4-33. The navigator can then determine the coordinates of a point on the chart. He should—
- Place a straightedge at the given point and parallel to a line of latitude.
- Determine the latitude where the straightedge crosses a meridian.
- Hold the straightedge in place, and set one point of the dividers at the given point and the other at the nearest interior meridians on longitude.
- Place the dividers on a longitude scale and read the longitude from this point, making sure not to change the spread.

DETERMINING DIRECTION

4-34. Direction is measured clockwise from 000 degrees to 360 degrees. When speaking of degrees in giving a course or heading, navigators should always use three digits; for example, "270 degrees" or "057 degrees." Directions can be in true degrees, magnetic degrees, or compass degrees. True direction uses the North Pole as a reference point. Magnetic direction uses the magnetic north. There are important differences between true and magnetic compass directions. True direction differs from magnetic direction by variation.

4-35. The compass rose is the primary means of determining direction. Nautical charts usually have one or more compass roses printed on them (Figure 4-8, page 4-10). These are similar in appearance to the compass card and are oriented with north at the top. A compass rose consists of two or three concentric circles several inches in diameter and accurately subdivided. Personnel use these circles to measure true and magnetic directions on the chart. True direction is printed around the outer circle with zero at true north; this is emphasized with a star. Magnetic direction is printed around the inside of the compass rose with an arrow pointing to magnetic north. Variation for the particular area covered by the chart is printed in the middle of the compass rose (as well as any annual change). The middle circle, if there are three, is magnetic direction expressed in degrees, with an arrow printed over the zero point to indicate magnetic north. The innermost circle is also magnetic direction, but is listed in terms of "points." There are 32 points at intervals of 11 1/4 degrees, further divided into half and quarter points. Stating the names of the points is called boxing the compass.

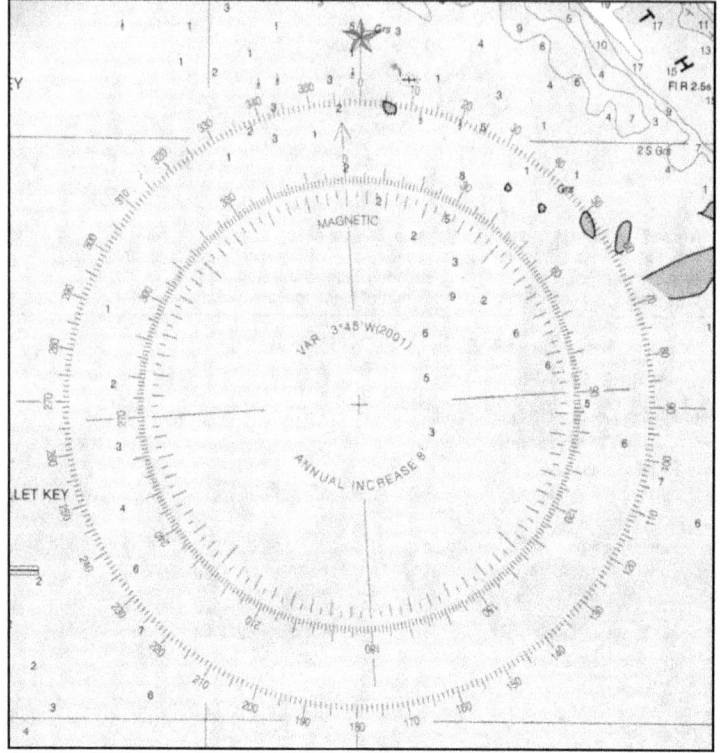

Figure 4-8. The compass rose

4-36. Each chart will have several compass roses printed on it at convenient locations where they will not conflict with navigational information. Roses printed on land areas may cause the elimination of typographical features in these regions.

4-37. Several cautions are necessary when measuring directions on charts. When large areas are covered, it is possible for the magnetic variation to differ for various portions of the chart. Navigators should check each chart before using it to be sure to always use the compass rose nearest the area for which they are plotting. Depending upon the type and scale of the chart, graduations on the compass rose circles may be for intervals of 1, 2, or 5 degrees. Navigators can use the following procedure for determining direction from point to point:

- Place parallel rulers so that the edge intersects both points.
- "Walk" rulers across the chart to the compass rose, without changing the angle of the rulers.
- Read degrees in the direction that the boat is heading (magnetic or true).

4-38. In summary, a navigator can express direction in three ways. He can refer to it as—
- *True*—differs from magnetic by variation when using the true (geographic) meridian.
- *Magnetic*—differs from compass by deviation when using the magnetic meridian.
- *Compass*—differs from true by compass error, the algebraic sum of deviation and variation, when using the axis of the compass card.

DETERMINING DISTANCE

4-39. The latitude bar is the primary scale used to measure distance. One degree of latitude is equal to 1 nautical mile. Latitude bars are located on the right and left side of the chart. The longitude scales are not used to measure distance due to the distortion of the Mercator projection. On large-scale charts, navigators should use the bar scale provided. The tool used to measure distance is the dividers. There are two different methods that can be used.

4-40. On large-scale charts, the navigator simply spreads the dividers and places the points on the distance to be measured. He then (without spreading the dividers further) places them on the bar scale provided in the upper and lower margin.

4-41. On small-scale charts, the navigator first measures off 5 to 10 nm on the latitude bar. Next, he places one point of the dividers on the desired point of origin. From this point, he walks the dividers along the plotted vessel track (counting the turns as he walks the dividers) to the end. If there is any distance left at the end of the track, he closes the dividers to the end of the vessel track and measures the remainder on the latitude scale (bar), adding this measurement to his distance (Figure 4-9).

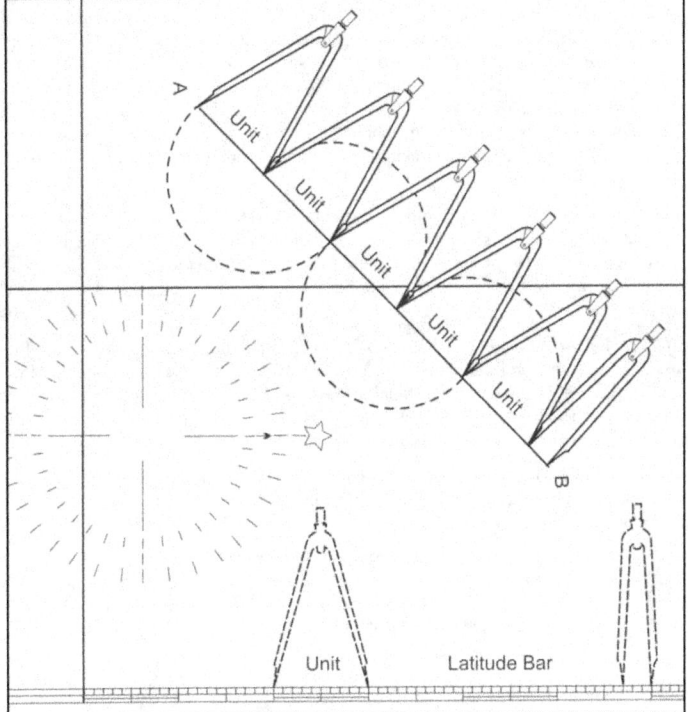

Figure 4-9. Determining distance on charts

Chapter 4

SOUNDINGS

4-42. Another important type of information that the navigator can learn from using the nautical chart is the bottom characteristics in the water. The chart provides this data by using a combination of numbers, color codes, underwater contour lines, and a system of standardized symbols and abbreviations.

4-43. Most of the numbers on the chart represent water depth soundings at mean low tide. Datum refers to the baseline from which a chart's vertical measurements are made. On the East and Gulf coasts, the tidal datum is mean low water (the average low tide). The tidal cycle on the East and Gulf coasts produces tides approximately equal in highness and lowness. Since the greatest danger to navigation is during low tide, a number of low-tide depths are averaged to produce the average low tide. On the Pacific coast, the datum is the mean lower low-water mark. The reason for using the mean lower of the two low tides for the West coast tidal datum is because the cycle of low tides may differ by several feet, thus making one lower than the other. In the interest of navigation safety, the mean or average of the lower of the two tides in the tidal cycles is used for soundings.

4-44. Contour lines, also called fathom curves, connect points of roughly equal depth and provide a profile of the bottom. These lines are either numbered or coded according to depth using particular combinations of dots and dashes. Generally, the shallow water is tinted darker blue on a chart, while deeper water is tinted light blue or white. Water depth may either be in feet, meters, or fathoms (a fathom equals 6 feet). The chart's legend will indicate which unit (feet, meters, or fathoms) is used. The nautical chart's water depth is measured downward from sea level at low water (soundings); heights or landmarks are given in feet above sea level.

SYMBOLS AND ABBREVIATIONS

4-45. Symbols and abbreviations appear in Chart No. 1 published jointly by the DMAHTC and the NOAA. They indicate the physical characteristics of the charted area, and details of the available aids to navigation. They are uniform and standardized, but are subject to variation, depending on the chart's scale or chart series. Generally speaking, man-made features will be shown in detail where they will reflect directly to waterborne traffic, such as piers, bridges, and power cables. Built-up areas will be determined or identified by their usefulness to navigation. Prominent isolated objects like tanks and stacks will be shown accurately so that they may be used for taking bearings. The following paragraphs explain these symbols and abbreviations.

4-46. **Color.** Nearly all charts employ color to distinguish various categories of information, such as shoal water, deep water, and land areas. Color is also used with aids to navigation to make them easier to locate and interpret. Nautical purple ink (magenta) is used for most information, as it is more easily read under red nighttime illumination normally used in the pilothouse of a small boat or bridge of a ship to avoid interference with night vision.

4-47. **Lettering.** Lettering on a chart provides valuable information. For example, slanted Roman lettering is used to label all information that is affected by tidal changes or current (with the exception of bottom soundings). All descriptive lettering for floating navaids is found in slanted lettering. Vertical Roman lettering is used to label all information that is not affected by tidal changes or current. Fixed aids such as lighthouses and ranges are indicated by vertical lettering.

4-48. **Lighthouses and Other Fixed Lights.** The basic symbol for these lights is a black dot with a magenta flare that looks like a large exclamation mark (Figure 4-10). Major lights are named and described; minor lights are described only.

All lights from lighthouses to a light on a piling are symbolized by a mark that looks like a magenta exclamation mark with a black dot as the bottom. Height, range, and rhythm are nearby. The number or name of the light is found in quotes.

Example: Fl.10s 12m 26M - means that the light is flashing at ten-second intervals, is 12 meters tall, and can be seen for 26 nautical miles.

Figure 4-10. Symbols for lighthouses and other fixed lights

Nautical Charts and Publications

4-49. **Ranges and Day Beacons.** Ranges are indicated on charts by symbols for the lights (if lighted) and a dashed line indicating the direction of the range. Day beacons are indicated by small triangles, which may be colored to match the aid. Day beacons, also commonly called day marks, are always fixed aids (that is, they are on a structure secured to the bottom or on the shore). They have many different shapes (Figure 4-11).

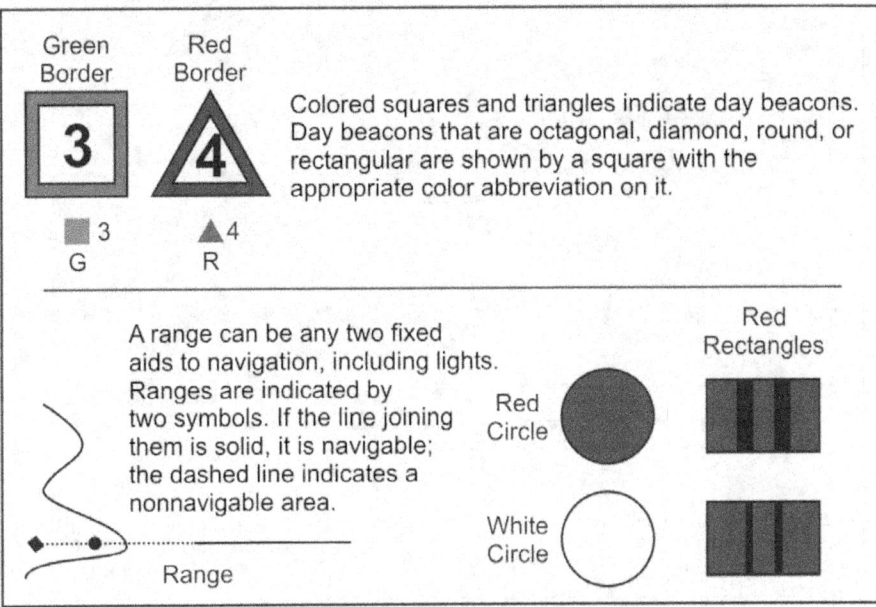

Figure 4-11. Symbols for ranges and day beacons

Note. Navigators should exercise caution during conditions of poor visibility, especially at night. If the observer is too close to the near range, his visual angle may make the near range appear to be higher than the far range. This perception is most likely in small craft, such as CRRCs. This confusion could result in navigational errors at a critical juncture.

4-50. Ranges are two man-made structures or natural features that are placed in a line to facilitate rapidly determining a line of position (LOP) when piloting a vessel in restricted waters. In port, ranges are set corresponding to the exact midchannel course, or turning points for course changes, to ensure safe navigation. Ranges may be either on the shore or in the water. The closest range mark will be shorter or lower than the far range mark. Ranges may use white, red, or green lights and will display various characteristics to differentiate from surrounding lights. As a general rule, the back range is higher than all other lights and is steady. The front is lower and usually a flashing light (Figure 4-12, page 4-14).

Note. Navigators should exercise caution during conditions of poor visibility, especially at night. If the observer is too close to the near range, his visual angle may make the near range appear to be higher than the far range. This perception is most likely in small craft, such as CRRCs. This confusion could result in navigational errors at a critical juncture.

Chapter 4

Figure 4-12. Range markers

4-51. **Prominent Landmarks.** Prominent landmarks such as water towers, stacks, and flagpoles are shown by a symbol of a dot surrounded by a circle (Figure 4-13). A notation next to the symbol defines the landmark's nature. The omission of the dot indicates the landmark's location is only an approximation.

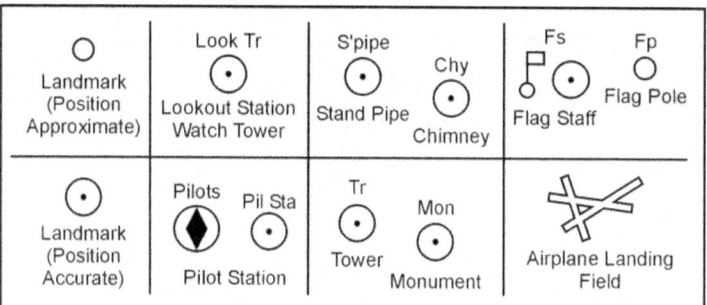

Figure 4-13. Prominent landmarks

4-52. **Wrecks, Rocks, and Reefs.** These features are marked with standardized symbols (Figure 4-14, page 4-15). For example, a sunken wreck may be shown either by a symbol or by an abbreviation plus a number that gives the wreck's depth at mean low or lower low water. A dotted line around any symbol calls special attention to its hazardous nature.

4-53. **Bottom Characteristics.** A system of abbreviations, used alone or in combination, describes the composition of the bottom, providing information for the navigator to select the best holding ground for anchoring (Figure 4-15, page 4-15).

Nautical Charts and Publications

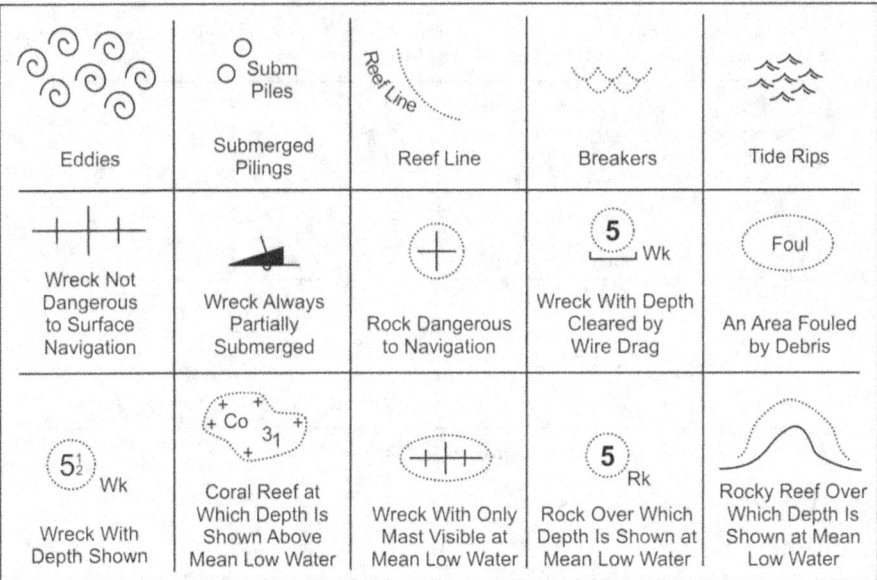

Figure 4-14. Symbols for wrecks, rocks, and reefs

Hrd	Hard	S	Sand	M	Mud; Muddy	Wd	Seaweed
Sft	Soft	Co; Cr	Coral	G	Gravel	Grs	Grass
Cy; Cl	Clay	Co Hd	Coral Head	Br	Brown	Oys	Oysters
St	Stone	Sh	Shells	Gy	Gray	Stk	Sticky

Figure 4-15. Bottom characteristics

4-54. **Structures.** For low-lying structures such as jetties, docks, drawbridges, and waterfront ramps, a shorthand representation has been developed and standardized. Such symbols are drawn to scale and viewed from overhead (Figure 4-16).

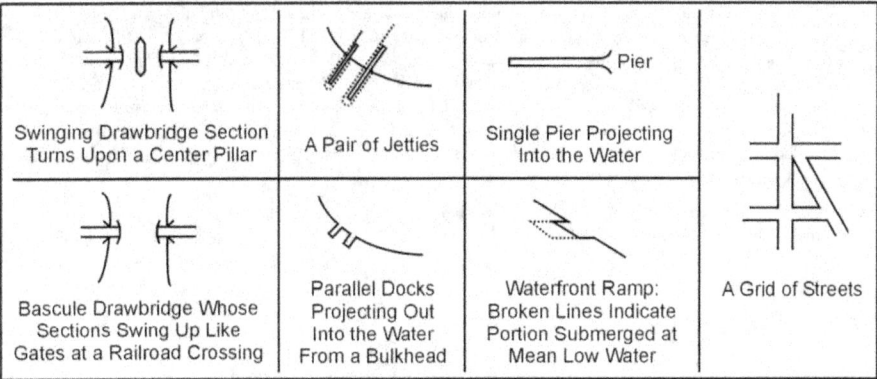

Figure 4-16. Structures

4-55. **Coastlines.** Coastlines are viewed at both low and high water. The navigator notes and labels any landmarks that may help him obtain a fix on his position (Figure 4-17).

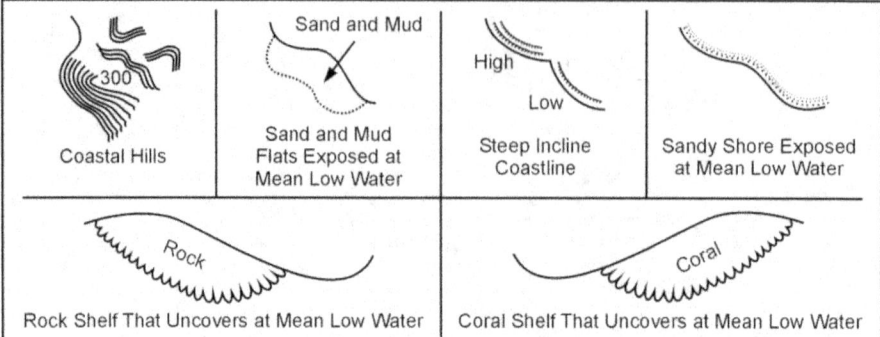

Figure 4-17. Coastlines

ACCURACY OF CHARTS

4-56. A chart is no more accurate than the survey on which it is based. All agencies try to keep their charts accurate and up-to-date. Major disturbances, such as hurricanes and earthquakes, cause sudden and extensive changes in the bottom contour. Even everyday wind and waves cause changes in channels and shoals. Because compromise is sometimes needed in chart production, the prudent navigator must be alert to potential changes in conditions and inaccuracies of charted information. Any information presented must be understood with ease and certainty. Various factors may prevent the presentation of all collected data for a given area. The navigator should always consider the following points to judge the accuracy and completeness of a survey.

4-57. The source and date of the chart are generally given in the title along with the changes that have taken place since the date of the survey. The earlier surveys were often made under circumstances that precluded great accuracy of detail. Until a chart based on such a survey is tested, it should be viewed with caution. Except in well-frequented waters, few surveys have been so thorough as to make certain that all dangers have been found.

4-58. Noting the fullness or scantiness of the soundings is another method of estimating the completeness of the survey. However, the navigator should remember that the chart seldom shows all soundings that were obtained. If the soundings are sparse or unevenly distributed, it should be taken for granted, as a precautionary measure, that the survey was not in great detail. Large or irregular blank spaces among soundings mean that no soundings were obtained in those areas. Where the nearby soundings are "deep," it may logically be assumed that the water in the blank areas is also deep. However, when the surrounding water is "shallow," or if the local charts show that reefs are present in the area, such blanks should be regarded with suspicion. This is especially true in coral areas and off rocky coasts. These areas should be given a wide berth.

4-59. The navigator or operator should ensure that he has the most recent chart for his AO. Before using a chart, he should first check the publication date, then check the Summary of Corrections, and finally check the local Notice for Mariners. After making the required annotations on the chart, he should record the changes on a Chart/Pub Correction Record (DMAHTC-86609).

PUBLICATIONS

4-60. Every detachment should acquire a basic library of charts and publications. At a minimum, it should include Chart No. 1; a collection of local charts sufficient to conduct training; Tide Tables and Tidal Current Tables for the local area; a nautical almanac; and Publication No. 9, *The American Practical*

Nautical Charts and Publications

Navigator. The detachment should also consider obtaining other publications on general seamanship. These references can be of considerable value when conducting training or missions, especially in areas where the detachment might encounter commercial marine traffic and aids to navigation. Because most publications are broken down by geographic region, it is only necessary to acquire those volumes pertinent to designated training and operational areas. Some of these publications will require periodic updates; the detachment should replace or update them as required to ensure their continued use and safety of navigation. Several of these publications are listed below.

4-61. **Chart No. 1.** This reference is also titled the *United States of America Nautical Chart Symbols, Abbreviations, and Terms* but is usually referred to as Chart No. 1. It is the only book that contains all symbols and abbreviations approved for use on nautical charts published by the United States. It is generally used worldwide. Chart No. 1 is divided into an introduction and five sections: general, topography, hydrography, aids and services, and alphabetical indexes. The table of contents is located on the back cover.

4-62. **Tide Tables.** Tide tables give daily predictions of the height of water at almost any place at any given time. They are published annually in the following volumes:

- Volume I. Europe and West Coast of Africa (including the Mediterranean Sea).
- Volume II. East Coast of North and South America (including Greenland).
- Volume III. West Coast of North and South America (including the Hawaiian Islands).
- Volume IV. Central and Western Pacific Ocean and Indian Ocean.

4-63. **Tidal Current Tables.** These tables provide the times of flood and ebb currents and times of the two slack waters when current direction reverses. They also tell the predicted strength of the current in kts. The time of slack water does not correspond to times of high and low tides; therefore, the navigator is unable to use the tide tables to predict current situations. These tables are published as follows:

- Volume I. Tidal Current Tables, Atlantic Coast.
- Volume II. Tidal Current Tables, Pacific Coast.

4-64. **Coast Pilots.** Available space and the system of symbols used limit the amount of information that can be printed on a nautical chart. Additional information is often needed for safe and convenient navigation. The NOAA publishes such information in the Coast Pilots. These are printed in book form covering the coastline and the Great Lakes in nine separate volumes. Each Coast Pilot contains sailing directions between points in its respective area including recommended courses and distances. It describes channels, their controlling depths, and all dangers and obstructions. Harbors and anchorages are listed with information on those points at which facilities are available for boat supplies and marine repairs. Information on canals, bridges, and docks is also included.

4-65. **Sailing Directions.** The Sailing Directions provide the same type of information as the Coast Pilots, except the Sailing Directions pertain to foreign coasts and coastal waters. They consist of geographically grouped volumes, which are shown in Figure 4-18, page 4-18.

4-66. **Light Lists.** This list provides more information about navigation aids than can be shown on charts. However, they are not intended to replace charts for navigation. The Light Lists consists of five volumes, and provides the same information for non-U.S. waters.

- Volume I. Atlantic Coast from St. River, Maine to Little River, South Carolina.
- Volume II. Atlantic and Gulf Coasts from Little River, South Carolina to Rio Grande, Texas.
- Volume III. Pacific Coast and Pacific Islands.
- Volume IV. Great Lakes.
- Volume V. Mississippi River System.

4-67. **List of Lights.** This is an annual publication (7 volumes) with similar information found in Light Lists but pertaining to foreign coastal areas.

Planning Guide

This guide consists of eight volumes and is printed annually. The volumes are divided into ocean basins that provide information about countries adjacent to that particular ocean basin. Contents include the following:
• Chapter 1. Countries.
  ▪ Government Regulations.
  ▪ Search and Rescue.
  ▪ Communications.
  ▪ Signals.
• Chapter 2. Ocean Basin.
  ▪ Oceanography.
  ▪ Environment.
  ▪ Magnetic Disturbances.
  ▪ Climatology.
• Chapter 3. Warning Areas.
  ▪ Operating Areas.
  ▪ Firing Areas.
  ▪ Reference Guide to Warnings and Cautions.
• Chapter 4. Ocean Routes.
  ▪ Route Chart and Text Traffic.
  ▪ Separation Schemes.
• Chapter 5. Navigation Aids.
  ▪ Systems.
  ▪ Electronic Navigation Systems.
  ▪ Systems of Lights and Buoys.

En Route Guide
This guide consists of 37 volumes and is printed annually. It is divided into subgeographical sectors that include detailed coastal and approach information. Each part contains the following information: • Pilotage. • Appearance of coastline (mountains, landmarks, visible foliage). • Navigation aids in general. • Local weather conditions. • Tides and currents. • Local rules of the road, if any. • Bridges, type and clearance. • Anchorage facilities. • Repair facilities. • Availability of fuel and provisions. • Transportation service ashore. • Industries.

Figure 4-18. Guides within the "Sailing Directions"

Notice to Mariners. The DMAHTC publications already mentioned are published at more or less widely separated intervals. As a result, provisions must be made for keeping mariners informed of changes in hydrographic conditions as soon as possible (ASAP) after they occur. The Notice to Mariners is the principal medium for distributing corrections to charts, light lists, and other DMAHTC publications. Each notice is divided into the following sections:
  • Section I. Chart Corrections.

- Section II. Light List Corrections.
- Section III. Broadcast Warnings and Miscellaneous Information.

4-68. **Local Notices to Mariners.** The local Coast Guard district publishes this information weekly. The notice provides corrections to nautical charts, coast pilots, and light lists. It also issues marine information particular to that district.

4-69. **Summary of Corrections.** Every 6 months, DMAHTC publishes this summary in two volumes. Volume I covers the Atlantic, Arctic, and Mediterranean areas. Volume II covers the Pacific and Indian Oceans and the Antarctic. These volumes cover the full list of all changes to charts, Coast Pilots, and Sailing Directions. The navigator uses the Summary of Corrections to supplement—not replace—the Notice to Mariners.

4-70. **Fleet Guides.** DMAHTC publishes Publication No. 940 (Atlantic) and Publication No. 941 (Pacific). Similar to the Sailing Directions and Coast Pilots, they are prepared to provide important command, navigational, repair, and logistic information for naval vessels. Both publications are restricted. They are a valuable source when conducting aid-to-navigation and harbor destruction missions.

AIDS TO NAVIGATION

4-71. Navigation aids assist the navigator in making landfalls when approaching from seaward positions overseas. They mark isolated dangers, make it possible for vessels to follow the natural and improved channels, and provide a continuous chain of charted marks for coastal piloting. All aids to navigation serve the same general purpose, and they differ only to meet the conditions and requirements for a certain location.

4-72. The prudent MAROPS detachment will prepare its navigation plan so that (as much as possible) it does not rely on artificial aids to navigation; for example, lights, buoys, or daymarks. During periods of hostility, these aids may not be present, accurately located, properly maintained, or lighted during periods of darkness. They are also most commonly emplaced in areas with concentrations of marine traffic. Detachments should seek to avoid these areas whenever possible.

INTERNATIONAL ASSOCIATION OF LIGHTHOUSES AUTHORITIES (IALA)

4-73. The IALA is a nongovernmental body that exchanges information and recommends improvements to aids to navigation. At present there are two systems termed IALA Maritime Buoyage System "A" Combined Cardinal and Lateral System (Red to Port) and IALA Maritime Buoyage System "B" Lateral System Only (Red to Starboard). Most European countries, including the countries of the former Soviet Union, have or will adopt System A. System B is still being developed. The United States uses the lateral system of buoyage (similar to but not the same as System B) recommended by the International Marine Conference of 1889. A graphic depiction of the different marking schemes as discussed below can be found in Chart No. 1.

BUOYS

4-74. The basic symbol for a buoy is a diamond and small circle. Older charts will show a dot instead of the circle. The diamond may be above, below, or alongside the circle or dot. The small circle or dot denotes the approximate position of the buoy mooring. Some charts will use the diamond to draw attention to the position of the circle or dot and to describe the aid. The various types of buoys are as follows:

- *Nun Buoys.* These are conical in shape, painted solid red, and mark the right side of the channel when one is entering from seaward.
- *Can Buoys.* These are cylindrical in shape and are painted solid green or black. They indicate the left side of the channel when one is entering from seaward (green) and mark the left side of rivers and intracoastal waterways (black).
- *Sound Buoys.* The four basic types are as follows:
    - **Bell** is sounded by the motion of sea.
    - **Gong** is similar to bell buoy but with sets of gongs that sound dissimilar tones.

Chapter 4

- **Whistle** is a tube mechanism that sounds by the rising and falling motion of the buoy at sea, making a loud, moaning sound.
- **Horn** has an electrically sounded horn at regular intervals.

4-75. Additional features on buoys include sound signals, radar reflectors, numbers or letters, or any combination of these features. Bells and horns are spelled out; radar reflectors are abbreviated (Ra Ref); whistles are abbreviated (WHIS); and numbers or letters painted on buoys are shown in quotation marks ("8").

Buoy Symbols

4-76. Nautical charts will show the buoy type by the initials of its shape; for example, nun buoys (N) and can buoys (C) (Figure 4-19). A mooring (anchor) buoy is the only one that is not indicated by the diamond and circle or dot. This symbol is a trapezoid (a figure having two parallel and two nonparallel sides) and a circle.

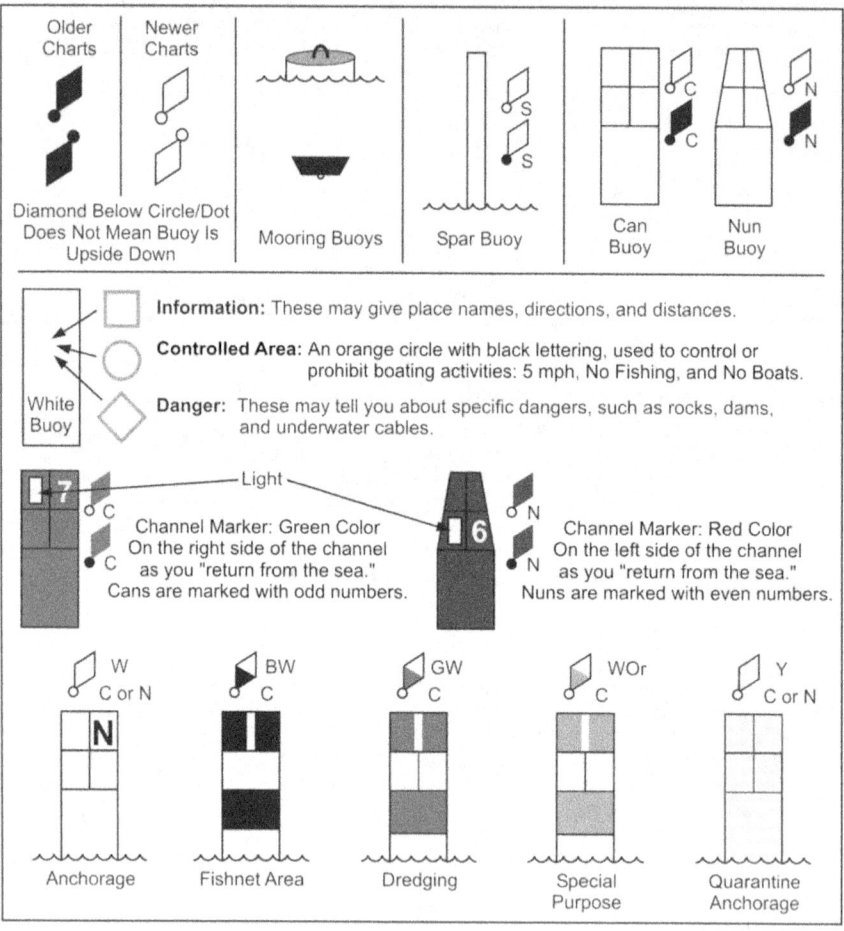

Figure 4-19. Buoys

Nautical Charts and Publications

4-77. If the aid is painted red, the diamond will usually be indicated in red on the chart; if the aid is painted black, the diamond will be black. There are five other color patterns used on buoys (Chart No. 1). These buoys have no lateral significance; that is they do not mark port or starboard. Although the buoys may not be numbered, they may be lettered (Figure 4-19, page 4-20).

4-78. The primary function of buoys is to warn the navigator of some danger, obstruction, or change in the bottom. A navigator may also use buoys to help mark his location on a chart, which aids in establishing his position. However, he should not rely solely on buoys or other floating objects for fixes because they are not immovable objects.

Buoy Lights

4-79. If a buoy is lighted, a magenta (nautical purple) disc will be overprinted on the circle. The characteristic of the light and its color will be indicated on the chart. Buoy lights can be red, green, or white. The letters R or G are used for red and green lights. The absence of a letter indicates a white light (Figure 4-19, page 4-20). The light phase characteristics and the meanings of abbreviations used to describe them are shown in Figure 4-20, below, and Figure 4-21, page 4-22. Each color is used as follows:

- *Red Lights.* They appear on red aids (nun buoys) or red and black horizontally banded aids with the topmost band red.
- *Green Lights.* These appear on black aids (can buoys) or red and black horizontally banded aids with the topmost band black.
- *White Lights.* These appear on any color buoy. The purpose of the aid being indicated by its color, number, or light-phase characteristic.

Abbreviations	Class of Light	Description	Illustration
F	Fixed	A continuous, nonblinking light.	
F. Fl.	Fixed and Flashing	A continuous light, varied at regular intervals by flashes of greater brilliance.	
F. Gp. Fl.	Fixed and Group Flashing	A continuous light, varied by groups of two or more flashes.	
Fl.	Flashing	A light that flashes at regular intervals of **not** less than 2 seconds and whose period of darkness **exceeds** the period of light.	
Gp. Fl.	Group Flashing	A light that sends out groups of two or more flashes at regular intervals.	
Gp. Fl. (1+2)	Composite Group Flashing	A flashing light in which the flashes are combined in alternating groups of different numbers.	
Mo. (A)	Morse Code	A flashing light which blinks signal letters in Morse Code. The letter "A" in Morse Code: (one short and one long flash).	
Qk. Fl.	Quick Flashing	A light that flashes 60 times or more a minute, used only on buoys and beacons.	
I. Qk. Fl	Interrupted Quick Flashing	A light in which 5 seconds of quick flashes is followed by 5 seconds of darkness.	
E. Int.	Equal Interval	A light with equal periods of light and darkness.	
Occ.	Occulting	A light that is eclipsed at regular intervals, but whose period of light is **always** greater than the duration of darkness.	
Gp. Occ.	Group Occulting	A light with regular spaced groups of two or more occultations.	
Gp. Occ (2+3)	Composite Group Occulting	A light whose combinations combine in alternate groups of different numbers.	

Figure 4-20. Light characteristics

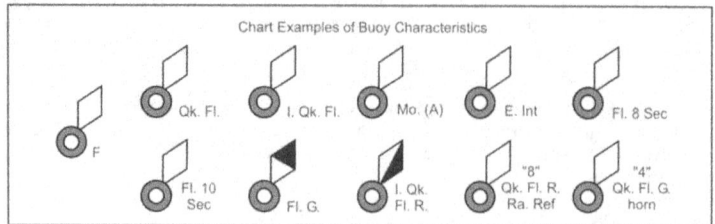

Figure 4-21. Additional light characteristics

Distance of Visibility of Lights

4-80. Distance of visibility is dependent on many factors. For all objects, the atmospheric conditions are the most important factors. The height of an object increases the distance from which it can be seen. Similarly, if one increases the height of one's view, one will increase the distance an object can be seen. Figure 4-22 explains the types of ranges that a navigator must consider when conducting small-boat operations.

Nominal Range	The maximum distance at which a light may be seen in clear weather (Figure 4-23, page 4-23). The Light Lists depicts the nominal range for lights that have a computed nominal range of 5 nm or more and can normally be found on the nautical chart next to the navigation aid, annotated with a large M.
Luminous Range	The maximum distance at which a light may be seen under existing visibility conditions (Figure 4-23, page 4-23). This range varies considerably with atmospheric conditions and the intensity of the light. The navigator may also determine the luminous range using the diagrams in the Light Lists.
Geographic Range	The maximum distance at which a light may be seen under conditions of perfect visibility, limited by the curvature of the earth only. Distance is expressed in nm for coastal waters. If the nominal range is less than the geographic range, the nominal becomes the geographic.

Figure 4-22. Types of ranges

4-81. To determine **geographic range**, the navigator first checks Table 12, Publication Number 9, Bowditch (Example 1), which is the distance to the horizon table. He will see that Table 12 is divided into four different columns—height in feet (of the navaid), nautical miles (seen from seaward), statute miles, and height in meters. The navigator first checks his nautical chart and finds the height of his navaid. In this case, the navigator uses the Point Loma Light that is 88 feet. He moves down the height column until he finds the height closest to his navaid's height (he should always round down; 88 feet = 85 feet on Table 12). The navigator then moves to the nm column and the figure there (10.8) will be the basic geographic range. He determines his height in feet and moves down the first column. For this height, he uses 4 feet, the height of a man sitting on a gunwale tube. The figure in the nm column for 4 feet is 2.3. The navigator then adds the range of the navaid to the range of his height (10.8 + 2.3 = 13.1), which results in his geographic range. He then plots a range ring on his course.

4-82. To determine **luminous range**, the navigator first checks the luminous range diagram in the Light Lists. The luminous range diagram is a graph that depicts nominal range on the bottom, visibility scale on the right side, and the luminous range on the left. The navigator requires two factors: an estimate of visibility and the nominal range of his navaid. He looks along the bottom of the diagram until he comes to his nominal range. He then looks up the scale to the center of the visibility tube that corresponds to his estimated condition. The navigator draws a straight line to the left side of the diagram and reads the luminous range for his navaid.

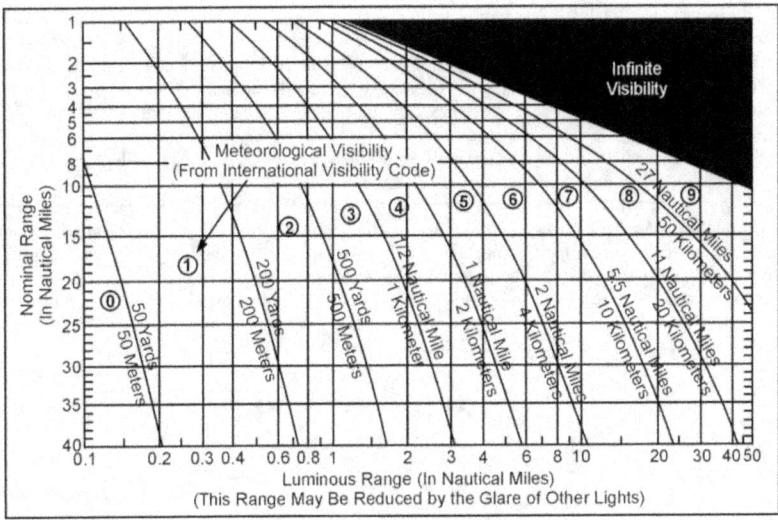

Figure 4-23. Nominal and luminous ranges

FOG SIGNALS

4-83. All lighthouses and light platforms or large naval buoys are equipped with fog signals that are operated by mechanical or electrical means and that are sounded on definite time schedules during periods of low visibility. For a fog signal to be an effective aid, the navigator must be able to identify the characteristic sound as coming from a certain point. This information is found on the chart and in the Light Lists. The characteristics of mechanized signals are varied blasts and silent periods. A definite time period is required for a complete signal. Various types of fog signals also differ in tone, which helps identify the respective stations. The various mechanisms are as follows:

- Diaphones produce sound by compressed air. They produce two tones of varied pitch.
- Diaphragm disc vibrates by air or electricity.
- Sirens are disc-shaped rotors that are activated by compressed air, steam, or electricity.
- Whistles are compressed air emitted through a circular slot into a cylindrical bell chamber.
- Bells are sounded by means of a hammer actuated by a descending weight, compressed gas, or electricity.

NAUTICAL COMPASS

4-84. The magnetic compass is the navigator's most important tool in determining his boat's heading. The purpose of a compass is to locate a reference direction and then provide a means for indicating other directions. A magnetic compass depends upon the magnetic fields of the earth for its directional properties; reference direction is called compass north or magnetic north. The compass generally has the following three components:

- *Compass Card.* It shows cardinal directions and degrees in 1-, 5-, or 10- degree increments.
- *Lubber Line.* This reference line indicates the direction of the ship's head.
- *Compass Bowl.* This container holds the compass card and a fluid not subject to freezing; for example, methanol or mineral spirits. The fluid is intended to dampen the oscillation of the compass card caused by the pitching and rolling of the boat.

4-85. For consistent and correct readings, the compass should be permanently affixed and aligned level to the vessel. To align the compass lubber line with the established centerline of the boat, the navigator must establish a second line that is perfectly parallel with the centerline. He then aligns the compass lubber line

Chapter 4

with the offset parallel line. The best way to mount a compass in CRRCs is to use a console that incorporates a knotmeter and integral lighting. There are current variants designed to mount on the main buoyancy tube that is forward of the coxswain's position. This position best enables the navigator to guide the vessel. To reduce magnetic interference (deviation), he should attempt to keep electronics and metallic (ferrous) objects at least one M16's length away from the compass when loading the boat.

Note. Coxswain fatigue and steering errors are best avoided by using only hard-mounting compasses of sufficient size that are designed for marine use.

4-86. The navigator will check the accuracy of the compass frequently. The magnetic compass is influenced not only by the earth's magnetic field, but also by fields radiating from magnetic materials aboard the boat. These two effects are referred to as "variation" and "deviation." A compass is also subject to unanticipated error resulting from violent movement as might be met in heavy weather and surf operations (Figure 4-24).

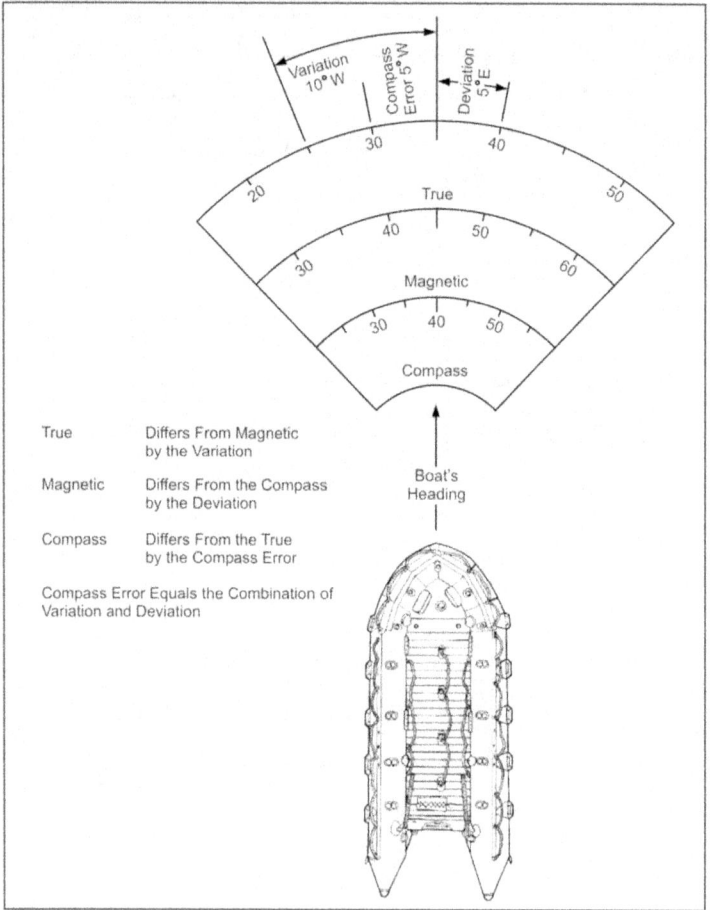

Figure 4-24. Compass direction

VARIATION

4-87. Variation is the angular difference between the magnetic and geographic meridians. It is measured as the difference in degrees between the directions of the magnetic and true North Poles. The amount of variation changes from one point to the next on the earth's surface. However, it will always be expressed in degrees in either an easterly or westerly direction.

4-88. A navigator can find the amount of variation for a given location on the compass rose of any chart for that area. Increases in variation may continue for many years (sometimes reaching large values), may remain nearly the same for a few years, and then may reverse (decrease). Navigators should use predictions of the changes for short-term planning, such as a few years.

4-89. Applying variation is similar to using a declination diagram on a topographical map. On a nautical chart, variation and its annual change can be found in the center of the compass rose. The navigator should always use the compass rose closest to the desired area, as variation changes from one area or compass rose to another. Not all charts will have a compass rose. Because variation changes daily (diurnal) and yearly (secular), the navigator should refer to NGA Chart 42 to find the correct variation in the intended objective area, especially if his chart is very old.

4-90. Navigators should always use the latest charts available. Normally, the compass rose will show the amount of predicted change. To determine the annual variation increase or decrease, the navigator should—

- Locate the compass rose nearest to the AO on the chart.
- Read the variation and annual increase or decrease from the center of the compass rose.
- Locate the year for which the information is given from the center of the compass rose.
- Subtract the year indicated in the compass rose from the present year.
- Multiply the number of year's difference by the annual increase or decrease.
- Add or subtract the sum of the previous step to the variation within the compass rose.

DEVIATION

4-91. Magnetic compasses function because of their tendency to align themselves with the earth's magnetic field. Unfortunately, the compass is subject to the magnetic influences of other metallic or electronic objects. The presence of these outside magnetic influences—for example, macheguns or radios—tends to deflect the compass from the magnetic meridian. This difference between the north-south axis of the compass card and the magnetic meridian is called deviation. If deviation is present and the north arrow of the compass points eastward of magnetic north, the deviation is named easterly and marked E. If it points westward of magnetic north, the deviation is called westerly and marked W.

4-92. The navigator can easily find the correct variation. However, deviation is not as simple to determine. It varies, not only on every boat, but also with the boat's heading and the equipment load on board. The navigator must swing (check) not only his compass but also each boat's configuration once it has been loaded. The most convenient method for small craft (CRRCs and kayaks) is to check the boat's compass every 15 degrees on a compass rose. Larger vessels use a process called "swinging the ship" that is outside the scope of this manual.

4-93. Most naval air stations or air bases have a compass rose laid out on the ground. It is "surveyed" so that accurate and repeatable measurements of the compass deviation caused by differences in equipment loading or configuration are possible. To record the deviation for his compass, the navigator should—

- Center the loaded boat with its mounted compass on the compass rose.
- Rotate the boat on the compass rose, stopping every 15 degrees.
- As he moves through the headings, compare the compass bearing with the surveyed bearing of the compass rose and record the number of degrees the compass is off on a deviation card. He should label it west if it is high and east if it is low.
- Apply the deviation when he actually conducts movement, if a magnetic heading is taken off the chart.

4-94. When compass error (the sum of variation and deviation) is removed, the correct true direction remains. Hence, the process of converting a compass direction to a magnetic or true direction, or of

converting a magnetic direction to a true direction, is one of "correcting" or removing errors. The opposite of correcting is "uncorrecting," or the process of applying errors. The following rules apply:
- When correcting, easterly errors are additive and westerly errors are subtractive.
- When uncorrecting, easterly errors are subtractive and westerly errors are additive.

4-95. To plot a course properly, the navigator must apply variation and deviation. Figure 4-25, below, and Figure 4-26, page 4-27, explain how to apply both of these differences.

Example

To correct the heading to a desired course, the navigator should set up his problem by taking the first letter from each of the following key words:

T = True
V = Variation
M = Magnetic
D = Deviation
C = Compass
CE = Compass Error (sum of deviation and variation)

The navigator then assigns each letter the known direction and error. For example, he is given a true course of 200 degrees, a variation of 14 degrees, and a deviation of 4 W. He needs to find out the magnetic heading, compass heading, and the compass error. This problem is uncorrecting, therefore—EAST IS LEAST (-) AND WEST IS BEST (+).

T	V	M	D	C	CE
200	14	(Step 1)	4	(Step 2)	(Step 3)

Step 1: Determine magnetic course by subtracting the variation from the true course.

200 (T)
- 14 (V)
Answer 186 degrees magnetic

Step 2: Determine the compass course by adding the deviation to the magnetic course.

186 (M)
+ 4 (D)
Answer 190 degrees compass

Step 3: Determine the compass error by algebraic method. Remember, if the signs are the same, add; if they are different, subtract; keep the sign for the highest number.

-14 (V)
(-) + 4 (D)
Answer 10 degrees (CE)

Figure 4-25. Guide to applying variation and deviation

Nautical Charts and Publications

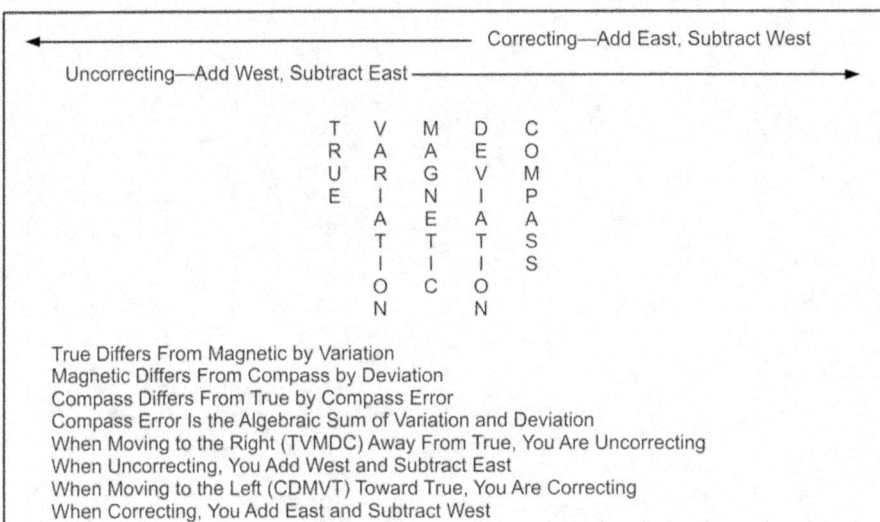

Figure 4-26. Determining variation and deviation

This page intentionally left blank.

Chapter 5

Small-Boat Navigation

The successful use of small craft by SOF requires accurate navigation. Traditional methods of small-craft navigation have been replaced by the global positioning system (GPS). The GPS has become the primary means of accurately locating a unit at sea. Unfortunately, like any other electronic device, the receivers are subject to failure. The cautious navigator will still use the old techniques, combining DR and piloting skills, to back up the input available from the GPS.

DR is the oldest form of navigation, based upon the projection of the unit's location from a known position along a course line using time and speed to determine the distance traveled. This information allows the navigator to determine an approximate position. Piloting is the art of directing a boat using landmarks, other navaids, and soundings to determine a more exact location. Piloting and DR are complementary skills.

Normally, the navigator will combine navigational techniques to ensure that the boat's location is accurately plotted. He must be consciously aware of his current location, as well as where he will be. He should always remember that all navigation deals with both present and future consequences.

NAVIGATION

5-1. Navigation is the process of directing the movement of a unit from one point to another. Both art and science are involved in navigating a unit accurately to a BLS or RP at sea. Art is involved in the proficient use of all available aids, methods, and the interpretation of the resulting data with good judgment to determine (or fix) a position. Science includes the mathematical computation for various navigational problems and the use of tables, almanacs, and publications intended to increase the accuracy of an intended route. Great progress has been made in advancing the science of navigation.

THE NAVIGATION TEAM

5-2. Accuracy in navigation demands that certain events take place simultaneously and in a timely manner. One man, of course, cannot perform all of the duties required. Consequently, a division of labor is necessary.

5-3. The piloting team consists of three individuals: the navigator, the assistant navigator, and the coxswain. All three should be well-versed in small-boat navigation and must become intimately involved in the planning, research, and development of the navigation plan. Once underway, the navigation team members perform the following duties:

- *Navigator.* The navigator orders course and speed to the coxswain and keeps total time of passage (TOP). He operates the plotting board by plotting bearings, currents, and DR plots. He notifies the assistant navigator when to take bearings and speed checks. The navigator oversees the entire navigation plan and is in charge of the safe piloting of the boat.
- *Assistant Navigator.* The assistant navigator renders aid to the navigator as directed and takes bearings, measures speed, and keeps time between legs. The assistant navigator must be alert and observant of sea conditions, aids to navigation, and land features.
- *Coxswain.* The coxswain maintains the ordered speed and course as well as avoids any deviation.

Chapter 5

NAVIGATION TERMS

5-4. Navigation is a "scientific" art with its own specialized vocabulary to ensure precision when communicating navigational information. All detachment personnel should thoroughly understand the terms and definitions listed in Figure 5-1, pages 5-2 and 5-3, if conducting MAROPS.

Actual Current	It is determined by measuring the displacement of a boat from a DR position to a fix.
Beach Landing Site	The area designated on shore where the boat is intended to land.
Bearing	The horizontal direction of one terrestrial (earthbound) point from another (the direction to an object from a position). It is expressed as the angular distance (degrees) from a reference direction (a direction used as a basis for comparison of other directions). A bearing is usually measured clockwise from 0 degrees through 360 degrees at the reference direction: true north, magnetic north, compass north, or relative to the boat's centerline.
Coast Piloting	Directing the movements of a boat near a coast or in harbors where the navigator can see charted geographic features for use as a reference. It involves the frequent (or continuous) determination of position (or lines of position) relative to the charted geographical position. It is usually done by shooting bearings at fixed land positions and using those bearings to fix the boat's position in a process that is very similar to resection as used in land navigation. This technique requires the greatest amount of area study and premission planning before launching.
Course	The intended horizontal direction of travel (the direction one intends to go) expressed as angular distance from a reference direction clockwise through 360 degrees. For marine navigation, the term applies to the direction to be steered. The course is often designated as true, magnetic, compass, or grid depending on the reference direction. A course can either be an accomplished or anticipated direction of travel. Small boats normally use magnetic or compass; large boats use true.
Course Line	The graphic image of a ship's course, normally used in the construction of a DR plot.
Course Made Good	The actual track of the boat over the ground. It is the resultant direction of movements from one point to another.
Current Triangle	A graphic vector diagram with one side representing ITR, another the set and drift of the current, and the third the course to steer (CTS). It is the heading to be taken to cause the boat to actually follow the ITR.
Dead Reckoning	The determination of approximate position by advancing a previous position using course and distance only, without regard to other factors such as wind, sea conditions, and current.
Drift	The speed of a current, usually stated in knots.
DR Plot	The graphic representation on the nautical chart of the line or series of lines showing the vectors of the ordered true courses. Distances run on these courses at the ordered speeds, proceeding from a fixed point. The DR plot originates at a fix or running fix; it is suitably leveled as to courses, speeds, and times of various DR positions. DR plots are made and confirmed every 15 minutes and when changing course or speed.
DR Position	A boat's location, determined by plotting a single or a series of consecutive course lines (vectors) using only the direction (course) and distance from the last fix, without consideration of current, wind, or other external forces acting on the boat.
Estimated Current	A prediction derived by the use of publications. It is also used for planning.
Estimated Position (EP)	The most probable location of a boat, determined from incomplete data or data of questionable accuracy. This point is also a DR position modified by additional information that in itself is insufficient to establish a fix.

Figure 5-1. Navigation terms

Estimated Time of Arrival	The best estimate of the arrival time at a specified location IAW a scheduled movement.
Estimated Time of Departure	The planned time of departure from a specified location IAW a schedule for movement.
Fix	An exact position determined at a given time from terrestrial, electronic, or celestial data or two or more lines of position from geographic navaids.
Heading	The actual direction the boat's bow is pointing at any given time. It is not necessarily the same direction as the boat's intended course.
Insert Point	The place where the mother ship or intermediate boat launches the CRRC or kayaks.
Intended Track	The anticipated path of a boat relative to the earth. The ITR is plotted during mission planning as part of course determination.
Line of Position	A line of bearing to a known object along which a boat is presumed to be. The intersection of two or more LOPs can result in a fix.
Off-Set Nautical Navigation	A method that allows for current when determining the course made good, or determining the effect of a current on the direction or motion of a boat.
Position	The actual geographic location of a boat. It may be expressed as coordinates of latitude and longitude or as the bearing and distance from an object whose position is known.
Range	Two types are used in piloting. A range can be two or more man-made structures or natural features, plotted on the appropriate charts, placed in a line. A range is used as an aid to piloting—to determine bearings, distance off, or course to steer. When two plotted objects line up to provide a bearing along which the boat's position can be estimated, these objects are said to be "in range." Ranges are also distance measured in a single direction or along a great circle. Distance ranges are measured by means of radar, range finders, or visually with a sextant. In small crafts, ranges can also be determined by using binoculars with a mil scale—measuring the angular height (or width) of an object of known size and using the formula (distance = height in yards x 1000/mil scale points).
Rendezvous Point	The linkup point at sea.
Running Fix	A position determined by crossing LOPs obtained at different times.
Set	The direction toward which a current is flowing, expressed in degrees true.
Speed	The rate measured in knots that a boat travels through the water. A knot is a unit of speed equal to 1 nautical mph. A nautical mile is 2,025.4 yards (2,000 yards for rough calculations over short distances) or 1 minute of latitude. External forces (for example, wind or current) acting on the boat cause the difference between the estimated average speed and the actual average speed.
Speed Made Good (SMG)	The vessel's speed over the surface of the earth. It differs from the vessel's speed through the water by the influence of any currents acting on the vessel. The effect of any currents is calculated by constructing a vector diagram. This data is then used to determine how the current influences the vessel's speed, which is sometimes called speed over ground (SOG).
Speed of Advance (SOA)	The average speed in kts that must be maintained to arrive at a destination at a preplanned time. The rounded speed in kts is plotted on the ITR.
Time of Passage	The time estimated for the CRRC to travel between specified points.
Track	The rhumb line or lines describing the path of a boat actually made good relative to the earth (sometimes called course over ground [COG]). The direction may be true or magnetic.

Figure 5-1. Navigation terms (continued)

Chapter 5

BASIC NAVIGATION EQUIPMENT

5-5. In determining position and safely conducting a boat from one position to another, the navigator uses a variety of piloting instruments. The navigator must be skilled in using these instruments and experienced at interpreting the information obtained from them. Piloting instruments must provide considerable accuracy.

5-6. One of the best-known navigational instruments is the magnetic compass, which is used for the measurement of direction. Aside from the compass, piloting equipment falls under the categories discussed below. Most of these instruments are readily available at battalion or group level as components of the diving equipment set.

BEARING-TAKING DEVICES

5-7. The navigator uses the **magnetic boat compass** to reference direction for accurate steering over a long track. It is the most important tool for quickly determining a boat's heading relative to the direction of magnetic north. A compass can also be incorporated into a navigation console that may include a knotmeter and some type of illumination system. A compass should be hard-mounted parallel to the centerline of the boat and have a light system for navigation in low-visibility conditions. The glow should be shielded from the view of all but the coxswain. Chemlights are the preferred method of illuminating the compass because of the potential problems with electrical systems in small craft.

5-8. The compass gives a constant report on the boat's heading. Users must make corrections for deviation and variance. A compass may also be used as a sighting instrument to determine bearings. A mark, called a "lubber line," is fixed to the inner surface of the compass housing. Similar marks, called 90-degree lubber lines, are usually mounted at 90-degree intervals around the compass card. The navigator can use these intervals to determine when an object is bearing directly abeam or astern. Centered on the compass card is a pin (longer than the lubber line pins) that enables the navigator to determine a position by taking bearings on visible objects.

5-9. The **hand-bearing compass** is a hand-held, battery-powered (illuminated) compass that has transparent forward and rear sight vanes with yellow cursor lines. The navigator uses this compass to determine relative bearings to navaids. It can also be an invaluable aid to piloting. Users must correct for the boat's heading by cross-referencing the bearing obtained from the hand compass with the boat's primary compass. A lensatic compass can be substituted if a hand-bearing compass is not available.

SPEED-MEASURING DEVICES

5-10. **Knotmeters** are speedometers for boats. They are usually incorporated into a navigation console that will also include the primary magnetic compass and some type of illumination system (preferably chemlight). The navigator can use either of the following knotmeters to accurately measure the speed of a small boat.

5-11. The *impeller-type knotmeter* measures speed by the use of an impeller (a rotating paddlewheel or propeller) mounted underneath or behind the boat. The knotmeter measures the boat's speed by counting the impeller's revolutions as the boat moves through the water. This measurement can be made with a direct mechanical connection to the gauge or indirectly using optical or magnetic sensors. Some systems may require a through-hull penetration.

5-12. The *pitot tube-type knotmeter* measures speed using an open-ended tube mounted to the stern or underneath the boat. The pitot tube is attached to the knotmeter at one end and submerged in the boat's slipstream at the other end. As the boat moves through the water, the open end of the tube creates a venturi effect, and draws a vacuum relative to the boat's speed that registers on the knotmeter as indicated speed.

5-13. Of the two types, the impeller is more accurate than the pitot tube. However, it is also at greater risk of being damaged, especially when beaching the boat. This risk increases significantly in a surf zone. Pitot tubes are mechanically simpler but are not as accurate as impellers. The tube is prone to being blocked by sand, debris, and algae. Damage from beaching is less likely to render the pitot tube useless, provided the mount is not destroyed and the navigator checks for blockages.

5-14. The **speed wand (tube)** measures speed through the water. It is a hollow, handheld tube with graduated markings that designate boat speed. Speed tubes are available in two types (speed ranges): 1.0 to

7 kts and 2.5 to 35 kts. Boat speed is measured by testing the pressure of the water mass against the moving tube. To use the speed wand, the navigator holds the large end and submerges the small end into the water with the inlet hole facing in the direction of travel. He lifts it out after 4 or 5 seconds (the metal ball will trap a specific volume of water in the tube). Speed is determined by comparing the trapped water level with the graduated lines on the tube. The navigator can then read mph and translate it into kts from the top of the tube.

Note. Most GPS receivers are capable of showing "speed made good" or speed over ground. This function is a byproduct of the continuous position update (fix) available from the GPS. It is most useful for projecting time of transit and fuel consumption when the boat's indicated (knotmeter) speed is being affected by the set and drift of an unknown current.

A motorized boat's speed may also be closely estimated by premarking throttle settings. This method is known as a timed-distance run and consists of noting the time needed to cover a known distance at different throttle settings. For this technique to be accurate, it is essential that the timed-run duplicate mission requirements or boat configuration and sea state conditions as closely as possible.

5-15. An accurate, reliable timepiece for clocking transit times between two points or at specified speeds is essential. A **stopwatch or navigational timer** that can be started and stopped at will is very useful. The navigator can use it for D/ST calculations when running a speed check. He can also use it to identify the lighted period or interval of a navigational aid, such as a flashing light. A waterproof digital watch works well for this requirement.

Depth-Measuring Devices

5-16. A **handheld lead line** is a graduated line with attached marks and fastened to a sounding lead, used for determining the depth of water when making soundings by hand. The leadsman takes a sounding from the bow of the boat by casting the lead (sounding weight) forward, allowing it to sink as the boat passes over the site. He then reads the graduated marking as the line comes taut. The leadsman usually uses the lead line in depths of less than 25 fathoms. An improvised lead line is most commonly used by swimmers to determine the bottom profile when conducting hydrographic surveys.

5-17. The **type fathometer or echo sounder** is a diver-held, sonar system that is available as part of the diving equipment set. It is a handheld (flashlight-sized), waterproof sonar system designed to measure depth or distance to submerged objects.

Plotting Instruments

5-18. The most basic of plotting instruments is the **pencil**, preferably a quality mechanical pencil with a 0.5 millimeter (mm) lead. During the mission-planning phase, it is best to use a fine-line mechanical pencil to maintain accuracy. As an alternative, the planner can use a No. 2 or No. 3 pencil, sharpening it regularly. He should keep all lines short, and print legibly and lightly for easy erasure. Art gum erasers are normally used for erasure since art gum is less destructive to chart surfaces than India red rubber erasers. Again, the planner should keep his pencils sharp; a dull pencil can cause considerable error in plotting a course due to the width of the lead. A grease pencil will be necessary during the transit for marking on the position, bearing, course (PBC) plotter, chart case, or acetated chart.

5-19. There are two **navigation sets** available in the diving equipment set. The sets are complementary and their components are designed to be used interchangeably. The navigator can use either of the following:
- NSN 6605-01-363-6346, PN: 00331, CAGE: 0CJM9 consists of an instruction book and the following parts:
    - Parallel rules.
    - A speed-time-distance computer (nautical slide rule).
    - A three-arm protractor.
    - Dividers.

Chapter 5

- NSN 6605-01-362-6327, PN: 00792, CAGE: 0CJM9 consists of an instruction book and the following parts:
    - A one-arm protractor.
    - A course plotter.
    - A protractor triangle.
    - A parallel rule.
    - Professional dividers.
    - A pencil.

5-20. **Dividers** are two hinged legs with pointed ends. They are used to measure distance on a chart. The **drafting compass** is an instrument similar to dividers. However, one leg has a pencil attached. The planner uses this tool for swinging arcs or circles when plotting coordinates and marking arcs of distance on the chart.

5-21. The **flat divider** is an improvised scale that is usually constructed by transferring distance increments (a latitude scale) from the chart being used onto a tongue depressor. The planner uses it to measure distances on laminated charts and plotting boards.

5-22. A navigator uses the **nautical slide rule** to determine speed, time, and distance problems. Using the slide rule provides greater speed and less chance of error than multiplication and division, especially during open-boat transits at night. There are several makes of nautical slide rules, but all operate on the same basic principle. The nautical slide rule has three clearly labeled scales: speed, time, and distance. By setting any two of the values on the opposite scales, the third is read from the appropriate index. More specifically, it consists of two dials on a base plate. These dials will turn together or independently. A nautical slide rule (NSN 6605-00-391-1110, PN: S2407-533638, CAGE: 80064) is available as part of the diving equipment set.

5-23. The PBC **plotting board** is a precision navigation instrument designed to accommodate small-boat navigation (Figure 5-2, page 5-7). The plotting board is a flat plastic board with an overlay containing a rotating wheel and a variation offset device that serves as a mount for the nautical chart. It enables the navigator to quickly convert bearings to LOP. The plotting board permits instant navigation in true or magnetic direction without protractors or parallel rules.

5-24. To set up the plotter, the planner places an appropriately sized map (folded or cut) on the plotter's base sheet with the map's longitude lines as nearly vertical as possible. He reassembles the rest of the plotter on top of the map and snaps or clips the assembly together. He orients the plotter to the map by turning the plotter's red grid to true north on the underlying map (he aligns a meridian of longitude on the map with the red lines on the plotter). To obtain magnetic variation (the preferred method), he then determines local variation and sets it by rotating the plotter's compass rose (adjustable azimuth ring) until the correct number of degrees of variation line up with the index line, and locks it in place with the setscrew.

5-25. A planner can use the plotter to obtain LOP and fixes by taking bearings of visible navaids with binoculars (integrated compass) or hand-bearing compasses. He can plot these magnetic bearings by turning the red pointer to the measured bearing (it is not necessary to calculate back azimuths as in resection). He then traces a grease pencil line parallel to the red lines starting at the navaid and extending backward through his probable position. He repeats this process with another navaid or does a running fix using the original navaid. If one of the red lines does not fall directly on top of both the start and finish points, he will have to interpolate. The planner simply rotates the dial until the red line is equidistant from (and on the same side of) the desired points. The intersection of the two LOPs represents a fix. His next course from any fix is determined by turning the red pointer so that the red parallel line connects (interpolates) both his position and his intended destination.

5-26. A plotting board, coordinate converter (NSN 5895-01-362-6329, PN: S10154, CAGE: 0H9W2) is available as part of the diving equipment set. It is not initially issued as a component of the Set Kit Outfit; however, it may be requisitioned "as required" when authorized by the commanding officer.

5-27. **Parallel rules** are simple devices for plotting direction. The rules consist of two parallel bars with parallel cross braces of equal length, which form equal opposite angles. The rules are laid on the compass rose (direction reference of a chart) with the leading edge aligning the center of the rose and the desired direction on the periphery of the rose. Holding first one bar and moving the second, then holding the second and moving the first, parallel motion is ensured. Firm pressure is required on one leg of the ruler while the other is being moved to prevent them from slipping off course. Lines representing direction may be plotted as desired on the chart. The primary use of the rulers is to transfer the direction of the boat's course line to the compass rose and vice versa.

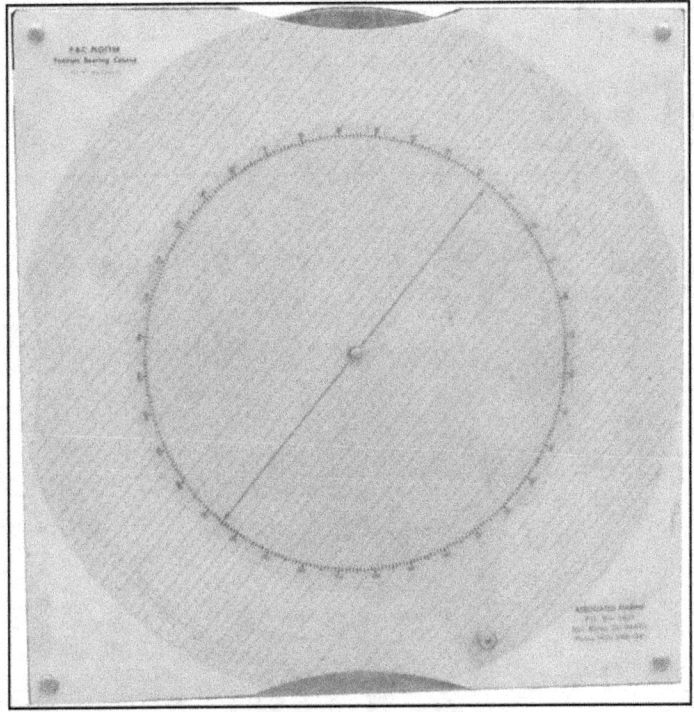

Figure 5-2. Plotting board

MISCELLANEOUS INSTRUMENTS

5-28. High-quality waterproof **binoculars** are extremely useful in waterborne operations. 7 x 50 (magnification x diameter of the objective lens) has been found to be the best compromise between light-gathering ability, magnification, and stability of the visual image. Some binoculars are equipped with an illuminated magnetic compass and a mil scale calibrated at 10 mils per index mark. These can be used to determine a magnetic bearing, measure the height/breadth, or determine range or height of any object. Short tactical airborne operation (STABO) binoculars work very well in a maritime environment.

Example: $\text{Distance} = \frac{\text{Height in Yards} \times 1000}{\text{Mil Scale Points}}$

5-29. Navigators use the **radar reflector** for linkup at sea with a mother ship. Small rubber boats generally have no discernible signal or signature in the water. By using the reflector on a telescopic pole, it will aid the mother ship in locating its position and assist in guiding the craft to the linkup point.

RECOMMENDED NAVIGATION EQUIPMENT

5-30. To assist in mission planning and execution, the navigation tools listed below and shown in Figure 5-3, page 5-8, are recommended:
- A—Binoculars (mil scale reticule, integral compass).
- B—Hand-bearing compass (hockey puck).

Chapter 5

- C—Stop watch.
- D—Parallel rule.
- E—Divider.
- F—Nautical slide rule.
- G—Compass.
- H—Depth finder.
- I—Pencil.
- J—Three-legged protractor.

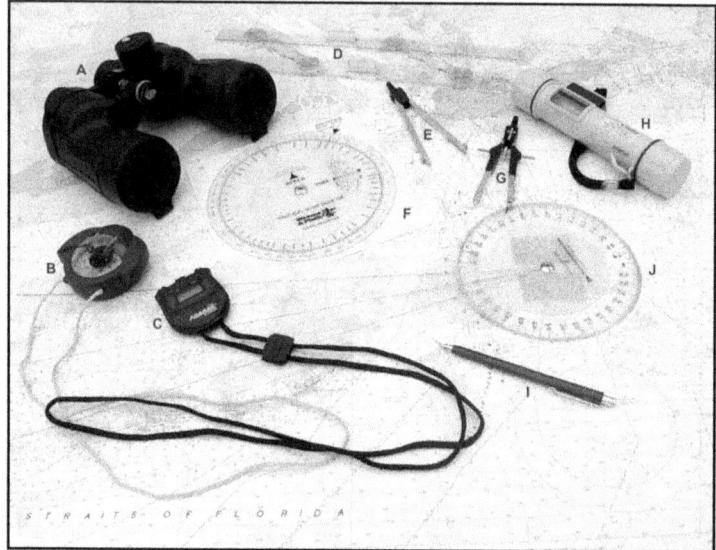

Figure 5-3. Navigation tools

TYPES OF NAVIGATION

5-31. There are four methods of navigation used in SF waterborne operations today. They are piloting, DR, and electronic or radio navigation. Each method is explained below.

Piloting

5-32. Piloting is the process for obtaining constant fixes (positions) by using geographical features, LOPs, and electronic devices. It involves the frequent (or continuous) determination of position (or LOPs) relative to the charted geographical position. Navigators most commonly use piloting near land or in harbors where they can see and use charted geographic features for reference. Piloting is usually done by shooting bearings at fixed land positions and using those bearings to fix the unit's position. This process is very similar to a resection as used in land navigation. Piloting requires the greatest amount of area study and premission planning before launching.

Dead Reckoning

5-33. DR enables the navigator to determine a unit's approximate position by moving the unit's position on a chart from a known point, using the course steered and the speed through the water. As long as the navigator can measure time, speed, and direction, he can always plot a DR position. This position is approximate but it is the principal method by which a basic fix can always be determined. It does not take into account the effects of wind, waves, currents, or poor steering. DR gives only an approximate position; the navigator uses it as a reference to locate his true position.

Current Sailing

5-34. Current sailing is the art of determining course and speed through the water, making due allowance for the effect of a predicted or estimated current, so that, upon the completion of travel, the ITR and the actual track will coincide. Current sailing may also include the determination of an existing current as measured by the displacement of the boat from its ITR. In essence, current sailing is applying the best information to the ITR to determine the course and speed to steer.

Electronic or Radio Navigation

5-35. In some circumstances, highly accurate electronic means and radio waves may be available to provide an exact location or fix. This method includes such equipment as radar, radio direction finder, Doppler and inertial systems, eLORAN, and the GPS. These systems are generally unaffected by environmental conditions or tidal current data. They are capable of greatly enhancing navigation accuracy. Therefore, their availability should be considered in the planning phase of the mission. Most of these systems are available on large ships. Detachments conducting infiltrations from a "mother ship" should access all onboard navigation systems during mission planning and conduct an accurate position check before debarkation.

5-36. The most viable form of electronic navigation for detachments to use in small craft is satellite navigation. The primary satellite navigation system is the NAVSTAR GPS. This system is an all-weather, worldwide system that will give pinpoint positioning through three-dimensional fixes. Given the advances in recreational, commercial, and military receivers since the introduction of the service, accuracy results routinely exceed 5 meters. Accuracy results have been enhanced by the implementation of the Wide Area Augmentation System (WAAS) and the Differential Global Positioning System (DGPS)—two commercial GPS supplement systems normally available on newer civilian GPS receivers. These systems use ground stations to augment the satellite signal and further refine the positional accuracy. They are not available in all areas. During periods of conflict, it is still possible for the U.S. Government to encrypt or "dither" the GPS signal to reduce its accuracy and subsequent use to unauthorized users.

5-37. The drawbacks of using electronic or radio navigation should be readily apparent. Despite the sophistication of electronic aids, they may become ineffective due to loss of power, propagation difficulties, and equipment malfunction. They are also subject to being jammed by the enemy. Therefore, it is necessary that navigators become skilled in and practice piloting and DR techniques.

NAVIGATION PLANNING

5-38. There are many factors to consider during the initial planning phase for insertion and extraction. Because time is top priority, the first step is to have all reference material on hand, to include pilot charts, sailing directions, light lists, and the latest nautical chart editions for the area (updated from the Summary of Corrections and the Notice to Mariners).

5-39. The navigator should conduct planning with the following information:

- *Weather.* The navigator should obtain information from the S2 or SWO to determine tide, current, wind, visibility due to natural phenomenon, sea state, wave height, and astronomical data.
- *Enemy Situation.* The navigator should obtain the visual and auditory threshold of detection; radar, satellite and hydrophone detection capabilities; and the effects of sea state and weather upon these thresholds. He should determine particular areas to avoid due to patrol boats, weapons, or natural hazards.
- *Insert/Rendezvous (RV) Boat.* The navigator should coordinate with the craft's master and quartermaster to determine boat capabilities; for example, speed, estimated time of departure (ETD) from ship, estimated time of arrival (ETA) to insert point (IP), and methods of fixing insert position. He should consult the ship's quartermaster for any additional navigational and landing area information.

5-40. After the navigator collects all the information possible, he should make a detailed chart or map for navaids that will assist in positioning himself once he is in piloting waters. Some navaids to look for are—
- Land forms (profiles).
- Towers.
- Lights.
- Buoys.
- Horns.

5-41. The navigation plan is a point-by-point detailing of the ITR. It should include all of the variables that could affect navigating the boat. The plan should include, but not be limited to, the following:
- Intended track (complete with labels).
- Current vector (included in ITR).
- Magnetic bearings to navaids.
- Range rings of luminous distances of lights.
- Geographic distance rings of other navaids.
- Danger bearings for course restrictions and arcs of detection.

5-42. When preparing the navigation plan, the navigator should not depend upon navaids to fix his position. Navaids are not always as depicted on the charts. They are continuously updated to reflect changes in hydrography and traffic patterns. Lights (lighthouses, range markers, and lighted buoys) are subject to being turned off during periods of hostilities. These changes are usually reflected in the Notice to Mariners, Sailing Directions, or other publications for navigators. The navigator should use tides and currents to his advantage. However, he must also fully understand the effects of tides and their ranges (for example, some places have a 9-meter tidal range).

LAYING THE COURSE

5-43. The navigator must follow specific steps to lay a course. He should—
- Draw a straight line from his departure point to the intended destination. This fix is his course line.
- Consider all navaids and other information presented on the chart along the course line.
- Lay one edge of his parallel rules along the course line. Walk the rules to the nearest compass rose on the chart, moving one rule while holding the other in place. He must ensure that the rules do not slip. If they do, the original line of direction will be lost.
- Walk the rules until one edge intersects the small plus (+) that marks the compass rose center. From the inside degree circle, he should read the course where the rule's edge intersects the center of the compass rose in the direction he is heading. This fix is his magnetic course.

PLOTTING AND LABELING COURSE LINES AND POSITIONS

5-44. All lines and points plotted on a chart must be labeled. Figure 5-4 shows the standardized symbols commonly used in marine navigation.

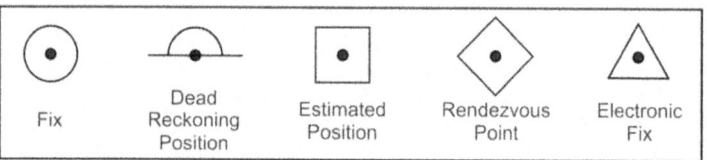

Figure 5-4. Standardized plotting symbols

CONSTRUCTING A DEAD-RECKONING PLOT

5-45. The following steps explain how to construct a DR plot. The navigator should—
- Plot the starting point and label it with A and the starting time.
- Plot the remaining turn points and the finish point, and label them B, C, D, and so on.

- Attach a rhumb line (magnetic-deviation compensation) from A to B, B to C, and so on.
- Find the heading in degrees magnetic for each leg of the transit.
- Measure distance of each leg using the latitude scale.
- Place speed of the transit for each leg below the distance for that leg.
- Calculate the time of transit for each of the legs using the D/ST formula.
- Label the arrival time at the beginning of each leg next to the letter designator.
- Calculate the distance between DR positions using the D/ST formula.
- Use the dividers to mark the DR position on the rhumb line. Label the DR positions with the proper symbol and with the time of arrival.
- Plot the DR positions (as nearly possible) at least every half hour, every course change, and every speed change.

LABELING A DEAD-RECKONING PLOT

5-46. The DR plot starts with the last known position (usually a fix). The navigator labels a DR plot using the steps below. He should label the—

- DR positions with a two-digit arrival time (minutes).
- First plot of a new hour with a four-digit time.
- Course with a three-digit heading and the letter M (magnetic) above the course line and as close to the departure point for each.
- Departure and arrival time in four digits rounded to the nearest minute. If the time falls exactly on a half minute, he should round to the nearest even minute.
- Distance rounded to the nearest tenth of an nm and placed below the course line as close to the departure point for each leg.
- Speed rounded to the nearest knot and placed below the distance for each leg (Figure 5-5).

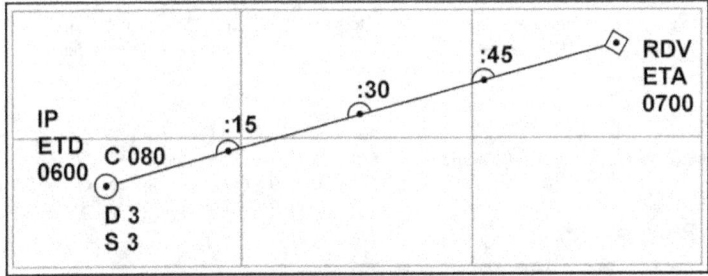

Figure 5-5. Example of labeling a dead-reckoning plot

5-47. There are various ways to label a DR plot; however, it is important that all involved can readily understand the method the navigator uses.

5-48. After plotting the ITR and the above information, the navigator writes the plan on a chart or plotting board in a noncritical area using the format shown in Table 5-1, page 5-12.

5-49. At least two navigation team members should inspect and validate all information. Once inspection is complete and final coordination is made, the navigator cuts the chart to a workable size to fit the plotting board. He also makes sure the portion of chart that remains includes the following features:

- Latitude and longitude bars.
- Logarithmic speed scale.
- Compass rose.

Table 5-1. Example of a navigation plan

Leg	Course (mag)	Distance (nm)	Time/Speed (min/kn)	Bearing to Navaid
IP – B	090	5.0	30/6	lighthouse 290 mag
B – C	171	6.75	40/6	
C – D	085	1.0	10/6	lighthouse 300 mag
D – C	265	1.0	10/6	
C – B	359	6.75	40/6	
B-IP/RDV	270	5.0	30/6	lighthouse 110 mag

5-50. The last step is to trace the ITR and navigation plan in black, highlight navaids, and laminate the chart front and back.

5-51. To ensure the navigation plan is thorough and complete, the navigator should consider the following planning hints:

- Include all members of the piloting team in planning and have them double-check all computations.
- Be careful and methodical during plotting and inspecting. Remember, one miscalculation will throw all of the plans awry.
- Check with every weather service and cross-reference their forecasts on a constant basis until departure.
- Remember the CRRC is a small, under-powered boat, and that the sea will always win if the crew tries to fight it. Always use the wind and current to the boat's advantage if possible.
- Avoid transiting with the moon (depending on percent of illumination) to the boat's rear. If this is not possible, remember to travel with the swell, staying in the trough. Use the swell as an advantage because not riding the waves will provide a lower silhouette.
- At a minimum, always plan for an alternate BLS. Think through the entire plan and ensure proper contingency plans are developed for any delays in transit (engine failure).

ESTABLISHING A FIX

5-52. Piloting involves continuous determination of the navigator's position by LOPs relative to geographical points. A single observation does not provide a position; it does provide the observers with a line, on some point of which the navigator is located. There is no connection between the DR course line and LOPs. The DR course line and DR positions are statements of intention; LOPs are statements of fact.

5-53. The simplest and most accurate LOP is determined by observing a range. If two fixed objects of known position appear to the observer to be in line, then at that instant, the boat's location is somewhere on the line passing through the objects.

5-54. It is not always possible to find two fixed objects in line at the time the navigator wishes to make an observation. Consequently, the LOP is usually obtained by plotting a bearing on a chart. The observer sights his hand-bearing compass or military binoculars (with integral compass) toward a fixed, known object and thus determines the direction of the LOS to the object. This point is called the bearing of the object. There are numerous possible positions on any single LOP. To fix the boat's position, the navigator must therefore plot at least two LOPs that intersect (no more than three), preferably at angles as near to 90 degrees as possible. This method is exactly like a resection in land navigation; however, the process is easier due to the functions of the plotting board. The ideal angle is 90 degrees. If the angle is small, a slight error in measuring or plotting either line results in a large error in position. Fixes less than 30 degrees should be regarded with caution. Closer navaids are preferable to those of considerable distance because the linear (distance) error resulting from an angular error increases with distance.

Note. Rule of sixty—the offset of the plotted bearing line from the observer's actual position is 1/6th the distance to the object observed for each degree of error. Therefore, a 1-degree error represents about 100 feet at a 1-mile distance and about 1,000 feet at 10 miles distance.

5-55. **Running Fix.** When the navigator cannot obtain two simultaneous observations, he must resort to a running fix. This point results by using two LOPs that are obtained by observations at different times. To plot a running fix, the navigator must make allowance for the exact time elapsed between the first observation and the second. He then advances the earlier LOP to the time of the second observation. The navigator assumes that, for the limited period between observations, the boat makes good via the ground a definite distance in a definite direction. He then plots this point by moving the earlier LOP, parallel to itself, to this advanced position. The new advanced line now represents the possible positions of the boat at the time of the second observation. A new DR is started from a running fix.

5-56. **Danger Bearing.** The navigator uses this factor to keep his boat clear of an outlying danger area, zone of detection, or other threats. He establishes a danger bearing between two fixed objects or a fixed object and a projected zone of danger.

5-57. If a distance to an object is known, the boat must lie somewhere on the perimeter of a circle centered on the object, with the known distance as the radius. This distance is called a distance circle of position. Distance (D) may be determined by binoculars and mil scale for the purpose of CRRC navigation (formula follows), or by observation of a precalculated threshold (range rings). Plotting concentric circles of position of various distances before entering piloting waters is recommended, rather than attempting computation while underway (as discussed in aids to navigation).

$$D = \frac{\text{Height in Yds} \times 1000}{\text{Mil Scale Points}} \qquad MSP = \frac{\text{Yds} \times 1000}{\text{Distance}}$$

DEAD RECKONING

5-58. The small-boat team, given the required resources, must be able to plan and execute the navigation required to conduct an accurate beach landing or rendezvous at sea at the appointed time after a minimum transit of 20 nm. Ideally, the team meets this requirement using a combination of all of the tools and techniques available. Small-craft navigation will frequently be centered on the use of a GPS. However, because no single system is fail-safe, DR and piloting navigation skills are essential to back up the GPS. DR enables the navigator to determine a boat's position at sea, any time, by advancing the last known position using speed, time, and distance computations. The navigator also uses DR to determine an intended course during premission planning. Besides providing a means for continuously establishing an approximate position, DR also has key applications in—

- Determining the availability of electronic aids to navigation.
- Predicting times of making landfall or sighting lights.
- Estimating time of arrival.

5-59. DR is assessing something dead in the water (a previously fixed position) and hence applying it to courses and speeds through the water. The navigator always starts from a known position and does not consider the effects of current in determining a DR position. Due to leeway caused by wind, inaccurate allowances for compass error, imperfect steering, or error in measuring speed, the actual motion through the water is seldom determined with complete accuracy. Because of this leeway, detachments must remember that a DR position is only an approximate position. It must be corrected (with piloting or electronic input) as the opportunity presents itself.

MEASUREMENTS

5-60. Navigators keep their DR by plotting directly on the chart, drawing lines to represent the direction and distance of travel, and indicating DR positions and estimated positions from time to time. The key elements of DR are discussed below.

Chapter 5

Direction

5-61. Magnetic courses are used to determine a DR position. Direction is the relationship of a point (known as the reference point) to another point. Direction, generally referred to as bearing, is measured in degrees from 000 through 360. The usual reference point is 000 degrees. The relationships between the reference points and reference directions are listed below:

- Measurement is always done from the closest compass rose to the individual's position. Compass roses for both true and magnetic directions may be given, but they must be labeled magnetic or true.
- The plotting board is a useful alternative to the compass rose. It can be oriented to any reference direction and can account for local variation, thereby enabling the navigator to always plot true or magnetic direction.

Distance

5-62. The distances traveled are obtained by multiplying a boat's speed in knots by an anticipated time or an accomplished time ($D = S \times T$). Once the distance run has been calculated, it must be transferred to the chart. The length of a line on a chart is usually measured in nautical miles to the nearest 0.1 mile. For this reason, navigators should use the latitude scale on a chart (1 minute of latitude = 1 nm). Navigators measure distance by marking it off with dividers and comparing the span of the dividers to the chart's scale or the closest latitude line. When the distance to be measured is greater than the span of the dividers, the dividers can be set at a minute or number of minutes of latitude from the scale and then "stepped off" between the points to be measured. The last span, if not equal to that setting on the dividers, must be measured separately. To do this, the navigator then steps the dividers once more, closing them to fit the distance. He measures this distance on the scale and adds it to the sum of the other measurements. He should use the latitude scale nearest the middle of the line to be measured. (The longitude scale is never used for measuring distance.) To measure short distances on a chart, the dividers can be opened to a span of a given distance, then compared to the nm or yard scale.

Time

5-63. Time is usually expressed in four figures denoting the 24-hour clock basis. When using the $D = S \times T$ formula, most individuals will find it easier to calculate times if they convert hours and minutes to straight minutes before they perform any calculations. A time expressed in decimal form may be converted to minutes by multiplying the decimal time by 60. Time expressed in minutes may be converted into decimal form by dividing the minutes by 60.

5-64. As previously discussed, the navigator cannot totally rely upon electronic systems and, even if they are available, he must still be able to navigate to the intended destination from a known (fixed) location. In DR, course and speed are generally computed without allowance for wind or current. The necessary components of DR are—

- *Course.* A course must be known and an exact heading maintained.
- *Speed.* The boat must maintain a consistent speed that is verified on a regular basis.
- *Distance.* The navigator must preplan the course distance (always plotted in nms) that he intends to travel.
- *Time.* Total time must be kept from each fixed position. DR will be used in planning the intended course before infiltration and then used to locate the CRRC's geographic position at sea, using speed, time, and distance. During the mission-planning phase, the navigator will plot an intended DR track. Using all of the information available, he will determine the potential BLS and IP. He will then plot a minimum-risk route between the IP and BLS, taking into account hazards to navigation and potential enemy threats. Once the detachment is inserted and underway, the navigator will track the detachment's position from the IP, as confirmed by the navigator of the delivery craft. The detachment is dependent on DR for its approximate position along the ITR until it reaches a point where navaids are available for determining LOPs.

COMPUTATIONS

5-65. Calculating time, distance, and speed requires a single basic mathematical formula. An easy way to remember how the formula works is to use the phrase "D street" or in mathematical terms, D/ST.

5-66. In the course of piloting the boat, both coxswain and crew members will make numerous calculations for time (minutes), speed (kts), and distance (nms). For these computations, speed is expressed to the nearest tenth of a knot, distance to the nearest tenth of a nautical mile, and time to the nearest minute. There are three basic equations for speed, time, and distance. In each case, two elements are known and used to find the third, which is unknown. When solving problems involving time, individuals should change time into hours and tenths of an hour. For example, 1:15 minutes = 1.25 tenths. Figure 5-6 shows the basic equations for making the necessary calculations.

T (time) = $\dfrac{60 \times D}{S}$	S (speed) = $\dfrac{60 \times D}{T}$	D (distance) = $\dfrac{S \times T}{60}$

Figure 5-6. Basic equations

Note. The nautical slide rule is made to solve speed, time, and distance problems. It provides greater speed and less chance of error than multiplication and division, especially during open-boat transits at night. The slide rule has three clearly labeled scales—one each for speed, time, and distance. By setting any two of the values on the opposite scales, the third is read from the appropriate index. Figure 5-7, pages 5-15 and 5-16, provides examples of using the D/ST formula.

Determining Distance When Speed and Time Are Known

A unit is running at 10 knots. How far will it travel in 20 minutes?

Step 1. D = $\dfrac{S \times T}{60}$

Step 2. D = $\dfrac{10 \times 20}{60}$

Step 3. D = $\dfrac{200}{60}$ = 3.3 nautical miles

Determining Distance When Speed and Time Are Known

A unit can travel from their station to the shipping channel in 3 hours and 45 minutes moving at a speed of 10 knots. What is the distance to the shipping channel?

Step 1. Use minutes to solve the time, distance, and speed equation. Convert the 3 hours to minutes (3 hours x 60 minutes = 180 minutes). Add the 45 minutes; time equals 225 minutes.

Step 2. Use the distance equation: D = $\dfrac{S \times T}{60}$

Step 3. Compute information opposite the appropriate letter.

D = $\dfrac{10 \text{ knots} \times 225 \text{ minutes}}{60}$

Step 4. D = $\dfrac{2,250}{60}$ = 37.5 nautical miles (nearest tenth)

Figure 5-7. D/ST formula examples

Determining Speed When Time and Distance Are Known

A unit assumes it will take 40 minutes to travel 12 nautical miles. What is its speed?

Step 1. $S = \dfrac{60 \times D}{T}$

Step 2. $S = \dfrac{60 \times 12}{40}$

Step 3. $S = \dfrac{720}{40} = 18$ knots

Determining Speed When Time and Distance Are Known

A unit's departure time is 2030. The distance to its destination is 30 nautical miles and it wants to arrive at 2400. What is the speed that it must maintain?

Step 1. Find the time interval between 2030 and 2400. Remember to subtract "hours and minutes" from "hours and minutes." Determine the time interval as follows:

    23 hours 60 minutes (23 hrs/60 min equals 24 hrs 00 min)
  − 20 hours 30 minutes
    3 hours 30 minutes

Step 2. REMEMBER to use minutes to solve the time, distance, and speed equation. Convert the 3 hours to minutes (3 hours x 60 minutes = 180 minutes). Add the 30 minutes remaining from Step 1. Time (T) is 210 minutes.

Step 3. Write down the speed equation: $S = \dfrac{60 \times D}{T}$

Step 4. Compute the information in the equation:

$$S = \dfrac{60 \times 30 \text{ miles}}{210 \text{ minutes}}$$

Step 5. $S = \dfrac{1,800}{210} = 8.6$ knots (nearest tenth)

Determining Time When Speed and Distance Are Known

A unit is cruising at 15 knots and has 12 nautical miles to cover before arriving on station. How long will it take before it arrives at his destination?

Step 1. $T = \dfrac{60 \times D}{T}$ D = 12 miles

                              S = 15 kn

Step 2. $T = \dfrac{60 \times 12}{15}$

Step 3. $T = \dfrac{720}{15}$

Step 4. T = 48 minutes

Figure 5-7. D/ST formula examples (continued)

OFFSET NAUTICAL NAVIGATION

5-67. The difference between the DR position and an accurate fix is the result of the action of various forces on a boat, plus any errors introduced when laying out course and speed. A boat is set off course by wind and current. The distance it is displaced results from the combination of set and drift, as well as steering error. The results can be disastrous if not corrected. There are also terms that navigators and crew members use that explain current sailing and help each person to understand what functions or actions may occur. A clear understanding of these terms is essential when conducting waterborne operations (Figure 5-8).

Estimated Current	Determined by estimating all the known forces that will contribute to the sum total of current effects in a given area.
Actual Current	Determined by the displacement of the boat from the DR position to a fix. This point can be determined when an accurate position can be obtained once underway. The difference is direction and distance between the fix and the DR position, for the time of the fix establishes the actual current.
Set	The direction toward which the current flows; expressed in degrees true.
Drift	Current velocity expressed in knots.
Estimated Position	The most probable position of a boat determined from all available data when a fix or running fix is unobtainable. The EP includes the effects of any estimated currents.
Current Triangle	A graphic vector diagram in which one side represents set and drift of the current, and the third side represents the track of the craft. If any two sides are known, the third can be determined by measurement.

Figure 5-8. Terms used in current sailing

5-68. The direction in which the current is moving, the direction a boat is moved by current and winds, or the combination of both is called set. When referring to the boat's movement, the term commonly used is leeway. This movement is the leeward (downwind or down current) motion of a boat caused by wind and sea action. The direction of set is usually expressed in degrees (true). The speed of the current and the speed at which a boat is being set is called drift, which is expressed in knots.

5-69. As previously discussed, a DR position represents the actual position of a boat only if the direction is accurate, the speed is accurate, and if no external forces have acted upon the craft. In navigation, many factors can cause a boat to depart from its intended course. These factors are referred to as currents. However, all of the currents that can affect a CRRC can be preplanned or estimated from several publications. Table 5-2 describes types and effects of currents that can cause navigational error and how to preplan for them.

Table 5-2. Current types, effects, and corrections for navigational errors

Type	Effect	Estimating Correction
Ocean Current	Is a well-defined current that extends over a considerable ocean area.	Is best planned using the Pilot Chart, Coast Pilot, and Sailing Directions. If planning onboard a ship, ocean current updates are usually available from the quarter-masters and navigators onboard.
Tidal Current	Is caused by tidal action. Effect is often seen in harbors or estuaries. Most often occurs along the coastline.	Is computed from the Tidal Current Tables. Tidal current is **not** computed with ocean current.
Wind Current	Is limited to an area affected by a strong wind blowing 12 hours or more. The force of the earth's rotation will cause the current to set to the right of the direction of the wind in the Northern Hemisphere and to the left in the Southern Hemisphere.	Is most often described in the Coast Pilot, Pilot Chart, and the bimonthly Mariners Weather Log. Local air or naval weather stations will also issue updated information about wind and weather.

5-70. The combined effects of these factors can cause a boat to depart from its intended course. Hence, the term *current* has two meanings—the horizontal movement of water due to ocean and wind currents, or the combined effect of all errors listed above.

Chapter 5

CURRENT EFFECTS/VECTOR

5-71. If a navigator can determine the effects of set and drift, he can obtain a better position by applying the correction to the ITR. In working problems to correct for set and drift, the navigator must allow for current and compensate for its effect using a vector diagram called a "current triangle." From this diagram, the navigator can find the course and speed his boat will make good when running a given course at a given speed, the course he must steer, and the speed at which he must run to make good a safe, desired course and itinerary.

5-72. He may solve the set and drift problem by constructing a current vector (or triangle) (Figure 5-9) to provide a graphic solution. He can construct the vector directly on the chart's compass rose, or set up with a protractor. Vector diagrams are most commonly constructed to determine course to steer (estimated current) and course made good (actual current). Information obtained from these vector diagrams will allow the navigator to anticipate course corrections so that he can adhere as closely as possible to his ITR. A discussion of both types of problems follows.

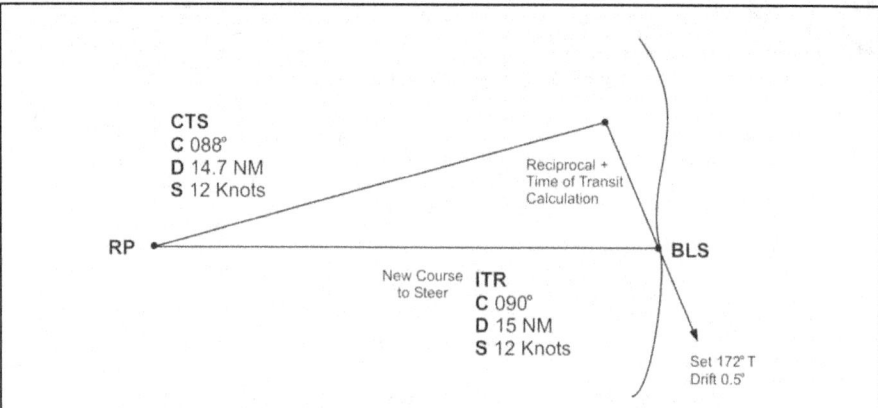

1. The navigator has an ITR of 15 nm, speed to make good of 12 knots, a course to make good of 90 degrees mag, and a TOP of 1:15 minutes (1.25).

2. From the pilot chart, he determines an anticipated set of 172 degrees and a drift of 0.5 kn.

3. From the BLS, the reciprocal course of 352 degrees T is plotted: 172°T + 180°T.

NOTE: Remember, if the set is less than 180 degrees, add 180; if it is more, subtract 180.

4. Multiply the drift of 0.5 by the TOP of 1.25: 1.25 x 0.5 = 0.625 (round down to 0.6).

NOTE: A distance of 0.6 nm is plotted on the reciprocal current line.

5. After plotting the current vector, a course to steer of 88 degrees mag is measured, and a distance of 14.7 nm is measured.

6. Time of passage of 1.25 is divided into the vector distance of 14.7 nm.

NOTE: Example of speed is rounded to the tenth place; therefore, the speed remains at 12 knots.

Figure 5-9. Steps required to prepare a vector diagram

Actual Current

5-73. Upon fixing a boat's position and finding that it does not correspond to the correct DR position on the DR plot, the navigator can compute and compensate for the effects of actual currents and forces being applied to the boat to push the boat off course. In short, he can fix the boat's position and compare his actual position to his preplanned position at any given time.

Constructing an Actual Current Triangle

5-74. The following steps explain how to construct an actual current triangle (Figure 5-10). The navigator—

- Plots the actual position of the craft.
- Draws a line through the actual position of the craft and the intended position. The intended position is along the DR track or the course-to-steer track if an estimated current triangle is in use.
- Determines the direction in degrees true from the intended position to the actual position using the parallel rules. This path is his set. He draws a line parallel to the set through the arrival point. He places the set above this line on the downstream side of the arrival point.
- Determines the drift using the D/ST formula. He measures the distance the craft drifted (the distance from the craft's actual position to the craft's intended position). The drift is equal to the distance the craft has drifted divided by the time it took to drift that far. He places the value of the drift under the set line at the arrival point on the downstream side.
- Calculates the distance the craft will drift for the remaining portion of the transit leg using the D/ST formula. Drift is equal to the speed of the current times the time remaining to reach the arrival point.
- Plots a point on the upstream side of the arrival point the same distance that the craft will drift along the set line.
- Draws a line from the craft's actual position to this point. This line is the new course-to-steer line. He writes the course to steer above the line.
- Calculates the new speed using the D/ST formula. The speed is equal to the distance of the new course-to-steer line divided by the time remaining in the transit. He writes this value below the new course-to-steer line.

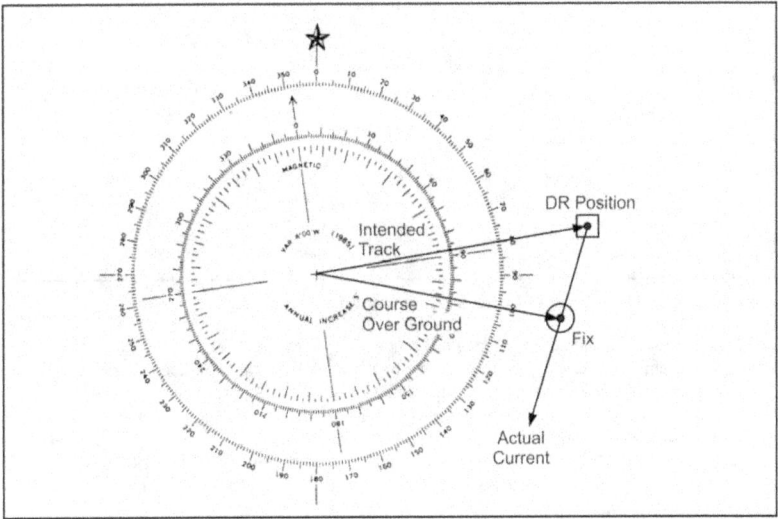

Figure 5-10. Example of actual current triangle with labeling

COMBINED CURRENTS

5-75. The small-craft navigator operating close inshore is likely to encounter multiple currents. The combined effects of all of the currents will have to be calculated to determine their impact on the boat's ITR.

Chapter 5

Currents in the Same Direction

5-76. When two or more currents set in the same direction, it is a simple matter to combine them. The resultant current will have a speed that is equal to the sum of all the currents, and it will set in the same direction. For example, a boat is near the former location of the Sand Key Light at a time when the tidal current is setting 345 degrees with a speed of 0.5 kts, and at the same time a wind of 50 mph is blowing from 150 degrees. The current that the boat will be subjected to is computed as follows: Since a wind of 50 miles from 150 degrees will give rise to a current setting about 345 degrees with a speed of 0.7 kts, the combined tidal and wind currents will set in the same direction (345 degrees) with a speed of 0.5 + 0.7 = 1.2 kts.

Currents in Opposite Directions

5-77. The combination of currents setting in opposite directions is likewise a simple matter. The speed of the smaller current is subtracted from the speed of the greater current, which gives the speed of the resultant current. The direction of the resultant current is the same as that of the greater current (Figure 5-11).

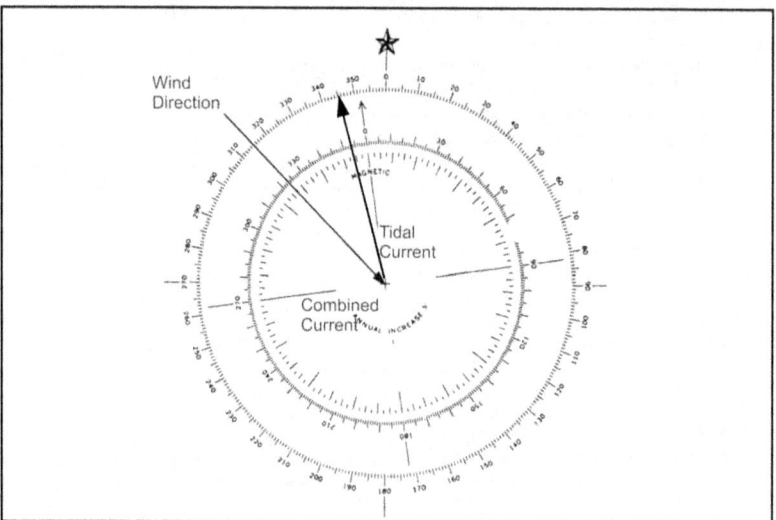

Figure 5-11. Opposing currents

5-78. As an example, suppose the navigator is required to determine the speed of the current at the former location of the San Francisco Lightship when the tidal current is setting 331 degrees with a speed of 0.5 kts and with a wind of 45 mph blowing from the northwest. The current produced by a wind of 45 mph from northwest would set 151 degrees with a speed of 0.6 kts. Therefore, the tidal and wind currents set in opposite directions, the wind current being the stronger. Hence, the resultant current will set in the direction of the wind current (151 degrees) with a speed of 0.6 kts minus 0.5 kts or 0.1 kts.

Currents in Different Directions

5-79. The combination of two or more currents setting neither in the same nor in opposite directions, while not as simple as in the previous cases, is, nevertheless, not difficult. The best solution is a graphic one (Figure 5-12, page 5-21). The navigator draws a line from the cross reference of the compass rose in the direction of one of the currents to be combined and whose length represents the speed of the current from the latitude bar scale. From the end of this line, he draws another line in the direction and the length of the other current to be combined. By joining the two lines together (the origin with the end of the second line), the navigator determines the direction and speed of the resultant current.

Small-Boat Navigation

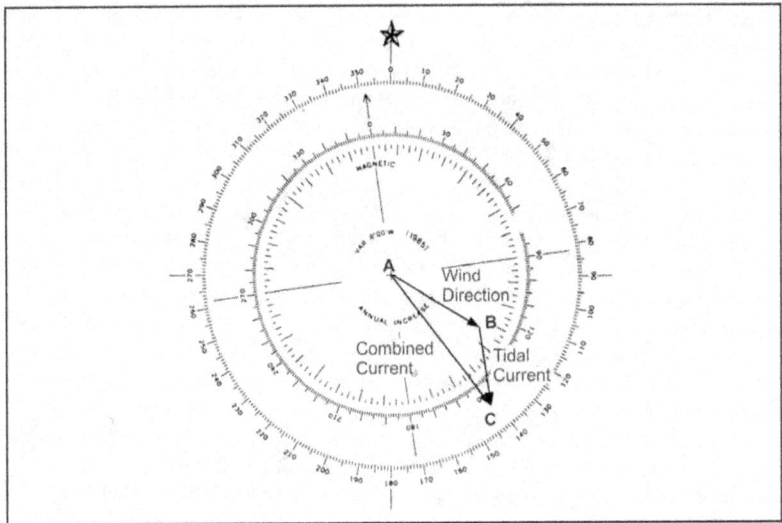

Figure 5-12. Currents in different directions

DETERMINATION OF ACTUAL COURSE AND SPEED MADE GOOD

5-80. This current triangle is a graphic vector diagram in which the direction of the first line drawn from the center of the compass rose indicates the boat's intended direction, course length in miles (or other convenient scale unit on the chart), and speed. The navigator draws a second line down from the end of the intended direction (first line) and along the set (direction) of the current. The length of this line in miles (or in the same units used to draw the intended course line) will be the current's drift (speed) in kts. He draws a third line from the center of the compass rose to the tip of the second line that provides the course the boat is actually making good. Its length in miles (or the units used to draw the first two lines) is the actual SMG.

5-81. The navigator should show the current direction in degrees relative to true north as noted in the Tidal Current Tables. He should also always plot the current using the compass rose outer circle (true direction).

5-82. As an example, the navigator's intended course to the destination is 120 degrees true and the desired speed is 5 kts. Checking the tidal current table for the operating area would show that the current will be setting the navigator's boat 265 degrees true and drift (speed) 3 kts (Figure 5-13, page 5-22). He will then obtain the actual course made good and SMG by following the steps below:

- *Step 1.* Lay out the chart. The center of the compass rose is the departure point. Draw the boat's intended direction (120 degrees) through the center of the compass rose (+). Make this line's length, in miles (or scale units), the same as the intended speed, in kts (5 kts). Put an arrowhead at its tip. This is the intended course and speed vector.
- *Step 2.* Draw the line for the direction of the current, intended course, and speed vector through 265 degrees true. Make this line 3 miles (or scale units) long, putting an arrowhead at the outer end. One kt is expressed as 1 nm per hour. Measurement can be made from the nm scale or the latitude scale running along both sides of the chart (or other convenient scale), using dividers.
- *Step 3.* Draw a straight line to connect the center of the compass rose to the arrow point of the direction and speed current line (set and drift vector). This line is the actual course and SMG vector. Measure the direction of this line as it crosses the outer circle of the compass rose. This line is the actual course made good. Measure the length of this line to obtain SMG.

Chapter 5

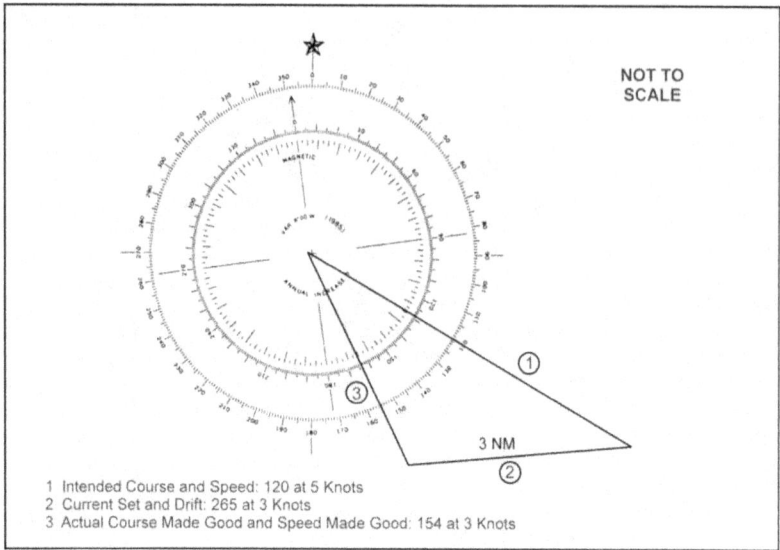

Figure 5-13. Determination of actual course and speed made good

DETERMINATION OF COURSE TO STEER AND SPEED TO RUN FOR ACCOMPLISHING INTENDED COURSE AND SPEED

5-83. This current triangle is a graphic vector diagram in which the navigator once again draws the first line from the center of the compass rose in the direction intended to be run. He draws its length in miles (or scale units) for the speed (kts) intended to be run. This line is the intended course and speed vector. The navigator puts an arrowhead at its end. Next, he lays out the current vector with its set from the center of the compass rose outward, to a length in miles equal to its drift and puts an arrowhead at its tip. Then, he draws the actual course and speed vector from the tip of the current vector to the tip of the intended course and speed vector. He puts an arrowhead at this tip. The length of this line is the actual speed to run. The direction of the line is the actual course to steer (Figure 5-14, page 5-23).

DETERMINATION OF COURSE TO STEER AND ACTUAL SPEED MADE GOOD USING INTENDED BOAT'S SPEED

5-84. The navigator intends to run at least 6 kts. He uses the graphic vector diagram and draws the first line from the center of the compass rose, which is the course he intends to run. At this time, he draws a line of indeterminate length.

5-85. Next, from the center of the compass rose, the navigator draws the current vector with its direction outward in the direction of the set and its length (in miles) that of the drift (kts). He puts an arrowhead at its tip. Using his drawing compass, he lays down a circle with its center at the end tip of the current vector and its radius (in miles) equal to the intended speed of his boat (6 kts). Where this circle crosses the first line (the intended course line), he makes a mark. He then draws a line from the tip of the current vector to this mark. The direction of this line is the course to steer at 6 kts. The length now established by the mark on his intended course line will be the actual speed he will make good. He should then use this SMG to calculate his estimated time en route (Figure 5-15, page 5-23).

Small-Boat Navigation

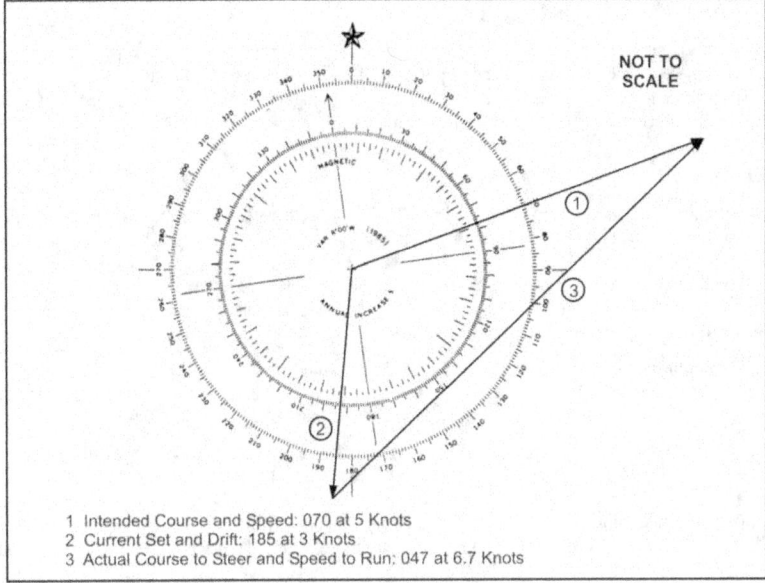

1. Intended Course and Speed: 070 at 5 Knots
2. Current Set and Drift: 185 at 3 Knots
3. Actual Course to Steer and Speed to Run: 047 at 6.7 Knots

Figure 5-14. Determination of course to steer and speed to run

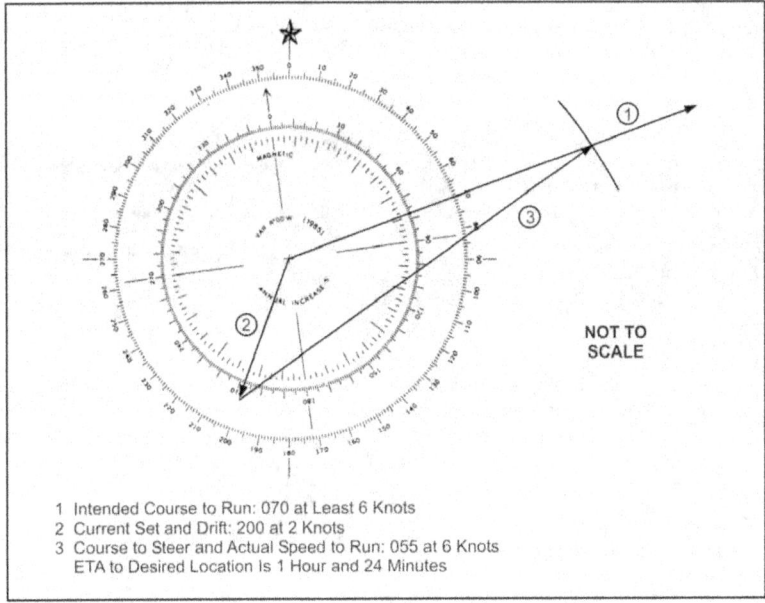

1. Intended Course to Run: 070 at Least 6 Knots
2. Current Set and Drift: 200 at 2 Knots
3. Course to Steer and Actual Speed to Run: 055 at 6 Knots
   ETA to Desired Location Is 1 Hour and 24 Minutes

Figure 5-15. Determination of course to steer and actual speed

Chapter 5

MADE GOOD USING BOAT'S INTENDED SPEED

5-86. It is possible for the navigator to determine set and drift by obtaining visual bearings and fixes and comparing these with computed positions from his DR track. However, conditions are not always favorable for visual sightings. Applying the information contained within the Tidal Current Tables enables the navigator to predict the effect currents will have on his boat, allowing him to apply corrections well in advance.

PILOTING

5-87. Piloting is directing a boat using landmarks, other navaids, and soundings. Piloting should be used in conjunction with DR and GPS navigation when determining a boat's exact position. The most important phase in piloting a boat is adequate preparation. Naturally, safe piloting requires the use of corrected, up-to-date charts. The navigator should always remember that piloting deals with both present and future consequences. He must be continuously aware of where he is as well as where he soon will be. The basic elements of piloting are direction, distance, and time.

5-88. **Direction** is the relationship of a point (known as the reference point) to another point. Direction, generally referred to as bearing, is measured in degrees from 000 through 360. The usual reference point is 000 degrees. The relationships between the reference points and reference directions are listed below.

5-89. When the **distance** to be measured is greater than the span of the dividers, the dividers can be set at a minute or number of minutes of latitude from the scale and then "stepped off" between the points to be measured. The last span, if not equal to that setting on the dividers, will require the navigator to measure it separately. He then steps the dividers once more, closing them to fit the distance. He measures this distance on the scale and adds it to the sum of the other measurements. He should use the latitude scale nearest the middle of the line to be measured. The longitude scale is never used for measuring distance. To measure short distances on a chart, the navigator can open the dividers to a span of a given distance, then compare them to the nm or yard scale.

5-90. **Time** is the third basic element of piloting. Time, distance, and speed are related. Therefore, if any two of the three quantities are known, the third can be found. The basic equations, the speed curve, the nautical slide rule, and their uses have been discussed earlier.

PLOTTING BOARD

5-91. Upon sighting the shoreline or the offshore navaids system, the navigator of a boat may make a transition from DR methods to piloting by using a plotting board and basic tools of navigation. Information gathered from the navigator's field of view can be quickly depicted graphically on the nautical chart. This data allows the navigator to frequently and more precisely determine his position in relation to geographic features.

5-92. The navigator follows a step-by-step procedure for setting up the plotting board. He should—
- Prepare the chart for the plotting board. He ensures that all plotting (DR plot, distance rings, corrections, enemy positions, preplanned LOPs) is completed.
- Laminate the chart and fold it to the size of the plotting board.
- Orient the plotting board to the chart. He—
  - Aligns the plotting board meridians to true north.
  - Sets the variation using the thumbscrew.
  - Follows all directions on the back of the board for setting up and using ABC MARK III plotting boards.

Note. The plotting techniques discussed below can be used without a plotting board. However, applying these techniques on the water would be considerably more difficult and time consuming, likely resulting in less-accurate piloting.

FIXING A POSITION

5-93. Techniques for fixing a boat's position on the water include LOPs, distance rings, ranges, running fixes, or various combinations of these techniques.

5-94. **Lines of Position.** An LOP is a bearing (azimuth) to an object depicted graphically on a chart. Simply stated, once the navigator's bearing to an object is drawn on the chart, he knows that his position lies somewhere along that LOP. Bearings must be to a known fixed position and converted to back azimuths before depicted on the chart (resection).

5-95. **Distance Rings.** Distance from an object can be determined if the height of that object is known by using the mil scale displayed in some binoculars. The navigator should—
- Locate a fully visible object of known height (found on the chart).
- Place the horizontal reticle line at the base of the object.
- Read the height of the object in mil scale points, 1 on the scale is equal to 10 mil scale points.
- Use the distance = height (in yards)/1000 mil scale points formula.
- With the bow compass, draw a circle (ring) around the object, using the determined distance as the radius.

The boat's position lies somewhere along the circle (ring) this distance from the object.

5-96. **Range.** A range is achieved when two objects of known position appear along a straight line of sight. The boat's position lies somewhere along this line. The three types of ranges are as follows:
- *Natural range* has two natural objects such as the edges of two islands.
- *Man-made range* has two man-made objects such as two lights.
- *Combination range* has both natural and man-made objects.

Note. When this line is drawn on the chart it becomes an LOP.

5-97. **Fixes.** A position fix can be determined by various methods as Figure 5-16, page 5-26, illustrates. The navigator can determine a fix by—
- Range and LOP—using the intersection of a range line and LOP, which is considered very reliable.
- Range and distance—using the distance/mil scale rings from a fixed object of known height.
- Distance and LOP—using the distance and LOP from the same object of known height.
- Two distances—where distance rings obtained from two different objects intersect.
- Two or more LOPs—using a minimum of two and a maximum of five LOPs, which give a fairly accurate position.

5-98. **Running Fix.** The navigator uses a running fix technique when only one navigational aid is available. He—
- Determines an LOP to an object and marks the DR plot accordingly. He notes the time.
- Takes a second LOP from the same object after a specific period of time. He plots it and notes the time.
- Calculates the distance run between the two LOPs.
- Advances the first LOP on the DR track. He draws a line parallel to the first LOP "advanced" along the ITR by the distance run since establishing the first LOP.

5-99. Where the advanced LOP intersects the second LOP is the running fix position. Constant speed and accurate time are necessary for a running fix.

Chapter 5

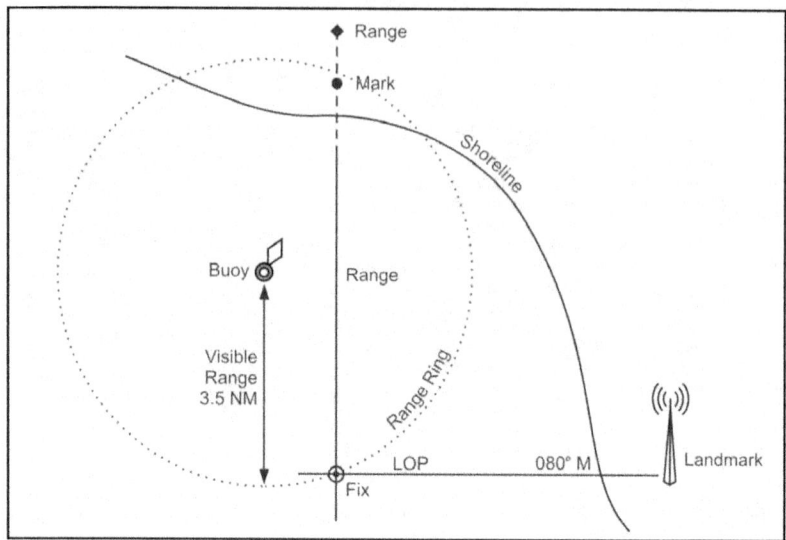

Figure 5-16. Determining a fix

Chapter 6
Combat Rubber Raiding Craft Operations

This chapter specifically addresses CRRC operations. It is intended to ensure interoperability among all U.S. special operations commands and to promote operational readiness and safety through uniform execution of policy and training. For the purposes of this chapter, CRRCs are noncommissioned, inflatable rubber boats, powered by outboard motors and capable of limited independent operations. They are not designated as service craft.

MARITIME PROPONENCY

6-1. The United States Special Operations Command (USSOCOM) Directive 10-1, *Terms of Reference*, gives the Commander, Naval Special Warfare Command (NAVSPECWARCOM) proponency for SO maritime (including underwater operations), riverine, and coastal operations. As such, NAVSPECWARCOM serves as USSOCOM's senior advisor on all matters pertaining to SF maritime, riverine, and coastal operations; training and doctrine; safety; equipment; and interoperability for USSOCOM's Active and Reserve forces. Accordingly, NAVSPECWARCOM shall—

- Standardize maritime, riverine, and coastal operations training, techniques, and procedures, to include basic skill qualifications for all USSOCOM components.
- Validate programs of instruction at all USSOCOM-component maritime, riverine, and coastal operations schools and training facilities.
- Test, validate, and standardize maritime, riverine, and coastal operations equipment.
- Develop, publish, and distribute safety messages, equipment bulletins, and quality deficiency reports.

RESPONSIBILITIES

6-2. This section describes the minimum duties of key personnel in the planning, preparation, and execution of CRRC operations by the following USSOCOM components.

UNIT COMMANDERS AND COMMANDING OFFICERS

6-3. Unit commanders and commanding officers implement and administer CRRC safety and training policies. Individual commands may issue supplementary regulations and instructions for the safe operations of assigned CRRC. In addition, commanders will—

- Ensure personnel involved in CRRC handling or operations are properly qualified and follow approved procedures.
- Ensure that all CRRCs are equipped with the required safety equipment.
- Develop and implement NAVSPECWARCOM-approved training plans to ensure adequate training of appropriate personnel in CRRC mission areas. Requalification and training will be documented in individual service records, as appropriate, and in command training files.
- Ensure accidents and mishaps resulting in death, lost time (24 hours or more), personnel injury, or significant material damage are reported IAW Service directives. Ensure that NAVSPECWARCOM is an "INFO ADDEE" on all CRRC accident reports.

Chapter 6

OFFICER IN CHARGE

6-4. All SF evolutions require a designated individual in charge. The officer in charge (OIC) will plan the assigned evolution using the CRRC officer, as appropriate, to supervise the CRRC portion of the operation. The OIC and the CRRC officer may be one and the same if the OIC is a qualified CRRC officer. The OIC is not required to remain in the CRRC during the entire mission.

CRRC OFFICER

6-5. A competent, qualified, and reliable E-5 or above will be designated as the CRRC officer. He is in charge of the operation and handling of the CRRC during the mission. Each CRRC officer should be experienced in underwater operations, Standard Maritime Practice, prudent seamanship, navigation, CRRC handling characteristics, sea-state limitations, surf-zone procedures, and communications with the chain of command. The CRRC officer—
- Should be the coxswain.
- Will assist the OIC in coordinating and planning all aspects of the CRRC operation, identify safety hazards, and develop emergency procedures during the planning phase of the mission. He should use checklists and guidance developed by his unit or provided in this manual.
- Will advise the operations OIC on all matters pertaining to the planning, execution, and safety of the CRRC portion of the mission.

THE NAVIGATION TEAM

6-6. The navigation (or piloting) team consists of the navigator and coxswain. Long-range transits and OTH operations would benefit from the assignment of an assistant navigator. All team members should be well-versed in small-boat navigation and be intimately involved in the development of the navigation plan (NAVPLAN). Development of the NAVPLAN should include input from indigenous sources, if available, familiar with the waters being navigated. Once underway, the duties of the navigation team are as follows:
- *Navigator.* He orders course and speed to the coxswain; keeps total TOP; and operates the plotting board by plotting bearings, currents, and DR plots. He notifies the assistant navigator when to take bearings and speed checks. He oversees the entire NAVPLAN and safe piloting of the CRRC.
- *Assistant Navigator.* He renders aid to the navigator as directed and shoots bearings, measures speed, and keeps time between legs. He must be alert and observant of sea conditions, aids to navigation, and land features.
- *Coxswain.* He maintains the ordered course and speed.

6-7. All personnel who embark in the CRRC will wear the equipment and clothing directed by the OIC or CRRC officer. They will be aware of the unit SOPs and the particular contingency plans associated with the mission or training evolution.

TRAINING

6-8. To ensure safe and successful operations, all personnel who participate in CRRC operations must be skilled and proficient. This section provides guidance on SF CRRC training, to include initial training qualification and the integration of CRRC training into unit training plans.

INITIAL CRRC TRAINING QUALIFICATION

6-9. Training for CRRC operations occurs primarily at the unit level. However, the Naval Special Warfare Center, Coronado, California; the Naval Special Warfare Center, Detachment Little Creek, Norfolk, Virginia; and the U.S. Army John F. Kennedy Special Warfare Center and School, Fort Bragg, North Carolina, and its school in Key West, Florida, have offered formal courses in long-range MAROPS and waterborne infiltration. United States Marine Corps (USMC) reconnaissance forces and assigned raid forces receive their initial training and familiarization in CRRC operations at one of the following sites:

The Basic Reconnaissance Course (selected reconnaissance personnel only), Atlantic Expeditionary Warfare Training Group, Pacific Expeditionary Warfare Training Group, or 1st, 2d, or 3d Special Operations Training Groups. Follow-on training and advanced OTH skills may be taught at a formal school or the unit level with oversight by the S-3 training sections. Initial training for all Services consists of classroom lectures, conferences, and practical exercises.

Classroom Lectures and Conferences

6-10. The lectures and conferences should familiarize personnel with—
- Unit SOPs.
- CRRC safety regulations.
- CRRC operating characteristics.
- Proper waterproofing of equipment and stowage procedures.
- Boat and engine maintenance requirements.
- Navigation techniques.
- Standard maritime practice.
- First aid.

Practical Exercises

6-11. Practical exercises (PEs) are conducted under the supervision of experienced personnel. PEs are designed to introduce trainees to the various aspects of CRRC operations. Areas covered in these exercises include the following:
- Practical navigation.
- Basic seamanship.
- Launch and recovery techniques from various host platforms.
- OTH operations.
- Surf passage.
- At-sea rendezvous techniques.
- OTB operations.
- Chart study and publications.
- Tides and currents.
- Planning navigation routes.
- Repair and maintenance of the CRRC and outboard motor (OBM).

SUSTAINMENT AND UNIT TRAINING

6-12. After initial qualification, it is necessary to sustain and build upon the skills acquired during initial training. Personnel should perform refresher and advanced training on a continuous basis to maintain proficiency.

Refresher or Sustainment Training

6-13. Training will consist of practical training and classroom instruction. Units should maintain lesson plans and review or update them annually. Regular command training for all personnel who participate in CRRC operations should include, but not be limited to, the following:
- Classroom instruction, to include the following:
  - General safety precautions.
  - Equipment maintenance.
  - Waterproofing techniques.
  - First aid.
  - Cardiopulmonary resuscitation.

- Navigation techniques.
- Standard maritime practice.
- Unit SOPs.
- Practical work, to include the following:
  - Practical navigation.
  - Repair and maintenance of the CRRC and OBM.
  - In-water emergency procedures.
  - Launch and recovery techniques from various host platforms.
  - Over the horizon operations.
  - Surf passage.
  - At-sea rendezvous techniques.
  - OTB operations.

6-14. Commanding officers, after completion of a unit-level CRRC/rigid hull inflatable boat (RHIB) refresher training package or Special Operations Training Group (USMC) small-boat refresher package, may approve and issue licenses that may be required by higher commands for forces to be properly designated to operate small boats.

Mission-Essential Task List Focus

6-15. Unit CRRC training will be mission-essential task list (METL)-focused to the maximum extent possible. CRRC training plans will be unit-specific. Training will be progressive to attain or maintain the skills required to conduct CRRC operations in support of assigned missions in projected operational environments. Collective CRRC training will be integrated with other METL-focused training as much as possible. The use of realistic field training exercises (FTXs), based on full mission profiles (FMPs), to train and evaluate the unit's CRRC capability should be the norm rather than the exception. Standards for various CRRC evolutions (for example, CRRC OTH navigation, CRRC paradrop insertion, and at-sea rendezvous) are delineated in U.S. Navy fleet exercise publications, with specific lesson guides contained in NAVSPECWARCOM instructions.

Advanced Training and Qualifications

6-16. Additional training and certification is necessary to perform certain technical and supervisory functions required in the conduct of SF CRRC operations. The functions and training are discussed below.

6-17. **OBM Technician.** Advanced training is offered at factory schools to train mechanics in the proper maintenance and repair of OBMs. On-the-job training (OJT) must also be available at the command level. Personnel who have demonstrated a mechanical aptitude can learn maintenance and repair techniques and procedures under the tutelage of experienced OBM technicians. Motor maintenance skills are critical to ensure mission completion.

6-18. **CRRC Officer.** Training that leads to written certification of individuals as CRRC officers is provided by the unit commander and must include detailed knowledge of the following subjects:
- *Navigation.* Personnel must demonstrate their abilities in the following areas:
  - Charts and publications, which include how to interpret, plot, use, order, and correct nautical charts.
  - Magnetic compass, which involves determining deviation, variation, and true direction.
  - Tide tables, which test the computation of the tides' state.
  - Current sailing, which applies the proper course corrections for currents of various speeds.
  - Tidal current sailing, which involves computing tidal currents and applying appropriate course corrections.
  - Tools of navigation, which involve using the GPS, a hand-bearing compass, a plotting board, and a sextant.

- Navigation planning, which includes laying out a track with the turn bearings, waypoints, and ranges.
- Piloting, which involves navigating from one location to another.
- DR, which involves determining position without using navaids.
- Electronic navigation, which involves using the GPS.
• *Seamanship.* Personnel must show proficiency in the following areas:
  - Standard maritime practice.
  - Small-boat safety regulations.
  - CRRC handling characteristics.
  - Sea-state limitations.
  - Surf-zone procedures.
• *Operational Planning.* Personnel must be well-versed in the following:
  - SF planning procedures.
  - Planning for CRRC operations.
  - Communications planning.

OPERATIONS AND PLANNING

6-19. A unit can conduct CRRC operations as infiltration and exfiltration operations. The organization for a particular operation is dependent on the nature of the mission and the unit's SOPs. When organizing for the mission, unit and individual qualifications or experience should be matched to the specific requirements of the mission or training event. Unit integrity is an important consideration when conducting CRRC operations; however, there are times when unit attachments may be necessary to fulfill the mission. In such cases, all participants must be briefed regarding the unit's tactical SOPs to be followed in an emergency.

6-20. Without preliminary planning, the entire operation may fail and, in extreme cases, lives may be endangered. The following information provides planning guidelines for a safe and successful CRRC operation. This information is not a guide to tactical mission planning. Specific mission planning factors to CRRC operations are discussed below.

TACTICAL LOADS

6-21. As a rule, the mission OIC should not load the CRRC to its maximum capacity when preparing for a mission. In a crowded CRRC, personnel in the forward positions are subjected to a greater degree of physical discomfort owing to the turbulence created by the effect of swell and wave activity. This concern is especially significant during long-transit periods. Increased physical stress may diminish an individual's ability to perform once he has arrived at the objective site. Furthermore, the OIC must take into account the possibility of prisoners, casualties, and the evacuation of friendly forces. Thus, it is always wise to have sufficient additional boat space for unforeseen contingencies. For example, although the F470 CRRC with a 55-horsepower (hp) engine is capable of transporting ten personnel, no more than six personnel should be embarked. The optimum weight ceiling for the F470 CRRC is 2,000 pounds (lb). Any weight above this ceiling significantly reduces the CRRC's efficiency. Experience has shown that six personnel, with mission-essential equipment, on average, come closest to this weight ceiling. In marginal sea states or when there are extended distances to be covered, consideration should be given to limiting embarked personnel to four.

COMMAND RELATIONSHIPS

6-22. The SF mission or unit OIC is responsible for every phase of the CRRC tactical mission. He conducts planning and coordinates any external support required to execute the mission. However, SF elements will frequently be tactically transported aboard host ships or aircraft. SF elements may also have personnel from other units attached. The following paragraphs provide additional guidance for these circumstances.

6-23. The ship or submarine commanding officer is generally the officer in tactical command (OTC) and the SF OIC is embarked under his command. Although the ship's commanding officer is in a supporting role to the SF OIC, he has ultimate responsibility for the safety of his ship and crew. Thus, his authority is absolute until the CRRC mission is launched. At the point of launch, tactical command of the mission shifts to the SF OIC. It is essential that the OIC coordinate every aspect of the operation with the host ship's commanding officer and discuss with him every phase of the mission that requires his participation.

Note. The above guidelines also apply when being transported aboard host aircraft.

6-24. After launch, the authority of the SF mission commander is absolute during the conduct of the mission. He is solely responsible for the successful execution of the mission, as well as the safety and well-being of assigned personnel. Any attached personnel, civilian or military, regardless of rank or position, cannot override the decision-making authority of the SF mission OIC. Any variance to this policy must be clearly addressed in the operation plan (OPLAN), with guidelines as to who is in tactical command at each phase of the operation.

6-25. The CRRC officer, generally the coxswain, will follow the directions of the SF mission commander. It is his responsibility to keep the mission commander informed as to location and status of the CRRC.

6-26. When planning an OTH CRRC operation, the SF OIC must be the focal point from the outset. The planning sequence must depict continuous parallel, concurrent, and detailed planning. Planning support and information will be needed from all sources. In addition, planning requires close cooperation and teamwork between the supported and supporting elements. As the plan develops, all aspects should be briefed, reviewed, and understood by planners and decision makers before becoming finalized.

6-27. As with the planning of any operation, provisions must be made to deal with uncertainty. Plans should be developed to address contingencies and emergencies during CRRC transits to include—

- *OBM Breakdown.* Unit should have a trained technician and bring tools and spare parts or spare motor, if space and time permit.
- *Navigational Error.* Unit should study permanent geographical features and known tides, currents, and winds; take into account the sea state; and intentionally steer left or right of the target BLS so once landfall is reached, a direction to target is already established. (When using the GPS, the CRRC should aim directly at the BLS; it should not offset left or right of target. Proper use of the GPS will take the CRRC directly and accurately to the BLS.)
- *Low Fuel or Empty.* Unit should run trials with a fully loaded CRRC in various sea states to calculate fuel consumption rates. The CRRC should take enough fuel for a worst-case scenario.
- *Emergency Medical Evacuation (MEDEVAC) Procedures.* Unit should know location of nearest medical treatment facility and develop primary and alternative plan for evacuating casualties.
- *CRRC Puncture.* Unit should carry expedient plugs.

OPERATIONAL CONTINGENCIES

6-28. Many other contingencies may arise during the conduct of CRRC operations. They may be caused by a wide variety of factors or events, and can occur at any time. When CRRCs operate with host platforms, operational contingencies fall into four general categories.

Mission Abort

6-29. This action occurs when a decision is made not to continue the mission as scheduled. It may lead to the cancellation of the mission or to a delay. Mission abort may result from a system or equipment failure that severely impacts performance, adverse weather conditions, or a changing tactical situation that jeopardizes mission success. The abort decision should be made after evaluating all the circumstances and their impact on the mission. Mission-abort criteria should be defined before the mission gets underway. Should the SF OIC decide to abort the mission while underway, he should communicate his decision to the supporting craft as soon as tactically feasible to initiate rendezvous, recovery, or casualty assistance operations. The decision to abort the mission is normally controlled by the on-scene tactical commander, or may be dictated by the operational chain of command.

Rendezvous Point Interference

6-30. Detailed rules of engagement must be developed to respond to both hostile and civilian interference at the primary and alternate RPs. The presence of hostile forces in the rendezvous areas may indicate compromise of the mission. Whatever the interference, a decision must be made to proceed with the rendezvous as planned or shift to an alternate RP.

Missed Rendezvous

6-31. A rendezvous is missed when the allotted time window has expired. The time window will vary depending on each mission. Mission planners define the time window and include it in the OPLAN. Generally, the time window will be approximately 1 hour. If late arrival is anticipated and emission control (EMCON) conditions permit, the CRRC should attempt to communicate with the host platform and coordinate an adjusted rendezvous time. In most cases, unless notified by the CRRC of a delay, the host platform will not remain in the rendezvous area longer than the scheduled time window. If the initial scheduled rendezvous and recovery do not succeed, subsequent efforts will be more difficult because—

- The CRRC crew will have extended exposure to the elements.
- Endurance (fuel) will be nearing exhaustion.
- The enemy will have a longer time to react.

6-32. Many factors can cause a missed rendezvous. For example, the CRRC may not make the RP on time. An equipment casualty, such as loss of motor propulsion or CRRC damage can prevent the CRRC from reaching the RP. Also, cumulative navigation error of the CRRC and the host platform can exceed the range of capability of the rendezvous locating system. This problem may not be obvious at the time. The navigation error may go unrecognized until the CRRC has exhausted its fuel. Bad weather can seriously reduce the range of the rendezvous locating systems or a failure of the rendezvous locating system may occur.

Search Considerations

6-33. Four items should be considered before starting a search for a CRRC after a missed rendezvous. Each should be carefully evaluated during planning and thoroughly briefed to the support personnel and host platform crew so they can make informed decisions as the tactical situation changes. The four questions are—

- How much time will be available to conduct a search, and what size search area can be covered before departing the rendezvous area?
- How much fuel will be expended during the search and how will it affect the host platform's ability to return to its base or host ship?
- Will the search effort increase the probability of detection?
- How long can the CRRC wait for the searching host platform?

6-34. The SF mission OIC will be faced with a "time-to-leave" decision based on endurance (fuel), remaining hours of darkness, and alternate COAs—for example, lay up ashore or return to sea for a combat search and rescue (CSAR) or other preplanned evasion and escape (E&E) procedures. If communications with the operational unit fail and a search is initiated, the search will continue at the discretion of the host platform's commander as preplanned E&E procedures are initiated. If the host platform is in positive communications with the operational element, the search will continue until the element is recovered or as long as feasible to prevent detection of the host platform or operational element.

SEA CONDITIONS

6-35. The reference for understanding and judging sea conditions is *The American Practical Navigator* (Bowditch), Chapter 37. The sea-keeping characteristics of the CRRC permit operations in Sea State 3, although rough weather greatly increases the fuel consumption and risks associated with launch or

Chapter 6

recovery of the CRRC. The maximum sea state for CRRC launch or recovery and ocean transits during training evolutions is Sea State 3. The high winds generally associated with higher sea states will adversely affect the maneuverability of the CRRC. High sea states and corresponding winds force slower CRRC speeds and result in longer transit times. An advantage offered by operating in higher sea states is a reduced vulnerability to detection. High seas degrade the detection range and effectiveness of electronic sensors and contribute to tactical surprise. Sea conditions can fluctuate rapidly. During the planning phase for an OTH operation, the navigator must consider the forecasted meteorological factors that could adversely affect the sea state during all mission times. For example, if the navigator anticipates that conditions at launch will be Sea State 3, he should, during the planning phase, determine if any forecasted meteorological factors exist that could cause the sea state to go to 4 or 5 during the operational window. Also, a low sea state does not mean a benign surf zone; sea and surf zone conditions must be considered independently.

Distance

6-36. The optimum CRRC launch occurs at least 20 nm from shore, but usually no more than 60 nm. In determining the actual distance to execute a launch, the navigator must consider the sea state, weather, transit times, and enemy electronic-detection capabilities. The objective is to keep the host platform undetected while minimizing the CRRC transit distance, thereby reducing the physical demands caused by long open-ocean transits.

Navigation

6-37. DR and coastal piloting are the primary methods of navigation for the CRRC. Passive electronic methods of navigation (for example, GPS) are a useful complement. However, electronics should not be totally relied upon due to problems in propagation, signals interruption in adverse weather conditions, and the general unreliability of electronic instruments in a saltwater environment.

> *Note.* The GPS is faster and much more accurate than DR. By naval convention, the GPS should be considered the secondary navigation method. Militarized GPS receivers are extremely rugged instruments—not fragile. The GPS was designed to minimize signal propagation problems due to adverse weather conditions.

6-38. Celestial navigation in CRRCs is normally impractical because of the extensive publications required, and the unstable nature of the CRRC as a platform from which to use a sextant. Whenever possible, development of the NAVPLAN should include input from indigenous sources familiar with the waters being navigated.

The Surf Zone

6-39. A key planning consideration for a CRRC operation is the characteristics of the surf zone near the BLS. The navigator will analyze key elements, planning and execution factors, and surf limits for inclusion in the NAVPLAN. Detailed information on the surf zone can be found in instructions from the Commander, Naval Surface Force, U.S. Pacific Fleet and the Commander, Naval Surface Force, U.S. Atlantic Fleet. OTB operations should always include scout swimmers to verify or monitor surf conditions, especially if the conditions are marginal or likely to change during the operation.

> *Note.* Surf observation (SUROBS) is an 8-line report; the North Atlantic Treaty Organization (NATO) format surf report (SURFREP) is a 10-line report, the first two lines being unit of measure and date-time group (DTG), the last eight lines corresponding to the 8-line SUROBS.

Key Elements

6-40. The critical considerations for CRRC surf-zone operations are the significant wave height (height of the highest one-third of the breakers observed), the period (the time interval between waves measured to the nearest half-second), and the breaker type (spilling, plunging, or surging). The danger factor in CRRC

Combat Rubber Raiding Craft Operations

surf-zone operations increases as wave height increases, the period between waves lessens, and waves begin plunging or surging. These critical factors must be considered, both individually and in combination, and their effect on CRRC operations carefully and prudently evaluated, especially in light of boat or engine maintenance conditions and coxswain or boat crew experience level. Personnel trained and experienced in CRRC operations—for example, designated CRRC officers—should conduct this evaluation. Although these three elements represent the most critical considerations, other information contained in the SUROBS is important for navigation and planning for negotiating the surf zone.

Planning and Execution

6-41. Surf conditions are critical in making the final determination to launch or not launch the CRRC. Once the CRRC is launched and arrives outside the surf zone at the BLS, the CRRC officer will advise the OIC regarding the final beaching decision. If prelaunch SUROBS, visual sighting, or mission necessity warrant further surf evaluation, the OIC may use scout swimmers for this purpose. He must consider the surf conditions as they exist at the time, his mission and command guidance, boat or engine maintenance conditions, and coxswain or boat crew safety. Further surf conditions, like sea states, can change rapidly. During planning, the forecasted meteorological factors that could adversely affect surf conditions should be evaluated. Swimmer scouts should always be used during OTB operations.

Surf Limits

6-42. The recommended CRRC surf passage table in Figure 6-1 outlines recommended operating limits for CRRC surf-zone operations. The table plots breaker height versus breaker period, and is applicable to both spilling and plunging waves. Plunging breakers are more dangerous to CRRC operations than spilling breakers, and greater care and judgment must be exercised as the percentage of plunging breakers increases. Surging breakers are not included in the table. The recommended surf limits are provided as a guide and are not intended to usurp the judgment of officers exercising command.

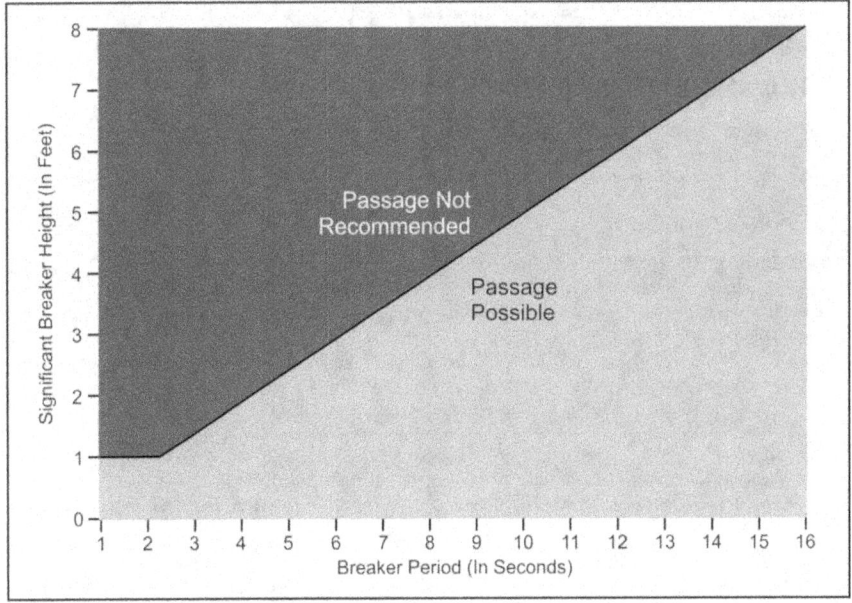

Figure 6-1. CRRC surf passage table

Mission Rehearsals

6-43. Rehearsals should be conducted for all phases of the CRRC operation, if time and the tactical situation permits. Rehearsals are as important for the crew of the host or supporting platform as they are the SF element. This practice occurs because supporting personnel may not be familiar with the support requirements for SF missions. Host platform crews should be given every chance to rehearse their portion of the mission with the SF unit they are to support. If embarked on a host ship, time may be available while in transit or during an in-port period. Special attention should be given to the time-critical elements in each phase and to the completion time for each task. Performance conditions and standards for success or failure should be determined to allow modification of the plan or training. Planners must balance the advantages of rehearsal experience against possible equipment degradation and personnel fatigue. As a minimum, rehearsal events for CRRC operations should include the following:

- Launch procedures.
- Rendezvous procedures.
- Recovery procedures.
- Communications procedures.
- Contingency plans.
- OTB insertion procedures.
- Insertion and extraction procedures.

6-44. All MAROPS have inherent risks. Proper training and in-depth preparation and planning will reduce the potential for accidents. Safety considerations will be fully addressed during planning and integrated into all CRRC training. Unit commanders will perform risk assessments before conducting the "crawl-walk-run" manner of training.

6-45. Safety is the responsibility of every individual participating in or supporting the CRRC operation. All personnel involved in a particular evolution must be constantly aware of the nature and progress of the operation and must remain alert for possible danger. Specific individual responsibilities for the safe conduct of SF CRRC operations are set forth in Chapter 2.

ENVIRONMENTAL PERSONNEL LIMITATIONS

6-46. During the planning process, consideration must be given to how weather, temperature, surf conditions, and sea state will affect the safety of the CRRC evolution. These planning considerations and minimum equipment requirements are discussed below.

Operational Procedures

6-47. Training and operations must be planned with consideration to individual and unit experience, personal fatigue and exposure factors, and equipment capabilities. Observance of the safety regulations prescribed below is mandatory unless waived by the appropriate level of command as determined by the operational risk assessment:

- All personnel involved in CRRC operations should wear an inherently buoyant personal flotation device (life jacket) or an inflatable life preserver; for example, an underwater demolition team (UDT) life jacket or a life preserver unit (LPU).

Note. LPU is a new type of inflatable vest that Navy SEALs and USMC reconnaissance forces wear while crossing over water by helicopter for visit, board, search, and seizure (VBSS) operations. The device is an inflatable collar that enables a person to be lifted up when in full mission gear.

Note. During CRRC training evolutions, personnel will not be bound in any manner (for example, handcuffs, flex-cuffs, or zip ties). Fatalities have occurred in the past when bound personnel were inadvertently thrown from the craft.

- For single CRRC operations, a minimum of two persons are required to be in the craft (towed craft are considered single-craft operations). For multiple CRRC operations, one person or coxswain per boat is acceptable, and one CRRC officer is sufficient for all boats. The SF OIC should be in the same CRRC as the CRRC officer. For administrative movement under good conditions and within confined, protected, or sheltered waters, it is acceptable to operate a CRRC with only one person in the craft, and a CRRC officer is not required.
- Before getting underway, the CRRC coxswain will—
  - Make certain that the CRRC is equipped with the required safety equipment.
  - Verify adequacy of boat fuel.
  - Establish and maintain positive communications as directed in the OPLAN or training plan.
  - Warm up the outboard engines at idle speed for 5 minutes.
- CRRC launch and recovery from underway platforms will not be routinely conducted in Sea State 3 or greater during training evolutions.
- CRRC transits should also not be conducted in Sea State 3 or equivalent sea conditions for routine training evolutions.
- CRRC will not be operated in surf zones with breaker heights greater than 8 feet. In this situation, the recommendation of the CRRC officer and judgment of the OIC will dictate alternate options for transiting the surf zone.
- Every CRRC will be equipped with a radio. Positive radio communications will be established before getting underway. Communications will be maintained with the parent command or operational commander as provided for in the training plan or the signal operating instructions (SOI).
- CRRC operations will be cleared with the immediate operational commander. He will approve intended and alternate navigation tracks.
- The CRRCs will be operated within the limits of the designated specifications for weight or personnel on board. (Chapter 7 provides information on the F470 CRRC and the CRRC [small]).
- Equipment will be secured in the CRRCs so as not to impede egress of personnel in emergency situations.
- Commanders or commanding officers and their appointed safety officers must ensure their personnel, particularly those directly involved with CRRC operations, are fully aware of the potential hazards inherent in all waterborne activity. The safety regulations and operating guidelines are not intended to usurp the judgment of officers exercising command.

Note. Minimum equipment requirements have been established. They are listed in the equipment section of this chapter and are considered essential for conducting safe CRRC operations.

Emergency Procedures

6-48. Emergency procedures are key collective skills that support the METL of all combat units. Planning and rehearsing these skills should be an integral part of all unit training. Unit SOPs should be developed to deal with anticipated emergencies that could arise during CRRC operations. An emergency action plan should be established and coordinated with supporting commands or units, as appropriate. At a minimum, the plan should include considerations for towing, capsized CRRC, loss of contact, man overboard, engine failure, and low fuel.

Towing Operations

6-49. All CRRCs must be rigged and equipped for tow or to-be-towed operations. When a CRRC fails, personnel should—
- Immediately prepare the craft for towing. Towlines should be long enough to sag under their own weight. Taut towlines result in accidents (Figure 7-1, page 7-4).
- Cross-deck a technician to repair the OBM.

Chapter 6

- Cross-deck personnel to alternate craft to lighten the load, if towing is required.
- Tow the inoperable CRRC with only the coxswain and technician onboard.
- Ensure the principles below are followed while towing. CRRC personnel should—
    - Tow at a slower speed than the CRRCs normally travel.
    - Not make quick turns.
    - Assign a lookout in the towing craft to observe the towed CRRC.
    - Not stop suddenly.
    - Maintain steering control of the towed CRRC. If the towing craft stops suddenly, the towed CRRC must be able to steer clear without ramming the towing boat.
    - Raise the engine on the towed craft.

Capsized CRRC

6-50. If a CRRC begins to capsize, the coxswain should warn the craft's occupants and attempt to stop the engine. When the passengers and coxswain are in the water, they should protect their heads with their arms. Actions taken after capsizing include the following:

- All personnel should inflate life jackets, orally if possible, when clear of the CRRC.
- The senior man or coxswain should conduct a head count.
- All personnel should swim to the seaward side of the craft to keep it from being pushed by swells over the top of them. They should remain in a group.
- Personnel should attempt to right the craft. If unable to do so, they should stay with the CRRC. If capsized in the surf zone, the CRRC will be pushed ashore. If in the near-shore area but outside the surf zone, personnel should try to swim the CRRC to shore. If at sea, it is easier for CSAR efforts to spot a capsized CRRC than individual swimmers.

Loss of Contact

6-51. If contact is lost between crafts when conducting operations involving multiple CRRCs, the usual procedure is for the lead craft to wait a preplanned period of time for the missing CRRC to catch up. Every attempt must be made to reestablish contact ASAP. Factors to consider include the following:

- CRRCs are most likely to lose contact at night and in reduced visibility (fog, rain, and heavy seas). To reduce the likelihood of lost contact, the following measures should be followed:
    - Formations are kept tight and the coxswain maintains visual contact with the other craft.
    - Other passengers in the CRRC are assigned to assist the coxswain in his duty.
- If a CRRC falls behind the formation or stops, the other craft will stop to render assistance. In the event that a CRRC cannot maintain the formation speed, the CRRCs will travel at the slow craft's best possible speed.
- If contact is lost, the lead CRRC should stop and wait for the lost craft to catch up. If after a short period the lost craft does not catch up, the lead CRRC should follow a reciprocal bearing until contact is made. Contingency plans include the following:
    - If contact is not made along the track, the lead CRRC will proceed to a predesignated rally point and wait until the lost craft arrives.
    - The coxswain of the lost craft should attempt to reestablish contact. If contact is not swiftly made, the coxswain should move to the predesignated rally point to await the rest of the force.
    - The OIC shall decide if the mission or the lost craft takes precedence, taking into account the time available for the mission and the preplanned abort criteria.

Man Overboard

6-52. If a man falls overboard, the coxswain or his designee will alert the OIC and other craft in the area using voice, radio, or visual signals. To facilitate recovery, each person will have a chemlight or strobe (these may be infrared [IR] depending on the tactical situation) attached to his life jacket. It will be activated upon falling overboard. Preventive measures include the following:

- Unless the CRRC is in a dangerous surf zone, where stopping would endanger the entire CRRC crew, the coxswain will maneuver to avoid the man and shift the engine into neutral until past the man to prevent the propellers from striking him.
- Depending on the tactical situation, a man overboard in the surf zone will either return to the beach where a designated boat will recover him, or he will swim through the surf zone and be recovered.
- To recover a man overboard into the CRRC, the following steps must be performed in sequence:
  - Observe the man in the water. Make a rapid turn toward the man, ensuring the propellers do not hit him. Keep the man in sight until he is recovered. Recover the man on the coxswain side to maintain visual contact by the coxswain throughout the evolution. Before reaching the man overboard, reduce speed and cut the engine off.
  - Approach the man facing into the current or wind, whichever is the stronger of the two. (By steering into the elements, the coxswain will maintain steerageway even at slow speed and can use the elements to slow and stop the CRRC.)
  - When close to the man, the coxswain will indicate to the crew whether he will recover him to port or starboard.
  - When the man is within reach, a line or paddle will be extended to assist him back into the CRRC. Only as a last resort should a crew member attempt a swimming rescue.

6-53. The CRRC from which the man fell will normally perform the recovery. If it cannot, the coxswain will request assistance from another craft and direct it to the vicinity of the man.

EQUIPMENT

6-54. For the purpose of this chapter, a CRRC is a noncommissioned inflatable rubber boat. It is powered by an OBM and capable of limited independent operations. CRRCs are not designated as service craft. Tables 7-1 and 7-2, page 7-13, and Table 7-3, page 7-17, outline the characteristics of CRRCs, associated OBMs, and fuel containers currently used by SF.

6-55. Equipment will be used IAW applicable Service regulations, instructions, and appropriate technical manuals. Information on the characteristics of specific craft and fuel consumption data is provided in Chapter 7 along with further guidance for preparing the CRRC and OBM for an operation.

6-56. A dangerous oversimplification exists when relying on any general minimum equipment list. The requirements of each operation will dictate the equipment necessary. The OIC and CRRC officer must take all facets of the mission profile into consideration before issuing the equipment list for a particular CRRC operation. The OIC must consider the following items for each operation:

- *Inherently Buoyant Personal Flotation Device (or UDT life jacket).* One device is required for each embarked member. A personal flotation device must be worn by each individual in the CRRC and must have sufficient lift to support the individual and his combat load. Each individual should conduct "float tests" to ensure that he has adequate flotation before the CRRC mission or training evolution. For flotation devices that use gas bladders for buoyancy, the activation mechanism (for example, gas cartridges and activators) should be inspected for proper functioning and ease of access while Soldiers are in full combat dress.
- *Radio Transceiver.* A compatible radio transceiver, with sufficient spare batteries, will be carried in each CRRC.
- *Navigation Equipment.* Passive electronic methods of navigation such as GPS are recommended. However, they should not be totally relied upon due to problems in propagation, signal interruption in adverse weather conditions, and the delicacy of electronic instruments. As a minimum, the following navigation equipment is recommended:
  - Binoculars (waterproof).
  - Hand-bearing compass.

Chapter 6

- Speed-measuring devices.
- Navigation chart of the AO.
- Nautical slide rule (speed wheel).
- Plotting board.
- Flat dividers.
- Boat compass.

● *Flares.* Only approved military flares will be used IAW current directives. Personnel should carry a minimum of four flares.

● *Paddles.* A sufficient number of paddles should be transported to provide propulsion and steering for the CRRC in case of OBM failure.

● *Repair Kit.* This kit should contain provisions for the CRRC and OBM.

● *Foot Pump.*

● *Strobe Light.*

● *Bow and Stern Lines.* They should be at least 15 feet long.

● *Towing Bridle.*

● *Protective Clothing.* Generally, protective clothing will be worn at the discretion of the OIC and CRRC officer as dictated by water and atmospheric temperatures. The planners should consider the desires of the individual crew. Planners should remember that crew members can suffer hypothermia during prolonged exposure to sea spray and wind even in relatively mild air temperatures.

● *Identification, Friend or Foe (IFF).* This transponder is similar to those used on aircraft, ships, and armored vehicles. It transmits its identification in response to a query by U.S. targeting systems.

● *Blue Force Tracker.* This transponder works through a series of satellites to pinpoint the location back to the crew's tactical operation center.

6-57. When planning for MAROPS, the "submariner's rule" should be applied when practical. It is a rule of thumb that provides an operational guide to equipment redundancy requirements when operating isolated from support facilities. This guideline is intended to foster the mindset that preparedness with redundant equipment is the only way to ensure mission success in case a critical piece of equipment fails. The submariner's rule states that three of any required item is actually only two that can be counted on, two is one, and one is none. For example, when conducting an OTH infiltration using a Special Operations Combat Expendable Platform (SOCEP) with two F-470s, prudent planning requires three OBMs.

Chapter 7

Inflatable Boats

This chapter provides information of value to detachments desiring introductory training on inflatable boats. Information is provided to familiarize personnel with boat team composition, organization, and duties. Basic boat-handling drills and emergency procedures are discussed. These procedures explain how to safely operate a CRRC in all conditions, with or without a motor. This chapter also provides information about CRRCs in general and the Zodiac F-470 in particular. The F-470 is the CRRC that detachments are most likely to use. It is the baseline by which all other CRRCs are judged.

Note. When coordinating with sister Services, detachments must be aware that the term CRRC only applies to the F-470.

Historically, inflatable boat training has been based on the inflatable boat, small (IBS)—a small (7-man) inflatable landing boat that personnel paddle onto the beach and cache or abandon. Much of today's team organization and duties, as well as the handling techniques, are still based on this craft even though SF now use vastly improved, motorized CRRCs to conduct MAROPS.

BASIC BOAT-HANDLING SKILLS

7-1. The extended distances that MAROPS teams must travel to avoid detection have made motors a necessity. Unfortunately, this need has resulted in a general degradation of basic seamanship and boat-handling skills. Motors can fail leaving the unprepared detachment at the mercy of an unforgiving ocean. Therefore, basic seamanship skills are still of great value.

BOAT TEAM COMPOSITION

7-2. A boat team consists of the crew and passengers. The number of crew and passengers will be dictated by the rated capacity of the boat and motor, the mission load, the distance to be traversed, the sea state, and the required speed over the water. Obviously, the more restrictive the qualifiers, the smaller the mission load and the fewer the number of personnel that can be transported in the boat.

7-3. During training, the boat team of an inflatable landing boat will normally consist of seven men—a coxswain and six paddlers. Three paddlers are located along each gunwale, and the coxswain is located in the stern.

7-4. There are two methods of numbering paddlers: the long count and the short count. When the numbering of the boat team is by pairs, it is known as the short count; when the numbering of the boat team is by individuals, it is known as the long count. When using the long count, starboard (right side) paddlers are numbered 1, 3, and 5; port (left side) paddlers are numbered 2, 4, and 6 counting from bow to stern. The coxswain is team member Number 7.

7-5. When using a short count, paddlers are numbered in pairs from bow to stern. If passengers are carried, they are numbered consecutively from bow to stern starting with Number 8. The coxswain will issue commands to all boat team members using their respective number. As an example, when the coxswain issues a command to an individual paddler, he will use that team member's number. When

Chapter 7

addressing a pair of paddlers, he will use the short count terms, "ones," "twos," or "threes," indicating team member Numbers 1 and 2, 3 and 4, and 5 and 6, respectively. When addressing either side of the boat, he will use the terms starboard side or port side.

BOAT TEAM ORGANIZATION

7-6. The coxswain forms the boat team by commanding, "Team, fall in." The boat team forms facing the coxswain in a column of twos. The team members assume the relative positions they will occupy in the boat, hereinafter referred to as boat stations. Passengers form at the rear of the two columns by number—even numbers in the left column and odd numbers in the right. After the team has formed, the coxswain commands, "Team, count off." All hands sound off with their position numbers in order, including the coxswain who is Number 7.

TEAM MEMBER RESPONSIBILITIES

7-7. The tactics and techniques used in boat handling are similar and generally apply to most operations. However, to operate efficiently, each boat team member must perform the following specific duties:
- *Number 1.* The stroke sets the rate of paddling and maintains the paddling rhythm as directed by the coxswain. He assists the coxswain in keeping the boat perpendicular to the breaker line when beaching or launching the boat. He assists the coxswain in avoiding obstacles in the water.
- *Number 2.* He also assists the coxswain in keeping the boat perpendicular to the breaker line and in avoiding obstacles. He handles the towline and quick-release line during towing operations.
- *Numbers 3 and 4.* They lash and unload equipment in the boat. They are used as scout swimmers during tactical boat landings if scout swimmers are not carried as passengers.
- *Numbers 5 and 6.* They assist in lashing and unloading equipment in the after section of the boat. They also assist the coxswain in maneuvering in swift currents, and rigging and handling the sea anchor.
- *Number 7.* The coxswain oversees the team's performance, the handling of the boat, and the distribution of equipment and passengers in the boat. He issues all commands to team members, maintains course and speed, and operates the OBM if it is used.

7-8. During boat training, all team members will not display the same aptitude for boat handling, nor do all members need to acquire the same degree of boat-handling proficiency. When a boat team is formed for an operation, as opposed to a boat training exercise, the man chosen as coxswain should be the one with the greatest ability as a boat handler. The one chosen as the stroke should be selected for his strength and ability to maintain a steady rhythm.

BOAT TEAM SETUP

7-9. Additional duties of the boat team members will be determined by the mission requirements. Factors that must be considered are as follows:
- The individual overall responsible for the mission should consider being the primary navigator located in the lead boat. Each boat will have its own coxswain whose only duties are driving and following a bearing.
- Each boat will also have a navigator who independently navigates that boat. Navigators should cross-check with the lead navigator or command craft. The assistant navigator (or bearing taker) takes bearings for the navigator and assists him as required.
- The boat watch monitors communications if applicable and observes the other boats to avoid separation and breaks in contact. Only the coxswain should sit on the buoyancy tubes. The remainder of the boat team sits inside the craft to reduce their exposure to the elements or enemy observation.

Inflatable Boats

INDIVIDUAL EQUIPMENT

7-10. The unit leader specifies the uniform and equipment carried by each boat team member. There are some general rules, the most important being to ensure that all personnel wear an approved life preserver. All boat team members must wear adequate flotation, either rigid or inflatable life jackets. Every man wears a knife and a distress light or marker on his life jacket during night operations. Inflatable flotation must have CO_2 cylinders for pull-type inflation. Inflatable life preservers are usually more comfortable and less restrictive. Unfortunately, they require additional maintenance, are more susceptible to damage that would render them useless, and require activation before they will provide flotation. The detachment must perform an "in-water" float test to determine if the life preserver provides enough flotation to support the swimmer and his combat load; for example, weapon, load-carrying equipment (LCE), and any additional exposure protection. This need will be particularly important when operating in a riverine environment where the team may be in close proximity to hostile forces and jettisoning the equipment may not be an option.

7-11. When team members wear LCE in the boat, they should prepare it as for any waterborne operation and ensure it is easy to jettison if the wearer falls overboard. They should sling the weapon diagonally across the back, muzzle up and pointing outboard. If the operator falls overboard and cannot remain afloat with the rifle, he must be able to quickly jettison it.

7-12. For cold-weather operations, personnel should wear suitable exposure protection. This cover may include neoprene wet suits or dry suits, either of which will provide considerable warmth and buoyancy while in the water. These suits somewhat restrict paddlers' movements and may cause chafing in the armpits. Personnel should wear rubber tennis shoes for foot protection. Each man should carry a field uniform and boots to be worn during extended operations after the landing.

RIGGING THE CRRC AND STOWING ORGANIZATIONAL EQUIPMENT

7-13. After inflating the CRRC, crew members must prepare it for operations by rigging towing bridles, capsize lines, lifting slings (if required), equipment tie-down lines, and setting up the navigation console.

Towing Bridles

7-14. Accident, combat damage, or mechanical failure can render a CRRC immobile in the water. Continuing with a mission or extracting a damaged vessel to a place of safety requires it to immediately be placed under tow. The detachment does this by prerigging the boat for towing, so that a tow can be initiated quickly. There are two types of towing rigs used with the CRRC.

7-15. The first type is the factory-supplied Zodiac towing harness consisting of two 15-foot nylon ropes with eye splices and shackles on each end. This rig is primarily used under relatively calm conditions and when towing short distances. To use it, the crew member attaches the ends of the ropes to the towed boat's bow towing D-rings with the included shackles. He then passes the ropes forward to the towing vessel and attaches to that boat's opposite stern towing D-rings (port bow to starboard stern, starboard bow to port stern). The ropes should cross (X-fashion) between the bow of the towed vessel and the stern of the towing vessel.

7-16. The second type is an expedient harness, permanently tied onto each vessel during mission preparation. This is the preferred towing rig for operational use. It distributes the tow stresses over the entire structure of the CRRC allowing towing at higher speeds, over extended distances, and in rougher sea states (Figure 7-1, page 7-4).

7-17. The expedient harness is made using approximately 60 feet of 7/16th climbing rope or 1-inch tubular nylon webbing. Two CRRCs can be rigged from one 120-foot rope. To tie the harness, the crew member cuts the 120-foot rope in half, one half per boat. For each boat, he doubles the rope and ties a figure-8 loop on the bight. He places the loop in front of the bow and extends one "leg" of the harness down each side of the CRRC. Working from the bow to the stern on each side of the CRRC, he threads the rope on that side straight through the bow towing D-ring and back through each carrying handle. The crew member passes the rope from the top of the carrying handle through the bottom, continuing through all handles to and through the rear towing D-ring. He takes the extreme end of the rope and ties a snug bowline around the tail cone of the main buoyancy tube immediately behind the rear towing D-ring. He secures the bowline onto the main buoyancy tube by tying it to the transom on the inboard side with a piece of suspension line (this keeps it from slipping off the tail cone if the harness should become slack). The crew member works any slack in the harness forward toward the bow, extends the figure-8 loop to the front of the CRRC, and ensures it is centered on the bow. If necessary, he adjusts the loop's position and fastens a snap link into it. He secures the excess harness by daisy chaining or coiling, and correctly stows it in the bow for rapid deployment without tangling.

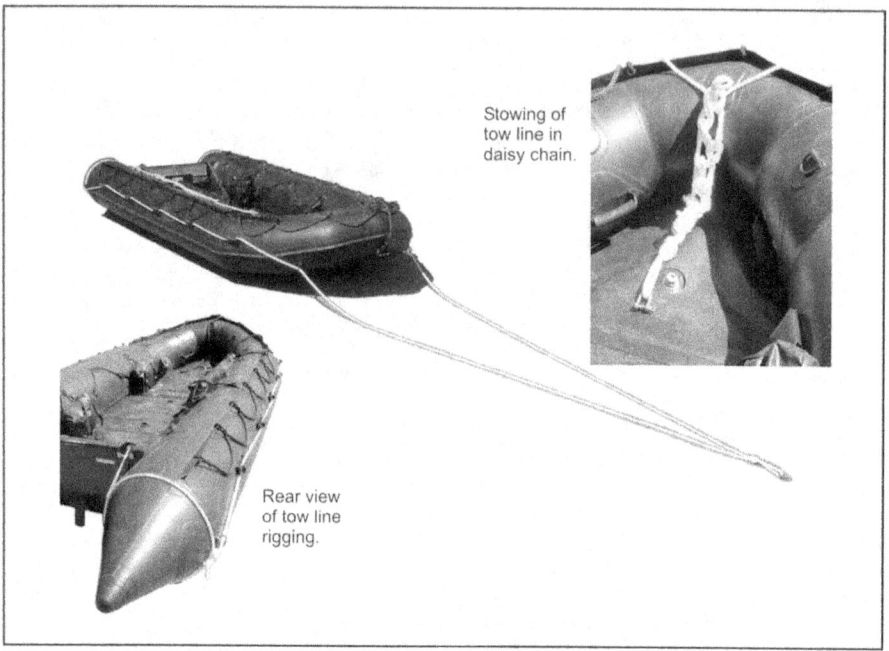

Figure 7-1. Expedient harness

7-18. The crew member prepares the towing component by tying (or splicing) small loops into the ends of a sling rope. He ensures the sling rope is long enough to pass around the rear of the CRRC's OBM without interfering with its operation. He puts a snap link into each loop and clips the snap links to the transom lifting rings. This bridle can be left in place by flipping it forward over the OBM and securing it with retaining bands on the inside of the transom. It can also be stowed in the towing vessel and emplaced only when required.

7-19. To initiate a tow, a crew member has the towing vessel back down on the bow of the towed vessel. He clips the snap link on the towed vessel's harness to the towing bridle on the towing vessel. As the towing vessel pulls away, the bow man on the towed vessel controls the deployment of the harness's slack to prevent entanglement. Throughout the tow, crew members in both vessels monitor the towing line to ensure that any slack is controlled before it can become entangled in the towing vessel's propeller. If the sea state requires extending the tow, crew members can easily add additional line (rope) between the towing harness and the towing bridle.

Inflatable Boats

Capsize Lines

7-20. If the CRRC capsizes, crew members must right it as quickly as possible. To help right the boat, crew members must rig the capsize lines. These lines help apply the leverage required to right an upside-down boat. There are two common methods of rigging capsize lines.

7-21. The preferred method is to attach a 20-foot line (rope or 1-inch tubular nylon) to the foremost and aftmost equipment lashing (50 mm) D-rings on the starboard main buoyancy tube. The crew member can either tie on or secure this line with snap links or shackles. He zigzags the slack between the two D-rings and secures it with retaining bands. When used, a crew member pulls the capsize line (Figure 7-2) loose from the retaining bands and passes it across the bottom of the boat to the other crew members assigned to right the boat.

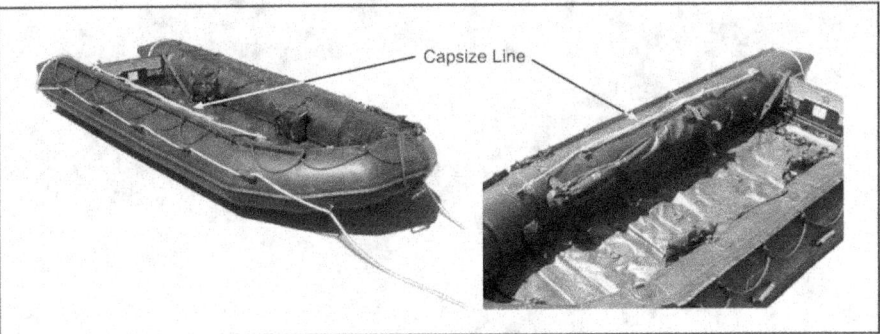

Figure 7-2. Capsize line

7-22. The alternate method is to attach sling ropes, one to each of three equipment tie-down (50 mm) D-rings, usually on the starboard side of the CRRC. The crew member stows the sling ropes by coiling and securing them with retaining bands. To use, the crew member pulls the ropes free of the retaining bands and tosses each capsize line across the bottom of the CRRC to a crew member waiting on the other side.

Lifting Slings

7-23. The Zodiac F-470 comes from the factory with an accessory lifting kit consisting of lifting points (eye bolts on the bow thrust board and lifting rings through-bolted on the transom) and a lifting harness (a center ring with a pair of long lines and a pair of short lines—all terminated in eye splices with shackles for attachment to the lifting points) (Figure 7-3, page 7-6). The lifting kit allows detachments engaged in mother ship operations to launch and recover fully rigged (but not loaded) CRRCs from the deck of a mother ship equipped with a deck crane or hoist. If the mother ship has the capability, hoisting the CRRC into or out of the water provides significant advantages, greatly increasing the safety and security of the operation and significantly reducing the physical exertion required.

7-24. In use, the CRRC is completely rigged with OBM mounted, fuel bladders attached and secured, and all operating equipment stowed. Individual and team equipment is staged on the deck of the mother ship. The lifting harness is attached to the CRRC with the short legs of the lifting harness shackled to the bow lifting points and the long legs to the stern. This ensures that when the CRRC is lowered over the side of the mother ship, the stern hits first causing the bow to "weathervane" and point into the waves and current.

7-25. To hoist the CRRC, the crew member places the center ring of the lifting harness in the hoist's hook. The mother ship's crew operates the hoist to launch or retrieve the CRRC. Crew members maneuver the CRRC using its bow and stern lines to maintain control throughout the hoisting operation. Once the CRRC is in the water, it is secured to the mother ship with the handling (bow and stern) lines, the lifting harness is disconnected from the hoist, and individual or detachment equipment is loaded and secured.

Chapter 7

Figure 7-3. Lifting sling

Equipment Tie Downs

7-26. All individual and detachment equipment that is not worn by crew members must be secured inside the CRRC. Crew members should place all organizational equipment that is not worn by the boat team, such as machine guns, radios, and demolitions, in waterproof bags (as required) or rucksacks. They should pad sharp corners and projections on equipment to prevent damage to the boat. They should also stow and lash the bags or rucksacks securely in the boat before launching.

7-27. There are numerous equipment tie-down procedures available to crew members. The most common ones include securing each bundle or rucksack to a fixed D-ring with snap links, and rigging a series of individual equipment tie-down lines athwartships between opposing D-rings to secure the bundles or rucksacks to these lines with snap links. When using the accessory rigid floorboard, crew members can rig equipment attachment loops by running lengths of 1/2-inch tubular nylon through holes drilled in the floorboards. This modification is required when rigging the CRRC for airdrop.

7-28. An alternate method that offers some advantages if the boat capsizes is for a crew member to rig a single equipment line from the lifting ring on the transom, forward to a D-ring on the opposite bow. He then ties a loop near the bow end of the line and snap links a rucksack (stopper) into the loop. He snap links the remainder of the bundles or rucksacks into the equipment line between the stopper and the transom. The equipment should be free to slide the length of the equipment line and to be retained on the line by the stopper. If the CRRC capsizes, the crew member releases the bow end of the line and allows the cargo to float free (secured at the transom), while the CRRC is righted. This method removes a significant portion of the weight from the CRRC and makes it easier to right the boat. Personnel can then recover the equipment over the stern (or towed, if the situation requires).

7-29. An alternative to individually securing each piece of equipment is to secure the entire load with a cargo net. Netting the cargo allows detachments to rapidly secure odd-shaped bundles or large quantities of

Inflatable Boats

smaller items, a characteristic that is especially useful when CRRCs are used for sustainment support missions. Purpose-built nets designed for the F-470 and similar craft are available through the General Services Administration (GSA) Catalog (NSN# 3940-01-477-7081) from various suppliers.

Navigation Console

7-30. CRRCs designated as navigation platforms should have a navigation console secured on the gunwale tube forward of the coxswain. The manufactured consoles consist of a boat compass (for example, Ritchie compass) and a marine knotmeter (for example, Outboard Marine Corporation [OMC] speedometer) mounted in a molded housing contoured to fit on the main buoyancy tube (Figure 7-4). Instrument lighting is provided by a chemlight. This equipment allows the coxswain to maintain course and monitor speed.

Figure 7-4. Boat navigation console

7-31. The console is strapped or tied to attachment points (D-rings, carrying handles) with 550 cord or straps laced through cutouts in the console. Care must be taken when routing the vacuum tube for the knotmeter to ensure that it is not cut or kinked. Running the hose to the inside, under the swell of the main buoyancy tube, and taping it in place with duct tape will maximize the available protection.

7-32. The boat must also be "swung" to account for deviation. This allows the navigator to account for compass errors caused by the magnetic field of the boat or any embarked equipment. In use, crew members ensure that large metal objects and electronics are kept far enough from the console so they do not affect the compass.

INFLATABLE BOAT COMMANDS

7-33. All boat personnel should be knowledgeable of operational terms to ensure they respond or react appropriately. The specific commands used when handling inflatable boats are as follows:
- *Preparatory Commands.* The coxswain issues preparatory commands to indicate who will execute the command to follow. In some instances the preparatory command includes the expression "stand by to...."

Chapter 7

- *Boat-Handling Commands.* These commands should be understood by all boat team members:
  - LOW CARRY. Team members form up at boat stations and face the bow of the boat. On command, they lift the boat by the carrying handles to about knee height.
  - HIGH CARRY. Team members form up at boat stations and face the stern of the boat. On command, they lift the boat to shoulder height while simultaneously rotating their bodies under the boat so that they face forward with the boat resting on their inboard shoulders. Team members maintain their grip on the carrying handle with the outboard hand.
  - GIVE WAY TOGETHER. Paddlers stroke in unison, following the rhythm set by Number 1.
  - HOLD WATER. Paddlers hold their paddles motionless in the water with the blade perpendicular to the direction of motion.
  - BACK WATER. Paddlers paddle backward in unison with Number 1.
  - REST PADDLES. Paddlers rest their paddles across their legs.

LAUNCHING THE BOAT

7-34. The boat team may launch the boat bow-first or stern-first. Personnel should launch the boat bow-first whenever the water is shallow enough for them to wade in carrying the boat at low carry. An example is launching the boat in surf or from the shore of any body of water that is shallow at its banks. The boat team should launch the boat stern-first when the water is too deep for wading; for example, launching the boat over the side of a larger boat or from the steeply sloping bank of a river. Personnel should waterproof, stow, and lash all equipment in the boat before launching. They should distribute the paddles in the boat at each boat station with the blades wedged under the cross tubes and the handles forward and slightly outboard. In both cases, the coxswain commands, CREW, BOAT STATIONS, and team members form alongside the boat in their relative boat positions facing forward.

Launching the Boat Bow-First

7-35. On the coxswain's preparatory command, STAND BY TO LAUNCH BOAT, the team members grasp the boat's carrying handles. On the command, LAUNCH BOAT, team members perform a low carry and move into the water at a fast walk. When the depth of the water is such that the boat floats free of the bottom, all hands continue pushing it seaward remaining in their relative positions alongside the boat. As the water reaches the knees of the first pair of paddlers, the coxswain commands, ONES IN. The ones climb into the boat, unstow their paddles, and give way together. The coxswain orders each pair of paddlers into the boat in succession by commanding, TWOS IN, and THREES IN. The pairs climb into the boat on command, break out their paddles, and pick up the stroke of Number 1. The coxswain orders the passengers into the boat after the paddlers by commanding, EIGHT IN, NINE IN, and so forth. Passengers board over the stern and move forward in the boat to their boat positions. The coxswain enters the boat last and sounds off, COXSWAIN IN.

Launching the Boat Stern-First

7-36. On the coxswain's preparatory command, STAND BY TO LAUNCH BOAT, the team members grasp the boat's carrying handles. On the command, LAUNCH BOAT, team members perform a low carry and carry the boat stern first to the water's edge. They launch the boat by passing it back along the line of team members. When Number 2 can no longer aid in passing the boat back, he moves to the bow of the boat and handles the bowline. Other team members follow suit, taking their places along the bowline between Number 2 and the boat. When the boat is in the water, the coxswain enters the boat and takes his boat station. The coxswain then commands the boat team to load by the long count; that is, SIX IN, FIVE IN, FOUR IN, THREE IN, AND ONE IN. Passengers are loaded next: Nine In, Eight In. When the coxswain is ready to cast off, he commands TWO IN. Number 2 casts off and climbs into the boat and takes his boat position.

Inflatable Boats

Paddling Rate

7-37. A paddling cadence of 30 strokes per minute is maintained for open-water paddling. A faster rate is necessary when launching a boat through surf; a slower rate for long distances. The coxswain may order a cadence of 10 to 20 strokes per minute by commanding, REST STROKE. The stroke decreases to about half of normal cadence and each paddler comes to the position of rest paddle between strokes. Well-conditioned paddlers can maintain a rest stroke cadence for hours.

Maintaining Course

7-38. Inflatable boats should be kept perpendicular to the waves in rough seas but may take waves off the bow or stern in a calm sea. Thus, the course may have to be maintained by tacking. Strong winds and currents may require the coxswain to steer a compensated course.

Turning the Boat Rapidly

7-39. The coxswain steers the boat with his paddle, called the sweep. If he wants to turn the boat rapidly, he orders the paddlers on one side to back water while those on the other side give way together. Thus, a rapid turn to starboard is executed by commanding, STARBOARD, BACKWATER. PORT, GIVE WAY TOGETHER.

TRANSIT FORMATIONS

7-40. Transit formations are used to control several boats operating together. Formations are determined based on the distance to be transited, the sea state, and the enemy situation. Mission commanders should select the formation that provides the greatest control while still ensuring the flexibility and mutual support necessary to react to enemy contact. Detachments should use prearranged hand signals to change formations (if required) without using communications. The mission planner should review the NAVPLAN and enemy situation to anticipate formation changes en route. The navigator uses the following formations when encountering the conditions described:
- File:
    - Tight areas.
    - When expecting contacts from flanks.
- Wedge:
    - Very dark and low visibility.
    - All-around defense.
    - Can move to more of a diamond with four boats.
    - Echelon (left or right).
    - Open areas.
    - High visibility.
    - Flexible response.
    - All-around defense.
    - Formed away from sea obstacles.

LANDING THE BOAT

7-41. The manner in which the coxswain orders the boat team to unload the boat when landing depends on the depth of the water at the landing point. Each depth and its differences are discussed below.

Shallow Water

7-42. As the boat nears the shore, the coxswain sounds with his sweep and orders the team to disembark when the water is waist deep, using the command "beach boat." On this order, the team stows their paddles and disembarks. The team members grasp the carrying handles and push the boat as far as it will float, then they carry it at low carry thereafter.

Chapter 7

Deep Water

7-43. When the boat is landed in deep water, such as when tying up to a submarine, the coxswain uses the following procedure to unload and land the boat. As the boat comes alongside the landing platform, the coxswain orders, "two out." Number 2 stows his paddle and, with bowline in hand, gets out of the boat onto the landing platform and pulls the boat up close to the landing. The other team members either stow their paddles in the boat or pass them up to personnel on the landing platform. The coxswain commands, "one out," and Number 1 climbs out of the boat followed in order by the passengers and Numbers 3, 4, 5, and 6. The coxswain is the last to leave the boat. As the team members climb out over the bow of the boat, they take their places along the bowline corresponding to their boat positions. After the coxswain gets out of the boat, the team pulls the boat out of the water and onto the landing platform. As the boat is pulled up, team members leave the bowline and move to positions where they can handle the boat by its carrying handles and pass it up between them. When boat team members get into or out of the boat, they maintain a low silhouette to prevent being thrown off balance overboard.

BEACHING THE BOAT IN SURF

7-44. The coxswain beaches his boat during one of the recurring periods of relatively light breakers that exist between series of heavy breakers. Before entering the surf zone, the coxswain orders all hands to shift aft to put as much weight as possible in the stern. Even small breakers will capsize a boat if they can raise the stern out of the water.

7-45. As the boat enters the surf zone, the coxswain keeps the boat perpendicular to the surf line, assisted by the ones who may back paddle as necessary. The paddlers take advantage of each wave's momentum by paddling faster as the wave raises the boat. The coxswain periodically observes the sea astern of the boat, but paddlers never look astern since they must concentrate on maintaining the cadence.

7-46. The coxswain sounds with his sweep and orders the team to disembark when the water is waist deep. The team grasps the carrying handles and pushes the boat as far as it will float, carrying it at low carry thereafter. The coxswain may order the boat overturned at the water's edge to empty it of water.

7-47. The greatest chance of capsizing or swamping exists just before a boat is beached. The turbulence caused by incoming waves meeting the backrush of water from the beach makes it hard to keep the boat perpendicular to the surf.

LAUNCHING THE BOAT IN SURF

7-48. The coxswain launches the boat in surf during a period of relatively light breakers. He takes advantage of rip currents if they are present. As each pair of paddlers embarks, they shift forward in the boat to put as much weight as possible in the bow. Number 1 sets a fast, strong cadence without waiting for orders from the coxswain. In moderate or heavy surf, the boat team has little chance of successfully transiting the surf zone if the boat is not through before the next series of relatively heavy breakers arrives.

OVERTURNING AND RIGHTING THE BOAT IN THE WATER

7-49. A coxswain may have to right a capsized boat or empty a swamped boat by overturning it and righting it in the water. A boat is overturned (broached or capsized) and righted by the same procedure. All but three members of the boat team and the coxswain (who holds all the paddles) stay in the water during the overturning or righting. Prerigged capsize lines attached on one side of the boat are passed across to the opposite side where they are grasped by the three remaining paddlers. These paddlers stand on the gunwale or main buoyancy tube and fall backward into the water, pulling the boat after them and overturning it. They are assisted as necessary by the three paddlers in the water. The command, BROACH BOAT, is used in training to cause the boat team to overturn the boat in water.

TACTICAL BOAT LANDINGS AND WITHDRAWALS

7-50. Boat teams conduct minor tactical exercises during training to demonstrate and improve their boat-handling ability under operational conditions. All members of the team are trained to perform the duties of all other members during these exercises.

Landings

7-51. As the boat approaches the surf zone, the team leader orders the coxswain to lie outside the surf zone and maintain position relative to the beach. The team leader then orders the coxswain to send in his scout swimmers. The coxswain commands, TWOS OUT, and Numbers 3 and 4 enter the water and swim to the beach.

7-52. Scout swimmers must avoid splitting a breaker line or foam line because doing so results in them being silhouetted against a white foam background. The scout swimmers determine the presence or absence of enemy in the landing area. They usually move singly about 50 yards in opposite directions after reaching the beach.

7-53. When the scout swimmers have determined that the landing area is free of enemy personnel, they advise the boat team using a predetermined signal that will be clearly visible and unmistakable to the detachment waiting at the boat pool. The selected signaling method must also take into consideration the threat of detection by the enemy. One method is for the scout swimmers to use a filtered (infrared), hooded flashlight with a prearranged light signal. The scout swimmers give the signal from the point on the beach that they have selected as the most suitable for landing. After signaling the boat, the scout swimmers move in opposite directions from the landing point to listening and observation positions. These security positions are at the limit of visibility from the landing point.

7-54. When the boat team receives the scout swimmers' signal, the team leader orders the coxswain to beach his boat at the point from which the signal originated. After the boat is beached, the team leader orders it hidden and camouflaged as previously covered in his fragmentary patrol order.

Withdrawals

7-55. The detachment conducts a security halt short of the withdrawal area or cache site. The team leader then orders a pair of scouts forward to determine if the withdrawal area is clear of enemy personnel. These scouts reconnoiter the cache to see if it has been disturbed and observe the beach for the presence of enemy personnel.

7-56. After the scouts report that the area is clear, the team leader orders the coxswain to prepare the boat for withdrawal. In some cases, the boat may have to be inflated from air bottles or hand and foot pumps. During this preparation, the team leader posts security just short of the limit of visibility from the cache to warn that the enemy is approaching. One observer is usually posted at each flank of the withdrawal area, while the team leader observes the route previously taken by the boat team on their way to the cache.

7-57. When the boat is prepared for withdrawal, the coxswain informs the team leader, who then orders the coxswain to launch the boat. Just before the boat enters the water, the team leader calls his security observers to the boat where they take their boat team positions, usually as twos. The coxswain then assumes control of all hands for the launching.

ZODIAC F-470 INFLATABLE BOAT

7-58. The Zodiac Marine Commando F-470 is the most frequently used assault boat (Figure 7-5, page 7-12) for all SOF. The 35-hp OBM is issued for use with the Zodiac F-470 inflatable boat (Figure 7-6, page 7-12). However, with minor modifications, the Zodiac F-470 can accommodate other engine packages, such as the 55-hp OBM, that will enhance performance.

Chapter 7

Figure 7-5. Zodiac F-470 inflatable boat

Figure 7-6. Outboard motor

7-59. The USMC's Small Craft Propulsion System (SCPS) is an optional power plant for the CRRC. It was developed and adopted to replace the USMC's OTH configuration that used twin 35-hp OBMs. The SCPS is a 55-hp submersible OBM. The principal difference between the SCPS and a standard 55-hp OBM is a replacement bottom-end that exchanges the standard propeller for a pump jet. The pump jet is a fully enclosed (shrouded) impeller. It provides an added degree of safety by reducing the possibility of a prop-strike. When compared to the twin 35-hp OTH configuration, the SCPS offers significant advantages for the USMC.

7-60. SF units will find that the SCPS offers detachments some limited safety advantages when operating in the vicinity of swimmers, in debris-laden waters, shallow-draft, or in case of a capsize (for example, heavy surf zones). The disadvantages of this system for SF units (when compared to the propeller-driven, 55-hp OBM) include a weight increase, reduced top speed, increased fuel consumption, increased mechanical complexity, and more expensive to purchase.

Inflatable Boats

CHARACTERISTICS

7-61. The Zodiac is a squad-sized inflatable craft capable of amphibious small-boat operations. The F-470 is made of neoprene-coated, tear-resistant nylon cloth. When fully inflated (240 millibars [mb] or 3.5 lb per square inch gauge [psig]), the F-470 weighs approximately 280 pounds without the engine. The maximum payload is 1,230 kilograms (kg) (2,710 lb). The overall length of the craft is 4.70 meters, or 15 feet and 5 inches. The overall width of the craft is 1.90 meters, or 6 feet and 3 inches. The F-470 should be loaded to a weight ceiling of 2,000 pounds to allow the craft to perform at its optimum. F-470s outfitted with aluminum decking ensure a higher rate of speed, mobility, and maneuverability. The aluminum deck is composed of four lightweight, self-locking aluminum sections and two aluminum stringers. Table 7-1 lists the F-470 specifications.

Table 7-1. Zodiac F-470 specifications

Feature	Size/Weight/Dimension
Overall length	4.70 m or 15 ft, 5 in.
Overall width	1.90 m or 6 ft, 3 in.
Inside length	3.30 m or 10 ft, 10 in.
Inside width	0.90 m or 3 ft
Tube diameter	0.50 m or 20 in.
Maximum number of passengers	10
Maximum payload	1,230 kg or 2,710 lb
Maximum hp with standard 40 hp	40 hp matted floor, short shaft
Maximum hp with optional 65 hp	65 hp aluminum floor, short shaft
Dimensions in bag	1.50 x 0.75 m or 59 in. x 29.5 in.
Weight with standard matted floor	120 kg or 280 lb
Weight with optional aluminum floor	120 kg or 280 lb
Number of airtight chambers	8
Weight of CO_2 charged bottle with manifold (U.S. DOT-approved)	20.45 kg or 45 lb

7-62. Detachments working with other Services may encounter additional types of small craft that fill the same operational niche as the Zodiac F-470. Knowing what is available is important, especially when requesting insertion or extraction by a sister Service, for example, a USMC raid or reconnaissance force. Table 7-2 shows different types and characteristics of inflatables used by other agencies.

Table 7-2. CRRC characteristics (inflatables)

Item	Weight (lb)	Length (ft)
Avon	450	15
Avon	460	15
Avon	520	17
Z-Bird	400	15
IBS	120	13
Zodiac F-470	280	15
OBMs Used With CRRCs		**Gas Bladder**	**Gas Can**
55 hp	202 lb	18 gallons (13.5 gallons maximum allowed on aircraft)	6 gallons (4.5 gallons maximum allowed on aircraft)
35 hp	118 lb		
15 hp	78 lb		

7-63. The F-470 is the CRRC in the Marine Corps supply inventory. The USMC uses it for both Service reconnaissance forces and amphibious raid operations. Marines also use the CRRC-LR and RHIBs. The CRRC-LR is a long-range, rigid hull version of the CRRC that can be taken apart and stored in nearly the same space as a CRRC. The RHIB is a one-piece rigid hull with an inflatable flotation collar (equivalent to

Chapter 7

the main buoyancy tube of the CRRC). Both of these boats have handling and load-carrying characteristics that are superior to the F-470, especially when operating in marginal sea states. They can also use more powerful OBMs.

7-64. The Zodiac offers numerous advantages when conducting waterborne infiltrations or exfiltrations. The Zodiac—
- Offers mobility and speed that surpasses scout swimming and scuba methods.
- Enables personnel to reach the BLS rested, dry, and ready to conduct the mission.
- Can be launched from submarine, aircraft, or mother craft.
- Has five compartments that are separately inflatable.
- Cannot be overinflated because of overpressure relief values.
- Has an inflatable keel that offers a smoother and more stable ride.
- Can oftentimes avoid radar detection due to its fabric construction.
- Has accessory rigid floorboards that prevent buckling and pitching with heavy loads during operations.
- Can adequately accommodate six men with equipment.

COMPONENTS

7-65. The Zodiac consists of the following components:
- Five airtight chambers in the hull.
- Two separate airtight chambers (speed skegs) under the main buoyancy tube.
- Overpressure relief valves.
- A plank-reinforced, roll-up floorboard that can be left installed when packing the boat.
- An accessory rigid floor that has four reinforced aluminum antislip floorboards.
- An inflatable keel.
- Eight carrying handles.
- A reinforced transom.
- Stainless steel towing rings.
- Lifeline and bow grab lines.
- Bow pouches with foot pumps, hoses, and a field repair kit.

7-66. The Zodiac F-470 (SW) CRRC variant for USMC use has a full-height transom modification designed specifically to accommodate a 55-hp OBM. New or replacement F-470s acquired for USASOC units will also conform to the USMC specifications. F-470s with the full-height transom require the use of long-shaft OBMs. Short-shaft engines will cavitate.

ASSEMBLY

7-67. Personnel follow specific steps to assemble the Zodiac. They should—
- Roll out the Zodiac and inspect it for serviceability.
- Turn the valves to the inflate position and remove the valve caps.
- Place the accessory rigid aluminum floorboards in the boat (1 through 4) and interlock them.
- Attach hoses with foot pumps to each valve and inflate the boat.
- Place stringers (short ones in front, long ones in the rear) on each side of the floorboards and lock them in place.
- Check the inflation pressure by placing a pressure gauge in a valve for the recommended (240 mb) reading when the boat is fully inflated.
- Turn all valves to the navigation position, thereby closing the valve and isolating the five separate air chambers in the main buoyancy tube.
- Secure the clamps on the connecting tubes between the main buoyancy tube and the speed skegs.
- Inflate the keel to the prescribed (220 mb) reading.

GENERAL MAINTENANCE

7-68. Personnel should follow the recommendations outlined below for the use and care of the Zodiac. They should—
- Never run the Zodiac underinflated.
- Wash down the Zodiac with freshwater after every use.
- Avoid dragging the Zodiac on rocky beaches, rough surfaces, coral reefs, and so forth.
- Ensure the Zodiac is thoroughly dry and free of grit and sand before deflating it and placing it in a carrying bag.
- Never lift the Zodiac by the lifeline.

EMERGENCY REPAIR AT OPERATOR LEVEL

7-69. CRRCs are extremely vulnerable. In any exchange of fire, it is probable that the CRRC will sustain damage. Punctures will cause the boat to deflate rapidly! Air pressure gives the boat its rigidity and, depending on the quantity and severity of the punctures, the detachment may only have seconds before the CRRC has lost enough air that it will fold up when the engine is used. The first priority of the team members is to clear the kill or danger zone. Then they attempt to patch the CRRC (and any wounded). If the CRRC has lost a significant amount of air, it may be easier and safer to tow it to a safer location rather than risk collapsing the boat by trying to motor to safety. To this end, CRRCs must always be rigged to facilitate rapid towing.

7-70. Detachment SOPs should specify quantities, types, and locations of emergency repair materials. The most effective emergency repairs are made with "lifeboat patches." There are two types available. The first is a series (graduated by size) of plugs resembling threaded cones that can be either plastic or wood. The team member should screw the plugs into the puncture until it is sealed. The second emergency patch is the "clamshell" type. It is a two-piece, gasketed, elliptical metal patch joined by a stud with a retained wing nut. The team member should separate the halves and insert the back half (point-first) through the hole, enlarging as required. He then rotates the patch to align its long axis with the hole and screws the two halves together. Both types of patches should be contained in easily accessible repair kits, strategically placed throughout the boat. The CRRC has pouches for repair kits and pumps located fore and aft on the main buoyancy tube. Second-echelon maintenance facilities can add extra pouches if desired.

7-71. After plugging any holes in the boat, personnel must reinflate it before continuing with or aborting the mission. The fastest, easiest method for reinflating is to use the CRRC's CO2 inflation system. It is a single-point, pressurized inflation system consisting of a CO2 bottle (original issue) or a 100-ft3 (updated to meet USN requirements) air cylinder with a low-pressure manifold attached to valves in the CRRC. To inflate the boat using this system, the team member should first open the interconnecting (I/C) Zodiac valves by rotating them from "navigate" to "inflate," and open the cylinder valve. He controls the rate of inflation by manipulating the cylinder valve to prevent blown baffles or overinflation. He only adds enough air to the CRRC to enable it to maintain its shape and resist the force of the outboard motor. The emergency repairs performed with the lifeboat patches are not secure enough to resist the full inflation pressure (240 mb in the main tube). The repairs will still leak, even at a reduced pressure. Any gas (air/CO2) remaining in the cylinder will be needed to top off the CRRC at regular intervals. To conserve the available gas, it may be possible to top off the CRRC using the foot pumps. These measures will allow the detachment to continue using a damaged CRRC until it can be patched using the emergency repair kit issued with the boat.

7-72. Once the detachment reaches an area of relative safety, personnel make temporary repairs that will allow them to use the CRRC. When repairing the CRRC, the team member should—
- Clean the patch and the area to be repaired with methyl ethyl ketone (MEK) and allow it to dry before application. The objective is to remove all contaminants such as oil, grease, or salt and to chemically prepare the surface for gluing.

Note. Alcohol or water may be used if MEK is not available, but the patch will not adhere as well.

- Cut patch material so that it extends a 1-inch minimum on all sides of the tear.

- Use the pumice stone included in the repair kit to rough up the surface of the material on the patch and on the boat.
- Apply a thin, even coat of the Zodiac glue to both the patch and the boat, making a total of 3 applications to each.
- Wait approximately 10 minutes between each application to allow the glue to set up.
- Carefully apply the patch, rolling it on evenly and smoothly to guard against wrinkles.
- Vigorously rub the patch over its entire surface with the boning tool (stainless steel spatula) included in the repair kit to ensure positive contact between the patch and the boat.
- Let the repair dry (cure) for 12 to 24 hours before inflating the boat to full pressure (if the situation permits).

Note. Detachments should seek additional training in the use of the emergency repair kit from qualified technical personnel.

FUEL CONSUMPTION DATA

7-73. To plan a maritime infiltration using any small craft, detachments must be able to accurately calculate the amount of fuel required to execute a transit from the debarkation point to the BLS. The prudent planner will then add in allowances for exfiltration and possible contingencies such as the requirement to move to an alternate BLS. He then uses this data as a baseline to determine an appropriate safety margin (at least 15 to 20 percent additional fuel) to allow for increased fuel consumption caused by wind, waves, currents, and variations in the boat's calculated fuel usage. These combined factors will help define the total amount of fuel required to safely conduct the mission. The planner also considers the following variables when determining overall fuel use:
- Type of boat.
- Speed maintained.
- Displacement and weight of personnel and cargo.
- Type of motor and propeller.
- Engine throttle setting.
- Wind speed and direction.
- Current, set, and drift.
- Sea state.

7-74. Fuel consumption is a critical element when planning mission requirements. The detachment computes fuel consumption to provide to the planner for inclusion in the NAVPLAN. Team members perform the following:
- Ensure fuel tanks are topped off.
- Measure and record total gallons in fuel tanks.
- Start engines and record the time.
- Set desired rpm for engines or throttle setting, and record.
- Record set rpm/throttle.
- Stop engines and record the time.
- Measure and record total gallons of fuel in tanks.
- Subtract total gallons in tanks after running 1 hour from total gallons recorded on boat at beginning of underway period.
- Record the difference.
- Measure the distance traveled, and record.
- Compute boat speed, and record.
- Apply the formula—time (T) x gallon per hour (GPH) = total fuel consumption (TFC).
- Recompute GPH after changing rpm or throttle setting and follow the procedures "record set rpm/throttle" through "apply the formula."

Inflatable Boats

7-75. The following data was assembled from test runs conducted by the Naval Special Warfare Center (NSWC) (Table 7-3, below, and Figure 7-7, page 7-18). It is presented for information only. To accurately calculate fuel usage for their CRRCs, detachments must conduct their own sea trials with mission loads and sea states as nearly identical to mission conditions as possible.

Table 7-3. Sample fuel usage for CRRCs

Payload (lb)	Maximum Speed (kn)	Nautical Miles Per Gallon
Configuration 1: Zodiac F-470, 35-hp Evinrude OBM, 11" X 13" Propeller		
1,800	6.5	2.5
1,600	6.7	2.8
1,400	8.2	4.1
1,200	16.4	4.3
Configuration 2: Zodiac F-470, Twin 35-hp Evinrude OBM, 11" X 13" Propeller		
2,200	7.9	3.9
2,000	17.7	6.5
1,800	18.8	6.7
Configuration 3: Zodiac F-470, 55-hp Evinrude OBM, 13 1/4" X 17" Propeller		
1,800	7.5	2.5
1,600	20.3	3.4
1,400	17.3	3.4
1,200	18.5	3.4
1,000	18.8	3.4
800	20.0	3.4
Configuration 4: Zodiac F-470, 55-hp Evinrude OBM, 13 3/4" X 15" Propeller		
2,200	8.8	1.7
2,100	18.8	3.1
2,000	18.9	3.1
1,800	19.3	4.2

7-76. The SCPS's carrying capability and speed estimates are slightly different from propeller-driven, 55-hp OBMs. It is rated by the USMC to carry a payload of 2,018 pounds, obtaining an average speed of 19 kts per hour at wide open throttle (WOT) in Sea State 1. It has a fuel consumption of 3.0 gallons per hour or 4.5 nm per gallon.

7-77. The following notes and recommendations are derived from the operational experience of units using the CRRC:
- Solid 7-foot stringers, as used by the USMC for their OTH configuration, provide extra deck plate rigidity to handle the increased weight and horsepower of the 55-hp OBM with or without a pump jet.
- The Zodiac 6-gallon (NSN 2910-01-365-2694) and F-gallon (NSN 2910-01-447-4893) fuel bladders are certified by the USAF at Wright Patterson Air Force Base for use aboard all USAF aircraft. This certification alleviates what has been a hazardous material shipping problem.

7-78. The use of Zodiac (PN N45451) heavy-duty fuel lines with dripless connections will further minimize fuel-line dripping and the attendant fumes aboard aircraft and naval vessels.

• Navigation Equipment:
  ■ Nautical Slide Rule
  ■ Hand-Held Compass
  ■ Steiner Binoculars
  ■ GPS
• Plotting Board:
  ■ Laminated
  ■ Bar Scale
  ■ 2 Grease Pencils
  ■ Flat Divider With Laminated Scale
• Cache Gear:
  ■ Camouflage Netting
  ■ Shovels
  ■ Rakes
• Communications Gear:
  ■ Radios
  ■ Signal Mirror
  ■ Strobe With Directional Shield/IR Cover
  ■ Para Flares: 3 White, 3 Red, 2 Green
  ■ Smoke Grenades (Floatable)
	■ Hand-Held Flares
	■ Pencil Flares
	• Poncho for Tactical Plotting
	• Boat Equipment
	• Bowline (Not to Reach Prop if Overboard)
	• Spare Water (Minimum of 5 Gallons/ 1 Jerry Can)
	• Boat Pumps/Hose
	• Floorboards With Tie-Down Loops Through Boards
	• Gear Tie-Down Straps
	• Towing Line/Capsize Lines
	• Paddles
	• Gas Cans or Bladders
	• Stern Line
	• Boat Compass and Lighting System
	• Lifting Slings (For Pickup by Hoist)
	• CO_2 Bottles and Inflation Fittings
	• Night Vision Equipment/Radios

Figure 7-7. Maritime operations gear

Chapter 8
Kayaks

The kayak originated thousands of years ago. It was developed by the indigenous people of the North American Arctic regions as a form of transportation and as a way to hunt the marine mammals that were a dietary staple. Arctic explorers returning home brought with them incredible stories of the "Eskimo skin boats." These stories excited the interest of the rest of the world and "kayaking" became very popular in Europe during the 19th century.

The most serious attempts to develop the Eskimo kayak into a craft suitable for European needs happened in Germany. Working from earlier efforts, Johann Klepper manufactured and marketed his first folding kayak in 1907. He chose the kayak design because it was the most universally successful small-boat type on earth and functioned well in flat water, waves, and on the open ocean. It was also the most efficient craft in weight-to-load ratio and overall performance per pound of material. This fact is still true today.

Since the kayak's discovery by the rest of the world, it has proven effective in all types of activities. It has been used in sports, in the Olympics, for fishing, as a means of exercising, in military operations, and in adventuring. Since World War II, military forces worldwide have opted for the folding, sea-touring kayak as a means of transport. Its load-carrying capacity and low signature, coupled with its portability and durability, have made it an ideal means of conducting clandestine movements over long distances in many regions of the world.

MILITARY BACKGROUND DATA

8-1. Kayaks are used as either a means of long-distance infiltration and exfiltration into a joint special operations area (JSOA) or as the primary method of infiltration and exfiltration in a target area. Planners consider using kayaks in MAROPS when the LP cannot be within a reasonable distance of the BLS or target, when exfiltration of the operatives or detachment is limited or critical, when equipment is excessively large or heavy, and when the use of motors is not tactically sound. Combat dive and scout swimmer operations can be supported directly from kayaks. Kayaks can be launched using AIROPS, SUBOPS, civilian or naval vessels, or from land.

8-2. Kayaks were used throughout World War II, in all theaters, in virtually all weather and sea conditions, with routine successes. In fact, some of the most successful SO missions conducted during the war used kayaks.

8-3. From the initial formation of 101 Troop (Special Boat Section) in 1940, several SO units based many of their missions on the use of kayaks. Combined Operation Assault Pilotage Parties (COPPS) served in Europe, the Mediterranean, and in F-month tours in Southeast Asia. They were primarily concerned with beach surveys before amphibious landings and as portable navaids during the landings. COPPS were used extensively before and during the Normandy Invasion on 6 June 1944. The Royal Marine Boom Patrol Detachment (RMBPD) operated throughout Europe and the Middle East. Primarily used as raiding and reconnaissance parties, they are the direct forerunners of the Royal Marine special boat section (SBS). Sea reconnaissance units (SRUs) trained with U.S. Navy UDTs and were used primarily in the Pacific theater. The Special Air Service (SAS), operating under the designation of the 1st Special Raiding Squadron, used

Chapter 8

kayaks in all their missions in the Mediterranean. The special operations executive (SOE) and the operational groups of the Office of Strategic Services (OSS) used kayaks in conducting many of their missions. The Germans and the Italians used kayaks as well.

8-4. After the war, the Allies disbanded most SO units. Of those that remained, the SBS carried on the kayak tradition. During the 1950s and 1960s, the SBS and SAS used kayaks during the Malaysian Emergency and the Borneo Conflict.

8-5. Today, virtually all major European armed forces use kayaks in some way. British SOF used kayaks to support operations during the Falklands War. The former Soviet Union's special purpose (Spetznaz) brigades routinely drafted Olympic-class kayakers into their ranks.

8-6. There are basically two types of kayaks—white-water and sea-touring. White-water kayaks are small, durable, highly maneuverable boats with limited load-carrying capacity designed primarily for negotiating river rapids. They are generally not suitable for military operations. The sea-touring kayaks are either folding or rigid, longer than white-water kayaks, and normally have a V-shaped hull. They are also usually wider (beamier) and have a greater load-carrying capacity. Sea-touring kayaks in general are designed for carrying large stores and for negotiating rough water.

8-7. Since World War II, the military has opted for folding sea-touring kayaks. They are stable in high seas, can be packed and assembled in limited spaces, can carry large stores, and can generally be field-repaired if damaged. However, they do require somewhat gentler handling and more maintenance than the hard-shell, white-water kayaks.

8-8. There are a number of folding sea-touring kayaks available to the general public. They cover the full spectrum from expedition-grade boats to relatively inexpensive recreational boats. For military use, the expedition-grade boats have proven to be the most durable, and in the long run, most cost-effective. Although Germany's Klepper Quattros and Aerius 2 Expeditions are the most widely used, there are several other similar kayaks suitable for use by military forces. These include the French Nautiraid and the American Long-Haul Commando (Figure 8-1, page 8-3). The characteristics of these boats are all very similar.

8-9. The open-sea folding kayak is very stable in high seas. Due to its hull shape, the kayak tracks well on a narrow keel. It derives its stability from its wide beam and a center of gravity that is at or below the waterline. Because of its nonrigid design, the kayak offers an elastic yield in the water that contributes to its ability to weather rough seas. Compared to a swim team, kayaks offer increased maneuverability and greater range. They present a lower silhouette and greater stealth than a CRRC. Kayaks can significantly increase the range of infiltration and exfiltration with a true OTH capability. They provide increased maneuverability in the target area and BLS and provide a means of exfiltration if the infiltration is aborted. Kayaks also reduce the reliance on sister Service support.

8-10. Kayaks provide increased flexibility in delivery systems. They can be assembled or disassembled rapidly to facilitate caching or portaging over most types of terrain. Kayaks can be cached in small areas for prolonged periods without affecting their performance. Their ability to be disassembled and transported in component bags offers increased flexibility in delivery methods. Kayaks can be airdropped as individual loads or delivered by surface craft or submarine.

8-11. Extensive training is required to complete long transits. However, with appropriate training and physical conditioning, detachments can paddle for long periods at an average speed of 3 kts. Kayaks can be outfitted with various sailing rigs to conserve the paddler's strength and greatly increase the operational range. Depending on the configuration, some kayaks can increase the operational load capability to as much as 1,000 pounds per two-man team. The speed and load-carrying capacity of the kayak is somewhat restricted by the weather, tides, and currents. High surf at the BLS can prevent landings and departures. Surf operations are limited to waves not greater than 6 feet. Kayakers must practice in rough surf if an operation is to be conducted under those conditions. Figure 8-2, page 8-3, identifies the advantages and disadvantages of folding kayaks.

Kayaks

Dimensions:
Length (without rudder): 17 feet
Length (with rudder): 18.5 feet
Beam: 2 feet, 9 inches
Depth: 1 foot, 4 inches
Weight: 100 pounds

Components and Capacity:
Thirty-one wooden parts with the hull.
Long wooden parts are ash.
Cross-ribs are marine-grade, nine-layer laminated Finnish birch.
Decking is cotton and hemp weave.
Hull is hypalon rubber.
Built-in air tubes, once inflated, make the kayak nearly unsinkable.
Spray skirts can be fitted over the two deck cockpits.
The rear man operates the rudder with foot pedals.
Carries up to 1,000 pounds of stores and equipment
 (normal operational loads are 200 pounds).
Average speed with 200 pounds of equipment is 3 knots.

Figure 8-1. Typical expedition-grade folding kayak

Advantages
- Is stable in high seas.
- Presents a low silhouette.
- Is extremely quiet or silent.
- Is seldom prone to mechanical failure.
- Can be routinely portaged over most types of terrain.
- Can be air-dropped (either platform or personal load).
- Can be deployed from submarines.
- Can be towed on the surface by submarines.
- Can be easily hidden and camouflaged (either made up or bagged).
- Can be paddled for long distances at speeds of 3 to 4 knots.

Disadvantages
- Its speed may depend on weather or tides.
- Must be very carefully prepared.
- Is very physically tiring to paddle for untrained personnel.
- Is relatively fragile.
- Its time of transit becomes a significant limitation on extended paddles.
- Requires additional training and physical conditioning to use effectively.
- Requires additional time and planning for loading and unloading equipment.

Figure 8-2. Characteristics of the folding kayak

Chapter 8

TERMINOLOGY

8-12. Many kayaking terms are not standardized and have different meanings in the United Kingdom and Europe. When this is so, the different meanings have been noted. Much of the terminology concerns marine navigation. The term kayak and canoe are used interchangeably in this list. Kayak is the traditional Inuit Eskimo name and has been used primarily in North America. Canoe has been used by both British and European kayakers since the last century. Because the bulk of all sea kayaking (or sea canoeing) has been developed by the British, and the SBS has been the guiding light in most military kayak operations since the end of World War II, the term canoe has been adopted by Europeans to imply military usage of the craft. Personnel use the terms listed in Figure 8-3, pages 8-4 and 8-5, during kayak operations.

Air Sponsons	Air tubes that are incorporated into the gunwales of kayaks to stabilize the boat when awash (filled with water), so that it can be reentered easily from the water and, if necessary, paddled while awash.
Bailers	Refers to bilge pumps, either hand-operated or foot-operated. It can also include any other device intended to remove water from the inside of a flooded kayak. All kayaks must have at least one bailer to remove water after capsizing or during rough seas. Bailers also have sponges that personnel can use to remove the remnants of water from the kayak's bilges.
Basic Kayak Strokes	The six basic kayak strokes are forward paddle, sweep stroke (bow and stern), slap support or recovery stroke, draw stroke, backward paddle, and sculling. A basic stroke is one that achieves the desired movement in the simplest manner.
Basic Ocean Items (BOI)	Equipment that is carried in all kayaks during paddles in the open ocean. Each kayak has its own assigned BOI that stay with the kayak at all times (Figure 8-4, page 8-5).
Beach Defense Posture (BDP)	Kayaker takes this posture when he first lands on a BLS, before securing it. A person "clears" a BDP and "secures" a BLS. (The BDP for kayak operations is taken from the UDT/SEAL BDP for combat swimmers, with modifications.) The BDP is based on fixed positions for the kayaks. Number 1 kayak "clears" the BDP (same as a scout swimmer) and signals the rest of the kayaks for movement to the beach at intervals. All kayaks beach to the left of the first kayak, turn bows to sea, and move to the BDP (which is normally near the waterline). The positions taken are always left, right, left, right, and so on, until the last kayak is in position. The first kayak that cleared the BDP takes up the rear.
Cockpit	The open space in the kayak where the kayakers sit. Number 1 is always in the front; Number 2 is always to the rear.
Cockpit Sock	A waterproof bag that is set in the cockpit after the boat is loaded and attached by an elastic rim to the coaming. The spray skirt fits on the top. If the boat capsizes, only the sock fills with water. It also offers increased warmth for the paddlers during operations in colder waters.
Cross-Paddle Turn	Movement that helps personnel turn a double kayak in a confined space or to turn on its own axis. For example, to make a sharp starboard turn, the rudder is held hard over (right rudder), and Number 2 executes a reverse sweep while, at the same time, Number 1 does a forward sweep on the port side. This procedure will normally turn a laden kayak on its own length.
Detachment Ocean Equipment (DOE)	This equipment is on paddles to support the paddle movement. The DOE normally consists of extra equipment and emergency equipment. The type and number of items depends on the paddle, the number of kayaks, and the mission. The DOE consists of the following, at a minimum, and is always cross-loaded: • One 120-foot rope (salamander/lizard line). • Two diving lights or any waterproof lighting system. • Ready-packed sleeping bag for easy access. • At least four extra lights and an extra repair kit. • Properly stored extra weather gear for easy access in open seas. • Two strobe lights. • Fins, mask, and dive tool (knife).

Figure 8-3. Selected kayaking terms

Kayaks

Downdrafts	Winds that result from mountains near the sea that can come from all directions as a result of the funneling effect down the valleys. Downdrafts can be extremely dangerous to all boats.
Drogue	Used mainly to reduce wind drift at sea and has been used by kayakers to stabilize entry through the surf. For a loaded double kayak, the effective size is about 36 inches long with an entrance 20 inches square and an apex aperture about 4 inches square. A swivel at the junction of the line and drogue reduces tangling. A sea chute is similar in purpose and function to the sea anchor, but is built exactly like a parachute. Both should be attached to a 50-foot white utility rope with 70 to 80 feet of 550 cord. The stretch (elasticity) of the 550 cord serves to absorb shock loads caused by wave and wind action. This arrangement places the anchor approximately 120 feet from the kayak and affords sufficient scope to allow the boat to "lie to" in inclement conditions. A float should also be attached to the crown of the chute by a 10-foot line to prevent the sea chute from sinking too deep (Figure 8-5, page 8-6).
Fade to Black (syndrome)	A modern sign of kayak angst or fear. It is usually seen in new kayakers in open water for prolonged periods. Victims become disoriented, usually under overcast or poor visibility. Dizziness may set in and there is an urge to find landfall ASAP.
Feathered Paddles	One paddle blade is set 60 to 90 degrees to the other paddle blade. Feathering reduces windage on the airborne blade when traveling into the wind.
Immersion Suit	For kayak operations, dry suits and wet suits can be used as a form of immersion suit. Care must be taken to avoid hyperthermia.
Lizard Line	The line used to tow IBS and kayaks from a mother craft. The line is larger in diameter (normally marine-grade rope) with hookup points fitted for quick release (Pelican hooks) that are staggered so the boats do not "ride" on each other. It is made at detachment level.
Loom	The area at the shaft of the paddle or objects that are normally below the horizon that become visible due to temperature inversion (increase in temperature with height). If this temperature change also includes a rapid decrease in humidity, the refraction will be greater than normal, causing objects to appear elevated and making the visible horizon seem further away.
Paddle Rate	Kayaker uses this paddling cadence for the changing sea conditions. In a head wind, the rate can be as low as 30 strokes per minute. With a wind pushing the kayak, rates of 60 or over can be maintained. With wind and waves abeam, the stroke is lowered and paddles lean into the wind.
Paddle Shaft Navigation	Navigation technique that allows the team member to use the shaft (or loom) of the paddle to check the uncompensated compass course. The bow of the kayak is held to course (as corrected for current). This technique is used to check the course set with the criticality of currents taken into account.
Painters	The bow and stern lines on a kayak. Normal length is at least as long as the kayak. Dismounted crew members use these lines as handling or docking lines and to control or maneuver the kayak in confined waters.
Rafting	Term refers to bringing kayaks together. It normally occurs when a meeting or a break for a meal in calm waters is needed. It is never used in kayak operations for storm procedure nor should it be used in rough water.
War Chest	A box that is maintained at B and C team levels and supports kayak operations when deployed. Although there are no fixed number of spares and tools in the chest, at a minimum it must be able to support six kayaks. The war chest is also capable of supporting major repairs on kayaks (Figure 8-6, page 8-6). It is not a field repair kit.

Figure 8-3. Selected kayaking terms (continued)

Kayak, operational	Extra double paddle
Bow and stern lifting straps	Bow and stern lines, with snap links
Compass board	Bail bucket
Repair kit	Drogue chute with 50-foot line
Flare, signal, two each	Sector single-unit panel, one each
UDT vest, two each	Bilge pump, one or more
Elastic carrying straps, two each	Bow light

Figure 8-4. Basic ocean items

Chapter 8

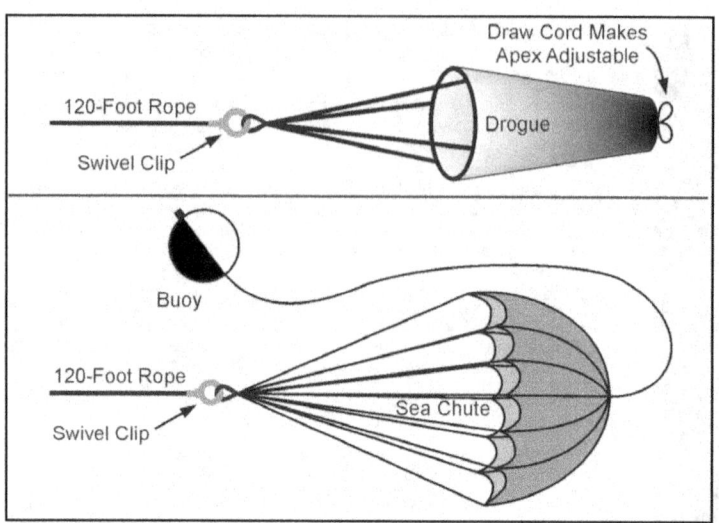

Figure 8-5. Drogue and sea chute

Spare Parts	Tools
Set of 8 rods (complete kayak)	2 seat backs (front)
3 rods with sliding sleeve	2 seat backs (rear)
10 extra black stoppers (for air sponsons)	Rivets
15 snap locks with rivets	2 steel chisels
10 rod holders with rivet (wash)	4 small hammers
10 keel board tongues	2 cutting pliers
3 pair of touring (double) paddles	1 hand drill (3- and 4-mm bits)
3 aprons	4 wooden carpenter clamps
1 gunwale tongue (flat)	2 round files
1 rudder blade with swivel head	2 bandage scissors
2 rudder pins	2 small flat-head screwdrivers
2 rudder yokes	2 medium Phillips screwdrivers
2 rudder cables	2 tubes of white glue
2-foot control pins with washers and nuts	2 cans marine varnish
1 yard of deck material	5 large needles with nylon thread
2 cans folding repair cement	Sandpaper (fine)
2 cans clear boat wax	Miscellaneous wood screws, 1/4 inch to 3/4 inch
1 rudder bracket	Air stop (plug for sponsons)
4 paddle bungies	Wood glue
4 rubber yokes	Punch
1 set gunwales (bow)	Rivet tools
1 set gunwales (stern)	3 rolls of folding tape
8 T-fittings	3 sewing awls
Note. A war chest includes only minimal tools and spare parts that will support six folding kayaks from a joint special operations task force (JSOTF) or special operations task force (SOTF).	

Figure 8-6. War chest

KAYAK OPERATIONAL TECHNIQUES

8-13. Techniques may vary with each kayak mission. Personnel must know how to assemble and disassemble the kayak, how to prepare the kayak for use at sea, launching and landing, paddling, righting a capsized kayak, and operating in the surf. The following paragraphs explain these procedures.

ASSEMBLY

8-14. Folding kayaks are generally similar in construction and assembly procedures. Trained kayak teams can assemble and disassemble a kayak in 15 minutes. Most boats will have two or three bags containing the long frame pieces, the ribs, and the hull. The bags are usually ample enough to store accessories like the spray skirt or bilge pump with the kayak components. The first step for the crew is to remove all of the components from their bags, lay them out, and inspect them (Figure 8-7).

8-15. Each kayak team member has specific responsibilities in assembling the kayak. Kayak team members are designated as Number 1 and Number 2, corresponding to their cockpit position. Number 1 handles the long bag and empties it on the ground while Number 2 empties the hull bag and the rib bag. Number 1 sorts the long wooden (or aluminum) parts into front and rear halves while the Number 2 lays out the hull from bow to stern lengthwise and places the ribs in sequence (from 1 to 7) alongside the hull.

8-16. Number 1 assembles the front half, and Number 2 assembles the rear half simultaneously (Figure 8-8, page 8-8). They do not assemble the cockpit ribs (usually 3 to 5) at this time. While folding and controlling the gunwales and keel boards, they insert the bow and stern assemblies one at a time into the hull. They make sure the inflatable sponsons are laying smoothly alongside the gunwales. Then they interlock the keel boards and press them down into place. They repeat this procedure for the two gunwales, pressing them outward. The team then installs the cockpit ribs and locks them in place. They connect the hull stringers and ensure that they are retained in their clips on the ribs.

8-17. Number 1 takes the coaming and inserts the masthead fitting into its position over rib Number 2 and the front deck bar. Working from front to rear, the crew secures the coaming to the deck and rib clips. The first man finished will attach the boomerang piece at the rear of the coaming or cockpit. Working on opposite sites, again from front to rear, the team will simultaneously secure the deck material to the coaming (usually by tucking a reinforcing strip into a slot in the top of the coaming). Each man will then install his own seat. Numbers 1 and 2 will then inflate by mouth one sponson two-thirds full.

Figure 8-7. Folding kayak components (Long-Haul Commando shown)

Chapter 8

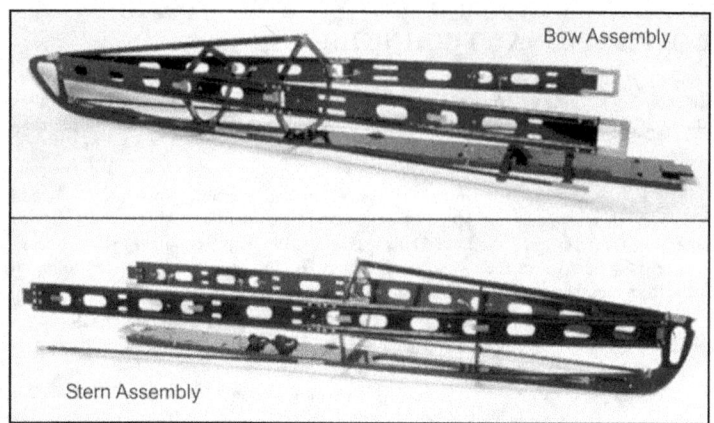

Figure 8-8. Bow and stern completed assemblies

8-18. Number 1 assembles the rudder's blade, yoke, pin, and cable system in the rear of the kayak. Number 2 inserts the foot pedal assembly for the rudder in its proper position and attaches the rudder cables to the pedals so they are comfortable for use.

8-19. Number 1 puts on the tuck-under spray cover (preferred) by placing the head of the cover over the masthead. Number 2 attaches the rear of the cover to the boomerang. Both team members then tuck the left and right sides of the cover under the coaming. With the spray cover in place, Numbers 1 and 2 simultaneously inflate the air sponsons completely. Finally, the team members assemble the paddles, stow their gear, and attach the spray skirts around the cockpits so the kayak is ready to launch.

DISASSEMBLY

8-20. The kayak team performs the disassembly simply by reversing the assembly steps. While Number 1 lets the air out of the sponsons, Number 2 removes the spray cover. Number 1 removes the rudder's blade, yoke, and pin, while Number 2 removes the foot pedal assembly. Both men remove the rudder cables.

8-21. Both men remove the material from the edge of the coaming and then remove the coaming. Then they remove the center ribs (3 to 5). Next they release the gunwales and keel boards and remove the front and rear halves of the frame from the hull. Number 1 disassembles the front while Number 2 disassembles the rear.

8-22. Number 1 puts the long parts in the long bag while Number 2 puts the hull, spray cover, spray skirts, seats, and repair kit in the hull bag. Both team members place the ribs and the rudder assembly into the rib bag. When the bags are properly cinched, they are ready for movement.

PREPARING THE KAYAK FOR THE SEA

8-23. Team members perform a prelaunch inspection when using the kayak at sea. They make sure—
- There are no holes in the hull.
- There are no holes or rips in the deck.
- The fore and aft lifting straps are present with two snap links.
- The bow and stern painters are present.
- There are two spray skirts, one with a spray cover.
- The rudder is complete with cocking line (used to raise the rudder).
- Sponson tubes are fully inflated.
- All paddles are interchangeable, including spares.
- There are spare paddles, three portage bags, a bailer, and a bow light (white only).

8-24. Team members must also stow various types of equipment on the kayak. To ensure balance and safety, personnel should keep to the following procedures:
- When stowing the gear, wisely balance the paddle (distance covered), sea conditions, tactical situation, and safety.
- During isolation, test-load the kayak by putting the entire operational load into an assembled frame (without the hull). This preliminary step will allow the detachment to gauge the effectiveness of their packing plan and make necessary adjustments.
- Ensure all equipment is waterproofed. Use commercial waterproof bags and waterproof *everything*. Bags make the cargo easier to store and easier to access. In the event of capsizing, the bagged gear displaces water that would otherwise have to be bailed, and it provides additional flotation.
- Load seldom-used equipment in the extremities of the kayak. Reserve the cockpit space around the paddlers for high-use items; for example, navigation, communications, safety, and immediate comfort (food, water, and thermal protection) items.
- Always pad and tape all-purpose, lightweight, individual carrying equipment (ALICE) rucksack frames if they are stored inside the hull. If the empty rucksack frame will be used as a seat, take care that the frame is not damaged by the paddler's weight, and that it does not damage the kayak.
- If the extra double paddle is stowed inside the hull, make sure it can be reached easily without removing other gear.
- Lash the kayak repair kit to the inside of the cockpit.
- Tie equipment to the two internal equipment lines, port and starboard. After securing equipment to the internal equipment lines, lash it with 1/2-inch tubular webbing to prevent movement. Ensure the knots are simple; untying wet knots with cold, cramped hands can become an impossible task.
- Keep deck loads to a minimum. Anything stored on deck increases windage and visual signature, and decreases stability. Deck loads are more likely to be lost if the kayak capsizes violently, and make righting the kayak more difficult. Ensure any gear stowed on the hull is free of the rudder cables and rudder cocking line.
- Load equipment for sea use last.
- Ensure navigational equipment is on the person or lashed to the cockpit.
- Lash the flashlight and flares to the inside of the cockpit.
- Store LCE and weapon last to even the trim of the kayak and to have it for ready use. If the weapon is carried on the outside of the hull, wear it on the person.
- Before departure, put the kayak into the water to make the final trim. Keep the heaviest equipment close to the cockpit, reserving the ends for lighter bundles. Head winds or seas require the bow to be heavier; following seas require more weight in the stern.

LAUNCHING AND LANDING

8-25. Kayaks are somewhat unstable and fragile. Care should be taken when embarking and disembarking to prevent capsizing or damage to the kayak. Figure 8-9, pages 8-10 and 8-11, explains various launching and landing methods that should be performed when using the kayak. In general, personnel should—
- Never step in the kayak when it is out of the water.
- Never step on the canvas (deck) portion of the kayak.
- Never drag the kayak (loaded or unloaded) across the ground.
- Never jump into the kayak.
- Always have the paddles ready when entering the water.
- Always place hands directly on the coaming above the open rib sections to prevent damage to the kayak.
- Always step directly onto the keel board when entering the kayak.

Chapter 8

Launching From a Beach
- Both team members move the kayak out bow-first, until it floats freely. They have their paddles in their hands.
- Both team members position themselves on the same side of the kayak.
- Number 1 stabilizes the bow for Number 2 to enter.
- Number 2 places his paddle on the opposite side and grabs the coaming with the same hand.
- Number 2 places one foot into the kayak.
- Number 2 grasps the coaming closest to him to support his weight and lifts his other leg into the kayak. He then lowers himself into the kayak, taking care not to sit on the seat back.
- Number 2 locates the rudder pedals with his feet and zips up (secures) the spray skirt.
- Number 2 stabilizes the kayak for Number 1.
- Number 2 places the end of his paddle on the sea bottom and leans his weight on the paddle shaft to stabilize the kayak.
- Number 1 enters the kayak in the same manner as Number 2.
- Both team members use the draw stroke to move away.

Landing on a Beach
- Number 2 cocks (raises) the rudder before landing.
- When in shallow water, Number 2 stabilizes the kayak with his paddle.
- Number 1 exits the kayak.
- Number 1 stabilizes the kayak for Number 2 to exit.

Launching From a Beach in Disturbed Water
- Number 2 carries the bow and Number 1 carries the stern.
- Both team members place the kayak in sufficient water to float.
- Number 2 holds the bow into the wind or sea.
- Number 1 gets in the front cockpit.
- Number 1 fits the spray cover.
- Number 1 stabilizes the kayak with his paddle.
- Number 2 gets in the rear cockpit and fits the spray cover.
- Both team members paddle into the wind or sea.

Landing on a Beach in Disturbed Water
- Team members reverse the steps above.
- Number 2 stabilizes the kayak for Number 1.

Launching From a Bank
- With Number 2 holding the bow, both team members place the kayak in the water with the bow upstream.
- Team members release the grip on the kayak, keeping the bow pointed toward the bank.
- Number 1 holds the kayak cockpit coaming with one hand and the bank with the other hand and puts one foot in the boat.
- Number 1 places his other foot alongside the first one, sits down, and adjusts his position.
- Number 1 stabilizes the kayak (with blade on bank) for Number 2.
- Number 2 gets in. Both team members use the draw stroke to move away.

Landing on a Bank
- Team members reverse the steps above.
- Number 2 stabilizes the kayak for Number 1.

Figure 8-9. Launching and landing methods

Kayaks

Launching From a Mother Craft
• Both team members place the kayak over the side using the bow and stern lines. • Both team members orient the bow in the same direction as the mother craft. • Number 2 holds the lines while Number 1 climbs in the kayak. • Number 2 passes the paddles down to Number 1. • Number 1 holds to the side of the mother-craft while Number 2 climbs down into it. • Both team members stow the bow and stern lines and use the draw stroke to move away.
Boarding a Mother Craft
• Team members reverse the steps above. • Number 2 stabilizes the kayak for Number 1.

Figure 8-9. Launching and landing methods (continued)

PADDLING

8-26. There are two basic paddling techniques: unfeathered and feathered. Both methods have their adherents and neither one is inherently superior. Unfeathered paddles have both blades on the same plane. This method makes paddling with a tailwind easier. Many people find this technique to be easier to learn and somewhat less fatiguing because it does not require a wrist roll. Feathered paddles are offset one from the other between 60 and 90 degrees. By using a control hand, the loom (paddle shaft) is rotated during the stroke so that, while one blade is in the water pulling, the other blade is feathered, or slicing into the wind. This method reduces wind resistance while paddling into the wind. Feathered paddling requires more practice to master and can cause physical problems because paddlers are constantly flexing their control hands.

8-27. When paddling, Number 1 sets the stroke or cadence. The cadence will vary from a sprint used to cross danger areas or shipping lanes to a slow stroke designed to conserve the team's energy during extended transits. Personnel use six basic strokes when conducting kayak operations (Figure 8-10, pages 8-11 and 8-12).

Forward Paddle	• Hold the paddle with a wide grip. • Make strokes as long as possible. • Lean forward. • Pull paddles. • Keep the paddle blade just barely covered by water.
Backward Paddle	• Take long strokes from the stern to the bow. • Keep the paddle blade at a flat angle as it enters the water.
Sweep Stroke (bow and stern)	• Make a large, sweeping arc with the paddles to turn the kayak hard right or hard left as desired. • Start the paddle blade either at the bow or stern.
Draw Stroke (kayak moves sideways on order)	• Keep the paddle blade parallel to the kayak. • Draw the paddle blade toward the kayak. • Turn the paddle blade 90 degrees, raising the wrist, and knife the blade out of the water.
Slap Support/ Recovery Stroke (recovers the kayak on the verge of capsizing)	• Keep the blade parallel to the kayak. • Draw the blade toward the kayak. • Turn the blade 90 degrees, raising the wrist, and knife it out of the water. Lean toward the side. • Keep the paddle blade out and horizontal to the water. • Slap the water with the flat side of the paddle blade and push down into the water forcing yourself upright.

Figure 8-10. Basic kayaking strokes

Sculling (for support or draw)	• When sculling for support:
	  ▪ Hold the paddle blade horizontal to the water.
	  ▪ Move the paddle back and forth in the water making a small arc.
	  ▪ Lean on the paddle.
	  ▪ Ensure the front edge of the paddle blade is raised slightly when moving forward, and that the back edge is raised slightly when moving backward.
	• When sculling for draw:
	  ▪ Use the same movements as for sculling for support, except hold the paddle vertical and do not lean on the paddle.

Figure 8-10. Basic kayaking strokes (continued)

RIGHTING A CAPSIZED KAYAK

8-28. As with any small-boat operation, capsize drills must be practiced in a controlled environment until they are mastered. Figure 8-11 explains the three basic capsize drills.

Cockpit Method of Self-Rescue
To right the kayak, each team member— • Keeps his paddle along the opposite side of the kayak. • Retains hold of the paddle. • Exits the kayak by rolling forward and out. If necessary, pull the thongs on the spray skirt to exit the kayak. Do not try to unzip the spray skirt to get out. Once out, surface and stow the paddles in the paddle bungies. • Checks his partner. • At night, checks his partner visually and by touch. • Retains grip on kayak. • Turns kayak into sea or wind. • Checks that equipment is secure. • Swims under the kayak and surfaces inside the cockpit. • Stows the paddles. • Throws the kayak up and over. To reenter the kayak— • Number 1 holds the bow into the waves while Number 2 climbs over the stern to the rear cockpit and bails out sufficient water to enter the kayak. • Number 2 on stern gets in and steadies the kayak with paddles. • Number 1 on bow gets in over bow. • Number 1 begins paddling while Number 2 continues bailing.
Pump and Reentry
• Complete the steps to right the kayak from the cockpit method. • Number 1 moves to the bow and holds it into the waves. • Number 2 swims up and over the bottom of the kayak and grasps the far sponson. • Number 2 pulls up and back on the far sponson and rights the kayak (Figure 8-12, page 8-13).
Raft Method
• Complete the steps to right the kayak from the cockpit method. • Assisting kayaks move alongside the capsized kayak. • Numbers 1 and 2 from the capsized kayak move to the assisting kayaks. Number 1 lies across the bow of the port kayak. Number 2 lies across the stern of the starboard kayak. • The assisting kayakers bail water out of the capsized kayak. • Once sufficient water is bailed out, Numbers 1 and 2 move across the assisting kayaks and into their kayak. Once in their kayak, they continue to bail out the remaining water.

Figure 8-11. Capsize drills

Kayaks

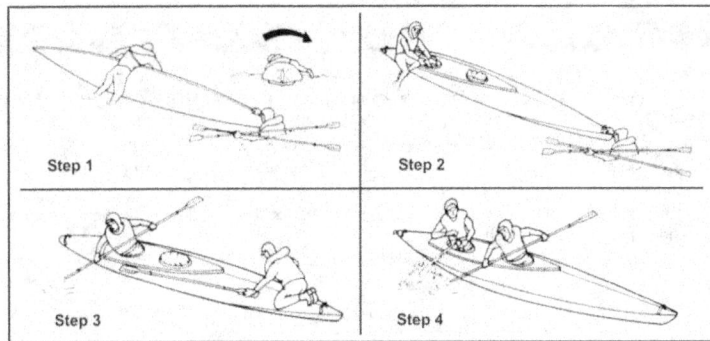

Figure 8-12. Pump and reentry method

8-29. Pump and reentry is a drill that mandates some type of pump (bailer). Personnel use this method to recover a loaded kayak. The disadvantage of this technique is that when rolling the kayak over, the cockpit takes in more water than normal. The raft method is used in heavy seas, or calm seas when the kayak is unstable or completely filled with water.

OPERATING IN THE SURF

8-30. There are many factors to remember when operating in surf conditions. Each technique discussed below requires specific movements that enable team members to properly maneuver the kayak (Figure 8-13, pages 8-13 and 8-14.).

Launching a Kayak in the Surf

- Team members stand beside the kayak on the beach one at the bow and one at the stern, each holding a paddle in the left hand.
- As soon as there is a lull in the surf action, they rapidly move the kayak beyond the beach line into sufficient water to float off.
- Number 1 gets in and fits the cockpit skirt while Number 2 stabilizes the kayak.
- Number 2 gets in and fits the cockpit skirt while Number 1 stabilizes the kayak.

Note. Speed is essential in fitting the skirts.

- Team members paddle rapidly straight into the waves.
- Before wave impact, the team members lean forward over the spray skirt, holding the paddle parallel to the kayak. Once the wave has passed, they resume paddling.

Note. This movement is only required if the wave is going to break on the kayak. If the wave breaks before, the team members continue to paddle through.

- Team members ensure that the kayak is going as fast as possible at the moment of wave impact to ensure that the kayak surmounts the crest.
- Once clear of the surf zone, the team members bail out and continue paddling.

Beaching a Kayak in the Surf

- Before the surf zone, the team members approach the beach stern-first, keeping the bow at right angles to the waves.
- They back paddle toward the beach during periods of lull.
- As waves approach the kayak, the team members paddle seaward into them.
- Before wave impact, they lean forward over the spray skirt holding the paddle parallel to the kayak.
- Immediately after the wave has passed, the team members paddle backward again and continue until the beach is reached.
- Number 2 gets out while Number 1 stabilizes the kayak; Number 1 gets out while Number 2 stabilizes the kayak.

Note. When landing in surf, the rudder assembly should be raised to prevent breakage.

Figure 8-13. Manuevering in surf conditions

Chapter 8

Surf Riding
• The team members choose an unbroken wave to ride on. • When the wave is astern of the kayak, they sprint toward the beach, keeping the kayak at right angle to the waves. Note: As the wave picks up the kayak, it should be going the speed of the wave. • The team members continue paddling to keep on the crest. • If the kayak gets off right angle to the wave, the team members lean over to the turning side and, with the flat of the paddle, hold on to the wave. • When the wave dies, the team members press the flat of the paddle on the dead wave to right the kayak and continue to the beach. • Number 1 gets out while Number 2 stabilizes the kayak. • Number 2 gets out while Number 1 stabilizes the kayak. **Note.** When landing in surf, the rudder assembly should be raised to prevent breakage.

Figure 8-13. Manuevering in surf conditions (continued)

KAYAK CONSIDERATIONS

8-31. The SFOD must consider many factors when preparing for kayak operations. Some of these considerations include planning factors, BLS procedures, combat diving and scout swimming operations, sailing, and hazards associated with kayaks.

PLANNING FACTORS

8-32. SFODs use kayaks solely as a means of transportation, normally during infiltration and exfiltration operations. The actual techniques used during kayak operations are influenced by the detachment's mission. DA and SR missions lend themselves readily to kayak operations. Regardless if the team has an initial mission after deployment, UW missions normally require the team to infiltrate onto an isolated BLS and to cache the kayaks.

8-33. When using the kayak for pure infiltrations, stores are packed so they can be unloaded in a minimum amount of time. DA and SR missions normally mandate that the BLS be within a reasonable distance to the target.

8-34. In isolation, personnel must decide either to cache the kayaks whole or in a bagged state. Kayaks can be camouflaged whole in a very secure manner, terrain permitting. Once the kayaks are disassembled, the reaction time and vulnerability are greatly increased if an abort is called. Therefore, the bagged state is a major disadvantage.

8-35. The cache site picked must be secure, must be able to withstand the elements, and must never be known to the resistance force. Prolonged periods in a bagged state will not affect the folding kayak. One of the most important advantages of the folding kayak is that it is a means of exfiltrating if the mission is aborted. It requires no support.

8-36. If the DA and SR mission is MAROPS (reconnaissance, harbor destruction, or beach reconnaissance), personnel may want to "live" in the kayaks. This technique was often used during World War II with complete success. However, they must remember that the mission revolves around the kayak. For long paddles and DA or SR missions where personnel will live in the kayak, they should use the sleeping pad or air mattress as a seat and backrest. When cooking inside the kayak, the personnel should make sure there is always one man on security, looking to the front. On long paddles, personnel should always have a relief bottle available.

8-37. When to portage must be carefully weighed against time, terrain, distance, and enemy activity. If the kayak must be portaged, Number 1 takes the hull with ribs securing them to his rucksack. Number 2 takes the long bag. The two painters are used to tie down the bags on the rucksack.

8-38. An empty kayak can be carried two ways. For short distances, it can be carried by the bow and stern carrying straps. If the straps are missing, personnel can carry the kayak under the opposite arms of each operator. Team members can carry the kayak for longer distances by lifting it onto opposite shoulders of each other. They can carry the disassembled kayak in the provided bags. For better balance, four persons should always move a loaded kayak up the beach. They should not carry a loaded kayak by the bow and stern lifting straps. Team members should fabricate two slings slightly longer than the width of the kayak with lifting loops at each end. They may place the slings under the kayak at the front and rear of the cockpit so that the weight is evenly supported. With one person on each lifting loop, a team can carry the kayak to an area where it can be unloaded safely.

8-39. Kayaks are extremely difficult to see, even at close quarters. The most exposed part of team members are their faces. Personnel should always camouflage exposed skin or use a cravat or cloth around their face when working near enemy positions.

8-40. When using scout swimmers or combat divers, they should come from different kayaks. This practice ensures that no kayak is without a kayak team member.

8-41. Two weapon systems may be taken on most missions. One for the kayak operations; the other for the land mission. In kayak operations, the weapon can be cached with the kayak and be used as a backup. The weapon, especially if working around enemy positions, must be small; otherwise it will interfere with the paddle strokes. It is normally slung over the shoulders and rides on the back. Examples of suitable weapons include an M4 carbine or an H&K MP-5, either one encased in a waterproof bag because of the requirement to fire in the water. When working close to enemy positions, team members should have their weapon and ammunition on hand (not inside). They should wear the LCE so that it does not interfere with wearing the UDT vest normally. A fatigue shirt can have pockets sewn onto the arms, back, or sides to facilitate access to equipment carried on the person's body.

8-42. Each kayaker must have adequate individual equipment. Although the items will depend on the paddle, type of training, and weather, at a minimum they will consist of the following:

- Army combat uniforms (ACUs).
- Pile or wool pullover.
- Foul weather jacket and pants.
- Running shoes or booties.
- Gloves.
- Wool cap.
- UDT vest with flare and whistle.
- Compass with overlay or chart.

8-43. Each kayak must always have a compass, map, and nautical chart. DA or SR missions can be completed by one solo kayak (and historically have been), but it must have a means of navigation. Although land maps can be used to identify prominent terrain features and man-made objects once the shoreline becomes visible, they do not have the necessary information for offshore navigation.

8-44. Personnel should know the tides and currents in the BLS and target area. They will affect the operator and he must compensate for them. He must offset to compensate for the tide and tidal current effects on his kayak.

8-45. Wind is one of the major determinants of the movement of water. Any strong wind—no matter in what direction—causes a net movement of water in the direction of its flow after a few hours. Once the wind dies, reverse flow occurs to redistribute the water that was initially moved. This "recovery effect" can be used for or against the operator.

8-46. Personnel should always do a nautical chart reconnaissance before infiltration and take notes, keeping security in mind. They should also consider that flow (currents, tidal movements, wind-driven currents) seeks the lesser friction of deeper water. Thus, current velocity is reduced over shoals, bars, and along shores, but it increases in channels. Nevertheless, with heavy stores, paddling near shore or in shallow water reduces the kayak's speed because of drag. Drag occurs with any vessel operating close to the ocean bottom; however, it becomes more noticeable with heavy stores.

8-47. Long paddles, or paddles through rough water, are mentally and physically demanding. Fatigue—no matter how experienced or fit the operators are—must always be considered when determining recovery time after beaching. The most dangerous time in kayak operations is after the paddle. It is at this time when most mistakes will occur.

8-48. Each kayak must have its own manufacturer's repair kit. It is designed to enable the crew to perform emergency field repairs. For specific mission purposes, it is not complete and select tools must be added to each kit. If maintenance is required in remote or denied areas, the repair kit with a spares kit will more than suffice for most major damages. This factor is another major advantage of the folding kayak. Long- and short-term maintenance procedures are relatively straightforward (Figure 8-14).

☐ **Short Term:**
- Bag the kayak for long storage periods only.
- Place the kayak in boxes for protection during transport.
- Wash all parts of the kayak with freshwater after seawater use.
- Remove all mud and grit from inside and outside the kayak after use.
- Dry the entire craft with a sponge and ensure all lines are dry.
- Lightly grease all metal parts.
- Store the kayak upside down on the coaming, with air sponsons inflated to allow the kayak to dry.
- Dry the kayak in the open air, if possible.
- Avoid exposing the kayak to direct heat when storing it.
- Avoid exposing the kayak to direct sunlight when drying it.

☐ **Long Term:**
- When bagging the kayak for an extended period, check its varnish, and revarnish as necessary.
- Store all lines (totally dry) with the kayak.
- Powder the inside of the hull with talcum powder when the kayak is stored indoors for long periods of time.
- Lightly wax the kayak's hull with neutral boat wax before storage and then at least once a year afterward. Never apply boat wax to the deck.
- When the kayak's skin starts to fade (usually after several years), apply marine canvas paint only.

Figure 8-14. Maintenance procedures for folding kayaks

BEACH LANDING SITE PROCEDURES

8-49. The team rafts up outside of small arms range. One kayak is sent in to conduct a reconnaissance of the beach. Before departing the boat pool, the scout team readies their weapons by slinging them across their torsos so that they are immediately available when the kayak reaches the beach. If the kayak capsizes in the surf zone or encounters enemy resistance, the scouts must be able to defend themselves until they can recover the kayak or exfiltrate back to the boat pool.

8-50. As the scout kayak approaches the beach, Number 1 has his weapon trained on the shore while Number 2 paddles in. Once the kayak is through the surf zone or into shallow water, Number 1 prepares to debark and stabilize the kayak so Number 2 can get out. After Number 2 debarks, Number 1 moves up on the beach, halts, and provides security while Number 2 turns the kayak around so the bow faces seaward. (The kayak is turned to facilitate a hasty departure.) Number 2 then faces the beach and provides cover for Number 1 as he returns to the kayak, removes the rudder, and stows it in the rear cockpit. Both men then carry the kayak onto the beach.

8-51. Once the kayak is secure from wave action, the kayak team then conducts a reconnaissance of the BLS and hinterland. After the area is determined to be secure, Number 1 maintains security while Number 2 returns to the beach and signals the rest of the detachment to enter.

8-52. The detachment comes in at 30-second intervals with Number 2 paddling and Number 1 with weapon ready. Number 2 on the beach can assist the first incoming kayak. As the kayaks land, they are placed in a beach-defensive posture with bows facing the water. When the team is assembled, it moves the kayaks off of the beach and into a concealed position.

COMBAT DIVING AND SCOUT SWIMMING OPERATIONS

8-53. Divers and scout swimmers perform specific tasks that enhance the security of the detachment using kayaks. Personnel should perform the following procedures when deploying divers or swimmers from kayaks:

- When launching scout swimmers from the kayak, the team rafts up beyond small arms range. The Number 1 men from the two center kayaks ease out of their cockpits and enter the water. The swim team swims to the BLS and conducts a reconnaissance. Once the BLS is cleared, they signal the remaining kayaks onto the beach. The Number 1 men in the remaining kayaks stow their paddles and get their weapons out for security. The kayaks with two personnel come in first so they can establish a secure BLS. They can then assist the remaining one-man kayaks onto the beach. Each kayak should maintain a 30-second interval (depending on wave period) when approaching the beach.
- If the BLS is unacceptable, the swimmers will have to return to their kayaks. They will reboard their respective kayaks using the same techniques as if the kayak had capsized.
- The kayak team must know where the pickup point is for the divers or scout swimmers. Not only is it extremely difficult for the kayak teams to see the divers or scout swimmers, it is also difficult for the divers or scout swimmers to see the kayaks.
- Once a diver or scout swimmer reenters the kayak, he will normally have a wet suit or dry suit on. He should avoid long-distance paddles without first taking the suit off. The suit can be removed a safe distance away from the pickup point.
- During training, open-circuit scuba may be used; however, it is far easier and safer to use closed-circuit scuba. Personnel should exercise extreme caution when handling divers' tanks around kayaks, especially the decks.

CAUTION

Never don open-circuit tanks in a kayak.

Note. Chapter 9 discusses generic scout swimmer and diver operations.

SAILING

8-54. Many folding kayaks can be modified for sailing. A keel, centerboard or leeboards, and substantial ballast are required. Most manufacturers of folding kayaks sell rigs and leeboards specifically designed for sea kayaking. In addition, a modified sailing rig can be made by using a parachute panel rigged to a spare paddle inserted in the aft section of the kayak.

8-55. The ability to rig a sail is an important option for the detachment considering longer-ranged kayak operations. Given favorable winds and sea conditions, a sail will conserve the strength of the detachment, increase the speed of transit, and greatly extend the range of a kayak mission. Unfortunately, effective use of a sail requires considerable training and continuous practice. Without a high degree of proficiency, sails can endanger the kayak, the kayakers, and the mission. When using sails on military operations, the increased visual signature must not compromise the mission. Personnel on an enemy patrol vessel can see a kayak's sail while still below a kayaker's visual horizon. Restricting the use of sails to night or low-visibility periods and selecting neutral-colored sails will reduce the visual signature.

8-56. There are a number of different sail kits or configurations available for use from various commercial sources. Some of them have been used very successfully for extended ocean voyages. These rigs are classified as upwind (against the wind) and downwind (with the wind) rigs. Because the wind hardly ever blows the direction that is needed, military users should concentrate on upwind rigs with their ability to sail into the wind.

8-57. An example of sails provided by the manufacturer is the S-4 sail ensemble (proprietary Klepper sailing rig). Other kayaks with similar mountings (Long Haul Commando) may use the same or similar rigs. The mast of the sail fits through the mast bracket in the coaming and is held in place by a raised wooden fitting on the keel. The entire ensemble consists of a gaff-mounted mainsail, a jib, and an M-1 drift sail. The M-1 can also be sailed up to 75 degrees off of the wind line. When the kayak is fitted with leeboards, it has the capability to tack upwind. The leeboards are mounted on a bar attached across the front of the coaming. In use, the downwind board is rotated into the water to prevent the kayak from slipping sideways while under sail.

8-58. Sailing technology for kayaks has advanced most rapidly in the aftermarket arena. One of the best assemblies available is the Balogh Batwing. This is a fully battened, camber-induced sailing rig that borrows much from sailboard sailing. With its patented outrigger system, it is a simple, powerful rig that has been proven under the most demanding conditions. Originally conceived as "training wheels," the outriggers have proven most useful while exploring the kayak's performance envelope (allows sailing in higher winds). This system has all the capabilities of older, more traditional rigs, with the added advantages of the outriggers for increased stability and safety.

8-59. Another sail system is the kite or parafoil, which is a recent introduction in kayaking. The parafoil resembles a small ram-air canopy. It is made of ripstop nylon and normally comes in three sizes: 7 1/2, 15, and 30 square feet. In use, it is flown like a kite and used to pull a kayak similar to a downwind sail. One of the greatest advantages over conventional sail rigs is its size and compactness. Unfortunately, because of the altitude at which it is flown, it has a correspondingly greater visual detection range.

8-60. When sailing a kayak, either the Number 1 or Number 2 can control the sail. The boom line should always be held and never tied down. If the boom line is tied down, a strong gust of wind may fill the sail and capsize the kayak. Power sailing is when the Number 1 controls the sail and Number 2 paddles. This conserves Number 1's strength and is a viable technique if used in light winds when a scheduled speed must be maintained. In rough seas, Number 1 can control the sail while Number 2 maintains his paddle at the "ready to do a recovery" stroke to prevent capsizing.

HAZARDS ASSOCIATED WITH KAYAK OPERATIONS

8-61. Environmental factors and medical hazards can also affect kayak operations. Therefore, all personnel should be aware of and alert to the following factors and hazards.

Environmental Factors

8-62. **Boomers.** These rocks are normally not exposed by the average swell, but when waves larger than normal pass through, they are exposed, usually by the wave crest exploding on the top of the rock. Boomers are commonly found on rocky lee (subjected to onshore winds) shores exposed to the open ocean.

8-63. **Clapotis.** On French charts, the term refers to riptides, a phenomenon in which waves striking a steep cliff cause rebounding waves, and where the incoming waves coincide with the rebounding waves to create high-standing waves. The power of the coinciding waves is vertical and can throw a laden kayak out of the water.

8-64. **Downdrafts.** These winds come from mountains near the sea in all directions as a result of the funneling effect down the valleys. Downdrafts are common in areas with ford-like terrain. They can be extremely dangerous to all boats.

8-65. **Surf.** It is extremely dangerous and should be avoided whenever possible. It becomes most critical at unprotected lee shores during limited visibility.

8-66. **Rocks and Coral.** Both submerged and exposed rocks are dangerous to the kayaker. He should be especially careful and alert for uncharted rock and coral formations.

Kayaks

8-67. **Rip Currents and Tides.** These can be useful for rapid egress but can be dangerous when beaching.

8-68. **River Entrances.** Kayakers should stay as close to the shore as possible when approaching any river entrance. These entrances are extremely hazardous at flood tide.

8-69. **Other Vessels in the Area.** Kayakers should be alert and stay clear of any vessel traffic. This practice should involve both visual alertness as well as concentrated listening during limited visibility. Kayakers should make sure they know the light patterns for various ships that may be in their area and, for training only, should make every effort to make their position known.

8-70. **Fog and Ice.** Fog presents a navigation problem. It also keeps others from locating a kayak. For training only, a radar reflector should be put up during dense fog. Hidden ice formations present serious problems to kayakers operating in dense fog.

8-71. **Marine Life.** Marine life does not present a big problem, but kayakers must be aware and alert. Large marine life such as sharks, porpoises, and dolphins may, out of curiosity, investigate a kayak and occasionally brush up against it. However, they generally present no problem. Jellyfish, such as the Portuguese man-of-war, occasionally may be encountered as they are swept over the kayaker by breaking waves. Jumping and flying fish, such as mullet and herring, can disrupt a kayaker paddling and he may be hit in the face, which can cause deep cuts. Seals and sea lions must not be agitated, as they are extremely dangerous and will attack a kayak if provoked.

Medical Hazards

8-72. **Back Injuries.** A lower back ache is common with long paddles. The kayaker should rest for short periods and take aspirin.

8-73. **Chafing and Rashes.** These are very common but also very serious if allowed to develop without attention. These normally occur in the arm and groin area. They can generally be prevented by using petroleum jelly or coconut oil. More severe cases can develop into boils or carbuncles—totally debilitating problems that require medical attention ASAP.

8-74. **Blisters.** These will normally develop at the start of a paddle and can generally be avoided by taping the hands or by wearing gloves.

8-75. **Seasickness.** This feeling must be treated seriously and considered during any planning and preparation phase. Motion-sickness pills may be taken before entering the water; however, they can cause drowsiness. Personnel should stay relaxed and paddle normally; aggressive paddling aggravates an upset stomach.

8-76. **Sunburn and Windburns.** These normally occur on long paddles and if not prevented can be catastrophic. Personnel can use zinc cream to protect their head and neck. They can use petroleum jelly and a cravat around the face to prevent serious windburn.

8-77. **Muscle, Joint, and Tendon Injuries (Tenosynovitis).** This injury is a painful inflammation of the tendons of the wrist and arm and most frequently occurs from using a feathered paddle. It can be prevented by gradually working up to longer paddles and by using a feathered paddle only when required. The pain can be treated with aspirin.

8-78. **Kayaker's Arm.** This pain is a more advanced case of tenosynovitis that can cause numbness of the entire arm. It is usually found during long-distance paddles where the kayaker uses feathered paddles. The only prevention is to use unfeathered paddles. Aspirin can be taken for the pain but a neurological exam is required following the mission.

8-79. **Kayaker's Elbow.** This very painful injury is characterized by sometimes severe swelling at the elbow. It is caused by too much paddling. The only treatment is complete rest for a period of at least 1 week.

Other Medical Conditions

8-80. **Hyperthermia (Heat Casualty).** This overheating of the body usually results from insufficient fluid intake or too much thermal protection. Symptoms are excessively high body temperature, breathlessness, dry mouth, dizziness, headaches, poor concentration, unconsciousness, and death. The best treatment is prevention. Kayakers should ventilate or remove excess clothing and drink lots of water. Victims should not be treated by immersing them in the water. This condition occurs in both hot and cold climates.

8-81. **Hypothermia.** This condition can occur in hot or cold water. Kayakers must rely on the buddy system and be aware of the standard signs and symptoms (uncontrollable shivering, erratic behavior, blurred vision, ashen face and hands, muscle rigidity replacing shivering, incoherence, and collapse). Kayakers must immediately "beach," get the victim warm, and seek immediate medical attention.

8-82. **Dehydration.** This condition is obviously caused by a serious shortage of body fluids. A sure sign of dehydration is dark urine. To prevent or treat dehydration, personnel should drink plenty of water.

8-83. **Hallucinations.** These are brought on by fatigue, usually associated with long periods of wakefulness. They normally occur at night. Rest is the only relief.

Chapter 9

Over-the-Beach Operations

The planning and execution that takes place during the initial phases of a waterborne operation are intended to move a detachment from home station to the debarkation point. Once the detachment has been "dropped off" at the debarkation point, they are committed to the mission and are essentially on their own. Aircraft making a static-line parachute drop cannot recover a detachment in trouble. Helicopters, surface vessels, and submarines have a limited loiter or recovery capability in the event the mission must be aborted.

Assembly on the water after delivery by aircraft, surface vessel, or submarine marks the start of the transit phase of the mission. At that point, the detachment's survival and mission success depend upon reaching land safely (seamanship and navigation) and without compromise. The transit plan must include a link-up plan in case there is a break in contact during the transit. Once the detachment comes within sight of land, the tactical portion of the mission commences. Getting the detachment ashore and inland is the most technically difficult aspect of a maritime operation. This chapter outlines how to aid a detachment in identifying requirements, formulating a plan or SOPs, and conducting training to prepare for the unique mission challenges.

BEACH LANDING SITE PROCEDURES

9-1. As the SFOD nears land, the transit portion of the mission is ending. General navigation (DR) gives way to the requirement for precision navigation (piloting). The detachment must place itself in the vicinity of the BLS and hold-up offshore in a "boat pool" to prepare for the landing operation. Although the use of a GPS receiver will ensure precision navigation, the detachment must have an alternate (independent) means of confirming its location. The preparation of a horizon sketch or profile during mission planning, with bearings to prominent terrain features, allows the detachment to confirm the BLS from the boat pool.

9-2. The boat pool is an assembly area or rallying point for the CRRCs, analogous to the last covered and concealed position before crossing the line of departure (LOD). The boat pool is located far enough offshore that the CRRC formation is reasonably protected from observation and direct fire, but still close enough for scout swimmers to reach the shore and perform beach reconnaissance in a reasonable amount of time. This distance will vary depending on meteorological (light, weather, and sea state) conditions. Actions to be accomplished in the boat pool mirror the actions taken in an objective rallying point (ORP) before and during a leader's reconnaissance of an objective.

9-3. The SFOD will assemble all of the CRRCs at the boat pool. When the boats assemble, personnel should be prepared to pass bow and stern lines if such action is required or desired to raft the CRRCs. If seas are not calm, bungee cords are an ideal substitute. CRRCs may be anchored to prevent or control drift while waiting in the boat pool. Whenever using an anchor, personnel should always secure it with a quick-release. In an emergency, it is easier to release the anchor than it is to recover it. A quick-release is also safer than attempting to cut an anchor line in adverse conditions.

9-4. The OIC should take time while in the boat pool to give last-minute orders, confirm signals, and review the rendezvous plan with the scout swimmers. If the boats are not rafted or anchored, the coxswains must work together to maintain the boats' positions and keep the boats' engines away from shore. This practice redirects and minimizes the exhaust noise signature. All personnel should have their equipment on but unbuckled, with weapons ready. Swimmers should be ready to enter the water. Everyone must keep a low profile to reduce the boats' outlines. The lower the boats' profiles, the closer the visual horizon of an individual standing on the beach and the less likely the detachment is to be compromised. After the OIC launches the scout swimmers, the detachment should motor (or paddle) the boats to a position that allows them to provide cover for the scout swimmers. This position should be left or right of the ITR of the swimmers so that any required covering fire does not endanger the swimmers. If necessary, the boats can be anchored to maintain this position without using the motors.

9-5. Whenever the SFOD elects to use scout swimmers, they should be a senior person (chief warrant officer [CWO], team sergeant) and the point man or rear security. Contingency plans will identify the revised chain of command on the affected boats. After the scout swimmers have negotiated the surf zone, they will conduct reconnaissance of the BLS, establish security, and conduct a surf observation report (SUROBSREP) to determine the wave sets. It will be the scout swimmer's call whether or not the boats can be brought in. If required, the scout swimmers will set up transit lights to mark and guide the boats in on a selected route that avoids obstacles.

Surf Passage

9-6. After the scout swimmers signal the boats, the detachment begins an orderly movement to the BLS. The first obstacle that the detachment is likely to encounter will be a surf zone. The planning criterion and operational restrictions for negotiating a surf zone are contained in Chapter 13.

9-7. The conduct of a safe and efficient CRRC surf passage is a critical part of any maritime operation. Surf passages place the detachment in a very vulnerable position. The chances of injury, loss of gear, or damage to the boats and motors are greatest in the surf zone. Because the detachment members are concentrating on safely passing through the surf zone, they become more vulnerable to enemy action.

9-8. The coxswain is in charge of the boat and crew during the surf passage. If the surf is expected to be high, the most experienced coxswains should take charge of the boats. They must ensure that all gear is securely stowed and tied into the boat. Gear also includes the motor, repair kits, gas cans, and crew-served weapons. Everything must be secured but ready to use. Detachments may also want to secure the engine cover with tape to prevent its possible loss if the boat capsizes. Whenever possible, personnel should conduct a surf observation to determine the high and low set. Finally, they check to make sure the engine kill switch works correctly.

9-9. If the surf is moderate to severe, the boats should execute their surf passages one at a time. This type of movement keeps the detachment from having all of its assets committed at the same time, and allows the uncommitted boats to provide cover or assistance as required. While advancing from the boat pool toward the surf zone, the coxswain will ensure the outboard motor is on tilt (unlocked) and that the crew members have their paddles ready. The coxswain must time high and low series and observe sets of waves in the surf zone. He should inch his way to the impact zone and position the CRRC behind a breaking wave. The CRRC should ride behind the white water as long as possible, watching for overtaking and double waves. Concern for follow-on waves depends on the height of the white water.

9-10. Personnel should remember the saying, "Be bold; he who hesitates is lost!" An outboard motor can outrun any wave unless the operator has placed himself in the wrong position. The objective is to position the CRRC on the backside of a wave and ride it all the way through to the beach. The key is to manipulate the throttle to follow the wave as closely as possible without overtaking it, or allowing the next wave to catch up. Following as closely behind a wave as possible keeps the greatest amount of water under the boat, keeps the boat concealed as long as possible, and is the easiest way to maintain control of the boat. Under no circumstances should the coxswain attempt to "surf" or ride a wave with a zodiac. A zodiac that gets ahead of a wave is in danger of broaching (getting sideways to the wave and being rolled), pitch-poling (catching the bow in the trough of the wave and flipping end over end), grounding out, or being overrun by the following wave. Once the entire detachment is ashore and the BLS procedures are complete, the detachment reorganizes and continues with the primary mission.

9-11. Surf zones are two-way obstacles. When the time comes for the detachment to exfiltrate, it must be prepared to negotiate the surf zone in reverse. The objective is to select a route having the smallest breakers and timing the passage so that the CRRCs are moving through the lowest-possible set. If conducting the surf passage following an OTB exfiltration, the coxswain should begin surf observations

ASAP. The surf observations allow him to determine the timing of the wave sets. High, medium, and low sets are usually present. Depending on the length of the surf zone, it may be desirable to start the passage at the end of a high set so that the CRRCs are not caught in the middle of the surf zone as the sets change. If the coxswain can determine the location of a riptide, it makes an excellent exfiltration route through the surf zone. The waves are usually lower, the risk of grounding on a sandbar (or meeting breaking waves on the bar) is minimized, and the outflowing current will assist.

9-12. Before embarkation, detachment members must ensure that the paddles are accessible and ready for use and that all gear is stowed and tied down. They should walk the CRRC into the water until it is deep enough to allow the coxswain to start the motor (if the detachment is not going to paddle out). After the motor is started, they maintain only the minimum number of people in the water to control the boat. At the coxswain's signal all personnel will get into the boat and start to go.

9-13. When the CRRC is breaching a wave, the coxswain should let the bow of the boat power through the wave before he lets up on the throttle. This maneuver will allow the bow of the boat to come down easily, and the weight of the bow will actually pull the rest of the boat through the wave.

9-14. The detachment should avoid overloading the bow of the boat to prevent plowing through the waves. If the boat does plow through a wave, everyone should attempt to stay in the boat. If it capsizes, they should attempt to return to and stay with the boat (to prevent possible separation at night). Wave action will carry it back to the beach. The boat OIC should assess the situation, carry out the necessary actions, and inform the detachment OIC ASAP of the situation. All craft should rendezvous at the back of the surf zone, then continue with the mission. The detachment OIC should always explain capsizing actions in detail within the patrol order.

BEACH LANDING

9-15. **All Going Well.** If the BLS is secure and it is safe to bring in the detachment, the scout swimmers will signal from the shore. The boats will cast off from each other and line up for the surf passage. All personnel will sit on the main buoyancy tube and maintain a low profile with one leg over the side. This routine will allow for a rapid, controlled exit when the boat has penetrated the surf line. The boats will move in one at a time adjusting their tactics (drive or paddle) depending on the characteristics of the surf zone. If using the outboard motor, the coxswain should not run it aground; however, he should keep it running as long as possible. He should make sure the motor tilt lock is released so that the motor can kick up to prevent damage if it does run aground. Everyone except the coxswain should exit the boat as soon as the water is shallow enough to wade. Designated members of the boat crew will move to augment BLS security. The remainder of the crew will act as gear pullers and move each CRRC and its components to the cache site individually to facilitate recovery after the mission. As each boat clears the beach, it should signal the next boat in. Care must be taken to prevent boats and equipment from piling up on the beach. A good point to remember is to "always think of security versus speed."

9-16. **All Not Going Well.** The detachment waiting in the boat pool depends upon the skills and judgment of the scout swimmers to determine if the BLS is suitable and safe. If the scout swimmers encounter unfavorable conditions (hydrography, topography, or enemy situation), they make the decision to abort the landing. They will signal out to the boats and activate the withdrawal plan. Personnel waiting in the boats must be prepared for contact or compromise. While the scouts are swimming back to the boat pool, the flank boats hold position and provide cover. The OIC or lead boat prepares to pick up the scout swimmers if required. The objective is to avoid compromise, move to the alternate BLS, and continue the mission.

9-17. **Enemy Contact.** If the boat pool or the scout swimmers are compromised from the shore before the detachment is committed to the landing operation, it may be safer to retreat out to sea. Unfortunately, this situation leaves the detachment separated with a requirement to recover, or cover the withdrawal of, the scout swimmers. The swimmers can either return to the CRRCs, get picked up by the CRRCs, or—if they have not been compromised—move to the shore and rendezvous with the detachment later. If the swimmers have been compromised, the boats must cover their withdrawal the best way possible. If a decision is made to recover the swimmers, the flank boats provide covering fires, and the center boat (normally the OIC) runs in to extract them. If possible, swimmers should attempt to get out past the surf line to make a pickup easier and safer.

9-18. The most difficult aspect of the problem is determining a method of marking the swimmers to facilitate recovery and protect them from friendly fire. The chosen method must mark the swimmers' location and aid in vectoring a recovery vessel to them without contributing to the risk of compromise or aiding the enemy's fire direction. The detachment must research available marking technologies and select one that will be visible in adverse conditions, that can be activated by disabled swimmers in the water, and that does not have an overly bright (compromising) signature. Most IR beacons, to include chemlights, will meet these requirements provided the detachment has adequate night vision goggles (NVGs) available.

9-19. If there is contact, the detachment's primary objective is to break off the engagement and withdraw. As soon as it is safe to do so, the mission commander must assess the situation and make a determination whether or not to continue the mission. If it is possible to continue, the detachment will regroup and move to the alternate BLS. If the contact resulted from a chance encounter with coastal patrol boats or aircraft, the detachment may be required to immediately get off of the water. Detachments can quickly find themselves outgunned with nowhere to hide. The only option is to evade to the nearest land and disperse until something can be done to salvage the mission or initiate an emergency recovery plan.

9-20. Detachments must have a sound rendezvous or linkup plan in case they have to split up to avoid contact or compromise. If there is enemy contact, the detachment should observe the same basic principles as a mounted detachment in land warfare—the coxswain (driver) does not shoot, he just drives; the remainder of the detachment works to gain superior firepower ASAP. If possible, the detachment will break contact and attempt to screen the withdrawal with smoke (bubble-wrapped or floating), then exit the area. The OIC must then assess the situation and take relevant action to continue or abort the mission.

9-21. Detachment members must maintain muzzle awareness and discipline. During an ocean infiltration, consideration should be given to having weapons without rounds in the chamber until just before the formation reaches the boat pool. The extended visual detection ranges make chance contacts unlikely. The same is not true for riverine or estuarine movement. Once contact has been initiated, it is important to consider the increased risk of fratricide caused by the unstable firing platform and the rapid changes in relative position during evasive maneuvers. After the engagement, detachment members must also be aware of the risks of a cook-off and the danger of a hot barrel making chance contact with the boat and puncturing it.

SCOUT SWIMMERS

9-22. Scout swimmers are surface or subsurface swimmers used in conjunction with small boat, combat swimmer, and combat diver infiltrations. They reconnoiter and secure the BLS before committing the entire team to the beach landing. Scout swimmers normally work in pairs. In addition to locating a suitable BLS, scout swimmers must also locate an assembly area, look for suitable cache sites, and select a location from which to signal the team. They may be used to confirm that the enemy does not occupy a proposed landing site, that a landing site is suitable and surf is passable, or to rendezvous with a beach reception party if one is present.

9-23. Scout swimmers are equipped with fins, flotation devices, exposure suits, and weights as needed. If the selected infiltration technique includes diving, the scout swimmers will require appropriate diving gear (O/C and C/C). Additionally, they carry LCE, an individual weapon, and the prearranged signaling device to be used to pass signals to the team from the BLS. Scout swimmers should be neutrally buoyant and armed with automatic rifles, pistols (sometimes suppressed), and grenades. Scout swimmers must be prepared to remain ashore if an unforeseen problem occurs. They must be able to survive for a reasonable amount of time if there is enemy contact, compromise, or any other separation where they cannot rejoin the main body. Experience has shown that a vest is the best way to carry the following equipment:

- Ammunition.
- Evasion and escape kit.
- Rations and water for 24 hours.
- Flashlight and signaling devices (and backup devices).
- Knife.

- Compass.
- Communications (for example, short-range COMM to main body or survival radio if required).

9-24. Scout swimmers should be selected during the planning phase to allow detailed plans to be formulated and drills rehearsed. Backup swimmers should be nominated in case of seasickness or loss. The composition of the team will depend on the situation, but may include a scout and the OIC, assistant officer in charge (AOIC), or noncommissioned officer in charge (NCOIC). Sometimes the swimmers may have to take in an agent, foreign national, or someone else who might be a weak swimmer. Contact drills, casualty recovery, and rendezvous drills must be rehearsed. A detailed and sound briefing must occur before the scouts are deployed. All contingencies must be covered and briefbacks conducted so there are no misunderstandings. When using scouts, a finite time must be placed on them to achieve their goals and "actions on" if that time limit is exceeded.

9-25. Scout swimmer deployment procedures begin on arrival in the vicinity of the BLS. The CRRC members form a boat pool a tactical distance from the shoreline and conduct a visual reconnaissance. The CRRC transporting the swimmers moves to the swimmer release point (SRP), fixes its position, and takes a bearing on an obvious shore identification point. The swimmers should move in and out on that bearing.

9-26. Before leaving the main body, the scout swimmers receive last minute instructions or adjustments to the original plan based on observations made during the infiltration thus far. Scout swimmers should leave their rucksacks with the main body. Normally, the main body halts in a holding area outside the surf zone and small arms fire (about 500 yards) from the beach.

9-27. Because the tactical (or hydrographic) situation may require the swimmers to return to the CRRCs, the current and drift at the SRP must be tested. Depending on the drift, the coxswain will decide whether to anchor or try to maintain position by paddling. The outboard motor can be left at idle to prevent the extra noise of trying to restart the motor. If the night is very still and quiet with no surf, paddles may be a better choice than the engine.

9-28. The swimmers, once released, should swim on a bearing to a previously identified point. To keep their direction, they use a dive compass or guide on prominent terrain features or lights on the beach. The raiding craft should be prepared to act as fire support to cover the swimmers if they are contacted. The swimmers should count the number of kicks and the time it takes to swim in. Most of the time it is impractical to do this through the surf zone. However, the swimmers should know how many kicks it will take them to get from the back of the surf zone to the RP or boat pool. Surface swimmers should swim and maintain 360 degrees visual observation around them. They can have this view by using the sidestroke. While approaching the surf zone or general vicinity of the BLS, swimmers should face each other and observe the area beyond the other swimmer.

9-29. As the scout swimmers reach the surf zone or when they get fairly close to the BLS, they use the breaststroke so they can observe the beach. It is imperative that the scout swimmers use stealth and caution while approaching the beach. They must keep a low profile in the water as well as when on the beach. The use of a camouflage head net is ideally suited for concealing scout swimmers as they approach the beach.

9-30. Swimmers should conduct a short security halt to look and listen before they commit to a surf zone passage. Once they reach waist-deep water and perform another stop, look, and listen, the swimmers decide whether to take off and secure their fins (but still keeping them accessible) while keeping as low as possible in the water. Depending on the beach gradient, there are two methods:
- If the water offshore is still relatively deep, one scout pulls security by observing the BLS, while the other scout removes and stows his own fins.
- If the water just offshore is relatively shallow and gently sloping up to the beach, one scout can low crawl to a position in front of the other. The man in front pulls security while the rear man removes the front man's fins for him. The scouts then change places and repeat the procedure. This method is much quieter in shallow water.

9-31. Before the swimmers leave the water, they should conduct a good visual reconnaissance to select the place they will cross the beach to ensure that the least amount of signs or tracks are left. Swimmers should

Chapter 9

not be hasty at this stage; there may be something on the beach that cannot be seen. They should use maximum stealth when approaching the BLS and leaving the water.

9-32. The swimmers should then move onto the beach maintaining as low a profile as possible. If the task is to clear the beach, the pair should stay together and use silent covered movement to conduct a reconnaissance of the landing site. They should be working to a set time. The two methods of moving across the beach, conducting reconnaissance, and securing the BLS are as follows:

- If the wood line (hinterland) can be seen easily from the waterline, one scout remains in the water just at the waterline and covers the other as he moves quickly across the beach. Once the inland scout has moved to the edge of the wood line, he covers his partner while he moves across the beach to the same position.
- If the beach topography is such that the hinterland cannot easily be observed from the waterline, the above method can be modified to include successive bounds, or both scouts can move quickly across the beach together. If both scouts move together, they must observe the beach area thoroughly before moving. It may be necessary to observe the beach from a certain distance offshore from the waterline to properly observe the hinterland.

9-33. Once both scout swimmers have moved inland, they conduct a reconnaissance and establish security. There are three techniques (cloverleaf, box, and modified box) used to reconnoiter and secure the BLS. The terrain, time, size of landing party, and vegetation (cover and concealment available) will determine which method the scout swimmers use. The scouts agree on a suitable assembly and cache site when they finish their reconnaissance. One scout then positions himself at the edge of the hinterland (or backshore, if suitable cover and concealment are available) to provide security for the main body's landing and from which he can guide the main body to the assembly area. The other scout positions himself where he can signal the main body. He must ensure that the signal is not masked by the waves and can be seen by the main body. As soon as he sees the main body, he moves to the waterline.

9-34. When the main body reaches the BLS, the scout at the waterline directs them to the other scout who guides them to the assembly area. After the last man has passed him, the scout at the waterline disguises any tracks left in the sand and then rejoins the main body.

9-35. If at all possible, scouts should not use an assembly area as a cache site. If the enemy discovers and follows the tracks or trail from the beach to the assembly area, he can easily determine the number of personnel involved in the operation by counting the swim gear. Also, the cached equipment may be needed to support exfiltration at another location.

9-36. If the swimmers are to return to the boats, they swim out from the previously identified point using the back bearing of their approach for the specified distance (set time and number of kicks). Divers should conduct a tactical peek after swimming a predetermined distance. They then signal over the agreed sector using red-lensed flashlights (or other appropriate mid-range signaling device) for identification and position location. The problem with this method is that flashlights offer very limited peripheral coverage and there is an increased risk of compromise.

9-37. If the swimmers are far enough offshore, the boats should move to the swimmers to effect a rendezvous. Having the boats move to the swimmers speeds the recovery and conserves the diver's strength. If the signal equipment fails and no pickup occurs, the swimmers should not swim about vaguely but return to the beach and try to fix the problem. If the time window has closed, the swimmers should move to the alternate swimmer recovery point. If they cannot, then they should carry out their rendezvous procedure as detailed in their operations order (OPORD).

9-38. If the swimmers make contact or are compromised, they must decide whether to move back through the water to the boat pool (swimmer recovery rendezvous) or across land to a land rendezvous (maybe the alternate BLS). If the swimmers are in contact, they should try to use grenades rather than rifle fire. Grenades cause confusion at night and do not pinpoint the swimmer's position. Illumination is a bad call in this situation; it will just compromise the boat pool and the swimmers. The use of darkness should be an advantage.

9-39. The swimmers require a means to identify their position tactically, so that if the raiding craft is in a position to provide fire support, the crew will know where the swimmers are to minimize risk to the swimmers. The CRRCs need to stay in visual distance of each other while loitering in the boat pool. The left and right flank boats need to be prepared to move out wide to provide fire support if required. The center boat may have to recover the swimmers while the OIC's boat is level with or forward providing fire support. Firing from the water is inaccurate. Boats located behind the swimmers may only be able to provide indirect fire with weapons like the M-203 or some other type of high-trajectory fire support. Boats to the flanks need to be out far enough so that they can fire small arms or light machine guns into the beach to support or assist the swimmers in making a clean break and withdrawing.

9-40. Actions on contact must be covered in-depth in the OPORD. The risk of fratricide is high. Any decision to compromise the boats and their location by providing fire support to the scout swimmers is a tactical decision that can only be made by the on-scene commander at that time. The other boats or crews must not give away their presence or position until the command to initiate fire support is given.

SIGNALS

9-41. Signals should be well thought-out, rehearsed, and coordinated with all necessary agencies. Personnel should plan for all possible contingencies. The various signaling devices, methods, and types are discussed below.

SIGNALING DEVICES

9-42. Personnel can use numerous signaling devices during waterborne operations. The capabilities, limitations, and uses of these signaling devices are discussed below.

9-43. **Infrared Signaling Devices.** IR emitting devices are probably the most secure and versatile signaling devices available. A good IR signaling device must possess the capability to send Morse code signals. Night vision devices (scopes or goggles) are a secure and available method of detecting IR signals.

9-44. **Flashlights.** When IR devices are not available, personnel can use flashlights. Personnel should always waterproof the flashlights and fit them with a directional cone to aid in limiting their signature. Personnel should consider using filters because white light will usually be too bright and attract undue attention. They should also avoid red and green filters where possible because these colors can easily be mistaken for red or green buoy lights that abound along coastlines of the world. Flashlights are the most commonly used signaling devices because of their availability and the fact that they can be used to send Morse code.

9-45. **Lightsticks (Chemlights).** When nothing else is available, a lightstick can be used for signaling. For correct use, it should be fitted with a directional cone. Morse signals can be sent by passing the hand back and forth over the cone. The primary drawbacks to using light sticks are that they are overly bright, cannot be turned off, and are difficult to adapt for directional and Morse code signals. Also, they float and can compromise the BLS if misplaced.

9-46. **Strobe Lights.** Strobe lights and other pulse-type devices are the least desirable signaling devices. Even when used in conjunction with IR filters or a directional cone, a strobe light's signal cannot be varied and Morse code cannot be used. Thus, only one signal can be sent as opposed to several: danger, abort, or delay. Personnel should use strobe lights as an absolute last resort only. However, a strobe fitted with an IR filter could serve as a good danger signal if another IR source were available to send other signals.

MISSION SIGNALS

9-47. The detachment should, as a minimum, plan for and coordinate the signals that will be used during the mission. All detachment members must understand and remember the signals that are established in the mission planning phase. Types of signals may include those discussed below.

9-48. **Safe Signal.** The scout swimmers send this signal to the main body upon completing their reconnaissance and determining the BLS area is clear. This signal should be a Morse code letter that does not consist of either all dots or dashes. This type message makes it readily distinguishable out at sea.

Chapter 9

9-49. **Delay Signal.** The scout swimmers send this signal to the main body to indicate that a temporary situation exists that requires the main body to delay its movement to the BLS. This signal can be sent before or after the safe signal has been given. It could also be given due to unanticipated activity on the beach. The signal should be Morse code and readily distinguishable from the safe signal.

9-50. **Abort Signal.** This is a signal from the scout swimmers to the main body to indicate a dangerous situation exists that will compromise the mission if the main body attempts to land. This signal is sent only if the scout swimmers can send it without compromising themselves. Although a Morse code letter can be used, it is best to use a quickly and easily sent signal such as a rapid series of either dots or dashes.

9-51. **Absence of Signal.** This signal is again from the scout swimmers to the main body. The absence of a signal indicates a condition exists on the beach that precludes safe landing. This is a time-driven signal. When none of the above signals are received within a certain time after releasing the scout swimmers, the main body uses a contingency plan that calls for meeting the scout swimmers at another location. The detachment must remember that the time required for scout swimmers to swim in, properly reconnoiter, and secure the BLS is extensive and must be stated in the contingency plans. For planning purposes, the detachment should allow 60 to 90 minutes before going to any contingency plans. The absence of a signal can also be used in conjunction with a delay signal. The OIC should activate a contingency plan if no other signal is seen for a specified time after receiving the delay signal.

9-52. **Signals Used With Pickup Craft.** When using an amphibious aircraft or surface craft for exfiltration or pickup, signals are normally exchanged in order for both parties to ensure that they are dealing with the correct people. Normally, the team initiates one signal and the pickup craft comes back with another. The team initiates because the pickup craft is considered more vulnerable and valuable at that point. Also, the team is generally signaling out to sea. The team—

- Heads out to sea on a specific azimuth.
- Sends a Morse signal after a specific time along that azimuth.
- Repeats this signal at a predetermined interval until a return signal is received from the pickup craft.

9-53. It is important to send the signals at an unhurried rate. The recommended rate is 1 second for a dot and 2 seconds for a dash with 1 second between dots and dashes. If signals are sent too quickly, the Morse letter that is intended may not be recognized and be misinterpreted as an abort or delay signal or as the wrong signal.

SIGNALING METHODS

9-54. Scout swimmers use specific signaling methods that are predesignated during mission planning. They normally use the one-lamp or two-lamp system.

9-55. **One-Lamp System.** The scout at the waterline positions himself on the beach where he can shine his light onto the water clear of the breakers. The scout should use the following procedures to ensure the main body sees the signal, whatever its angle of approach. He first sends the agreed upon signal at the proper time and interval—two times straight out to sea at the expected approach azimuth. After that, he sends—

- Two times at an angle of 010 degrees to the left of the approach azimuth.
- Two times at an angle of 010 degrees to the right of the approach azimuth.

9-56. **Two-Lamp System.** A reception committee normally uses this system. Personnel first position two weak white lights (different colors can be used as a positive identification technique) on the beach. Then they align the lights (exactly like a Range Mark) to provide a line of approach for the main body and to indicate the BLS. The committee will—

- Position the lower light as in the one-lamp system and use it in the same manner. Hold the light about 1.5 meters above the sea. In a variation of the system, stake it in place at that height.
- Secure the second light (which normally remains switched on continuously) to a stake. Position it 6 meters behind the lower light and 1 meter above it. Normally, this is a weak light that can be a different color from the lower one.

Note. If the distance between the two lights is more than 6 meters, personnel should ensure the gradient between the lights is as close to 1:6 as possible.

DETACHMENT INFILTRATION SWIMMING

9-57. Detachments may occasionally encounter circumstances that preclude landing on the beach with CRRCs or kayaks. In case the detachment cannot land, they must be prepared to swim from the launch point to the beach. Detachment infiltration swimming should only occur as a last resort. Infiltration swimming requires allowing additional transit time, higher physical exertion levels, greater environmental exposure, and reduced quantities of mission-essential equipment. It is the least favorable infiltration technique.

9-58. To maintain team integrity, combat diver and combat surface swims are conducted using a swim line. Once the infiltration team has arrived at the launch point, it is recommended that all swimmers hook up on a swim line before beginning the swim. This practice is a control measure designed to maintain unit integrity. An advantage of the swim line is that even if the environment adversely affects the team, the entire team and its equipment will remain together. There are many varieties of swim lines, each offering advantages and disadvantages. The most common techniques are discussed below.

SWIM LINES

9-59. There are generally two types of swim lines for infiltration swimming—on-line or column variation. The column method can also be conducted in two ways (Figure 9-1). The swim line can be constructed of climbing rope, tubular nylon, suspension line, or any other strong material. Tubular nylon is preferred because it is light but strong and is far less bulky than climbing rope. Swimmers should remember never to use climbing rope for actual climbing or rappelling after it has been exposed to salt water.

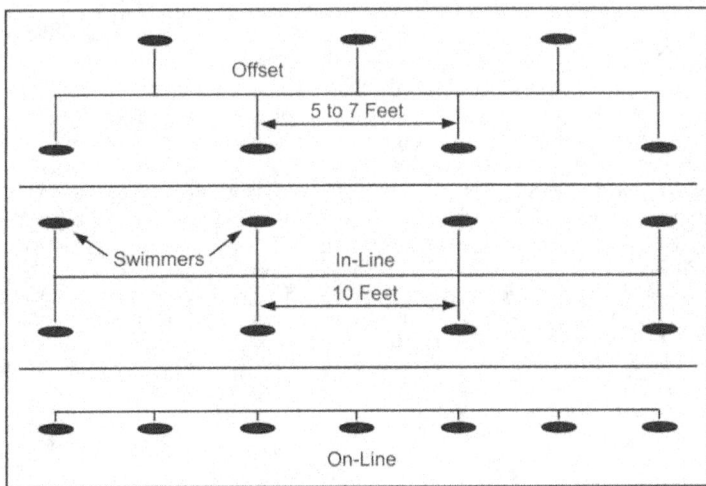

Figure 9-1. Swim line formations

9-60. **Column (Offset).** This method is one of many variations of the standard in-line column formation. Its major advantage is that the length of the line can be reduced to half the normal size while still allowing 10 feet between swim teams on the same side of the line. Swimmers primarily use this formation for underwater swims because it eases control of the swim team underwater.

9-61. **Column (In-Line).** This formation is used for combat surface or subsurface swimming. The line can be a solid straight rope with loops placed about every 10 feet. It can also be 10-foot sections of line with end loop splices at each end, then fastened together with snap links. The swim team buddy lines are then routed through the loops or snap links on the centerline. All swimmers then follow the azimuth and pace of the lead swim team.

9-62. **On-Line.** Swimmers most frequently use the on-line swimming formation for combat surface swimming. The actual line is formed by using a solid straight rope with hand loops spaced 5 to 7 feet apart. Each team member positions himself at a hand loop location. The team then swims on-line toward its target area.

SURFACE SWIMMING

9-63. Regardless of the swim formation that the team uses, certain surface swimming concepts remain constant. The team members swim on their backs or sides using the flutter kick. They wear enough weights to ensure their flutter kick remains subsurface, thereby maintaining a low profile and minimizing surface disturbances. Alternating the swimmers on a swim line or when swimming in two-man teams, allows each team member to observe across the other's back area and to the front. Swimmers can switch sides during the swim for partial physical recovery, because one leg tends to work harder than the other. This gives the body a partial rest when sides are switched and allows maintaining a steady swim pace. Equipment bundles are towed on lines held by the swimmers. The lines must be long enough not to interfere with the swimmer's flutter kick.

9-64. An alternative swimming style incorporating the use of the swimmer's arms is the combat sidestroke with fins method. World War II amphibious reconnaissance platoons sent in from submarines to reconnaissance islands in the South Pacific originally developed the technique. It consists of swimming with fins on feet, while executing the sidestroke. The swimmer should be on his side with arm extended out in front and one side down as a normal sidestroke; he then pulls the arms into the body's center, meeting at the chest. He can extend his arms back out in an underwater recovery stroke method as to not disturb the water, while the feet fin in a normal flutter-kick fashion maintaining position with one side down. Sprint speeds in excess of 1 kt are achievable by an individual who is proficient in the technique. This speed aids in overcoming littoral currents that often run parallel or diagonal to the beach.

9-65. Swimmers waterproof- and buoyancy-test any bundles to make sure they are neutral buoyant and likewise maintain a low profile in the water. They can wear LCE in the normal manner and their individual weapons are heavily lubed and slung across the chest with an over-the-shoulder sling. Very small rucksacks or equipment bags can be worn over the chest with the arms through the shoulder straps. If the swimmer desires, he can carry his weapon by attaching it to the rucksack; however, this is not recommended.

9-66. All individual swimmers must wear flotation devices, fins, appropriate exposure suits, knife with flares, and weights as required. Some type of swim or water sports goggle can be used to increase the surface swimmer's comfort level, especially when swimming in salt water. Diving facemasks with their large lenses are not normally worn when surface swimming due to the high probability of reflected light being seen. If the facemask is worn, it must definitely be removed well outside the surf zone and at a safe distance from the BLS. Designated swimmers should be equipped with watches and compasses. If a compass is required, the swimmer should remember that when swimming beyond the horizon on his back, he must use a back azimuth.

DIVING

9-67. For underwater operations, the team swim line can remain the same except that buddy teams must remain close together to be able to readily aid a diver in trouble. Normally, the column formation is used (Figure 9-1, page 9-9). To maintain the proper depth, pace, and azimuth of the lead team, at least one member of the buddy team must keep his hand on the main centerline. This hold enables the diver to alter his pace (speed up or slow down) by feeling the amount of slack or tautness in the centerline. He can also maintain direction and depth of the lead team.

WATERPROOFING

9-68. Waterborne operations, other than limited reconnaissance missions, require the infiltrating detachment or swimmer to transport equipment to and from the objective. Depending on the mission, this equipment will vary from light and compact to bulky. In every case, equipment must be limited to the absolute minimum necessary to accomplish the mission.

PROCEDURES

9-69. The size, weight, and quantity of bundles are dependent on the size of the team, distance to swim, and method of insertion into the objective area. Swimmers should use the following guidelines:

- *Size.* Bundle size should be limited to a cube no larger than 14" x 30" x 18." This size enhances diver comfort, reduces drag in the water, allows easy handling, and presents minimal problems for the swimmer if held correctly. Whatever container the equipment is transported in must also be carried on land to the ultimate inland objective or cached.
- *Weight.* Before swimming the bundle, the swimmer will adjust its buoyancy by either adding weight or putting floatation devices on it to make it neutral or slightly positive. Each bundle must be man-portable.
- *Quantity.* The number of bundles will be determined by dividing the mission-essential equipment into the size and weight restrictions previously identified.

PRINCIPLES

9-70. Waterproofing is necessary to ensure items will operate once they are used on the objective. Electronic equipment is particularly susceptible to water damage. Using the following principles when waterproofing equipment will make sure it functions reliably when needed:

- Do not waterproof anything that does not require it.
- Cushion all sharp edges of equipment in a bundle to prevent punctures.
- Try to use waterproof plastic or rubber sacks. Place item in it, withdraw all air, and seal.
- Disassemble equipment into component parts. Small items are easier to waterproof.
- Identify all items with tags or tape on the outside of the bag to aid in quick assembly.
- Waterproof items more than once. (Triple-consecutive containment is recommended.)
- Swim at shallow depths to minimize pressure and reduce potential water infiltration.
- Maintain equipment integrity within the bundle to facilitate putting the equipment into service once on shore.

9-71. Waterproofing of materials and equipment is an essential task when preparing for waterborne operations. The procedures discussed herein are not all inclusive; however, they do provide a basis from which to develop additional techniques peculiar to unit operations. All items of individual and table of organization and equipment are waterproofed basically in the same manner.

9-72. Waterproofing materials are divided into two groups: improvised and issued. Improvised items can be any type of plastic bag, ammunition can, tape, wax, or prophylactics. Issued items include commercial off-the-shelf or purpose-built items and waterproof bags available through the GSA Catalog. Examples include rucksack or daypack liners (NSN: 8465-01-487-3183), transport and shoot-through weapons bags, radio bags, and general purpose kit bags. They are superior in all ways to improvised systems based on the olive drab waterproof bags (issued with the rucksack) or waterproof bags issued with magazines, radio battery bags, mortar round canisters, and the waterproof bag for medical gear. These bags also offer simplicity of use and durability over improvised waterproofing systems.

9-73. Personnel should remove waterproofing one layer at a time. They should check to ensure that no water has leaked in any of the layers. If freshwater has leaked in, it should be wiped off and all metal parts should be oiled. If salt water has leaked in, it should be rinsed off with freshwater ASAP and should then be dried and oiled. If freshwater is not available, the article should be oiled well and checked for rust. If ropes or clothing are wet from salt water, they must be washed with freshwater ASAP. Salt water will cause material to rot relatively quickly and the dried salt cuts into rope fibers.

Chapter 9

RADIOS

9-74. To waterproof a radio, a swimmer first identifies all sharp edges. He removes the rubber boot from the handset plugs and checks to make sure the radio is operational. He should always install new batteries. The swimmer pads all sharp edges using the clothing in the rucksack; he does not use separate padding material unless absolutely necessary. This reduces the sterilization problem once in the assembly area. He leaves enough room so that the controls can be reached and used. Once he ties or tapes the padding in place, the radio is then put into the first waterproof bag with the controls and antenna or handset plugs in the bottom of the bag. The swimmer removes all air by either sucking it out by mouth or by using a small boat pump in the suction mode. Once the air is removed, he twists the remainder of the bag into a gooseneck, bends it back on itself, and ties it securely. He makes sure there is enough room at the bottom of the bag to turn the radio on and off with the switch between preset frequencies without breaking the bag. He repeats the same steps with the second and third waterproof bags. For best results, the first two bags should be heavy-duty plastic bags and the last one should be an issue bag so that the strongest and most durable bag is on the outside. The third bag provides abrasion and puncture protection for the first two bags. The swimmer then waterproofs the handset by placing it inside a small waterproof bag, such as a plastic battery bag or a plastic sandwich bag, and repeats the steps outlined above.

WEAPONS

9-75. The individual weapon must be waterproofed so that it is functional. The primary reason for waterproofing weapons is to keep sand and salt out of the working parts (a significant problem when swimming through heavy surf), and not necessarily to keep water out. The two primary techniques that can be modified for crew-served weapons are discussed next.

9-76. The simplest method is to place the weapon in a purpose-built weapons bag. There are two types: a shoot-through bag designed for individual weapons and a transport bag designed for crew-served or special-purpose weapons. These are available through the supply system or the GSA Catalog. The swimmer pads the weapon as required to protect the weapon and the bag. If the weapon is going into a shoot-through bag, he makes sure the weapon functioning is not compromised and the operator can still manipulate the weapon's controls.

9-77. If weapons bags are not available, then the swimmer must protect the weapon with improvised packaging. He tapes the muzzle or tapes a muzzle cap in place and places the weapon into a waterproof plastic bag. He then twists the bag into a gooseneck and ties it off as described above. He tapes the bag to the weapon so that it conforms essentially to the shape of the weapon, but allows the moving parts to function. If the weapon must be fired, the swimmer pushes his index finger through the bag and manipulates the selector switch from the outside. He fires the weapon through the bag. When taping the bag to the weapon, the swimmer leaves enough space around the ejection port so that ejection can occur and no stoppages are caused.

9-78. If the swimmer decides not to waterproof the weapon, he must still make sure it can resist the infiltration of sand into the mechanism. He tapes a muzzle cap in place or tapes the muzzle itself to prevent water from entering the barrel. He wraps plastic bag material around the receiver group allowing room for the charging handle selector switch and trigger to operate. He also leaves room for expended rounds to eject.

BUNDLE TRANSPORTING

9-79. It is recommended that the bundle be towed or carried in the swimmer's hands. It should not be attached to his body except in parascuba operations and then only for the duration of the jump. As a safety precaution, the swimmer should be able to quickly jettison heavy bundles that would hinder a safe ascent to the surface in an emergency situation. An attack board should be used when swimming bundles underwater.

Chapter 10

Combat Diving and Swimming Operations

This chapter addresses duties and safety for personnel and units conducting combat swimming and diving operations. Its purpose is to ensure interoperability among all USSOCOM elements and to promote operational readiness and safety through a uniform execution of policy and training.

Note. The U.S. Army and the U.S. Air Force use the term combat divers; the Marine Corps uses the term combatant diver. These terms describe the diving requirements of SOF personnel primarily engaged in infiltration swimming. The U.S. Navy uses the terms combat swimming and combat swimmer.

ORGANIZATION AND DUTIES

10-1. Every SF individual involved in combat swimmer and diving operations must attain and maintain a high state of mental and physical alertness and readiness for the safe and successful execution of each operation. This chapter outlines the minimum duties of key personnel in the planning, preparation, and execution of combat swimmer and diving operations by USSOCOM components.

THE NAVY DIVING ORGANIZATION

10-2. Combat diving is fundamental to all SF diving. The Navy trains the majority of military personnel in combat diving skills and has therefore been assigned proponency for diving throughout the Department of Defense (DOD). Accordingly, USSOCOM assigned proponency for combat diving operations to the NAVSPECWARCOM in USSOCOM Directive 10-1, *Terms of Reference*. Because combat diving is an inherently maritime activity, clear understanding of certain Navy instructions (the Navy uses "instructions" rather than regulations or orders) and organization is the key to full joint integration and standardization of combat diving within SF. The following paragraphs outline the manner by which the Navy has assigned duties with regard to the planning and conduct of diving operations.

Commander, Naval Special Warfare Command

10-3. The NAVSPECWARCOM is located in Coronado, California, and is the naval component of USSOCOM. The USSOCOM commander has designated the Commander, NAVSPECWARCOM as the proponent for maritime SO.

Supervisor of Diving

10-4. The Navy supervisor of diving (SUPDIVE) is located in Arlington, Virginia, and works for the Director of Ocean Engineering within the Naval Sea Systems Command. This is the primary Navy office that is concerned with diving operations and publishes manuals on the subject.

Naval Sea Systems Command

10-5. The Naval Sea Systems Command (NAVSEA) is located in Arlington, Virginia, and is the major Navy organization (called a Systems Command) which oversees all naval diving matters ashore and afloat except for shore-based diving facilities (for example, hyperbaric chambers). The NAVSEA command certifies dive systems that are used afloat. The SUPDIVE is an office of the NAVSEA.

Naval Facilities Engineering Command

10-6. The Naval Facilities (NAVFAC) Engineering Command is the major Navy organization responsible for design, construction, and maintenance of naval shore facilities. NAVFAC certifies dive systems that are used ashore (for example, shore-located hyperbaric chambers).

Naval Special Warfare Center

10-7. The Naval Special Warfare Center (NAVSPECWARCEN) is located in Coronado, California, and is the Navy's school for special operations. Among other courses of instruction, it teaches the Basic Underwater Demolition/SEAL (BUD/S) course. The NAVSPECWARCOM commander has designated NAVSPECWARCEN as the executive agent for SF open- and closed-circuit diving issues. These duties are comprehensive and include issues such as open-, closed-, and semiclosed circuit diving; combat diver operations; diving equipment; dive policies; dive planning; diving mobile training teams; and diving systems certification as directed by NAVSPECWARCOM instructions.

Naval Safety Center

10-8. The Naval Safety Center (NAVSAFECEN) is located in Norfolk, Virginia, and works closely with the NAVSPECWARCEN concerning safety aspects of combat diving and certification procedures for all SO diving equipment. The NAVSAFECEN conducts comprehensive safety inspections of USSOCOM Divers Life Support Maintenance Facilities (DLSMFs).

Navy Experimental Diving Unit

10-9. The Navy experimental diving unit (NEDU) is located in Panama City, Florida. Before recommending that diving equipment be approved as Authorized for Navy Use (ANU), the Navy SUPDIVE refers the request to the NEDU for testing and recommendations.

SPECIFIC DUTIES

10-10. The NAVSPECWARCOM commander is designated as the proponent for USSOCOM combat swimmer and diving matters. As such, NAVSPECWARCOM develops USSOCOM policy as it relates to combat swimming and diving matters and serves as the SOC's focal point for these issues. The personnel who participate in these operations are explained below.

Unit Commanders and Commanding Officers

10-11. The commanders and commanding officers of USSOCOM subordinate commands that participate in diving operations will—

- Ensure all diver life support systems, support equipment, and diving life support maintenance facilities are maintained IAW instructions from the Chief of Naval Operations and adhere to Navy certification standards.
- Ensure that divers use only systems and equipment that have been certified or ANU, unless a waiver has been obtained through the Chief of Naval Operations. Naval Sea System Command instructions (NAVSEAINSTs).
- Maintain diving system certification IAW NAVSEAINSTs.
- Ensure that diving is conducted only by qualified personnel following approved procedures.
- Ensure all assigned diving personnel meet the physical standards set forth in appropriate Service directives, regulations, and instructions.
- Develop and implement command training plans to ensure adequate training of divers in all unit diving mission areas. Requalification and training will be appropriately documented in individual service records, as appropriate, and in unit training files.
- Ensure a Dive Log (DD Form 2544) is maintained for all dives conducted by the command. This log is an official record and will be retained for three years. Ensure the dive log is completed for all dives and submitted to the NAVSAFECEN IAW Office of the Chief of Naval Operations

isntructions (OPNAVINSTs) and applicable supplemental instructions. In addition, individual divers will maintain a personal dive log.
- Ensure diving mishaps resulting in death, lost time (24 hours or more), personnel injury, recompression treatment, or significant material damage are reported within 24 hours to the NAVSAFECEN IAW OPNAVINSTs. The appropriate USSOCOM component command will also be notified ASAP following the incident via telephone and followed by a written report, as required.
- Establish procedures to ensure that diving equipment (which may have contributed to an accident) is secured, not tampered with, and shipped by the fastest traceable means to the NEDU for analysis. This procedure should specify that the equipment shall not be dismantled, cleaned, or altered in any way before shipment.
- Conduct annual operational, administrative, and material inspections of diving units or elements to verify compliance with this directive and other appropriate diving regulations or instructions.
- Designate, in writing, diving officers, supervisors, and other personnel as required by appropriate regulations or instructions.

Diving Officer

10-12. The commander or commanding officer designates the diving officer to oversee the safe and efficient conduct of diving operations. In addition to his duties outlined in U.S. Navy diving instructions and amplifying Service directives, regulations, and instructions, the diving officer will—
- Monitor diving operations to ensure compliance with established policies and procedures. Be responsible for the safe conduct of all diving operations.
- Ensure that the qualifications of all personnel participating in diving operations remain current IAW Service directives.
- Establish a dive training program. Monitor lesson plans and lectures to ensure correct and proper information is being disseminated.
- Ensure, by observation and routine inspection, that safe operation and maintenance procedures are performed on all diver life support systems, maintenance facilities, and associated equipment.
- Ensure that diving supervisors complete all diving forms, logs, and checklists, as required by OPNAV instructions, current U.S. Navy regulations, technical manuals for the type of diving equipment used, this manual, and amplifying regulations or instructions.

Diving Medical Officer

10-13. The diving medical officer (DMO) is a physician or physician assistant trained at the Diving Medical Officer School at Panama City, Florida, who maintains currency IAW appropriate Service directives. The DMO will—
- Review all diving physicals and determine if individuals are physically qualified to perform diving duties.
- Be present during buoyant ascents, free ascents, submarine lockout and lockin training, and pressure and oxygen tolerance testing.

Officer-in-Charge and Noncommissioned Officer-in-Charge

10-14. All diving evolutions require a designated individual in charge. An E-7 or above normally oversees SF operations that include diving as part of a larger operation, exercise, or training. The OIC or NCOIC will plan the evolution using the assigned diving supervisor to supervise the actual diving portion of the operation. The OIC or NCOIC and the diving supervisor may be the same individual if the OIC or NCOIC is a qualified diving supervisor. The divers will receive all commands and direction from the diving supervisor during his period of responsibility. The OIC or NCOIC may relieve the diving supervisor of his duties if the OIC or NCOIC is himself a qualified diving supervisor or if another qualified diving supervisor is immediately available. The OIC or NCOIC is not required to remain on the surface.

Diving Supervisor

10-15. The diving supervisor must be an E-5 or above who is thoroughly familiar with the equipment, conditions, safety precautions, and hazards inherent to diving operations. Diving supervisors will be designated in writing by his commander or commanding officer. He will be a qualified diver and have successfully completed an approved NAVSPECWARCEN or USAJFKSWCS diving supervisor course. Waivers to the above qualifications must be approved by the appropriate USSOCOM component commander after policy coordination with the NAVSPECWARCOM commander. The diving supervisor will perform the duties outlined in the current U.S. Navy regulations governing diving operations and amplifying Service directives, regulations, and instructions. The diving supervisor will—

- Be responsible to the diving officer for the safe and efficient conduct of diving evolutions. He will be in charge of the actual diving operation; no diving operations will be conducted in his absence and he will remain on-scene until all divers are out of the water.
- Coordinate and plan all aspects of the diving operation, ensure proper clearance is obtained from appropriate higher authority, identify safety hazards, and develop emergency procedures during the planning phase to ensure safety and the success of the mission. When planning the dive, the diving supervisor will use the diving safety checklist, emergency assistance checklist, diving boat safety checklist, and the ship repair safety checklist for diving found in the U.S. Navy diving regulations. Other derivative and locally prepared checklists may be used.
- Advise the OIC or NCOIC of the operation on all matters pertaining to the planning, execution, and safety of the dive.
- Conduct the dive brief.
- Assess the physical readiness and qualifications of the divers with the assistance of the DMO and the DMT, as available. All divers will have current diving physicals as prescribed by appropriate Service directives.
- Ensure that support personnel are present for all briefings and are adequate to support the dive. Inspect all support and special equipment to ensure that it is in proper working condition and able to perform its intended function.
- Personally supervise the setup and testing of all diving equipment.
- Ensure all divers receive a diving supervisor check before entering the water.
- Know the location and operational condition of the nearest recompression chamber and status of the chamber crew; for example, chamber supervisor, outside tender, inside tender, and log keeper. Ensure that arrangements have been made for expeditious transportation of any casualties and for contacting a DMO.
- Have the requisite equipment needed to perform his duties and responsibilities, to include meeting any likely emergency situation.
- Assign one or more assistant diving supervisors to perform his duties, as directed, when the nature of the operation requires more than one diving platform; for example, launch and recovery from different locations or a large number of divers.
- Ensure proper diving signals (day or night) are displayed when dive training operations are in progress.
- Ensure that divers maintain a safe distance from energized sonar IAW NAVSEAINSTs and that sonars that present a potential hazard are tagged-out and in the passive mode.
- On repetitive dives, check each diver's remaining air and oxygen supply and residual nitrogen times before divers leave the surface. When diving closed-circuit scuba (for example, MK 25), ensure the CO_2 absorbent is changed per Navy and manufacturer's technical manuals and NAVSEA directives.
- Be the single point of contact during any diving emergency.
- Supervise postdive cleaning and storage of all dive and support equipment.
- Ensure that all dive logs are properly completed, signed, and submitted to the diving officer.

Assistant Diving Supervisor

10-16. When assigned, the assistant diving supervisor will—
- Be a qualified diving supervisor, or a member who has completed an approved diving supervisor course and is under the supervision of a qualified diving supervisor for training purposes.
- Perform those functions directed by the diving supervisor.

Diving Medical Technician

10-17. The DMT must be a graduate of a recognized course of instruction for qualification as a DMT or special operations technician (SOT). He will—
- Be present for all mixed gas and closed-circuit dives. His presence during open-circuit dives will be at the discretion of the diving supervisor, unless otherwise required by amplifying Service directive.
- Provide medical care and treatment to divers on a routine basis or in case of a diving emergency. Accompany any injured party to the treatment facility if additional treatment is necessary.
- Have the proper equipment, as prescribed in current U.S. Navy diving regulations, to resuscitate a diver and provide first aid.

Standby or Safety Diver

10-18. The standby or safety diver will be a fully qualified diver whose functions are to provide emergency assistance to divers and to perform the duties outlined in current U.S. Navy diving regulations. Additional duties require him to be—
- Present during the dive briefing and operation. The standby or safety diver should be qualified in the type of scuba being used by the divers in the mission or training evolution.
- Knowledgeable of the rescue procedures for the type of scuba equipment being used by the divers.

Note. The standby diver will wear open-circuit scuba equipment regardless of what equipment the divers are using, unless directed otherwise by the diving supervisor. Current U.S. Navy diving regulations outline the minimum equipment required for scuba diving.

- Positioned as best possible to render assistance in the event of an emergency. He must be fully dressed for immediate entry into the water after receiving instructions to do so from the diving supervisor. (The regulations or instructions of the USSOCOM component commanders may authorize the standby diver to be fully dressed with the exception of scuba, fins, and face mask, which will be staged and ready for quick donning.)

Note. Single standby divers or safety divers will be tended by a surface crew with a tending line.

Divers

10-19. A diver is qualified for the type of scuba to be used during the dive. Basic qualifications and physical standards are described in current U.S. Navy diving regulations. He will also—
- Be present at the dive brief.
- Be physically and mentally prepared for each dive.
- Have personal equipment prepared for each dive.
- Properly set up scuba equipment, using appropriate check sheets.
- Promptly obey all diving signals and instructions received from the diving supervisor.
- Promptly report equipment malfunction or damage to the diving supervisor.
- Observe the buddy system as outlined in current U.S. Navy diving regulations and oversee the safety and welfare of his buddy.

Ship Safety Observer

10-20. A ship safety observer is assigned for simulated combat diver ship attacks. He ensures the target ship is safe to dive under and completes the Ship Safety Checklist (found in current U.S. Navy diving regulations). He acts as liaison between the target ship and the diving supervisor, and informs—
- The diving supervisor of any discrepancies in the Ship Safety Checklist.
- Other ships and craft nested with or along the same pier as the target ship of the dive plan.
- Ships on adjacent piers and pier sentries of the dive plan.
- All ship participants when the operation is secured.

TRAINING

10-21. To ensure safe diving operations, all divers must be skilled and proficient. This section provides broad guidance on the conduct of SF diving training to include initial training and qualification and the integration of diver training into unit training plans.

INITIAL TRAINING AND QUALIFICATION

10-22. Details concerning qualification standards and requirements are contained in the appropriate Service directives. Each level of training is discussed below.

Diver

10-23. A diver must be a graduate from a formal military dive course. Service divers who are not designated as combat divers will not normally participate in SF diving training or operations. Exceptions will be authorized by the appropriate USSOCOM component commander for the purpose of rendering specialized support (for example, dry-deck shelter [DDS] operation), documentation of training (for example, combat camera teams), or observing SF diving operations. A member is qualified as a combat swimmer or diver by completing one of the following:
- Graduate from the USAJFKSWCS Combat Diver Qualification Course.
- Graduate from the BUD/S Course.
- Graduate from a military dive course and subsequent attendance at NAVSPECWARCEN or USAJFKSWCS Combat Diver Course for training on the Draeger MK 25 scuba.
- Graduate from the Marine Combatant Diver Course (MCDC).

Sustainment Training or Unit Training

10-24. After initial qualification, it is necessary to sustain and build upon the skills acquired as individuals during diver qualification courses. Additional training is discussed below.

10-25. Refresher and sustainment training will consist of practical training, lectures, and classroom instruction. Units should maintain lesson plans and review or update them annually (Appendix B). Regular command training for all divers should include, but not be limited to, the following:
- Lecture. Information covered includes the following:
    - Underwater physics.
    - Underwater physiology.
    - Basic diving procedures for scuba.
    - Diving air decompression tables.
    - Diving hazards.
    - General safety precautions.
    - Reports and logs.
    - Standard diving equipment.
    - Open-circuit scuba.
    - Closed-circuit scuba.
    - First aid.

- Cardiopulmonary resuscitation.
- High-pressure air compressor (stationary and portable).
- Air sampling program.
- System certification procedures.
- Hyperbaric chamber operation.
- Hyperbaric treatment tables.
- Decompression sickness (Type I and II).
• Practical Work. Tasks include the following:
  - Maintenance and repair of all types of scuba used by the unit.
  - Repair and maintenance of diving air compressors.
  - Demonstrate working knowledge of decompression tables and, where appropriate, combat swimmer multilevel diving procedures.
• Diving. Operations might include the following:
  - Subsurface infiltrations and exfiltrations using open- or closed-circuit systems during day and night.
  - Underwater search and recovery operations.
  - In-water emergency procedures.

10-26. Details concerning diver requalification standards and requirements for Army divers are contained in AR 611-75, *Management of Army Divers*, and for Marine combatant divers in MCO 3150.4. A brief summary follows:
- Each Service prescribes the type, number, and periodicity of dives required to maintain currency.
- If a diver's currency has lapsed for 6 months or less, he may regain currency by fulfilling the requirements prescribed for requalification. Dive pay stops the date currency lapses; it restarts the date the diver becomes current again.
- If a diver's currency has lapsed for longer than 6 months, he must be retrained and recertified IAW established Service procedures.

MISSION-ESSENTIAL TASK LIST FOCUS

10-27. Unit dive training will be METL-focused to the maximum extent possible. Dive training plans will be unit-specific. Training will be progressive to attain and maintain the skills required to conduct dive operations in support of assigned missions in projected operational environments. Collective dive training will be integrated with other METL-focused training as much as possible. The use of realistic FTXs based on FMPs to train and evaluate the unit's divers should be the norm rather than the exception. Standards for various dive evolutions are delineated in U.S. Navy fleet exercise publications.

ADVANCED TRAINING QUALIFICATIONS

10-28. Additional training and certification is necessary to perform certain supervisory and medical support functions required to conduct SF diving operations. The training and qualifications for these positions are prescribed by appropriate Service directives. The positions are as follows:
- *Diving Supervisor.* He must be an E-5 or above who is thoroughly familiar with the equipment, conditions, safety precautions, and hazards inherent to SO diving. He will be a qualified diver and have successfully completed an approved NAVSPECWARCEN or USAJFKSWCS Diving Supervisor course.
- *Diving Medical Technician.* He will be a qualified diver and a graduate of a recognized course of instruction for qualification as a DMT or SOT.

OPERATIONS

10-29. SF diving, with the exception of specific missions assigned to NSW forces, can be described as an insertion or exfiltration technique. U.S. Army, Air Force SOF, and USMC divers use diving as a means of

clandestine infiltration and exfiltration and, as such, treat diving as a phase of a larger operation. NSW forces also use diving as an infiltration and exfiltration technique; however, as USSOCOM's maritime component, they are tasked with a myriad of maritime missions that require more sophisticated diving skills. Other NSW diving operations include the following:
- Combat swimmer ship attack (Limpeteer attack).
- Underwater demolition raids.
- Harbor reconnaissance.
- Submerged hydrographic reconnaissance.
- Underwater obstacle demolition or mine countermeasures.

ORGANIZATION

10-30. The organization for a particular operation is dependent on the nature of the mission, the type of diving to be conducted, and the unit's SOPs. The minimum supervisory and support personnel required by current U.S. Navy diving regulations will be present for all SF diving operations. When organizing for the dive, it is important to match unit and individual qualifications and experience to the specific requirements of the operation or mission. Unit integrity is an important consideration when conducting diving operations. However, there are times when attachments to a unit may be necessary to fulfill its mission. In such cases, it is imperative to fully brief all participants in the operation regarding unit SOPs to ensure that everyone knows what procedures to follow in the event of an emergency.

PLANNING

10-31. Preliminary planning, as outlined in current U.S. Navy diving regulations, is vital for a successful dive. Without adequate planning, the entire operation may fail and, in extreme cases, the lives of the divers may be endangered. The following information describes plans for a safe and successful dive operation and is appropriate for training as well. It is not a guide to tactical mission planning. The planning phase of a diving operation consists of the following tasks:
- *Surveying the Problem.* The first step is to study all aspects of the diving operation to be conducted and how it fits into the broader plan for the entire mission. After framing the entire diving operation and defining the objectives, select an approach to the problem.
- *Choosing Diving Equipment.* Planners should consider the requirements to solve the problem—range, depth, thermal protection, communication, decompression, and similar needs. Factors that must be considered before making a decision are—
  - Underwater visibility.
  - Tides and currents.
  - Water temperature.
  - Water contamination.
  - Bioluminescence.
  - Surface conditions (wave or surf action, air temperature, and visibility).
  - Qualifications and experience level of divers.
  - Available diving equipment (safety boats, recompression chamber, resuscitator, personal flotation devices, diver propulsion device).
  - Adequate diving platform (can the divers enter or return aboard without difficulty?).
  - Ability to recover a diver in distress.
  - The necessary items to mark the location of a missing diver.
  - Any other conditions that are required to safely support the particular diving operation.
- *Selecting Equipment.* Equipment selection depends on the type of evolution to be undertaken and the availability of specific equipment. A Diving Equipment Checklist is provided in current U.S. Navy diving regulations.
- *Establishing Safety Precautions.* The applicable safety precautions must be determined by considering the—

- General precautions for scuba diving.
- Specific precautions for the particular type of scuba to be used, to include hazardous materials such as oxygen and O2 absorbents (soda lime).
- Safety precautions particular to the operation.

Note. If time and the nature of the operation permits, a local Notice to Mariners should be issued. Furthermore, when routine diving operations are to be conducted in the vicinity of other ships, they must be properly notified by message.

- *Briefing the Divers.* A thorough briefing will be given for each dive evolution. All proper precautions against foreseeable contingencies should be thoroughly covered in the briefing. It must be remembered that a great percentage of the work involved in a diving operation should be completed topside through proper planning. Inadequate briefings and supervision can, and have, caused fatal accidents.

DIVING

10-32. SF combat diving will be conducted IAW the guidance provided in current U.S. Navy diving regulations and the safety considerations listed in USASOC Reg 350-20. Additional guidance for specific diving techniques and types of operations can be found in Chapter 11 of this manual.

SAFETY

10-33. Diving operations are inherently hazardous. Proper training and in-depth preparation and planning will reduce the potential for accidents. Safety considerations will be fully addressed during planning and integrated into all dive training. Unit commanders will conduct risk assessments before training and training will be conducted in a progressive crawl-walk-run manner. Specific guidance for conducting risk assessments is contained in FM 100-14, *Risk Management.* USASOC units must also reference USASOC Reg 385-1, *Accident Prevention and Reporting.* The arbiter of safety questions is the USASOC Safety Office.

RESPONSIBILITIES

10-34. Safety is every individual's duty if participating in or supporting the diving operation. All personnel involved in a particular dive evolution, operator and support personnel alike, must be constantly aware of the nature and progress of the operation, and must remain alert for possible danger.

ENVIRONMENTAL PERSONNEL LIMITATIONS

10-35. During the planning process, consideration must be given to how weather, temperature, surf conditions, and sea state will affect the safety of the dive evolution. Chapter 6 outlines the minimum equipment requirements. Environmental limitations on divers and equipment are found in current U.S. Navy diving regulations.

OPERATIONAL PROCEDURES

10-36. The element of danger is always present during SF diving operations. Training and operations must be planned with consideration given to individual and unit experience, swimmer fatigue factors, and repetitive diving times. General dive procedures are provided below as guidelines for the safe conduct of SF dive operations. (Specific guidance for operations such as combat swimmer ship attack, hull searches, turtlebacking and tactical purge procedures, and diving with foreign units is provided in USSOCOM Reg 350-4.) Adherence to these safety procedures will enhance the probability of success during combat operations and reduce the incident of diving casualties:
- All diving conducted by USSOCOM forces will follow the procedures of the *U.S. Navy Diving Manual* and the amplifying guidance set forth in AR 611-75 and USASOC Reg 350-20, without exception.
- Only equipment that is certified or ANU IAW NAVSEAINSTs will be used during diving operations, unless a waiver has been granted by the appropriate authority.

- Every dive will be preceded by a dive brief to be attended by all personnel involved in the dive. If key support personnel are unavailable to attend the dive brief, then the diving supervisor will ensure they are briefed separately.
- It is mandatory-that each diver use the appropriate predive check sheets for setting up his scuba. Each diver will sign the check sheet and ensure that his check sheet is signed by the diving supervisor. Sample check sheets are provided in USASOC Reg 350-20.
- A standby diver is mandatory for any diving operation. He need not wear the same scuba as the divers, but he must have the same depth capabilities and be able to enter the water immediately after being briefed by the diving supervisor. He should be qualified to dive the scuba being used by the divers in the mission or training evolution.
- No dives will be conducted through a surf zone if the surf zone contains more than three lines of waves with a wave height of more than three feet, unless otherwise directed by the unit commander. Dives conducted through plunging surf of any size require careful consideration of diver safety.
- A diver recall device is mandatory equipment for all dives. Lists of approved recall devices are published by NAVSEA.
- A buddy line will be used between divers as required by current U.S. Navy diving regulations, unless otherwise authorized by the unit commander.
- A marking float should be used whenever possible to mark the location of divers in the water. At the discretion of the diving supervisor, a light source may be attached for easier location during night dives.
- A life jacket with a whistle attached to the oral inflation hose and the proper CO_2 cartridges or air bottles in place will be worn during all diving operations. There will be no quick releases in the body straps of the life jacket. CO_2 cartridges will weigh no less than 3 grams below their original weight and will be weighed before each use.
- All scuba cylinders used during diving operations will be charged to at least 75 percent of the working pressure. The diving supervisor will ensure that sufficient diver's breathing medium remains to conduct any repetitive dive.
- Both individuals in a dive pair will use the same type of scuba.
- The diving supervisor and corpsman must have flashlights or diving light when conducting diving operations between sunset and sunrise. Chemlites, dive lights, or flares are mandatory for each diver.
- Radio communications will be maintained on the scene between all safety boats, the diving supervisor, ship safety observer, if designated, and the parent command for immediate notification in case of emergency.
- Before allowing divers to enter the water in the vicinity of or beneath surface vessels or submarines, the diving supervisor shall make contact with, and obtain clearance from, the ship safety observer aboard the vessel.
- Free swimming ascent (FSA) and buoyant ascent (BA) are considered to be emergency procedures and should not be included in a diving evolution, unless—
    - The sole purpose is to train divers in FSA and BA.
    - Properly trained instructors are on station and in control of the divers' situation from air source to the surface.
    - A certified recompression chamber is on station and immediately available for use.
    - A DMO is on scene and can be summoned to the recompression chamber within 5 minutes.
    - Divers are checked by a DMT or SOT immediately upon surfacing.
- Divers should not exceed the normal rate of descent (75 feet per minute [fpm]) or ascent (30 fpm).
- If a diver senses any adverse physiological symptoms or mechanical malfunctions, he should surface immediately. Never force continuation of a dive during training evolutions.

- A weighted ascent or descent line is required for all dives deeper than 100 feet. The following procedures will apply:
    - When at depth, remain within visual sight of the descending line unless using a circling line, then contact must be maintained with the circling line. In case the current is in excess of 0.5 kts, contact will be maintained with the descending or ascending line.
    - If the descending or circling line is lost, surface and report to the diving supervisor.
- Divers should breathe normally and continuously at all times to avoid lung-over pressurization or CO2 buildup.
- Supervisor should not turn off the main control valve for the breathing medium until the diver is out of the water.
- Diving training operations will be suspended for any of the following reasons:
    - Small craft warnings, when sea state makes it difficult to recover divers, thunderstorm conditions, or similar warnings.
    - When, in the opinion of the diving supervisor, the current or tides present an unsafe diving condition.
    - Restricted surface visibility of less than 500 yards, due to rain, snow, fog, or other meteorological conditions.

Note. Personnel should be careful when considering lost diver procedures and the possible hazard of unseen ship or boat traffic.

- Any unsafe condition exists (for example, environmental problems, water pollution, red tide, civilian agencies closing water areas, or unexpected or unresolved operational concerns).
- Any circumstance that requires one of the mandatory dive supervisory or support personnel to leave the diving scene.

EMERGENCY PROCEDURES

10-37. Emergency procedures are key collective skills that support the METL of all combat dive units. Planning and rehearsing these skills will be an integral part of all unit dive training. Units should establish SOPs to deal with any diving emergencies. The following procedures apply with each type of emergency:
- *Separation From a Dive Partner.* If divers become separated and lose visual contact, each should—
    - Stop, look (look up, down, and 360 degrees around; in poor visibility extend an arm in the direction he is looking), and listen—then surface immediately.
    - If the dive buddy is on the surface, regroup, connect buddy line, and descend to continue the dive.
    - If the buddy is not on the surface after a reasonable wait, signal the safety boat (signal by waving arms during the day or a chemlite at night; if there is not an immediate response from the safety boat, use smoke during the day or a flare at night) and wait for assistance.
- *Diver in Need of Assistance.* Functions should include the following:
    - In an emergency, render assistance to the swim partner by inflating his flotation device, dropping his weight belt if necessary, and bringing him to the surface. Use particular caution when surfacing from under a ship, floating pier, or other overhead obstruction. The diver rendering assistance should come off dive status; for example, remove his regulator mouthpiece only after the man in distress is safely aboard the safety boat or if required to administer first aid or CPR on the surface.

CAUTION
If diving the MK 25, drop the weight belt only as a last resort.

Chapter 10

- In an emergency, ignite a flare or signal with a flashlight in a circular motion at night. During the day, ignite smoke or wave one or both hands over the head. For nonemergency assistance, shine a flashlight toward the safety boat at night; during the day, hold one arm out of the water, palm toward the safety boat.
- Render first-aid, CPR, or mouth-to-mouth resuscitation as appropriate.
* *Loss of a Diver.* Functions should include the following:
  - Initiate diver recall immediately upon determining that a diver is lost.
  - Mark the last known location of a diver with a buoy.
  - Contact the parent command via radio. **At no time will the name of any lost diver be passed over the radio!** Reference may be made to the dive pair's number if appropriate. Pass the word that a diver is missing and that a search is being initiated.
  - Organize divers on hand and decide if there is adequate bottom time to conduct a search. If so, commence a search of the immediate area using the standby diver or other available divers. Use the buoy as the center of the search area. **Do not move the buoy!** If no bottom time is available with divers on hand or if it is suggested that available bottom time is inadequate for the task at hand, contact the unit for additional resources.

EQUIPMENT

10-38. All equipment used for any type of diving operation must be approved by the Navy. The U.S. Navy publishes an ANU list that contains all Navy-approved diving equipment. The procurement and use of unauthorized scuba and associated equipment as well as unauthorized modification of approved equipment, is expressly prohibited. Equipment will be used IAW current U.S. Navy diving regulations, applicable Service regulations and instructions, and appropriate technical manuals.

MINIMUM EQUIPMENT

10-39. current U.S. Navy diving regulations list the minimum equipment for open- and closed-circuit scuba diving. Equipment that does not appear on the approved Navy diving equipment list will normally not be used except for authorized research, development, test, and evaluation (RDT&E) purposes. However, a dangerous oversimplification exists when relying on any general minimum equipment list.

Diver Equipment

10-40. The requirements of each dive will dictate the equipment necessary. The diving supervisor must take all facets of the dive profile into consideration before issuing the equipment list for a particular dive. Specific items of consideration are discussed below:

- *Personal Flotation Device.* The Navy-approved personal flotation devices used by SF units are as follows:
  - The MK-4 flotation device is suitable for open circuit, MK 16, and MK 16 scuba.
  - The Secumar TSK 2/42 flotation device is suitable for all dives using the MK 25 Draeger scuba.

Note. Past experience has shown that CO_2 cylinders can lose their charge due to faulty manufacture. All unexpended CO_2 cylinders shall be weighed individually before the start of each dive and per appropriate Preventative Maintenance System schedules. Quick releases will not be used on flotation devices. CO_2 cylinders will weigh no less than 3 grams below their original weight.

- *Knife.* Carrying a knife is a safety requirement. The knife will be attached to the diver's web belt or the leg or body strap of the flotation device, but never to the weight belt. The knife will be kept sharp and free from rust. The dive buddy will be aware of the location of his partner's knife.
- *Depth Gauge.* A depth gauge will be worn on the arm or on the attack board of the diver.
- *Wristwatch.* A wristwatch will be worn by each diver.

- *Compass.* Compasses will be used at the discretion of the OIC or NCOIC of the dive, depending upon the type of evolution planned. For compass accuracy dives, at least one compass will be carried by each pair of divers. Whenever possible, an attack board with compass, depth gauge, and wristwatch will be used for those dive profiles requiring compass accuracy (for example, infiltration, exfiltration, and ship attacks). The attack board with the compass is to be attached to the diver's wrist or flotation device by a lanyard.
- *Signal Flare.* Only approved flares will be used IAW current directives.
- *Light Sources.* Chemlights or dive lights are mandatory for each diver when diving between the hours of sunset and sunrise.
- *Diver's Slate.* A slate will be worn when directed by the OIC or NCOIC and is used for recording information or for diver-to-diver communication.
- *Life Lines.* The two types are—
  - **Buddy lines.** The buddy line will be 6 to 10 feet long. It will be used to connect dive pairs at night or in other conditions of poor visibility and for all closed-circuit or semiclosed-circuit dives. It should be strong and have a neutral or slightly positive buoyancy. Nylon, Dacron, and manila are suitable materials.
  - **Tending lines.** If only one diver is available, he will be tended by a surface crew with a line. The line is to be of sufficient strength to support the diver's weight while pulling him to the surface and clear of the water. However, it should not be so heavy as to interfere with the diver's ability to perform his assigned duties. The line will be tended IAW procedures set forth in current U.S. Navy diving regulations.
- *Protective Clothing.* Generally, protective clothing will be worn at the discretion of the diving supervisor or the OIC or NCOIC as dictated by the water and atmospheric temperatures. Protective clothing also offers protection against scrapes and cuts from underwater obstacles. The desires of the individual diver should be considered by the dive planners. Even in warm water, a diver can suffer hypothermia during prolonged exposure.
- *Weight Belt.* A weight belt with an approved quick-release mechanism will be worn as required by each diver. It must be worn outside of all equipment so that it can be ditched in case of emergency.
- *Whistle.* One whistle per diver is required for all diving operations and should be attached to the oral inflation tube of the flotation device.
- *Dive Buoys.* A dive buoy for each swim pair may be used at the discretion of the OIC or NCOIC and diving supervisor. A light source (light or chemlite) may be attached to the buoy for easier location during night dives.

Breathing Apparatus

10-41. SF elements use three types of scuba units—open circuit, closed circuit (MK 25), and semiclosed circuit (MK 15 and MK 16). All scuba cylinders used during diving operations will be charged to at least 75 percent of their working pressure. The diving supervisor will ensure that sufficient breathing medium remains to conduct any repetitive dives. Equipment checklists for each of the three types of scuba can be found in USSOCOM Reg 350-4.

SUPPORT EQUIPMENT

10-42. Divers require specific support equipment for conducting all diving operations. The items discussed below reflect the minimum support equipment that is used by SF units.

10-43. **Safety Boat.** Personnel use a power boat during all administrative and training dives and for picking up divers in case of an emergency. They may use a safety boat on operational dives as circumstances permit. A safety boat is mandatory for any training dive conducted in open water. It must be highly maneuverable and must be ready to rapidly render assistance to a diver in distress. If the probability exists that divers will become widely dispersed (such as on dives or swims of 2 or 3,000 yards, around a point of land, or with inexperienced personnel), two or more safety boats should be used as deemed

necessary by the OIC or NCOIC and the diving supervisor. The following minimum equipment will be on board each safety boat:
- *High Intensity Light.* Divers should use a high intensity light during all night dives.
- *First Aid Kit and an AMBU-Type Resuscitator.* A DMT will have in his possession the proper equipment, as prescribed in current U.S. Navy diving regulations to resuscitate a diver and provide first aid.
- *Diving Safety Boat Checklist.* A sample checklist can be found in current U.S. Navy diving regulations.
- *Diving Flag or Diving Navigational Lights.* Navigation rules require the following visuals be displayed when diving evolutions are conducted:
  - **Daytime.** The international signal Code pennant and the Alpha flag.
  - **Night.** In addition to the normal running lights, three 360-degree lights in a vertical line (red over white over red) will be displayed.

10-44. **Light Sources.** A flashlight or diving light is required for the diving supervisor and DMT or SOT during all night dives.

10-45. **Diver Recall Device.** A diver recall device is mandatory equipment for all dives. Lists of approved recall devices are published by NAVSEA.

10-46. **Communications Equipment.** Radio communications will be maintained on the scene between all safety boats, the diving supervisor, safety observer (if assigned), the parent unit, and other units as required (for example, harbor patrol) for immediate notification in the event of an emergency.

SPECIFIC MANNING REQUIREMENTS

10-47. The minimum manning requirements for conducting dive operations are specified in AR 611-75. The dive supervisor oversees the safe conduct of SF diving operations. To assist the diving supervisor and render emergency medical support in case of a diving accident, the minimum requirements for on-scene medical support would be a DMO and a DMT or SOT:
- *Diving Medical Officer.* The DMO is required to be present during buoyant ascents, free ascents, submarine lockout and lockin training, and pressure and oxygen tolerance testing. The DMO is responsible for developing a medical plan to monitor all divers who are involved in the diving evolutions listed above. He has the responsibility to recommend that a diver not participate in an evolution due to adverse health or physical condition. He personally monitors diving operations in progress and stands ready to provide medical treatment if necessary.
- *Diving Medical Technician or Special Operations Technician.* He is required to be present for all mixed gas and closed-circuit dives. His presence during open-circuit dives will be at the discretion of the diving supervisor, unless otherwise required by amplifying Service directive. The DMT or SOT should be positioned where he can best monitor the progress of the dive and provide medical care and treatment to divers on a routine basis or in case of a diving emergency. He will have in his possession the proper equipment, as prescribed in current U.S. Navy diving regulations, to resuscitate a diver and provide first aid. He will accompany any injured party to the treatment facility if additional treatment is necessary.

Chapter 11
Open-Circuit Diving

Combat swimming and diving operations are a small piece of the total spectrum of SF waterborne capabilities. Combat diving operations are highly decentralized in both the planning and execution phases. It is because diving operations are specialized and require advanced training of all personnel involved. These skills are rarely required; however, when they are, no other skill set will meet mission requirements. To ensure combat divers are successful in their operations, the team must become involved early in the planning phase to ensure mission requirements and proper safety considerations are met.

The objective of this chapter is to provide the dive supervisor, dive detachments, commanders, and staffs with essential information necessary to identify planning considerations required to conduct and supervise horizontal dives, vertical dives, search and rescue dives, contaminated water dives, cold water or ice dives, and limited-visibility dives. In support of this objective, this chapter contains excerpts from relevant reference material. Where appropriate, that material has been expanded to emphasize key elements peculiar to SF diving. The definitive material that is required to actually conduct diving operations is contained in the following references:

- AR 611-75, *Management of Army Divers*.
- Most current U.S. Navy regulations and manuals pertaining to dive operations.
- USASOC Reg 350-20.

Note. Detachments conducting diving operations are required to have these references on-hand throughout the operation.

OPEN-CIRCUIT DIVE OPERATIONS

11-1. Open-circuit diving is an ideal medium for training and support. It is the simplest method to get underwater and stay there long enough to do useful work. It may be used for any operation not requiring secrecy or when underwater detection capabilities are very limited. Planning ranges for infiltration swims are approximately 1,500 yards depending on equipment loads and diver conditioning.

Note. When conducting combat diving operations, the planning considerations need to be calculated in U. S. standards of measure. Dive planners use a 1-kt swim speed as the goal. This establishes a time/distance standard for planning purposes of 100 yards per 3 minutes. If using the metric system, distances covered would be less than planned (if using the 1-kt standard or 30.8209 meters per minute, after 3 minutes the diver will travel 92.4627 meters vice 101.118 yards in the same time) causing the combat divers to surface early or miss the intended navigation point if using time.

11-2. SF teams are most likely to use open-circuit diving for initial diver training, detachment training and training support, underwater reconnaissance in a permissive environment, ship bottom search, and underwater search and recovery. It must be used whenever the operational depth exceeds the capabilities of the closed-circuit (O2) rebreathers (20 FSW with excursions to 50 FSW). Depth limits for open-circuit (diving) are 130 FSW in training and 190 FSW (IAW regulatory guidance) in exceptional circumstances. Complete guidance for operational depths may be found in NAVSEAINSTs.

DIVE PLAN AND BRIEFING

11-3. All diving operations require meticulous planning. The designated primary diving supervisor will prepare the dive plan and present the dive briefing.

11-4. The CDS prepares a dive plan to ensure the proper and timely execution of all required support functions. The plan is prepared in the normal five-paragraph OPORD format and serves to notify commanders and staff of the impending operation. It is required to ensure that all diving operations are coordinated, controlled, and conducted safely. Detachments will find that major portions of these formats do not change from dive to dive. They should be incorporated as a permanent part of the detachment's diving SOP.

11-5. A dive briefing precedes all combat diving operations. Dive briefings are a compact synopsis of the overall dive plan that is tailored to include only the information necessary for the completion of the dive. It is given to the participants (divers and support personnel) as close to the time of the dive as possible. Each dive briefing is different.

STANDARDS FOR COMPRESSED AIR

11-6. Internal training, external training, mission support, and combat diving operations all require that a dive detachment be capable of filling and refilling diving tanks. SFODs have the option of deploying with organic portable air compressors or obtaining air from host nation (HN) or commercial sources. Air purity standards and testing procedures are detailed in current U.S. Navy diving regulations. SFODs conducting diving operations must, as a minimum, adhere to the published Navy standards.

11-7. The quality of compressed air is critically important and must be strictly monitored and controlled. The air used in diving operations must meet the standards of purity as established by the Commander of the Naval Medicine Command. This standard applies to all sources of air or methods used for charging the cylinders. Table 11-1, page 11-3, outlines the required minimum standards.

11-8. Contamination of divers' air can cause illness, unconsciousness, or even death. The NAVSEA that an air sample be taken semiannually from each operational air source. The Naval Support Activity in Panama City, Florida, is the central authority to schedule the sampling of divers' air sources. All field units should coordinate air sampling requirements through that center. The Navy provides this service, semiannually, for each registered air source at no cost to the unit. Alternatively, diving detachments may use a commercial sampling service provided by a certified laboratory.

11-9. Detachments deploying with organic assets that have been maintained and tested IAW USN standards can be assured that their equipment and the air supply it produces meets or exceeds the minimum standards. Detachments that are forced to obtain air from other sources must exercise reasonable precautions such as examining the facility's physical plant (compressors, air banks, distribution panel) and the air analysis certificate to ensure that it meets the minimum standards. Diver's air procured from commercial sources shall be certified in writing by the vendor as meeting the purity standards of FED SPEC BB-A-1034 Grade A Source I (pressurized container) or Source II (compressor) air. When examining a commercial air analysis certificate, detachments should verify that the certificate is current (within previous 6 months) and the tested sample met or exceeded Grade E.

11-10. Units deploying to locations without certified air sources must be prepared to conduct their own tests of breathing gas purity before using air from unknown sources. Air quality test kits using disposable chemical test strips or test vials are available through the supply system, the GSA Catalog, or issued as a component of the unit's portable compressors. Personnel should ensure the kit's chemicals are fresh and they will not expire during the mission. Procedures vary between the types of test kits. Detachments must review the operating instructions whenever they conduct air quality sampling.

Table 11-1. Minimum standards of compressed air

U.S. Military Diver's Compressed Air Breathing Purity Requirements for ANU-Approved or Certified Sources	
Constituent	Specification
Oxygen concentration	20 to 22 percent by volume
Carbon dioxide	1,000 parts per million (ppm) maximum
Carbon monoxide	20 ppm maximum
Total hydrocarbons	25 ppm other than methane
Particulates and oil mist	5 mg/m^3 maximum
Odor and taste	Not objectionable
Diver's Compressed Air Breathing Requirements if From a Commercial Source	
Constituent	Specification
Oxygen concentration	20 to 22 percent by volume
Carbon dioxide	500 ppm maximum
Carbon monoxide	10 ppm maximum
Total hydrocarbons as methane (CH4) by volume	25 ppm maximum
Particulates and oil mist	.005 mg/l maximum
Odor and taste	Not objectionable
Separated water	None
Total water	0.02 mg/l maximum
Halogenated compounds (by volume): Solvents	0.2 ppm maximum
Reference: FED SPEC BB-A-1034 A and B	

DIVING INJURIES AND MEDICINE

11-11. Current U.S. Navy diving regulations contain detailed information about anatomy and physiology as it relates to diving. They also contain extremely detailed presentations of diving injuries and the required medical treatment, including recompression therapy.

ACCIDENT MANAGEMENT

11-12. Tables 11-2 through 11-4, pages 11-4 through 11-6, are extracts from the NOAA Diving Manual. They serve as an excellent guide for preparing and executing a diving accident management plan. The importance of evacuation planning cannot be overemphasized. Planning must be thorough and complete, keeping in mind that it will involve several modes of transportation (for example, boat, truck, aircraft, and ambulance) and communications (for example, voice, radio, and telephone).

Table 11-2. Injuries during descent

Type Injury	Cause	Symptoms	Treatment	Prevention
Sinus Squeeze	1. Blocked sinus opening. 2. Too rapid a descent.	1. Pain in facial sinus area. 2. Bleeding from nose in severe cases. 3. Blood in nasal mucous discharge in all cases.	1. Provided there are no complications, time is the only treatment required.	1. Do not dive with head cold. 2. Valsalva maneuver. 3. Bounce dive. 4. Use nasal spray and drops.
Ear Squeeze	1. Blocked eustachian tube or external ear canal preventing pressure equalization. 2. Too rapid a descent.	1. Pain, increasing with depth. 2. Rupture of tympanic membrane, relief from pain. 3. Blood from ear, nose, or throat.	1. Valsalva maneuver or bounce dive. 2. With bleeding or severe pain see medical doctor.	1. Do not dive with head cold. 2. Valsalva maneuver or bounce dive. 3. Use nose drops or spray. 4. Do not dive with tight-fitting hoods. 5. Stop dive if pain persists.
Lung Squeeze	1. Breath-holding while diving to a depth where lungs collapse. 2. Underwater explosions.	1. Pain, usually quite severe in chest. 2. Difficulty in breathing when diver returns to surface.	1. Artificial respiration if breathing has stopped. 2. Possible surgical drainage may be required. 3. If no complications, allow time for tissues in lungs to heal. 4. In all cases consult the medical officer.	1. Breathe normally during descent. 2. Do not make breath-hold dive to a depth sufficient to compress the lungs below their residual volume.
Minor Squeezes (Face, Intestine, and Tooth)	1. Unequal pressure to some part of the body. 2. Failure or inability to equalize.	1. Pain, sometimes severe. 2. Swelling, redness of tissues, bleeding in severe cases.	1. Equalize pressure. 2. Cold packs. 3. In case of tooth, report to a dental officer.	1. Equalize pressure in mask. 2. Proper dental hygiene. 3. No diving after major dental filling work.

Table 11-3. Injuries during ascent

Type Injury	Cause	Symptoms	Treatment	Prevention
Mediastinal Emphysema (Rupture of Lung, Overexpansion and Air Escaping Into the Mediastinal Space)	Holding breath on ascent produces overexpansion.	1. Slight pain under breastbone. 2. Possible shortness of breath or fainting. 3. Possible slight blueness of lips and face.	Usually none. Dissipates with time.	Do not hold breath on ascent.
Subcutaneous Emphysema	1. Holding breath on ascent. 2. Possibly from air embolism, pneumothorax, or mediastinal emphysema.	1. Swelling of neck tissues. 2. Change in voice. 3. Crepitation.	Usually none. Dissipates with time.	Do not hold breath on ascent.
Pneumothorax (Air Escaping the Lungs and Entering the Pleural Space Between the Lung Lining and Chest Wall)	1. Holding breath on ascent. 2. Weak pleura.	1. Severe chest pain made worse by deep breathing. 2. Sudden shortness of breath. 3. Irregular pulse. 4. Possible shock or cyanosis.	1. See medical doctor ASAP. 2. Air may have to be surgically removed from the pleural cavity.	Do not hold breath on ascent.
Air Gas Embolism (AGE)	Holding breath on ascent.	1. Sudden onset of bloody, frothy sputum. 2. Dizziness. 3. Paralysis. 4. Visual disturbances. 5. Unconsciousness (normally within 15 minutes of surfacing). *Note.* If a diver sustains an AGE, he most likely will also have the other injuries on this table.	1. Immediate recompression. 2. Administer 100 percent oxygen. 3. Contact a DMO ASAP.	Do not hold breath on ascent.

BLAST PRESSURE

11-13. Combat divers or swimmers may be exposed to the effects of underwater explosions in the course of their missions. Blast overpressure is the sudden increase in pressure, above the normal pressure, caused by an explosion. Because water is 800 times denser than air, it transmits this increased pressure very efficiently in the form of a shock wave. Ideally, swimmers should be out of the water whenever underwater explosives are detonated. Divers or swimmers exposed to blast pressure risk serious injury or death. An overpressure of 50 psig or greater may cause injury to organs and body cavities containing air. An overpressure of 300 psig may cause severe injury to a fully submerged diver, and 500 psig may cause death.

Table 11-4. Injuries caused by indirect effects of pressure

Type Injury	Cause	Symptoms	Treatment	Prevention

Decompression Sickness (Bends or Caisson Disease, N_2 Bubbles in Blood)	1. Inadequate decompression.
	2. Overstaying bottom time.
	3. Ascending too fast.
	4. Exceeding planned depth.
	5. Flying after diving.
		1. Brain or spinal cord involvement:
		  Unconsciousness.
		  Convulsions.
		  Inability to speak.
		  Muscular paralysis.
		  Nausea, vomiting, visual problems.
		  Dizziness, vertigo.
		  Paralysis of lower body.
		  Loss of bladder and bowel control.
		2. Lung involvement (chokes):
		  Shortness of breath.
		  Coughing and shallow breathing.
		3. Pain only:
		  Muscle and joint area bubbles.
		  Severe pain in joints.
		  Localized pain anywhere.
		  Swelling (associated with pain).
		  Skin bends.
		  Red itching rash.
		  Skin welts.
			Immediate recompression.	1. Adhere to dive plan.
				2. Do not fly for 12 hours after diving.
				3. Good physical condition.
				4. Avoid over-exertion.

Oxygen Toxicity	Generally unknown but normally caused by breathing 100-percent oxygen at 33 feet or greater.	Use letters VENTIDC: V—Visual problem (tunnel vision) E—Ringing in ears N—Nausea T—Twitching/tingling I—Irritability D—Dizziness C—Convulsions	1. Surface. 2. Remove oxygen source.	1. Know symptoms. 2. Observe operational limits. 3. Use buddy system.
Nitrogen Narcosis	Partial pressure of nitrogen gas has a narcotic effect at 99 feet (4 atmosphere absolute)	1. Disorientation. 2. Confusion. 3. Unusual behavior. 4. Loss of skill. 5. Lack of concern. 6. Drunkenness.	Controlled ascent.	1. Recognize symptoms. 2. Use buddy system. 3. Avoid deep dives.

Open-Circuit Diving

11-14. When divers cannot exit the water, detachments must calculate a minimum safe distance so as to reduce the effect of the blast overpressure on the divers. The objective is to ensure that they are not exposed to a blast pressure in excess of 50 psig. The detachment uses the following formula to determine the pressure exerted at a given distance by an underwater explosion of tetryl or TNT:

$$\text{Pressure (P)} = \frac{13{,}000 \times \sqrt[3]{\text{Weight of Explosive}}}{\text{Distance in Feet}}$$

> **WARNING**
>
> The above formula is for TNT. It is not applicable to other explosives.

11-15. Distance from the explosion is the primary factor serving to mitigate the effects of blast pressure. Once a safe distance has been computed, planners must calculate the amount of time the divers will require to reach it. Planners should use the DxS=T formula to determine the fuzing requirements of the charge.

Minimizing the Effects of an Explosion

11-16. When expecting an underwater blast, the diver should get out of the water and out of range of the blast whenever possible. If he has to be in the water, it is prudent to limit the pressure he experiences from the explosion to less than 50 pounds per square inch (psi). To minimize the effects, the diver can position himself with feet pointing toward and head directly away from the explosion. The head and upper section of the body should be out of the water or the diver should float on his back with his head out of the water. Divers should be aware of bottom topography, composition, underwater structures, and how these factors may influence the shock waves through reflection, refraction, and absorption. Generally, high brisance (REF) explosives generate a high-level shock wave of short duration over a limited area. Low brisance (high power) explosives create a less intense shock and pressure wave of long duration over a greater area.

DIVE TABLES

11-17. The USN is the proponent agency for all matters concerning military diving. The USN decompression tables are special tables and rules that have been developed to prevent decompression sickness and enhance the diver's safety. These tables and rules take into consideration the amount of nitrogen absorbed for any given depth and time period, and allow for the elimination of the inert gases through normal processes. With rapid changes and advances in diving medicine, the most current U.S. Navy diving regulations should always be available at the dive site for reference.

> *Note.* The tables must be rigidly followed to ensure maximum safety. Combat diving operations will normally be performed within the no-decompression limits IAW AR 611-75 and current U.S. Navy diving regulations.

Air Decompression Tables

11-18. To properly use USN air decompression tables, divers must understand specific terms. Each term is explained below:
- *Depth.* This term indicates the depth of a dive. Divers should always use the maximum depth attained during the dive. Depth is always measured in feet.
- *Bottom Time.* This term refers to the total elapsed time from where the diver leaves the surface until he begins his ascent. This time is rounded up to the next whole minute. It is always expressed in minutes.

Chapter 11

- *Decompression Schedule.* It is the specific decompression procedure for a given combination of depth and bottom time. It is indicated in feet and minutes.
- *Single Dive.* This dive is conducted at least 12 hours after the most recent previous dive.
- *Repetitive Dive.* This dive is conducted 10 minutes after and within 12 hours of a previous dive.
- *Decompression Stop.* This stop occurs at a specified depth where the diver remains for a specified time to eliminate inert gases from body tissues.
- *Surface Interval.* The diver spends this amount of time on the surface following a dive. It begins as soon as the diver surfaces and ends as soon as he begins his next descent.
- *Residual Nitrogen.* This term refers to excess nitrogen gas that is still present in the diver's tissues from a previous dive.
- *Repetitive Group Designation.* This letter designation relates directly to the amount of residual nitrogen in a diver's body.
- *Residual Nitrogen Time.* This amount of time must be added to the bottom time of a repetitive dive to compensate for the nitrogen still in solution following a previous dive.
- *Equivalent Single Dive Time.* This term refers to the sum of residual nitrogen time and bottom time of the repetitive dive. It is measured in minutes and used to select the decompression schedule and repetitive group designator on a repetitive dive.

Selection of Decompression Schedule

11-19. There are four USN decompression tables. They are as follows:
- USN Standard Air Decompression Table.
- No-Decompression Limits and Repetitive Group Designation Table.
- Surface Decompression Table Using Oxygen.
- Surface Decompression Table Using Air.

Note. Scuba or combat divers do not use the surface decompression tables; therefore, their use will not be discussed. They are mentioned for information purposes only.

11-20. Divers should select the appropriate decompression table based on the dive's parameters. These factors are depth and duration of the dive, availability of a recompression chamber, availability of an oxygen breathing system within the chamber, and the specific environmental conditions (sea state, water temperature).

11-21. All USN decompression schedules have general guidelines that divers must follow to account for slight differences in the actual dive and the manner in which the decompression schedules of all tables are arranged. Above all, the decompression schedules provide for maximum safety of the diver. Divers should consider the following principles when selecting a decompression schedule:
- Compute schedules for all tables in 10-foot depth increments.
- Make sure bottom times for all schedules are in 10-minute increments.
- Combine depth and total bottom time to form actual dives.
- Always select exact or next greater depth.
- Always select exact or next greater time.
- Do not interpolate between decompression schedules.
- Do not alter or modify decompression schedules without prior approval of a DMO.
- Ensure the diver's chest is located as close as possible to the stop depth.
- Begin decompression stop times when the diver reaches the stop depth.
- Do not include ascent time as part of stop time.

- If the diver is delayed in reaching the decompression stop, add the delay time to the bottom time and recompute the dive for the longer bottom time.
- Select the next longer decompression schedule, if the diver was exceptionally cold, showed signs of extreme fatigue during the dive, or if his workload was relatively strenuous.

No-Decompression Limits and Repetitive Group Designators Table for No-Decompression Air Dives

11-22. SF divers use this table most often. This table outlines the maximum amount of time that a diver may remain at a given depth without incurring a decompression obligation. Each depth listed has corresponding no-decompression time limits given in minutes. Should a diver need to make a repetitive dive, this table provides the group designator for no-decompression dives (Table 11-5).

Table 11-5. No-decompression limits and repetitive group designators for no-decompression air dives

Depth (fsw)	No-Stop Limit (min)	Repetitive Group Designation													
		A	B	C	D	E	F	G	H	I	J	K	L	M	N	O	Z
10	Unlimited	57	101	158	245	426	*										
15	Unlimited	36	60	88	121	163	217	297	449	*							
20	Unlimited	26	43	61	82	106	133	165	205	256	330	461	*				
25	595	20	33	47	62	78	97	117	140	166	198	236	285	354	469	595	
30	371	17	27	38	50	62	76	91	107	125	145	167	193	223	260	307	371
35	232	14	23	32	42	52	63	74	87	100	115	131	148	168	190	215	232
40	163	12	20	27	36	44	53	63	73	84	95	108	121	135	151	163	
45	125	11	17	24	31	39	46	55	63	72	82	92	102	114	125		
50	92	9	15	21	28	34	41	48	56	63	71	80	89	92			
55	74	8	14	19	25	31	37	43	50	56	63	71	74				
60	60	7	12	17	22	28	33	39	45	51	57	60					
70	48	6	10	14	19	23	28	32	37	42	47	48					
80	39	5	9	12	16	20	24	28	32	36	39						
90	30	4	7	11	14	17	21	24	28	30							
100	25	4	6	9	12	15	18	21	25								
110	20	3	6	8	11	14	16	19	20								
120	15	3	5	7	10	12	15										
130	10	2	4	6	9	10											
140	10	2	4	6	8	10											
150	5	2	3	5													
160	5		3	5													
170	5			4	5												
180	5			4	5												
190	5			3	5												

REPETITIVE DIVES

11-23. Any dive performed within 12 hours of a previous dive is a repetitive dive. The period between dives is known as the surface interval (SI). During the SI, divers are **off-gassing**, or reducing through normal respiration, the level of nitrogen absorbed into their bodies during previous dives. Keeping track of their SI allows divers to recalculate their residual nitrogen group to take advantage of the reduction (over time) of the absorbed nitrogen in their bodies.

11-24. Surface interval times are expressed in hours and minutes and range from a minimum of 10 minutes to a maximum of 12 hours. Any surface interval of less than 10 minutes is ignored and the bottom times of the two dives are combined and treated as one continuous dive. It is only necessary to track SI for

Chapter 11

12 hours. After 12 hours, the diver is considered **desaturated** (no remaining residual nitrogen from the previous dive) and any further dives start from scratch.

11-25. The residual nitrogen timetable for repetitive air dives provides information relating to the planning of repetitive dives. Divers should follow the steps below when using the timetable (Figure 11-1, page 11-11):

- Refer to the repetitive group designator (RGD) from the previous dive assigned by either the standard air or no-decompression table.
- Determine the residual nitrogen time by using the residual nitrogen timetable for repetitive air dives.
- Enter the SI table horizontally to select the appropriate surface interval.
- From the appropriate SI, read down to the bottom of the table to obtain the new repetitive group letter.
- Determine the equivalent single dive time. Add the residual nitrogen time and the total bottom time to get the equivalent single dive time.
- Obtain the residual nitrogen time corresponding to the planned depth of the repetitive dive as follows:
  - Read down from the repetitive group letter to the row that represents the depth of the repetitive dive.
  - Add the time shown at the intersection (the residual nitrogen time in minutes) to the repetitive dive to obtain total nitrogen time for the repetitive dive.
- Select the decompression schedule according to the new equivalent single dive time.
- Use the equivalent single dive time from the preceding repetitive dive to determine the proper schedule, if additional dives are needed.

Note. There is one exception to this table. In some instances, when the repetitive dive is to the same or greater depth than the previous dive, the residual nitrogen time may exceed the actual bottom time of the previous dive. In this case, the diver should use the actual total bottom time of the previous dive as the residual nitrogen time in determining equivalent single dive time.

11-26. The principles outlined above are illustrated in the repetitive dive flowchart (Figure 11-2, page 11-12). The flowchart converts into the repetitive dive profile shown in Figure 11-3, page 11-12.

AIR DIVING TABLES

11-27. These tables consist of the USN standard air decompression table and the exceptional exposure table. SF divers are usually limited to 130 feet and a no-decompression profile. For this reason, they normally do not use these tables. The type of diving that does require their use is outside the scope of this manual. If the SFOD has a requirement to conduct decompression dives, it must seek and obtain the appropriate waivers and ensure that only personnel trained in decompression procedures participate in the dives. Detachments must strictly adhere to the procedures laid out in current U.S. Navy diving regulations.

Note. Normally, combat diving operations do not involve decompression stops. Sometimes they are unavoidable; for example, when an entanglement inadvertently extends bottom time. If decompression stops are required, the diver must be prepared to safely execute them using the USN standard air decompression table.

Open-Circuit Diving

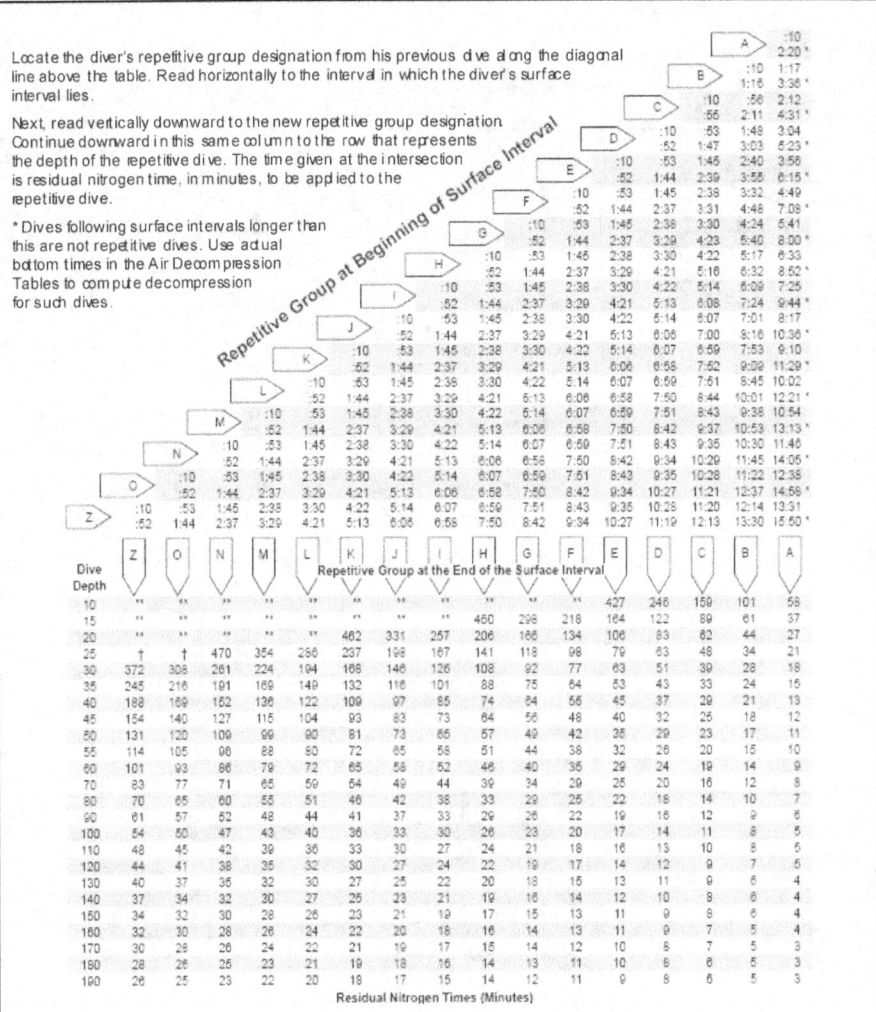

Figure 11-1. Residual nitrogen timetable

Chapter 11

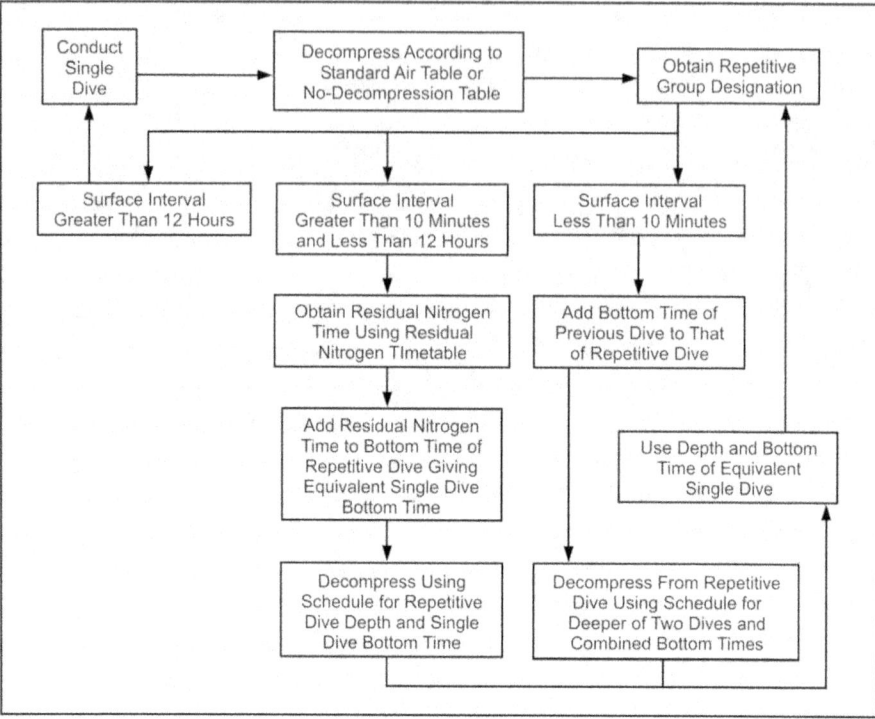

Figure 11-2. Repetitive dive flowchart

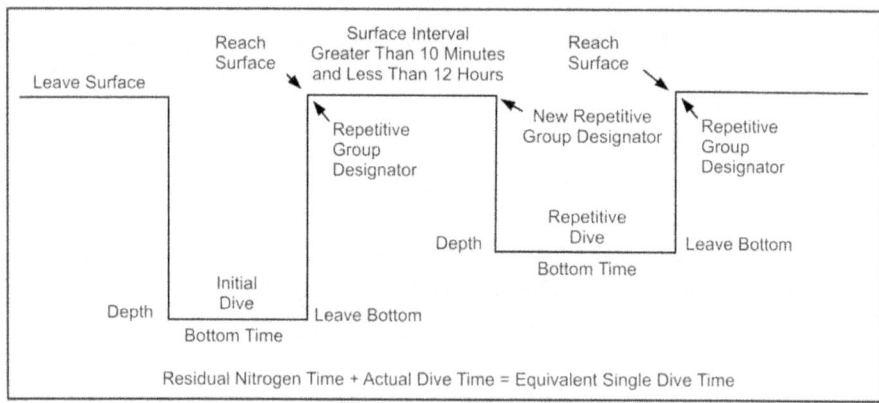

Figure 11-3. Repetitive dive profile

EQUIPMENT AND BASIC TECHNIQUES

11-28. The types of equipment available to SF divers are changing and improving rapidly as more equipment is tested and approved by the USN. The ANU list available from NAVSEA specifies the equipment that units and divers are allowed to purchase or use. SF units can use this list to tailor their equipment purchases to more closely satisfy requirements unique to their particular environment or mission. Because diving places extra stress on the human body, the one-size-fits-all approach to equipment acquisition cannot adequately meet all the unique requirements of SF divers and missions.

EQUIPMENT USE AND MAINTENANCE

11-29. Military diving places Soldiers and Marines in a potentially hazardous environment. The hazards found in diving are made safe through personnel selection, training, safety guidelines, supervision, and properly maintained equipment. The equipment provided to combat divers is extremely reliable and safe if properly maintained and used. Specific maintenance requirements for each piece of equipment are identified on that equipment's maintenance requirements card (MRC). These requirements derive from the USN's maintenance procedures. U.S. Army SF divers and Marine Corps divers are required to follow them. Most of them apply only to the servicing DLSMF. The maintenance allocation charts (MACs), coupled with the manufacturer's instructions, provide the framework to properly maintain equipment. For certain types of equipment additional formalized training needs to be given to certain individuals. For example, servicing compressors or regulators involves working on life support systems that must be maintained by qualified individuals. Appendix C contains an outline of mandatory maintenance procedures.

WEARING OF EQUIPMENT

11-30. The variety of equipment available to SF divers as well as the diver's personal preferences can result in a number of different equipment configurations. Units must establish an SOP that specifies how critical items of equipment should be worn. Examples would include the placement of an alternate air source, emergency signaling devices, depth gauges, watches, or the fastening of a weight belt. Other items may be left to diver preference for comfort and efficiency. Examples include determining the degree of exposure protection worn and placement of dive tools, lights, and nonessential accessories. Individual divers should rig and precheck their own and their buddy's gear before reporting to the CDS for the formal predive inspection. Individual divers should—

- Gauge their own cylinders and record pressure.
- Attach the regulator to the manifold keeping saltwater off the filter.
- Check for air leaks, check high-pressure gauge reading, and ensure the regulator is functioning properly.
- Adjust the harness for proper fit.
- Ensure the reserve lever is in the up position.

11-31. Before entering the water, all divers must be given a predive inspection by a diving supervisor. The CDS will inspect each diver for safety and ensure the diver meets mission requirements. Dive supervisors receive specialized training in performing predive personnel inspections (Appendix D) as part of their initial diving supervisor training. They must ensure that every diver participating in the operation is properly equipped and correctly rigged before entering the water. The CDS is the final determiner of a diver's readiness to enter the water. Points of inspection include the following:

- Air cylinders are worn so that the manifold will not strike the diver's head during the dive. They must be positioned so that the diver can manipulate the on/off valve while submerged and activate the reserve lever when needed.
- The primary regulator hose is routed over the diver's right shoulder.
- The alternate (octopus) regulator (if worn) must be routed to the front of the diver's body and secured so that it is visible and accessible to the diver and his dive partner. Units must establish an SOP that specifies how the octopus is to be routed and secured.

- The low-pressure inflator is routed (depending on the equipment configuration) over or under the diver's arm and firmly attached to the buoyancy compensator military (BCM) inflation device.
- The submersible pressure gauge is routed under the diver's arm (as desired) for easy access.
- All hoses and accessories are secured so as not to present an entanglement hazard.
- A shirt or protective dress (for example, wet suit, dry suit, or mission uniform) is worn for environmental protection.

PREDIVE PREPARATION

11-32. Before filling the diving cylinders with high-pressure air, divers must—
- Check the O-ring and manifold.
- Inspect the general condition of the cylinders. Ensure they have not been exposed to extreme heat, as it weakens the metal significantly. If heat exposure has occurred, ensure the cylinders undergo a hydrostatic test.
- Check for the Department of Transportation (DOT) or Interstate Commerce Committee (ICC) regulation stamp that indicates the type and grade of metal and working pressure of the cylinders.
- Check to ensure the hydrostatic test date has not expired. Cylinders must be hydrostatically tested every 5 years and visually inspected annually IAW the MRC.
- Ensure the reserve handle is down so that cylinders are filled completely.
- Check for proper strap adjustments and that straps are not missing, frayed, or torn.
- Check for proper band position and tightness.
- Check for air leaks with soapy solution or underwater.

Note. Divers should not fill cylinders faster than 200 psig per minute. This will prevent heat buildup and loss of psig when the cylinder cools down. Do not overfill!

11-33. Military divers currently use the single-hose, two-stage regulator fitted with an additional low-pressure hose for inflating the buoyancy compensator and a submersible pressure gauge to monitor the quantity of air remaining in the diver's cylinders. Divers use this type of regulator for training and it adequately fits most mission profiles. Safety divers should be equipped with an additional second stage regulator (octopus) on a longer hose so they can provide an alternate air source to a distressed diver. Navy guidance currently allows commanders to decide if all divers within their commands should be equipped with octopus regulators.

11-34. Specific regulator functions must be checked before use. The diver must—
- Ensure that hoses, plugs, and the mouthpiece are all correctly attached and serviceable.
- Inspect hoses for tears or breaks.
- Inspect the sintered filter on the first stage for corrosion.
- Attach regulator to manifold and turn air on. Inspect for air leakage and free flow.
- Check purge flow by pushing purge button and ensure that regulator functions correctly by taking at least three normal breaths.

11-35. Divers must also check the regulator after each use. They should—
- Rinse regulators with freshwater after each dive. Periodically wash off with warm water and mild germicidal solution. **Never purge the second stage while washing the regulator.**
- Allow regulators to dry and store in cool, dry area out of direct sunlight. Store regulator with dust cap assembly firmly in place. If possible, regulators should be stored flat with hoses loosely coiled.
- Periodically, lightly lubricate neoprene parts with breathable silicone lubricant.
- Inspect and maintain regulator body and internal parts at least once a year. Repair and adjust as stated in manufacturer's repair manuals. Adjust intermediate pressures as required.

DIVER COMMUNICATIONS

11-36. Safety and the efficient management of divers and diving operations require reliable communications. There are three types of diving signals— standard, emergency, and special. These may be either visual, audible, or by physical contact. All dive plans must include a subparagraph that details the designated communications methods.

11-37. Visual signals are the primary means of communications between submerged divers. Divers also use visual signals for longer distance surface communications. The simplest visual signals are hand-and-arm signals.

11-38. Current U.S. Navy diving regulations depict the standardized signals. Additional hand-and-arm signals are easily adapted from detachment SOPs to meet mission requirements. Other visual signals are more appropriate to long-distance signaling. They include smoke or handheld flares commonly worn attached to the dive tool, the safety sausage (a surface marker buoy), and the USAF survival mirror. All of these can be invaluable when attempting to locate and recover separated and distressed divers, especially at night in open water.

11-39. Audible signals are most valuable when communicating instructions from the surface to submerged divers. The diver recall system (DRS), a component part of the Support Set B, is the primary audible signaling device. It consists of a splash-proof, battery-operated surface unit with a microphone and a transducer that is suspended in the water. It is most effective when used in the tone mode. The dive supervisor can toggle continuous or intermittent tones (similar to a sonar pulse) and communicate with the divers using a prearranged code. The surface unit can also be used to give verbal instructions. Unfortunately, the unit has considerable distortion and limited effective range when used in the voice mode. Surface personnel must make sure to enunciate clearly, speak slowly, and keep instructions simple.

11-40. The simplest method of communicating with submerged divers is to bang two metal objects together. The clanking noise produced by the metal-to-metal contact is distinct from any other sound found in nature and it propagates well. This technique is especially useful when rendezvousing with a submarine at sea. Other acoustic signaling devices include the whistle attached to each diver's BCM, the sonic alert (an air-powered horn attached between the low-pressure inflator hose and the BCM), and the USN's pyrotechnic diver recall device.

11-41. Physical contact usually refers to line-pull signals. These signals are particularly useful when conducting surface-tended search operations. They are also useful between buddy pairs when swimming in limited-visibility water. Current U.S. Navy diving regulations give a list of standard signals. Special signals may be devised between the CDS and the diver to meet particular mission requirements. When a diver uses line-pull signals, he should first remove all slack from the tending line, then give a series of sharp distinct pulls. Even hand signals can become a contact sport when diving in limited-visibility conditions. By exaggerating the movements and feeling the partner's hand as the signal is being formed, it is possible to communicate in blackout conditions with reasonable accuracy. Signals are acknowledged by repeating them as received. All signals must be answered promptly. Failure of a diver or partner to respond to a signal should be considered an emergency.

11-42. Underwater communications systems are the most precise means of communicating for divers and support personnel. These systems require special equipment and training to use and maintain. They include hard-wired and wireless systems. Hard-wired systems consist of a surface unit, a helmet or full facemask for the diver, and a communications wire that is embedded in the diver's umbilical. Wireless systems are AM or single side band (SSB) transceivers, usually mounted in a full-face mask, that allow three-way communications from the surface, to the surface, and between divers. All of these systems allow a greater degree of control, especially when the diver is engaged in extremely precise operations underwater. This factor is especially useful in search and recovery operations. Before any through-water communications system is used, planners should consult the ANU list to ensure that it is approved.

Free Swimming Ascent

11-43. The FSA is a controlled ascent, from any given depth to the surface, by a diver that has been breathing compressed gas at ambient (depth) pressure. The diver executes the ascent by starting a controlled exhalation and swimming upwards from his start depth. The diver must exhale continuously until after his head breaks the surface. The entire ascent is completed without breathing additional compressed gas.

11-44. FSAs are commonly executed in training, or operationally, when practicing the ditching and donning of dive equipment, or when executing an underwater cache and recovery of diving equipment. During initial dive training, FSAs are normally conducted in a controlled environment (50-foot FSA tower) and the students only use facemasks. Later, training is conducted using surface swim gear. If the training is focused on "ditch and don" or underwater cache and recovery, the weight belt and diving equipment will be ditched before the ascent.

11-45. Because of the potential threat of a lung over-expansion injury, FSAs are considered a high-risk training event. Actual procedures for conducting FSA training in a training tower are detailed in NAVSPECWARCEN Instruction 1540.5C. This instruction has the force of regulation and must be adhered to by USSOCOM units conducting this type of training. In addition to the above instruction, trainers must review and adhere to the additional regulatory requirements in current U.S. Navy diving regulations, AR 611-75, and USASOC Reg. 350-20.

11-46. Combat divers receive FSA training during their initial dive course. Because of the risks, this initial training is required to be conducted in a controlled environment under the direct supervision of certified instructors. It should be repeated or refreshed before any operation requiring a deliberate FSA evolution; for example, submarine lockout/lockin (LO/LI) training. NSWC and USAJFKSWCS maintain training towers and adequate cadre to support units requiring this type of training, provided the units make timely coordination so as not to interfere with the schools' primary mission. An FSA supervisor should oversee FSAs conducted operationally.

11-47. The ideal rate of ascent for a planned FSA is no faster than the ascent rate of the diver's smallest exhaled bubbles (a vertical flow of bubbles must be visible above the student's face mask). This practice results in an ascent that is faster than normal (normal = 30 fpm); however, the bubbles are a readily identifiable reference that the diver can use to gauge his ascent rate under almost all conditions. To reduce the risk of a lung overexpansion injury, the diver must exhale continuously as he ascends while looking to the surface, with a hand extended over his head. The diver may fin slowly to reach the surface. He must not activate the life jacket firing mechanisms. If the ascent results from an out-of-air emergency, the diver should retain the weight belt until he surfaces if possible. (Dropping the weight belt unnecessarily will result in an uncontrolled ascent.)

11-48. If an FSA is required as the result of an out-of-air emergency, the diver should keep the dive gear until he surfaces. If he has difficulty reaching the surface, then he should ditch the weight belt and inflate the buoyancy compensator (execute a buoyant ascent). Once the diver has surfaced, a decision can be made whether or not to jettison the dive equipment.

Buoyant Ascent

11-49. Buoyant ascent training is no longer conducted at the initial entry level in combat diving. Buoyant ascents are an emergency procedure performed by ditching the diver's weight belt and inflating the buoyancy compensator. This movement results in an uncontrolled ascent to the surface. The diver is at significant risk of suffering a lung overexpansion injury and possible decompression sickness (DCS). Divers forced to execute a buoyant ascent should start the procedure with a forceful exhalation and make a conscious effort to continue exhaling all the way to the surface.

Open-Circuit Diving

> **WARNING**
>
> This method is an emergency procedure; present doctrine precludes practicing it in training!

11-50. Accidental buoyant ascents are normally the result of a loss of buoyancy control. These are usually caused by inattention, the loss of a weight belt, accidental activation of the buoyancy compensator, or problems with a dry suit. All divers training must include techniques for dealing with a loss of buoyancy control. Divers must also be aware of the dangers presented by overhead obstacles when deciding to execute a buoyant ascent. A fully inflated buoyancy compensator (BC) (or dry suit) is capable of pinning a diver to the underside of an overhead obstacle and holding him there immobile so he must be recovered by outside assistance.

DITCHING AND DONNING SCUBA EQUIPMENT

11-51. Scout swimmers or entire detachments approaching a BLS have the option of removing and caching their dive equipment in the water or wearing it across the beach. The MK 25 with its smaller, lighter case is usually worn until the detachment occupies the assembly area where it can be cached. Open-circuit equipment because it is heavier and bulkier is more likely to impede the movement of the detachment as it crosses the beach. Consideration should be given to caching it in the swimmer holding area behind the surf line. The technique for removing and replacing the equipment is referred to as "ditching and donning."

DIVING CONSIDERATIONS

11-52. This section discusses some of the more important considerations and training requirements associated with combat diving. Information is provided concerning air consumption, underwater navigation techniques, and limited-visibility operations.

DURATION OF AIR

11-53. All combat diver missions require planning for the diver's air supply. Duration of air depends upon the diver's consumption rate, depth of the dive, and capacity of the cylinders. No mission can be effectively accomplished if the diver runs out of air prematurely. The diver must compute his air consumption rate. There is a standard formula that must be applied; current U.S. Navy diving regulations explain this formula. The formula is quite simple; however, it requires gathering actual "in-water" air consumption information. That information is then used to calculate expected consumption requirements at any given depth and exertion rate.

11-54. The CDS must designate a "turnaround pressure" for each dive. Turnaround pressure is what is left in the dive tanks when the dive must stop and a return to the surface begins. This amount is especially important when conducting a dive where there is an "overhead environment" and the diver cannot ascend directly to the surface. An overhead environment can occur by an obstruction (ship bottom search), physiology (decompression obligation), or the threat of enemy activity (observation or fire). Rules to plan turnaround are based on an expected reserve requirement such as having enough air to return to the start point and still react to emergencies. The most common rules from most to least conservative are as follows:
- Thirds: 1/3 in, 1/3 out, and 1/3 for contingencies.
- 1/2 + (x) psig – 1/2 of total tank pressure plus a designated reserve (usually 200 psig).
- Designated minimum remaining pressure (usually 300 to 500 psig at end of dive).

Chapter 11

> **WARNING**
>
> Diving in a true overhead environment involving penetration into an enclosed space is beyond the scope of activity for combat divers. It is extremely hazardous and requires specialized training and equipment that is not normally available to SF divers.

UNDERWATER NAVIGATION

11-55. The success of any type of infiltration is directly dependent upon the unit's ability to arrive at its objective. Underwater operations are no different. Underwater navigation poses additional challenges because of the limited opportunities for terrain association. These problems are compounded if the infiltration route is complicated by "doglegs" or course changes en route. The most reliable navigation technique is DR with a magnetic compass.

11-56. DR requires an accurate determination of distance and direction. The accuracy of this method increases as divers gain experience conducting practice swims. Accurate distance calculations are especially critical if the dive mission includes "doglegs" or other precision navigation requirements. Divers determine distance using two methods—pace (or kick) count and timed distance runs. Both methods require the diver to swim a measured course a number of times so that a valid average can be determined. It is imperative that the swim be conducted with the equipment load that will be worn on the mission so the equipment's drag can be factored in. If the divers are swimming a relatively short distance, the pace count method will be marginally more accurate. For longer distances, the timed method will prove more useful. Neither one of these methods allow for the effects of contrary currents. These must be taken into account for whenever calculating pace/time required to cover a given distance.

11-57. A good operational swimming speed (goal) for slick divers is 100 yards every 3 minutes. This pace was determined based on the requirements for swimming the C/C MK 25 UBA. Well-conditioned divers swimming O/C equipment slick (without extra equipment causing drag) may find this pace to be slower than optimal. This pace must be practiced repeatedly using a measured course until the swim teams develop a consistent speed. Divers burdened with mission equipment will have to conduct numerous practice swims to determine their actual (sustainable) swim speed.

Compasses

11-58. Directional information or control is provided by the compass. There are two basic types of compasses—those designed to be worn around the wrist or handheld and those ball-type compasses that are secured to some type of board.

11-59. A wrist compass consists of a fluid-filled capsule containing a stationary lubber line and a north-seeking arrow. Two stadia lines on a rotating bezel enable the user to maintain alignment of the north-seeking arrow with the course bearing. The azimuth may be referenced by 5-degree graduations marked on the rotating bezel surrounding the capsule. All features are luminous to enable the diver to operate the compass in limited visibility. The compass is most frequently worn on the wrist but can also be handheld or strapped to an improvised tactical (TAC) board or to the rucksack frame. To use the compass, a diver simply points the compass's lubber line toward the target. He then turns the bezel ring until the two stadia lines straddle the north-seeking arrow. As long as the diver swims keeping the compass level and the north-seeking arrow between the stadia lines, he will arrive on target. It is important that a diver ensure that the north-seeking arrow does not get stuck or frozen as may happen if the compass is held at an angle.

11-60. The ball-type compasses are similar to ship-mounted binnacle ones that are of a much higher quality than the aforementioned wrist compass. They are usually more accurate, easier to read, and easier to use. The diver should point the compass at the target, observe the desired azimuth, and swim the desired azimuth to the target, as it is read or set. It is best to use these compasses in a TAC board-mounted role. The TAC board is highly recommended for all operations. It is a form of console usually set up with the compass center-mounted, with watch and depth gauge mounted above. Field-expedient TAC boards can be made of wood or Plexiglas and are normally held in the diver's hands. It is recommended that units use the TAC board in the console mode, as it keeps essential instrumentation organized and readily visible to the diver.

Other Techniques

11-61. If pinpoint accuracy is absolutely essential, the tactical peek should be incorporated into the swim. Under these circumstances, one diver, on azimuth, exhales all his air and very slowly breaks the surface, just to the point where he can see the objective and verify his position. He then returns subsurface as quickly as possible without unduly disturbing the surface of the water. His dive partner stands by underneath him, ready to pull the surface diver back underwater if he loses buoyancy control and rises too high out of the water. This method requires extensive buddy team practice.

11-62. Underwater navigation can also be facilitated with acoustic beacons. Beacons consist of two parts—a sending unit (the beacon) placed in the water at the objective and a receiver unit used by the divers to hone in on the beacon. They are a component of the Diver's Support Set B. Beacons work like a sonar direction-finder, indicating signal strength and bearing, usually by a series of light emitting diodes (LEDs) that illuminate sequentially as signal strength and receiver alignment improve. Obviously this is only viable when returning to (or going toward) a previously emplaced beacon. This is especially useful when returning to a cache, a work/search site, or an extraction point. Unfortunately, the sending units have limited range, are susceptible to enemy intercept, and require prior emplacement of the sending unit.

11-63. Research and development is underway to provide SO divers with other effective methods of underwater navigation. These efforts include handheld inertial or Doppler navigation systems and submersible GPSs. Obviously underwater navigation is a critically important skill that all combat divers must practice frequently.

LIMITED-VISIBILITY DIVING OPERATIONS

11-64. Planners define limited visibility as diving under conditions where a diver cannot distinguish objects at a distance of 10 feet or less. Limited visibility is caused by particles suspended in the water or a lack of ambient light. Turbidity is the description or measurement of the amount of suspended particles in the water and their effect on visibility. Turbidity is caused or exacerbated by surface runoff, high winds, a rolling surf, upwellings, and extreme tidal changes. Divers can expect better visibility at high tide than at low tide; they should use tide tables in dive planning. A minus tide (a low tide lower than the mean low tide) will greatly reduce visibility. Rivers and their outflows, harbors, bays, and other near-shore areas within the littoral zone are all subject to increased siltation and other conditions that contribute to reduced visibility.

11-65. Training for diving under limited-visibility conditions is especially important. Diving in these conditions can be very disorienting to an unprepared diver. The lack of visual references exacerbates latent tendencies to claustrophobia, which may manifest themselves as significant psychological stumbling blocks. Ideally, dive teams should have considerable experience under good conditions before they begin limited-visibility operations. Maintaining contact between dive buddies and maintaining precise navigation are much more difficult. Practicing in a pool at night with the lights out, or with masks blacked out, will improve skills.

11-66. Equipment that should be available and used on limited-visibility dives includes the following:

- Buddy lines.
- Diving lights.
- Diving knives.
- UDT vest or BC.
- One compass per dive team.
- Heavy gloves regardless of temperature.
- Descending lines, safety lines, and surface floats if the tactical situation permits and the mission so requires.

11-67. Procedures for limited-visibility diving operations are simple common sense. If possible, the diver should make the first dive in a new area when visibility conditions are good (before diver's actions silt up the site). He then should work upstream facing the current and allowing the current to wash silt downstream away from the work area. He must proceed slowly and cautiously when visibility is poor. Unseen debris on the bottom (or in the water column) can cause impalements, punctures, cuts, lacerations, and entanglements. If the diver becomes entangled, he should work slowly and carefully to clear the entanglement. He should attempt to back out; the direction of entry is probably clear. He should dive in dirty water with one hand extended in the direction of movement. When swimming parallel to the bottom, the diver should keep one hand held underneath his body slightly touching the bottom to assist in maintaining a reference. He must always be conscious of surge and currents to avoid being carried into unseen rocks.

11-68. A diver should verify direction frequently. He should make an effort to maintain a definite course or swim pattern and to know his position at all times. If not attached by a buddy line, he should check his swim buddy more often than usual and maintain close contact. If possible, he should use an anchor line or descent line for reference to assist in maintaining orientation. Judging the rate of ascent may be difficult. A diver should always use the watch and depth gauge in combination and maintain buddy contact.

11-69. Underwater searches in limited visibility require strict control measures to ensure diver safety and adequate coverage of the search area. The smaller the object to be found or recovered, the greater the degree of control required. Tended line or jack stay searches are the most effective. All must be practiced dry before divers are committed to the water. When conducting working dives (for example, search or recovery), the detachment should consider using a single diver on a tether, tended from shore or a boat. The safety diver (also tended on a tether) would then be able to clip onto the stricken diver's tether and follow it directly to him without concern for visibility issues.

11-70. In coastal areas, limited visibility in water may be caused by man-made pollution. Divers should pay particular attention to disinfecting their ears and ensuring that any cuts or abrasions are treated to prevent infection.

CONTAMINATED WATER DIVING

11-71. River mouths, estuaries, harbors, and bays located along an enemy coastline are areas of military interest where SFODs may find themselves operating. These areas are usually economically important and have a high probability of human activity that would result in significant pollution. Special characteristics of the local topography and hydrography may combine to further concentrate existing pollutants. In addition to tactical requirements, SFODs performing other missions may occasionally be called upon to dive in the vicinity of harbors, sewers, or industrial outlets that discharge contaminated waters.

11-72. The most common pollutants are sewage, surface or agricultural runoff, chemical or manufacturing wastes, and petrochemicals (usually as a result of spills or military operations). They present special problems when viewed from a military operations perspective. Working in or around these pollutants will expose detachments to disease or chemical hazards. Either threat has the potential to incapacitate personnel and destroy mission-essential equipment. MAROPS detachments must be prepared to deal with the event of operating in contaminated waters.

11-73. Divers conducting operations in polluted water are especially vulnerable to skin irritations and ear infections. Other more virulent pathogens (for example, hepatitis, E. Coli, salmonella) pose serious, potentially life-threatening risks to unprepared or unprotected divers. When planning a dive operation in waters that are known to be polluted, protective clothing must be used and appropriate preventive medicine procedures taken. The most effective medical protective measures are an active vaccination program with emphasis on blood and waterborne pathogens and a rigorously performed postdive decontamination program.

11-74. Physical protection is only afforded by a complete barrier system, usually a suitable commercial-grade, vulcanized rubber or polyurethane-coated, hooded dry suit with a full-face mask. Wet suits do not provide adequate protection for diving in contaminated waters. Full-face masks are essential to prevent the accidental ingestion of polluted water. A diver who gets polluted materials into his mouth may have both physiological and psychological problems.

11-75. Divers wearing full body protection (usually dry suits) or diving in waters with an elevated temperature are subject to hyperthermia—an elevated body core temperature (Figure 11-4).

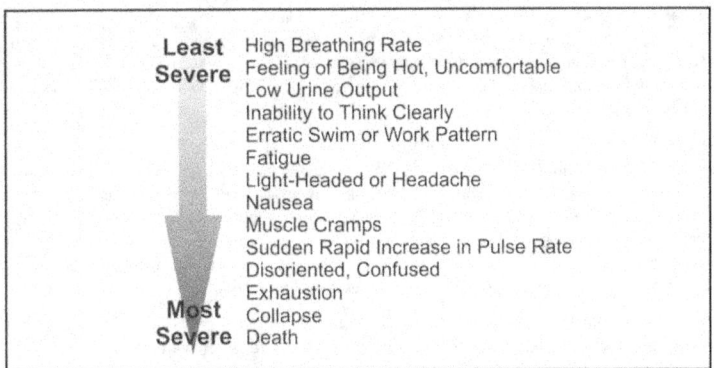

Figure 11-4. Hyperthermia symptoms

11-76. Toxic materials or volatile fuels leaking from barges or tanks can irritate a diver's skin and damage his equipment. Oil leaking from underwater wellheads or damaged fuel tanks can cause fouling of equipment. Aviation fuel is particularly damaging to dive equipment. It (and most other petrochemical-based pollutants) will actually dissolve rubber or neoprene components. Before diving in polluted water, all dive team members must be thoroughly briefed on the possible hazards.

11-77. Physical decontamination of the diver and his equipment is essential. All decontamination procedures should start with the diver still encapsulated in his gear and breathing compressed air. Personnel should not break the seal on the diver's gear until an initial decontamination is completed. The best and most thorough decontamination method is to adopt procedures similar to those used in a full-blown chemical decontamination. Obviously, this method works best in a nontactical environment where sufficient supply support is available. Operating in the field (nontactical) without a dedicated decontamination team requires some improvisation. The safest, simplest decontamination method is to use multiple washes with a strong detergent (emulsifies petrochemicals and provides a degree of sterilization), and multiple rinses preferably using hot water. Equipment that cannot be decontaminated should be bagged for disposal. If in an environmentally sensitive area, personnel should make sure that runoff from the decontamination process does not further contaminate the affected waterway.

11-78. When operating in a tactical environment, options are more limited. Divers should at least attempt a freshwater rinse before removing their dry suit and equipment. They should carefully remove the equipment while attempting to avoid unnecessary contact with contaminated surfaces, bag the equipment for disposal, and then wash exposed skin with a disinfectant and dry. They should always be prepared to discard contaminated equipment, especially if it has been exposed to petrochemicals or solvents. The deterioration of the materials in the dive gear cannot be reversed and it may render the equipment unsafe for further use.

11-79. Accidental exposure to nuclear radiation may occur due to proximity to weapons systems, or occasionally, substances in the natural state. Exposure to radiation can result in serious damage to the body and its systems. NAVSEA publishes a list of safe tolerance levels. Before diving on a vessel with nuclear capability, the SFOD leader must consult the radiological control officer. All divers will wear a thermoluminescence dosimeter (TLD) or similar device and should be made aware of the locations of possible exposure. The detachment must collect and process TLDs postdive IAW the supported vessel's SOPs, so that diver exposure can be accurately monitored.

ALTITUDE DIVING

11-80. Current U.S. Navy diving regulations contain the most recent information pertaining to altitude diving. The following information provides general guidance to SFODs and commanders required to conduct or supervise high altitude (>999 feet above sea level) diving operations. This information is intended to inform divers of the limitations imposed by the increased altitude. Any unit engaged in high-altitude diving must possess hard copies of current U.S. Navy diving regulations on hand whenever conducting operations.

11-81. Because of the reduced atmospheric pressure, dives conducted at altitude require more decompression than identical dives conducted at sea level. Standard air decompression tables, therefore, cannot be used as written. Some organizations calculate specific decompression tables for use at altitude. An alternative approach is to correct the altitude dive to obtain an equivalent sea-level dive, and then determine the decompression requirement using standard tables. This procedure is commonly known as the cross correction technique and always yields a sea-level dive that is deeper than the actual dive at altitude. A deeper sea-level equivalent dive provides the extra decompression needed to offset effects of diving at altitude.

11-82. No correction is required for dives conducted at altitudes between sea level and 300 feet. The additional risk associated with these dives is minimal. At altitudes between 300 and 1,000 feet, correction is required for dives deeper than 145 fsw (actual depth). At altitudes above 1,000 feet, correction is required for all dives.

11-83. To simplify calculations, current U.S. Navy diving regulations give corrected sea-level equivalent depths and equivalent stop depths for dives from 10 to 190 feet and for altitudes from 1,000 to 10,000 feet in 1,000-foot increments.

11-84. Upon ascent to altitude, two things happen. The body off-gasses excess nitrogen to come into equilibrium with the lower partial pressure of nitrogen in the atmosphere. It also begins a series of complicated adjustments to the lower partial pressure of oxygen. The first is called equilibration; the second is called acclimatization. Twelve hours at altitude is required for equilibration. A longer period is required for full acclimatization.

WARNING

Altitudes above 10,000 feet can impose serious stress on the body resulting in significant medical problems while the acclimatization process takes place. Ascents to these altitudes must be slow to allow acclimatization to occur and prophylactic drugs may be required. These exposures should always be planned in consultation with a DMO. Commands conducting diving operations above 10,000 feet may obtain the appropriate decompression procedures from NAVSEA.

11-85. If a diver begins a dive at altitude within 12 hours of arrival, the residual nitrogen left over from sea level must be taken into account. In effect, the initial dive at altitude can be considered a repetitive dive, with the first dive being the ascent from sea level to altitude. Current U.S. Navy diving regulations give the repetitive group associated with an initial ascent to altitude. Using this group and the time at altitude before diving, the diver must check the Residual Nitrogen Timetable to determine a new repetitive group designator associated with that period of equilibration. He determines sea-level equivalent depth for his planned dive using the current U.S. Navy diving regulations. From his new repetitive group and sea-level equivalent depth, the diver determines the residual nitrogen time associated with the dive. He then adds this time to the actual bottom time of the dive.

11-86. The exact procedures for altitude diving, to include work sheets and tables, are in the current U.S. Navy diving regulations. SFODs should not conduct altitude dives without referencing it.

COLD WEATHER DIVING

11-87. SF divers rarely have the luxury of diving in tropical environments, and mission requirements usually do not allow the detachment to wait until the water is warmer. The SF diver that finds himself forced to dive in cold water must be aware of the potential risks and knowledgeable in the techniques and equipment available to mitigate those risks and accomplish the assigned mission.

11-88. Water conducts heat 25 times faster than air. Because of its greater conductivity, water does not have to be extremely cold for it to adversely affect the diver's ability to conduct his mission. Even relatively warm (70 degrees F) water will eventually cause the unprotected diver to suffer hypothermia (Figure 11-5, page 11-24). The major effect of cold water diving is body heat loss or hypothermia. Other and sometimes related effects include cramps, increased fatigue, loss of strength, increased air consumption, and frequent urination. More serious effects are the loss of ability to recall information and to concentrate, increased stress, increased susceptibility to bends, occasional loss of muscle control, and symptoms similar to those of nitrogen narcosis.

SHORT-DURATION DIVING IN EXTREMELY COLD WATER

11-89. This section provides general guidelines to assist dive personnel in planning and conducting short-duration diving operations in cold water situations where the surface water temperature is 37 degrees F and below. Typical missions may include search or reconnaissance dives; inspecting ice conditions for cracks, pressure ridges, or thin spots before crossing by vehicles; and supporting amphibious operations.

11-90. Ice diving normally is not required of SF divers. SFODs that may find it necessary to conduct ice dives should refer to the specialized procedures outlined in current U.S. Navy diving regulations. They must conduct internal (detachment) training to ensure that all personnel are adequately prepared to perform both the diving and the support tasks required to conduct a safe operation. The dive supervisor should keep the following general rules in mind when planning extreme cold weather diving operations:

- Divers should have protected areas for dressing and undressing. After suiting up, a diver should not delay his water entry. Time on surface must be kept to a minimum.
- A chilled diver should never go into the water. He must be completely rewarmed before reentering the water. A diver should not remain in the water after he begins to shiver.
- A diver should make sure that equipment does not restrict circulation.
- Mittens should be large enough for protection, but not so large as to decrease dexterity. Upon surfacing, a diver should not remove his mittens until he is able to dry his hands and put on warm gloves.

11-91. The primary concern is to dive warm. **A diver should start warm and stay warm.** Cold divers start with a heat deficit that cannot be recovered from. This deficit only gets worse as diving operations continue. Factors that should be considered are the general state of the divers' health, physical and psychological conditioning, and the degree of training for the specific environment. Divers should be well rested and fed; they should generally consume a high protein meal 2 hours before the dive. The protein will supply energy gradually over a long period of time. Consuming simple sugars shortly before diving and between dives to provide quick energy also helps. Energy can be conserved and thus warmth retained by keeping warm before diving. Wearing a jacket and staying out of the wind en route to the dive site are simple but often overlooked precautions. Physical protection from the cold is achieved through the use of a wet or dry suit during the dive.

GENERAL EQUIPMENT CONSIDERATIONS

11-92. When selecting equipment for a cold weather diving operation, divers should use only equipment with proven cold weather reliability. They should also consider the specific climatic conditions of the dive site, the probable lack of facilities to support the operation, the lack of communications, and the distance over which support must be provided.

Chapter 11

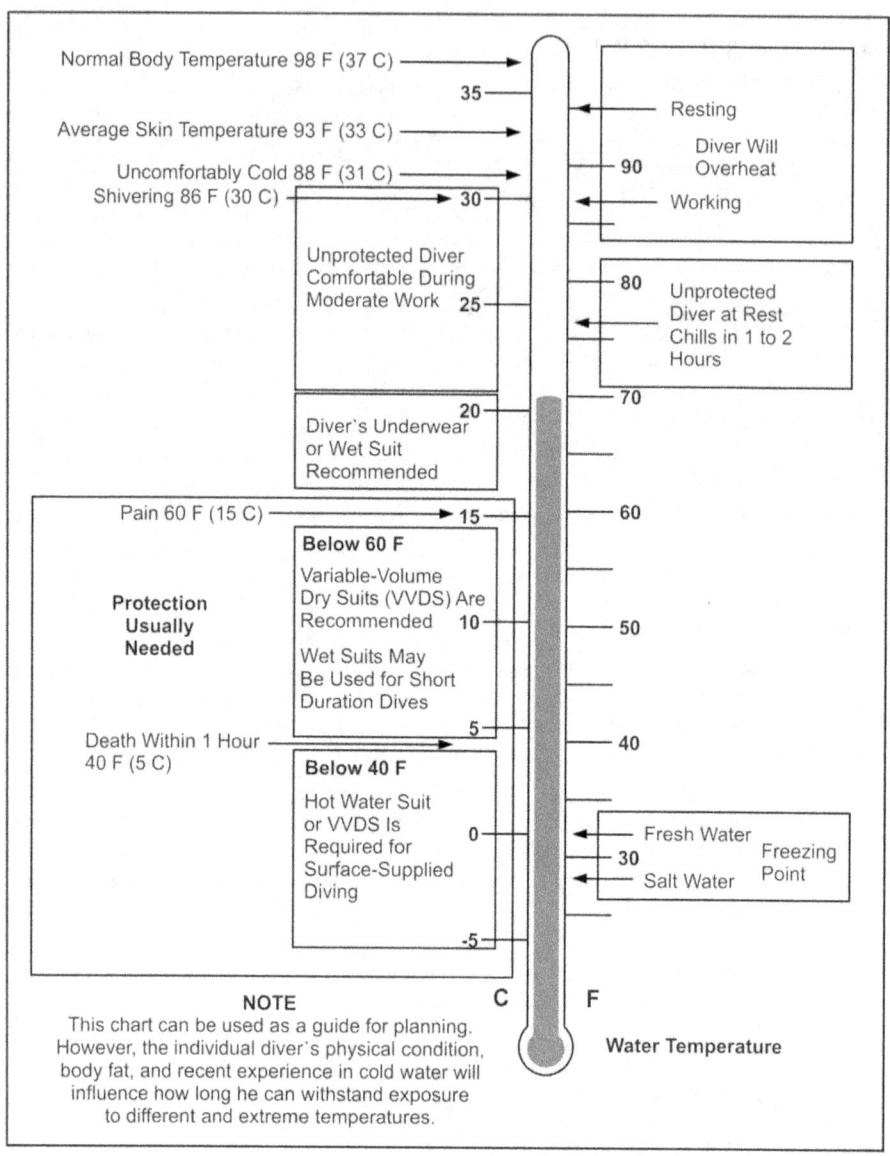

Figure 11-5. Water temperature protection chart

11-93. The selection of the regulator must be made very carefully for cold weather diving operations. The ANU list specifies which regulators have been found suitable for use in water temperatures less than 37 degrees F. This cutoff is used because at that temperature the expansion of compressed air inside the regulator can result in sufficient cooling to cause icing in the first stage and a potential free-flow.

11-94. Physical protection from the cold is provided by wet or dry suits. They protect the diver by providing an insulating layer between the diver and the surrounding water. Figure 11-6 depicts the differences in wet and dry suit functions.

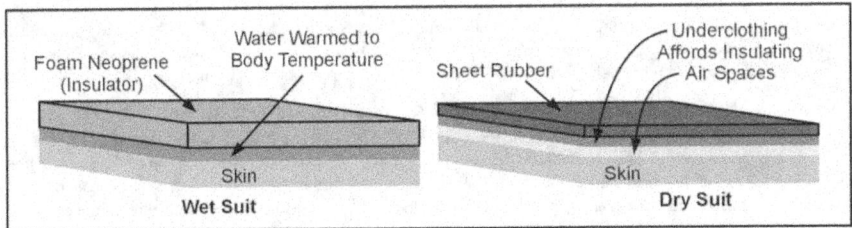

Figure 11-6. Wet and dry suits

WET SUIT CHARACTERISTICS

11-95. A wet suit works by restricting the flow of water into the suit and preventing the exchange of water inside the suit (warm) with water outside the suit (cold). Any water that infiltrates the suit should remain trapped in a thin layer next to the diver's body where it is warmed by body heat. Good seals at the wrists, neck, and ankles, and a sealing surface behind the zipper and on any of the suit's mating surfaces help to prevent or restrict this exchange of water. These features coupled with quality neoprene formulations make for wet suits that are better insulators, have better recovery from compression, are more flexible and durable, and conform to the diver's body (fit) more readily.

11-96. How a wetsuit fits the diver is its most important characteristic. A thin suit that has been properly fitted will provide far more protection for the diver than a thicker suit that does not fit properly. An excessively tight suit restricts circulation, especially in the extremities and contributes to the loss of body heat. However, there should not be any areas where the suit does not contact the diver's body. A properly fitted suit is one that envelops the diver snuggly without restricting movement or circulation. Any place where a gap exists between the diver and his wet suit is a potential point of water circulation or exchange. Each inhalation and exhalation, kick, or arm movement serves to pump warm water out of the wet suit and siphon cold water into it. The diver will rapidly become chilled or hypothermic as his body struggles to warm the cold water that is continuously flooding his suit.

11-97. The wet suit has been satisfactorily used in extremely cold water. However, it compresses at depth and greatly reduces its insulating capability. In the thicker suits, a diver's movement is restricted and the additional buoyancy becomes problematic. Other problems include the diver chilling after leaving the water and difficulty in drying the suit. In addition to the advantages of availability and simplicity, wet suits are unlikely to have material failure, allowing sudden flooding as may happen with all dry suits. Generally, for short-duration dives of limited depth, the wet suit is satisfactory, even in water temperatures below 32 degrees F. A wet suit consisting of the following items has been satisfactorily used in water at a temperature of 30 to 32 degrees F:

- Pants—5-mm neoprene.
- Jacket—5-mm neoprene.
- Hood—5-mm neoprene.
- Vest or hood combination—3-mm neoprene.
- Gauntlet type mitten—5-mm neoprene.
- Inner sock—3-mm neoprene.
- Booties—5-mm neoprene

THE DIVER AND WET SUIT USE

11-98. To be effective, each diver should have an assigned wet suit that has been individually measured and fitted. His most effective dress consists of the various pieces identified in the previous paragraph. They

Chapter 11

allow the diver the flexibility to choose the appropriate level of protection based on the water temperature, the expected level of exertion, and personal comfort. Semicustom wet suits are the preferred solution because generic sizing does not provide an adequate range of sizes to fit and protect all divers in extremely cold water. To meet this requirement, commands may purchase off-the-shelf suits from reputable manufacturers or commercial sources IAW the guidance promulgated in the ANU list. After the suit has been sized and issued to the diver, he is responsible for its use and maintenance. Because of issues like individual sizing and personal hygiene, suits should not be swapped off between divers.

11-99. When divers are using wet suits, they should consider the following points:

- It is especially important to insulate the head when diving in cold water. The unprotected head can account for more than 50 percent of the body's entire heat loss. The use of cold water hoods with chin cups, neoprene booties, and mittens or gloves will also help prevent heat loss at the extremities. A diver should not cut ear holes in the hood, for that will cause an unnecessary loss of body heat. He should also glue shut all small holes in the suit.
- A one-piece suit helps conserve heat. An attached hood adds more protection. Both features serve to restrict water exchange in the suit. Pullover jackets without zippers are much warmer than jackets with front and cuff zippers, but they are less convenient. Hooded vests worn under conventional zipped jackets are a good compromise between convenience and warmth.
- Custom or semi-custom suit modifications can add significantly to warmth. A diver can increase his warmth by adding a spine pad down the back to reduce water seepage and putting a backing flap behind zippers. Wrist and ankle cuffs of smooth neoprene that seal against the skin significantly reduce water circulation to help keep the body warm. Eliminating the nylon lining at the cuffs prevents wicking cold water into the suit.
- A diver should not exercise in an attempt to get warm. He will stimulate circulation to the extremities where heat loss is greatest. He should keep the breathing rate down if possible. Heating cold air inhaled from the tank requires a great deal of energy. If a diver begins to shiver uncontrollably, it is time to terminate the dive.
- A diver should preheat the suit by filling it with warm (not hot) water immediately before diving in cold water. He should avoid a high entry into cold water, particularly on repetitive dives. If he eases into the water, any warm water in the suit can be retained. Boots should be worn inside the pant legs. If worn outside, each kick will pump warm water out of the boot and suck cold water in.
- The SFOD should provide a sheltered area protected from the wind (heated, if possible) for divers to reduce evaporative cooling from wet wet suits.

DRY SUIT CHARACTERISTICS

11-100. Dry suits are the other option for protecting divers in extremely cold water. They provide superior thermal protection both in and out of the water. Modern dry suits are generally one piece with attached boots and a hood. Entry into the suits is usually through a zippered opening. They function by encapsulating the diver in a dry microenvironment that is maintained by watertight seals at the wrist, neck, and zipper.

11-101. Dry suits can be worn with a variety of clothing underneath. Depending on water temperature and expected exertion levels, the diver can wear various types and amounts of long thermal underwear to keep warm. Underwear specifically designed for use with the suit is best. The type used will vary with the mission and individual preference. Too much underwear will be bulky and can cause the diver to overheat before entering the water, thus perspiring and later chilling. For scout swimming purposes, dry suits proved satisfactory when used with a standard military uniform underneath. Used this way, they can greatly reduce the time required on the beach after infiltration and before entering the land phase of the operation.

11-102. Dry suits are available in various weights and constructions. While all suits work on the same basic principle, they differ greatly in design and quality. Dry suits appropriate for diving are usually of the variable volume type. Nondiving applications can use the somewhat simpler nonvariable volume suits.

11-103. The three suit materials or construction methods most commonly encountered are vulcanized rubber, trilaminate, and neoprene. The most durable are the vulcanized rubber suits. However, they are heavier and less flexible. Divers performing heavy work, diving in contaminated waters, or diving where there are significant risks of entanglement and snagging or tearing the dry suit should consider using the commercial weights of the vulcanized rubber suit. They are extremely durable, easily repaired by trained personnel, resistant to many chemical or biological contaminants, and readily decontaminated.

11-104. Trilaminate suits are lighter weight and more flexible, but they are not as durable. They are constructed of multiple (usually three) layers of material that have been laminated (bonded) together—normally an outer wear layer, a middle waterproof membrane, and an inner wear layer. Their flexibility and relative comfort make them more acceptable when the diver is expected to be extremely active, such as during an infiltration swim. Trilaminates also make ideal exposure suits for personnel conducting surface support operations or CRRC infiltrations. Unfortunately, most trilaminate suits are not as puncture resistant as the other types.

11-105. Neoprene suits are constructed of the same material as wet suits. Because the neoprene is itself an insulating material, neoprene suits do not necessarily require additional insulating undergarments. However, because it is constructed of neoprene, the suit is subject to the same compression at depth as wet suits. This factor will reduce its insulating properties somewhat. A neoprene suit that floods out (loses its integrity and fills with water) will still provide some thermal protection. Newer models take advantage of neoprene's stretchiness with closer tailoring to make a suit that conforms more closely to the diver. Older models were often very bulky. Neoprene suits, because of their inherent buoyancy, also require more weight to make the diver neutrally buoyant. Neoprene suits are not as readily decontaminated as the other types of dry suit nor are they as resistant to some pollutants.

VARIABLE-VOLUME DRY SUITS

11-106. Variable-volume dry suits (VVDSs) have been adapted for diving by the addition of a low-pressure inflator and various exhaust valves. As the diver descends, the suit material is subjected to normal compression. In a variable-volume suit, introducing additional air into the suit through the low-pressure inflator equalizes pressure to counteract the effects of suit squeeze. A diver must take care to prevent the suit from becoming overly buoyant, especially on ascent.

> *Note.* Divers must receive proper training in and be thoroughly familiar with the dry suit before they attempt to use it operationally. Any diver planning to use a VVDS in any water, especially while diving under ice, should be thoroughly familiar with the functioning of the suit and the manufacturer's operational literature. Divers should have experience with the suit before deployment to the dive site. In addition to gaining experience using the suit, the diver should get a thorough checkout on the proper method of donning and doffing, suit care, and maintenance. Three dives to increasing depths are normally sufficient for an experienced diver to become familiar with the suit.

11-107. The two greatest risks to the diver using a dry suit are floodout and blowup. Floodout is the sudden compromise of the suit's watertight integrity usually caused by a tear or the failure of one of the watertight seals. When the suit fills with (cold) water, it suddenly loses its buoyancy and quickly becomes very negative. The diver may also experience thermal shock caused by the sudden exposure to the very cold water. Blowup is the sudden loss of buoyancy control caused by the rapid expansion of air trapped inside the suit. It is most commonly caused by an inadvertent ascent without venting the suit, a stuck (often frozen) low-pressure inflator, or a closed or jammed exhaust valve. Proper training prepares the diver to deal effectively with these emergencies. Cause of death for untrained divers is usually drowning, air embolism, DCS, or a combination of these factors.

11-108. The VVDS should not be subjected to an ambient outside temperature below 32 degrees F before a dive. Such exposure can result in super cooling of the inlet and exhaust valves and can cause icing on immersion. If it is necessary to expose the suit to extreme temperatures before diving, the diver should lubricate the valves with silicone. He should attempt to rewarm the valves before entering the water.

11-109. Nonvariable-volume dry suits (usually trilaminate construction) are satisfactory for surface swimmers. They provide exceptional exposure protection for detachments conducting long-range surface transits in CRRCs. They can also be used for underwater swimming if precautions are taken to prevent suit squeeze.

11-110. Use of dry suits is limited by the following:
- Horizontal swims may be fatiguing due to suit bulk.
- If the diver is horizontal or head down, air can migrate into the foot area and become trapped there causing the diver to lose attitudinal control. Because there are no exhaust valves in the legs, the diver may find himself ascending feet first and out of control. The expanding air in the legs can cause overinflation, loss of buoyancy control, and a rapid uncontrolled ascent. The expanding air trapped in the feet may also cause the diver's fins to pop off.
- Inlet and exhaust valves can malfunction on variable-volume suits.
- A collapsing or parting seam or zipper or a rip in the suit can result in sudden and drastic loss of buoyancy and in thermal shock.
- With variable-volume suits, extra weight is required to achieve neutral buoyancy. It is generally best to use oversized weights.

11-111. The SFOD should keep the following spare parts and repair items on hand for suit maintenance and repair:
- Sharp, heavy-duty scissors for cutting neoprene.
- Needle and thread for seam repairs (15-pound test nylon fishing line works well).
- Neoprene rubber cement.
- Extra neoprene material for cuff, face seal, and suit repair.
- Spare exhaust and inlet valves for variable-volume suits.
- Spare low-pressure inflator hoses for variable-volume suits.
- Large amount of assorted O-rings, silicone spray, and zipper wax.

SAFETY CONSIDERATIONS

11-112. Personnel taking part in cold or ice water diving operations should know the emergency procedures for the following situations:
- *Surface Personnel Falling Into the Water.* Prevention is the key. Buddy teams should work around the entry hole and keep unnecessary personnel away from the hole.
- *Stricken Diver.* The buddy diver should get the stricken diver to the surface ASAP.
- *Breathing System Failure.* Diver should switch to backup system, notify partner, and surface together.
- *Suit Failure.* Diver should surface immediately.
- *Uncontrolled Ascent.* Diver should exhale continuously during ascent, relax against the ice, and relieve pressure from the suit. If applicable, he should signal on the tether and proceed to the entry hole or wait for assistance from his buddy diver.
- *Lost Diver.* Diver should ascend to the surface immediately, relax as much as possible, and attempt to return to the entry hole or wait for assistance.

DIVE OPERATIONS PLANNING

11-113. For convenience, the planning of SF dive operations is normally divided into vertical and horizontal dives. Vertical dives are usually open-circuit such as working dives that range from annual requalification to reconnaissance or search and recovery dives. Horizontal dives are normally closed-circuit infiltration or exfiltration and training dives. Each type of dive has its own unique requirements. Pertinent information is included in the Dive Safety and Planning Checklist, and current U.S. Navy diving regulations. Personnel should follow specific steps for conducting a vertical (deep) training dive. They should—
- Prepare the dive site.

- Update the situation.
- Conduct the dive supervisor personnel inspection.
- Supervise the entry of the dive team into the water.
- Get an OK from all divers. Get an OK from the group leader when his team is at the dive buoy and is prepared to descend.
- Begin the dive time as soon as the first diver's head leaves the surface of the water.
- Record all times and maintain elapsed times.
- Ensure normal descent procedures are followed. The entire group will maintain contact with the ascent or descent line and will descend no faster than 75 fpm.
- Ensure that the total time of the dive does not exceed the no-decompression limits.
- Record the total time of the dive. The dive is complete when the last diver's head breaks the surface of the water. Receive the total time of the dive and the maximum depth reached from the group leader.
- Supervise the water exit.
- Ensure that all personnel are accounted for before leaving the dive site.

11-114. Special equipment required for deep or vertical dives includes the following:

- An ascent and descent line.
- A safety line from the surface craft to the dive buoy.
- A safety line from the dive buoy to the safety boat.
- A safety line from the dive platform for divers.
- A securing line for the ladder.
- An inflatable safety buoy or pumpkin.
- Snap links.
- Lead weights or anchor system (weight clump sufficient to keep the bottom of the descent line at the desired depth).
- Small lift bag (50 to 100 pounds to facilitate the recovery of the weight clump).
- Backup regulators.
- A dive flag.
- A safety boat.
- Extra equipment.
- A diver's recall system.

11-115. Personnel should follow specific steps for conducting horizontal dives. The steps are as follows:

- Before loading the boat with divers and equipment, the dive supervisor will inspect the boat. Current U.S. Navy diving regulations contain a diving boat safety checklist.
- Divers will enter the boat when instructed to do so by the dive supervisor. They will maintain three points of contact when entering the boat without stepping on the gunwale. The buddy team will assist each other.
- Seating will be arranged with equal weight distribution in mind and with the buddy teams seated facing each other. Divers will don their fins.
- All boats will be under the control of the dive supervisor from the loading point to the dive site. While en route, the dive supervisor will initiate purge procedures when using a closed-circuit UBA. Upon reaching the dive site, the divers will signal the dive supervisor that they are ready to enter the water.
- The dive supervisor will command, "prepare to mount the gunwale," at which time each diver will don his mask and place his regulator in his mouth. (Closed-circuit divers will stay alert for the next command.)
- On the command, "mount the gunwale," the divers will carefully sit on the gunwale and direct their attention to the dive supervisor.

- The dive supervisor will direct the boat driver to "back it down" (the driver will place the boat in slow reverse). Then, from bow to stern, the CDS will dispatch the divers into the water with the commands "prepare to enter the water" (buoy man tosses the buoy rearward, and each diver looks right or left over his shoulders to ensure his water entry point is clear) and "enter the water" (each diver executes a proper backroll into the water).
- After entering the water and executing a proper ascent, each diver will exchange the OK signal with the dive supervisor. Divers will exchange the descend signal with the dive supervisor who will record the time the divers go subsurface. He will monitor the dive until all divers are safely ashore.
- The dive supervisor will ensure that he has proper accountability of all personnel and equipment. He will then supervise postdive operations, which include debriefing, maintenance of equipment, and record posting.

11-116. Special equipment for horizontal dives might include the following:
- Buoys.
- Buddy lines.
- Team swim lines.
- Snap links.
- Chemlites.
- A diver recall system.
- Extra equipment.
- Spotlights.
- Communications.
- A vehicle at the BLS.

11-117. The preceding guidelines are not intended to be all-inclusive. They should be seen as tools to aid the CDS in planning and executing his assigned duties. Obviously, requirements will change based on the mission and the situation. The prudent dive supervisor will conduct a thorough risk analysis and incorporate all of the pertinent information when he prepares and presents his dive plan and brief.

Chapter 12

Closed-Circuit Diving

The U.S. Army, the USN, and the USMC have adopted the MK 25 UBA for use by SOF. The MK 25 UBA is light, easy to operate, and eminently suited for the conduct of maritime special operations. The information presented in this chapter contains diving information from the U.S. Navy and NAVSEA. The procedures outlined must be followed, as they constitute the official doctrine concerning use of the MK 25 UBA.

SAFETY CONSIDERATIONS

12-1. Divers must make sure they adhere to and follow the data provided below. Diving safety and successful operations of the MK 25 UBA depend upon the following:
- Competence and performance of operation.
- Operations planning.
- Adherence to approved operating, emergency, and maintenance procedures.

12-2. The safety guidelines stated in Figure 12-1, pages 12-1 through 12-3, apply to operation and maintenance procedures. Personnel must thoroughly understand and comply with them to ensure the MK 25 UBA operates safely and efficiently.

General Warnings
Monitor the gas supply for the MK 25 UBA because it has no positive reserve. Use the pressure gauge that is provided for this check.
Monitor the MK 25 UBA breathing bag because it does not have a dump valve and overinflation may occur. Recharge the breathing bag at minimum possible depth. Do not overcharge.
Stay aware that the MK 25 UBA pressure reducer has an operating range of 10 to 207 bar (145 to 3,000 psig).
Do not charge oxygen cylinder above 207 bar (3,000 psig).
Do not exceed fill rate of 14 bar (200 psig) per minute.
Canister and Scrubber Warnings
Use only NAVSEA-authorized carbon dioxide absorbents with the MK 25 UBA. Use of other hydroxide chemicals is not currently authorized.
Thoroughly settle bed of carbon dioxide-absorbent granules. If improperly filled, channels that permit gas to bypass the absorbent may form, causing elevated levels of carbon dioxide in the breathing loop. Do not overfill canister past fill mark.
Fill the canister outdoors (or in a well-ventilated space) over a container suitable for disposal of the absorbent.
Do not mix different brands of absorbent in the same canister. Granules could pulverize, leading to channeling of the absorbent.
Avoid contact with carbon dioxide-absorbent dust; it will irritate the eyes, throat, and skin. Take appropriate precautions to avoid breathing absorbent dust or getting it into your eyes or on your skin. Do not stand downwind of canister while filling or emptying.

Figure 12-1. Safety guidelines

Chapter 12

Canister and Scrubber Warnings (continued)

- Do not use the last 1 inch of carbon dioxide absorbent in the container because absorbent dust accumulates in the bottom of the container.
- Do not overfill the canister. Overfilling can lead to canister flooding.
- Comply with hazardous material regulations for the appropriate state.
- Refer to NAVSEA regulations for MK 25 UBA 0/1/2 canister duration limits.

Oxygen Handling Warnings

- Ensure the cylinder valve on the oxygen cylinder is shut. Before working on pneumatic components, vent pressure from the pneumatic subsystem.
- Never mix different brands of MIL-G-27617, Type III greases. Ensure all old grease is removed before applying new grease.
- Do not allow oil, grease, or any other foreign material to come in contact with high-pressure oxygen. Such material exposed to oxygen under high pressure may explode or ignite.
- Keep sparks and flames away from oxygen systems. Secure electrical equipment in the immediate area during maintenance of oxygen systems.
- Purge cylinders by pressuring with oxygen to 60 to 120 psig because empty cylinders may contain residual nitrogen. Relieve pressure and purge again to charge the cylinder.
- Check the one-way valves in both hoses before each operation.
- Install cylinder valve cap and reducer plug to prevent dirt or water from entering the reducer. This practice will prevent the potential for an oxygen fire and possibly degrading performance of the oxygen reducer.

Maintenance Warnings

- Do not install expired components.
- Ensure poloxamer-iodine cleansing solutions are carefully diluted. Use of solutions with greater concentrations of iodine will cause degradation of the rubber components of this equipment. Failure to thoroughly rinse all poloxamer-iodine solution from the equipment may result in lung irritation or long-term degradation of the equipment's rubber components.
- Wear goggles to prevent chemicals from splashing in eyes. If sanitizing agent splashes in eyes, rinse with large amounts of water and consult medical personnel.
- Open cylinder valve slowly to avoid heat generation that may burn the lower spindle seat.
- Always assume pneumatic subsystem is pressurized. Before assembly or disassembly, vent pneumatic subsystem by depressing manual bypass button on the front of the unit.

General Cautions

- Ensure only qualified MK 25 UBA technicians perform maintenance. However, qualified divers may perform premission, predive, and postdive actions under the guidance of a technician.
- Do not interchange the rotary valve (barrel) and housing. If either part needs repair, replace the entire mouthpiece valve.
- When inspecting breathing bag connecting pieces for secure attachment, do not lift or pull bag material from glued seams or use fingernails to pry connecting pieces from the breathing bag.
- To avoid damaging canister lid seal, do not overtighten canister lid to canister housing.
- Do not soak the urethane canister housing, lid, or associated soft goods in a vinegar-water solution.
- Attach breathing bag to demand valve with care; connecting piece cap nut can be easily cross-threaded.
- Do not overcharge the breathing bag. Damage to the system may occur. The breathing bag should be firm, but not stretched.

Figure 12-1. Safety guidelines (continued)

Closed-Circuit Diving

General Cautions (continued)
▫ While rinsing the canister components, keep them separate from the other UBA components. When mixed with water, the absorbent produces a caustic base mixture that can damage rubber and plastic components.
▫ Before proceeding, read and understand the purpose and precautions of the purge procedures.
▫ Make sure all MK 25 UBA components are completely dry before reassembly and storage to prevent mildew formation, bacteria growth, and material rot.
▫ Ensure proper handling of valve body to avoid leakage. Handle carefully to avoid scratching body.
▫ Use extreme care not to damage thread on gauge line or connecting line while removing reducer.
▫ Do not allow pressure to exceed 200 psig while testing relief valve.
▫ Do not use sharp tools on the boot.

Figure 12-1. Safety guidelines (continued)

DESCRIPTION

12-3. The general description of the system provides an overview of the UBA and its principal components. The gas flow pattern, adding oxygen to the breathing bag, and operational duration of the MK 25 UBA are also addressed.

> **DANGER**
> Omission of any operating procedures may result in equipment failure and possible injury or death to operating personnel.

12-4. In addition to standard predive guidelines, planners must consider mission requirements for using the MK 25 UBA. SF, USMC reconnaissance units, and USN SEALs use the MK 25 UBA in shallow-water operations. They must also use it in conjunction with an approved life preserver. Personnel should only use MK 25 UBA-approved life preservers IAW with NAVSEA guidance.

12-5. The MK 25 UBA is worn in the front of the swimmer and is attached with two harness straps (neck and waist). The mouthpiece is held in place with a head strap. All system components are attached to or contained in an equipment housing. The oxygen cylinder is secured in place at the bottom of the equipment housing with two straps and is connected through a cylinder valve to a reducer, which reduces cylinder gas pressure. Gas flows from the reducer to the demand valve assembly. This valve controls oxygen flow into a single breathing bag. The swimmer receives breathing on demand from the breathing bag through the inhalation hose and the mouthpiece. The mouthpiece contains a rotary valve that, in the dive (up) position, supplies the breathing gas and, in the surface (down) position, isolates the gas flow loop from the ambient atmosphere or water conditions. The exhalation hose also connects to the mouthpiece and takes exhaled gas to the soda lime canister (carbon dioxide scrubber). Both the inhalation and exhalation hoses are fitted with one-way valves (discs) to ensure the correct flow path of the gas. The canister contains a soda lime material (hydroxide chemical) that absorbs carbon dioxide. Gas flows from the canister to the breathing bag, completing the loop.

12-6. Additional oxygen is metered into the breathing bag from the oxygen cylinder by the demand valve assembly as required. A pressure gauge mounted on the top of the equipment housing indicates the amount of oxygen remaining. On the front of the equipment housing is a bypass button that permits the operator to bypass the (automatic) demand valve and manually add oxygen to the breathing bag. The MK 25 UBA is equipped with a removable canister insulator package. When installed, the insulator package helps retain heat inside the canister, extending the duration of the carbon dioxide absorbent in cold water (below 60 degrees F). Using the insulator in warm water (60 degrees F and above) may decrease canister absorbent duration. When the canister insulator is used, the diver must fit weight pouches around the demand valve to counter the buoyancy effects of the insulator. The weight pouches fit into pockets installed at the demand valve and weigh a total of 3.18 pounds.

FUNCTION

12-7. Table 12-1, page 12-5, and Figure 12-2, page 12-6, depict the functional description of the MK 25 UBA. From the oxygen cylinder (15), high-pressure oxygen passes through the cylinder valve (14) to the reducer (11), where the high-pressure gas is reduced to an average intermediate working pressure of 3.3 to 3.7 bar (47.8 to 53.6 psig) over ambient pressure. The gas is then piped through the low (intermediate) pressure line (12) to the demand valve assembly (10), which is adjustable for an actuation pressure of 8.5 + 1.5 inches of water (0.31 + 0.05 psig). High-pressure gas is also piped through the high pressure line (13) to the 0 to 350 bar (0 to 5,075 psig) pressure gauge (9) located on top of the equipment housing (16). The demand valve assembly (10), secured to the equipment housing (16) and fitted to the breathing bag (7), functions each time the bag is emptied on inhalation and a negative pressure occurs. On inhalation, the one-way valve (disc) (5), located in the inhalation hose (6), opens and the diver receives gas from the breathing bag (7). If not enough gas is available, the demand valve actuates due to negative pressure, adding more oxygen to the system. As the diver exhales, the exhalation one-way valve (disc) (2) opens, the inhalation one-way valve (disc) (5) closes, and the exhaled gas flows through the exhalation hose (1) to the soda lime canister (carbon dioxide scrubber) (8). The gas then filters through the soda lime canister with the next inhalation. During descent or to purge the unit, the diver depresses the demand bypass button located in the front center of the equipment housing (10) to manually add oxygen to the system.

COMPONENTS

12-8. The MK 25 UBA has three major subsystems—the recirculation subsystem, the pneumatic subsystem, and the equipment housing and UBA harnesses. Each is described in the following paragraphs.

RECIRCULATION SUBSYSTEM

12-9. The recirculation subsystem consists of the breathing bag, soda lime canister, mouthpiece valve assembly, and inhalation and exhalation hose assemblies. During routine and postdive maintenance procedures, personnel should sanitize all components, except the soda lime canister, to ensure germ-free cleanliness.

Breathing Bag

12-10. The breathing bag is made of rubber-coated fabric. During normal operations, the breathing bag holds approximately 4 liters (0.141 cubic feet) of breathing gas. When it is fully expanded, the bag holds 7 liters (0.247 cubic feet). The breathing bag contains three connection points to other MK 25 UBA components. The bayonet connecting piece fits to the soda lime canister and locks with a 90-degree rotation. One plastic connecting piece fits to the connecting piece of the inhalation hose, and the second plastic connecting piece with cap nuts fits to the metal connection of the demand valve. The connecting pieces are mounted to the bag using sewing thread with a parallel whipping knot and covered with a clear waterproofing. The bayonet connecting piece and the demand valve-connecting piece with cap nut seals with O-rings.

12-11. Located inside the breathing bag is a spring that prevents complete collapse of the breathing bag. The safety cord, attached to the outside of the bag, helps to secure the breathing bag to the equipment housing and prevents the bag from rising up on the diver's chest. The bag acts as a flexible gas reservoir or counterlung and at the same time helps to provide underwater neutral buoyancy for the system.

Table 12-1. MK 25 UBA components list

#	Component
Recirculation Subsystem	
1	Exhalation Hose
2	One-Way Exhaust Valve (Disc Valve)
3	Mouthpiece Valve Assembly (with Rotary Valve)
4	Head Strap
5	One-Way Inhalation Valve (Disc Valve)
6	Inhalation Hose
7	**Breathing Bag Assembly**
7A	Connecting Piece, Inhalation
7B	Connecting Piece, Demand Valve
7C	Connecting Piece, Canister
8	**Soda Lime Canister**
8A	Canister Intake Port
8B	Canister Outlet Port
8C	Canister Lid
Pneumatic Subsystem	
9	Pressure Gauge
10	Demand Valve/Bypass Knob (reverse, not shown)
11	**Reducer**
11A	Reducer Hand Wheel/Hand Grip (rubber ring)
11B	Safety Valve Assembly
12	Connecting Line (Low-Pressure)
13	Pressure Gauge Line (High-Pressure)
14	**Oxygen Cylinder Valve**
14A	Oxygen Cylinder Valve Safety Burst Disc Oxygen Cylinder
15	Oxygen Cylinder
Equipment Housing and UBA Harnesses	
16	**Equipment Housing**
16A	Exhalation Hose Slotted Indent
16B	Inhalation Hose Slotted Indent
17	Cylinder/Canister Harness
18	Positioning Strap
19	Waist Harness
20	Neck Harness
21	Triglide
22	Lead Shot Pouch Kit
23	Canister Insulator

Chapter 12

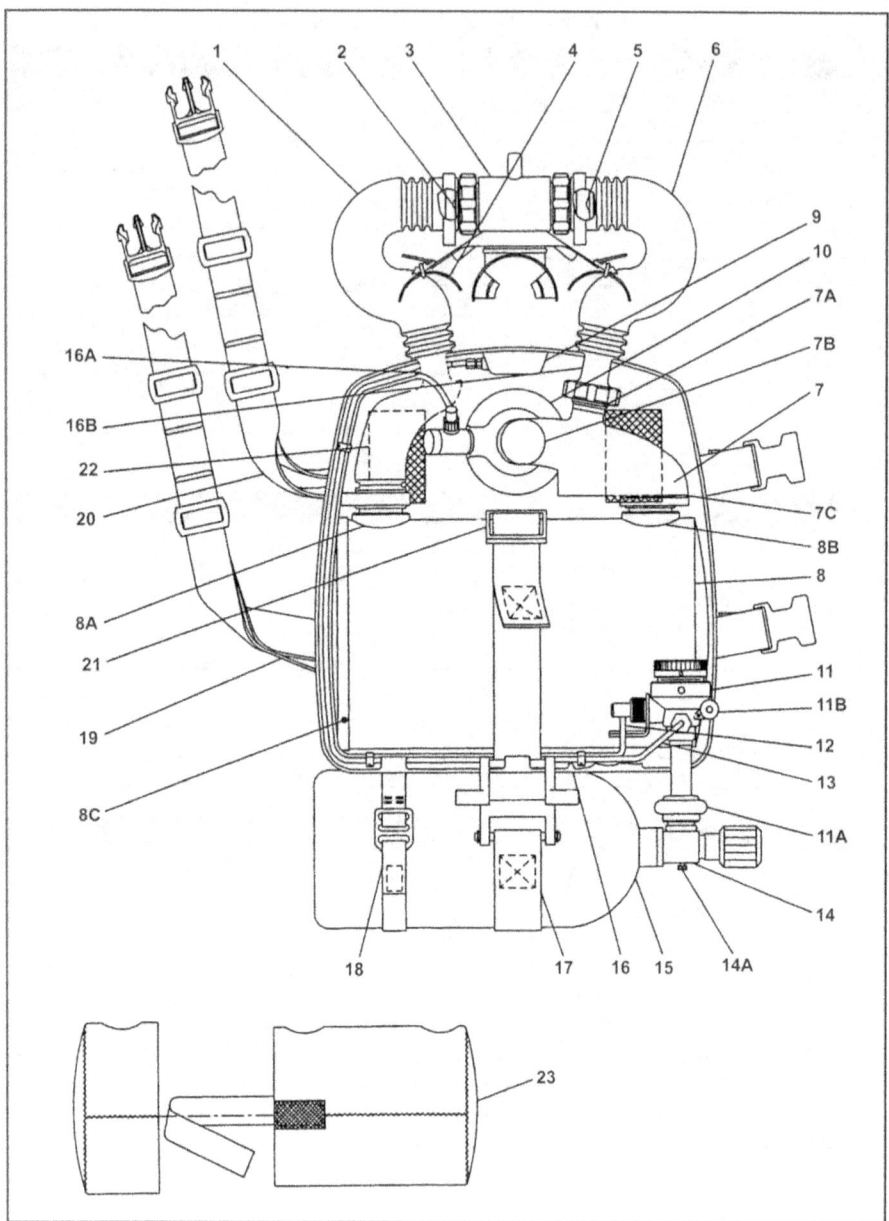

Figure 12-2. MK 25 UBA component locations (as worn)

Inhalation and Exhalation Hoses

12-12. The corrugated inhalation and exhalation hoses are fabricated from neoprene rubber and are very flexible. The exhalation hose is longer than the inhalation hose, as the distance from the soda lime canister to the mouthpiece is longer than that from the breathing bag to the mouthpiece. The exhalation hose is easily identified by a red ring located at the one-way valve end of the hose. The difference in length also helps eliminate confusion on installation.

12-13. The shorter inhalation hose assembly consists of the hose, one-way valve (disc), and two connecting pieces. The connecting piece with cap nut attaches to the breathing bag; the second connecting piece attaches to the mouthpiece. The connecting piece that fits to the breathing bag is smaller than the connecting piece to the mouthpiece, thereby eliminating incorrect installation of the inhalation hose.

12-14. The longer exhalation hose assembly consists of the hose, one-way valve (disc), and two connecting pieces. The bayonet connecting piece locks to the soda lime canister with a 90-degree rotation. The plastic connecting piece attaches to the mouthpiece. The collar connecting piece on the exhalation hose is larger than the connection on the breathing bag, thereby eliminating any chance of incorrectly installing the exhalation hose to the breathing bag.

12-15. Both hoses contain one-way valves (discs) at the mouthpiece end to ensure correct gas flow when the MK 25 UBA is used. The exhalation hose has a red ring at the mouthpiece connection which, when installed correctly, corresponds to a red dot on the mouthpiece to signify the exhalation side of the breathing loop. The inhalation side is not color-coded. All four hose-connecting pieces are mounted to the hoses by hose clamps. All four connecting pieces contain O-rings.

Mouthpiece Assembly

12-16. The mouthpiece valve provides the means for passing gas to and from the diver. It is secured in the diver's mouth by two rubber bite pieces and is held in place by an adjustable head strap. The mouthpiece contains a rotary valve, which opens and closes the breathing loop. The rotary valve consists of a barrel and housing. The barrel and housing are not interchangeable; each rotary valve is a matched barrel and housing set. The rotary valve is opened and closed with the rotary-valve knob. Correctly installed, the knob will point downward when the rotary valve is closed and outward when the valve is open. When the rotary valve is closed (down, surface position), the breathing loop is isolated from the surrounding environment. A water blowout hole, located on the bottom of the rotary-valve housing, permits the diver to clear the mouthpiece of water before opening the valve. When the rotary valve is open (up, dive position), the MK 25 UBA is ready for breathing. The mouthpiece bite pieces are mounted onto the rotary-valve housing using monofilament cord with a parallel whipping knot. The rotary-valve knob contains one O-ring.

Soda Lime Canister (Carbon Dioxide Scrubber)

12-17. The soda lime canister removes both moisture and carbon dioxide from the gas in the breathing loop. The canister is molded from urethane or fiberglass and contains two separate chambers. (The fiberglass canister is still in use and requires a different type of absorbent as described in NAVSEA guidance.) Gas flows from the exhalation hose into the first chamber, which contains carbon dioxide absorbent, and then into a smaller second chamber, which acts as a moisture trap. The absorbent chamber holds approximately 5.75 pounds of absorbent. The moisture trap can hold up to 200 cubic centimeters (cc) of liquid before any liquid will enter the absorbent material. After the carbon dioxide and moisture are removed, the gas flows into the breathing bag. The soda lime canister has two ports, both of which are bayonet couplings and contain spiral waves to ensure a snug fit. The port closest to the canister lid connects to the exhalation hose fitting; the second port connects to the breathing bag. The port for the breathing bag connecting piece is slightly larger than the port for the hose-connecting piece, which eliminates the possibility of incorrect installation. When the bayonet connecting pieces are locked to the canister, a snug fit results with no play in the connections.

12-18. The components of the canister can be completely disassembled. Should carbon dioxide-absorbent residue clog the screens and canister rod threads, these components may be soaked in a solution of vinegar water, scrubbed lightly with a nylon bristle brush to remove grit, and then rinsed with freshwater.

12-19. The canister lid seal should be inspected often for wear and tear. It is not lubricated. The soda lime canister contains two O-rings, one on the lid pin and the other on the nut at the opposite end of the canister rod.

PNEUMATIC SUBSYSTEM

12-20. The pneumatic subsystem of the MK 25 UBA consists of the oxygen cylinder and cylinder valve, reducer, demand valve, pressure gauge, and pressure lines. Each component of the pneumatic subsystem, except the pressure gauge, when disassembled for any reason, must be cleaned using NAVSEA-approved oxygen-safe cleaning procedures before reassembly. With the exception of the cylinder valve to reducer connection, any entry into the pneumatic subsystem requires that reentry control (REC) procedures be followed to maintain system certification.

Oxygen Cylinder

12-21. The MK 25 UBA oxygen cylinder is made of aluminum alloy 6061-T6. The cylinder has a basic internal volume of 1.9 liters (0.067 cubic feet) and may be charged to 207 bar (3,000 psig). When fully charged, the cylinder holds 410 liters (14.49 cubic feet) of oxygen at standard conditions. The cylinder has an average weight of 6.6 pounds (without the cylinder valve). The oxygen cylinder mouth has internal straight threads to accept the oxygen cylinder valve. The cylinder is held in place on the equipment housing with two straps: a self-tensioning cylinder or canister strap (harness) and a polypropylene Velcro positioning strap. The oxygen bottle connects to the oxygen reducer with hard piping and is equipped with a supply (on/off) valve. The MK 25 UBA 3AL aluminum oxygen cylinder is DOT approved. Regulatory guidance mandates that the cylinder undergo visual inspection annually and hydrostatic testing every 5 years.

Cylinder Valve

12-22. The cylinder valve is connected to the oxygen cylinder with a straight thread connection and is sealed with an O-ring. The outlet connection of the valve is a CGTA 540 connection that assembles directly to the reducer. The cylinder valve has a rated service pressure of 207 bar (3,000 psig), and is equipped with a burst disc that eliminates hazardous overpressurization of the cylinder. The burst disc ruptures when the cylinder pressure exceeds 5,000 (nominal) psig.

12-23. The cylinder valve has two O-rings. The O-ring between the valve and the oxygen cylinder is always replaced during the annual inspection of the cylinder or as required. The internal O-ring is replaced during the 5-year overhaul of the cylinder valve.

Reducer

12-24. This reducer regulates the cylinder gas pressure to 3.3 to 3.7 bar (47.8 to 53.6 psig) over ambient pressure. The reducer is designed to operate with an oxygen supply pressure of 10 to 207 bar (145 to 3,000 psig). The reducer contains two outlet ports. The high-pressure outlet port is connected by hard piping to the pressure gauge. The seal between the high-pressure outlet port and the piping is a ridged, metal sealing ring. The intermediate pressure outlet port is connected by hard piping to the demand valve. The connection is sealed with an O-ring.

12-25. The reducer has two safety features. The velocity reducer, located in the high-pressure inlet of the reducer, slows the high flow rate of incoming oxygen. Excessive oxygen flow rate increases the potential for combustion of any foreign particles inside the reducer. The second safety feature is the safety valve assembly, connected to the reducer body. The safety valve precludes overpressurization of the reducer's inner chamber that may result from a high-pressure oxygen leak. Excessive intermediate pressure from the reducer will cause an increase in back pressure in the demand valve and increased breathing resistance. The safety valve is set to vent a 9 to 13 bar (130 to 190 psig). The safety valve reseats at 8 bar (116 psig).

12-26. The MK 25 UBA reducer is an unbalanced pressure regulator. It means the actual pressure provided to the demand valve is dependent on oxygen cylinder pressure. As the cylinder pressure decreases, the intermediate pressure supplied to the demand valve increases the breathing necessary to actuate the demand valve increases. For this reason, the reducer intermediate pressure must be set at 3.3 to 3.7 bar (47.8 to 53.6 psig) using a supply pressure of 100 + 5 bar (1450 + 72 psig). The intermediate pressure output of the reducer must be checked annually. The reducer must be overhauled every 6 years or when flooded, whichever comes first.

Demand Valve Assembly

12-27. The demand valve assembly meters the oxygen supply into the breathing bag. This demand valve is located below the breathing bag and is attached to the equipment housing. The demand valve receives low (intermediate) pressure oxygen from the reducer via hard piping. The connection is sealed with an O-ring. The breathing bag plastic connecting piece with cap nut attaches to the demand valve's metal housing.

12-28. Negative pressure inside the breathing bag actuates the demand valve rocker, which allows oxygen to flow into the breathing bag. As negative pressure inside the breathing bag decreases, the valve rocker reseats, shutting off the flow of oxygen. The demand valve actuation point is set to 8.5 + 1.5 inches of water (0.31 + 0.05 psig).

12-29. A manual bypass button is located on the front of the demand valve. This button is used by the diver to override the automatic metering of the demand valve to add oxygen to the breathing bag. The bypass button allows oxygen to enter the breathing bag at a rate of 60 liters per minute. Adjustments to the demand valve actuation point must be in the range of 8.5 + 1.5 inches of water (0.31 + 0.05 psig). However, low actuation points may result in free flow of the demand valve due to depth or diving attitude changes. Adjustments must be made with the correct reducer intermediate pressure output using a 100 + 5 bar (1450 + 72 psig) oxygen supply. Increased intermediate supply pressure, either by incorrect intermediate pressure setting or depletion of oxygen in the cylinder, will increase the actuation point of the demand valve. The demand valve actuation point must be checked annually and must be checked using correct output pressure from the reducer. The demand valve must be overhauled every 2 years.

Pressure Gauge

12-30. The pressure gauge, the diver's only indicator on the MK 25 UBA, displays oxygen cylinder pressure. The gauge fits within a boot in the top of the equipment housing. When the MK 25 UBA is donned, the diver can see the gauge by glancing downward. The pressure gauge is connected to the reducer via hard piping. The connection at the gauge is sealed with a metal sealing ring. This flat sealing ring should not be confused with the ridged sealing ring connecting the piping to the reducer. The pressure gauge has a range of 0 to 350 bar (0 to 5,075 psig) and is graduated in bar; conversion to psig may be made by multiplying the gauge reading (in bar) by 14.5. The gauge must be tested for accuracy every 18 months.

Pressure Lines

12-31. Two pressure lines are used in the MK 25 UBA to carry oxygen. Both lines are made of chrome-plated copper tubing. The pressure-gauge line assembly provides oxygen cylinder pressure to the pressure gauge from the high-pressure port of the reducer. Connection seals on both ends are made by metal sealing rings. (The ridged sealing ring is used in the reducer port; the flat sealing ring is used in the gauge connection.) These sealing rings are replaced every time the connections are broken. The connecting line assembly carries low-pressure oxygen (3.3 to 3.7 bar/47.8 to 53.6 psig) to the demand valve from the low-pressure port of the reducer. Seals on both ends of this line are made with O-rings. These connections are hand-tightened.

EQUIPMENT HOUSING AND UBA HARNESSES

12-32. This subsystem consists of the equipment housing, associated hardware, and harness assemblies. These components need only be cleaned with soapy water when necessary.

Equipment Housing

12-33. The equipment housing is made up of a single unit from reinforced fiberglass and provides an attachment point for all components of the MK 25 UBA. Slots are provided for the exhalation and inhalation hoses, reducer, and harness and strap assemblies. Circular holes are provided in the housing for the demand valve and the pressure gauge. The housing is fitted to the diver by two harnesses. Two retaining straps are incorporated into the housing. The cylinder or canister strap (harness) that holds both the oxygen cylinder and canister in place is made of 1 1/2-inch, polypropylene or neoprene, self-tensioning material. It attaches to the housing via a plate retainer (located under the canister) and hooks to the bottom of the housing, after passing over the canister and encircling the oxygen cylinder, by means of a retaining hook. A triglide, located at the canister end of the strap, allows for adjustment. A positioning strap, made of 1-inch polypropylene webbing containing hook and pile (Velcro) tabs and a ladder buckle, passes through the bottom of the housing (passing under the pressure lines) and encircles the oxygen cylinder, acting as a second securing strap. The positioning strap may be tightened using the ladder buckle and Velcro tabs. The equipment housing also provides an attachment point for the reducer. The equipment housing holder secures the reducer in place; the holder is attached to the housing with two screws. Three sets of retaining clamps secure the pressure lines to the equipment housing.

Harnesses

12-34. The MK 25 UBA is fitted to the diver using two harnesses. The neck and waist harnesses are made of 1 1/2-inch polypropylene webbing that is fitted with quick-release side buckles. Adjustments in the harnesses are made with the incorporated triglide adjusters. When donned properly, the MK 25 UBA fits over the lung area of the chest, with the top of the UBA 7 to 10 inches below the chin. The waist strap fits loosely to allow expansion room for the breathing bag.

Canister Insulator Package

12-35. The MK 25 UBA canister is equipped with a removable neoprene insulator package that may be used in cold water (below 60 degrees F) to extend the duration of the carbon dioxide absorbent. The insulator has two components. The larger insulator slides onto the canister housing. After aligning the insulator holes with the canister ports, the insulator is held in place by a Velcro strap that fits over the canister lid. The smaller insulator fits on to the canister lid. The hole in the small insulator must align with the exhalation hose port in the canister. When installed, the insulation may be lifted to check for leaks and release any trapped air. To offset the positive buoyancy effects created by the canister insulator, a set of two lead-shot weight pouches are included in the equipment housing. The lead-shot pouches are loaded with 1.59 pounds of lead shot each (3.18 lb total) and are inserted into pockets that surround the demand valve. These weights are used only if the canister insulator package is installed.

Gas Flow Path

12-36. The gas is exhaled by the diver and directed by the mouthpiece one-way valves into the exhalation hose. The gas then enters the carbon dioxide-absorbent canister, which is packed with a NAVSEA-approved carbon dioxide-absorbent material. The carbon dioxide is removed by passing through the CO_2-absorbent bed and chemically combining with the CO_2-absorbent material in the canister. Upon leaving the canister, the used oxygen enters the breathing bag. When the diver inhales, the gas is drawn from the breathing bag through the inhalation hose and back to the diver's lungs. The gas flow described is entirely breath-activated. As the diver exhales, the gas in the UBA is pushed forward by the exhaled gas, and upon inhalation the one-way valves in the hoses allow fresh gas to be pulled into the diver's lungs from the breathing bag (Figure 12-3, page 12-11).

Breathing Loop

12-37. The demand valve adds oxygen to the breathing bag of the UBA from the oxygen cylinder only when the diver empties the bag on inhalation. The demand valve also contains a manual bypass button to allow for manual filling of the breathing bag during rig setup and as required. There is no constant flow of fresh oxygen to the diver.

Closed-Circuit Diving

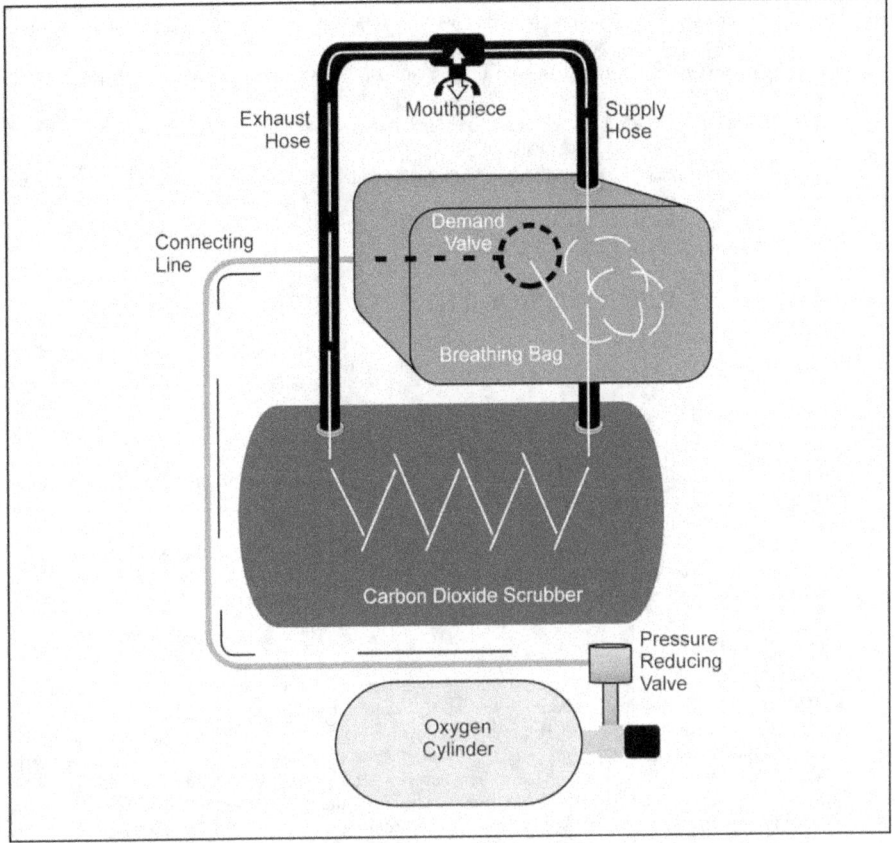

Figure 12-3. Gas flow path

CHARACTERISTICS

12-38. The MK 25 UBA has significant specifications that distinguish it from other models. They are as follows:

- *Weight*—27 pounds (with full canister and charged O2 bottle, and no insulation package). Neutral in water with approximately 2 liters of oxygen in the breathing bag.
- *Dimensions*—Length: 18.25 inches, width: 13.25 inches; height: 7 inches.
- *Operating Time*—4 hours based upon canister duration limits and an oxygen cylinder charged to full capacity.
- *Diving Depth*—0–50 FSW.
- *Temperature*—Diving: 35 to 90 degrees F/Transport: –22 to 120 degrees F.
- *Oxygen Cylinder*—1.9-liter volume, charged to 207 bar (3,000 psig) with 99.5 percent pure aviator's oxygen.
- *Reducer*—Reduces cylinder pressure to 3.3–3.7 bar (47.8–53.6 psig).

Chapter 12

- *Demand Valve*—Adds oxygen to breathing bag only when bag is emptied on inhalation or when the manual bypass button is manually bypassed. Opening pressure is adjustable from 9.0 to 11.5 inches H2O.
- *CO2 Scrubber Canister*—Volume of 0.09 cubic feet (2.55 liters) with approved carbon dioxide absorbent.
- *Breathing Bag*—Maximum volume is approximately 7.0 liters. Maximum volume, when UBA is donned by diver, is approximately 4 liters.
- *Manual Bypass Valve*—Allows oxygen to be added manually to the breathing bag via the demand valve.
- *Equipment Housing*—Material made of fiberglass. Lead-shot weight pouches (3.18 lb) at demand valve offsets neoprene insulation.

OPERATIONAL DURATION OF THE MK 25 UBA

12-39. The operational duration of the MK 25 UBA may be limited by either the oxygen supply or the canister duration. These two constraining factors are discussed below.

Oxygen Supply

12-40. The MK 25 UBA oxygen bottle is charged to 207 bar (3,000 psig). The oxygen supply may be depleted in two ways—by the diver's metabolic consumption or by the loss of gas from the UBA. A key factor in maximizing the duration of the oxygen supply is for the diver to swim at a relaxed, comfortable pace. A diver swimming at a high exercise rate may have oxygen consumption of 2 liters per minute (oxygen supply duration = 205 minutes) whereas one swimming at a relaxed pace may have an oxygen consumption of 1 liter per minute (oxygen supply duration = 410 minutes). Figure 12-4, page 12-13, illustrates oxygen consumption rates from which the diver can estimate his cylinder duration.

Canister Duration

12-41. The canister duration is dependent on the water temperature, exercise rate, and the mesh size of the NAVSEA-approved carbon dioxide absorbent. Typically, absorbent duration will increase as water temperatures decrease. The MK 25 UBA is equipped with a removable canister insulator package for use in water temperatures below 60 degrees F. Use of the insulator increases the duration of the canister in colder waters by holding heat within the absorbent material. However, use of the canister insulator in water temperatures above 60 degrees F can actually decrease canister duration. Divers should refer to NAVSEA guidance and applicable diving advisories for authorized canister duration for the MK 25 UBA. Exceeding authorized canister duration limits for given temperatures can result in very high levels of carbon dioxide in the breathing gas. The canister will function adequately for the time period as long as the UBA has been properly set up.

PREDIVE PROCEDURES

12-42. The MK 25 UBA is set up using DA Form 7532-R (MK 25 MOD 2 Predive Checklist) (Appendix E). The diver should pay special attention to the following details:
- Gauge oxygen bottle for minimum operating pressure; normal minimum is 207 bar (3,000 psig).
- Inspect and fill scrubber canister with absorbent (Figure 12-5, page 12-14). Use only approved CO2 absorbent. Check upper and lower screens for clogging. If screens are clogged, refer to MRC R-4. Fill canister to scribe line. Do not overfill. Inspect bayonet connector circle springs, ensuring they are free to move in their grooves and that no deformity is present. When installing top cover, do not overtighten. Overtightening may cause the support rod to rip from the fiberglass housing or to crack the bottom of the canister.
- Inspect and check the operation of the one-way check valves in the inhalation and exhalation hoses. Do not confuse the supply and exhaust hoses. Check the O-rings on the hose connectors; replace or lubricate if necessary.

Closed-Circuit Diving

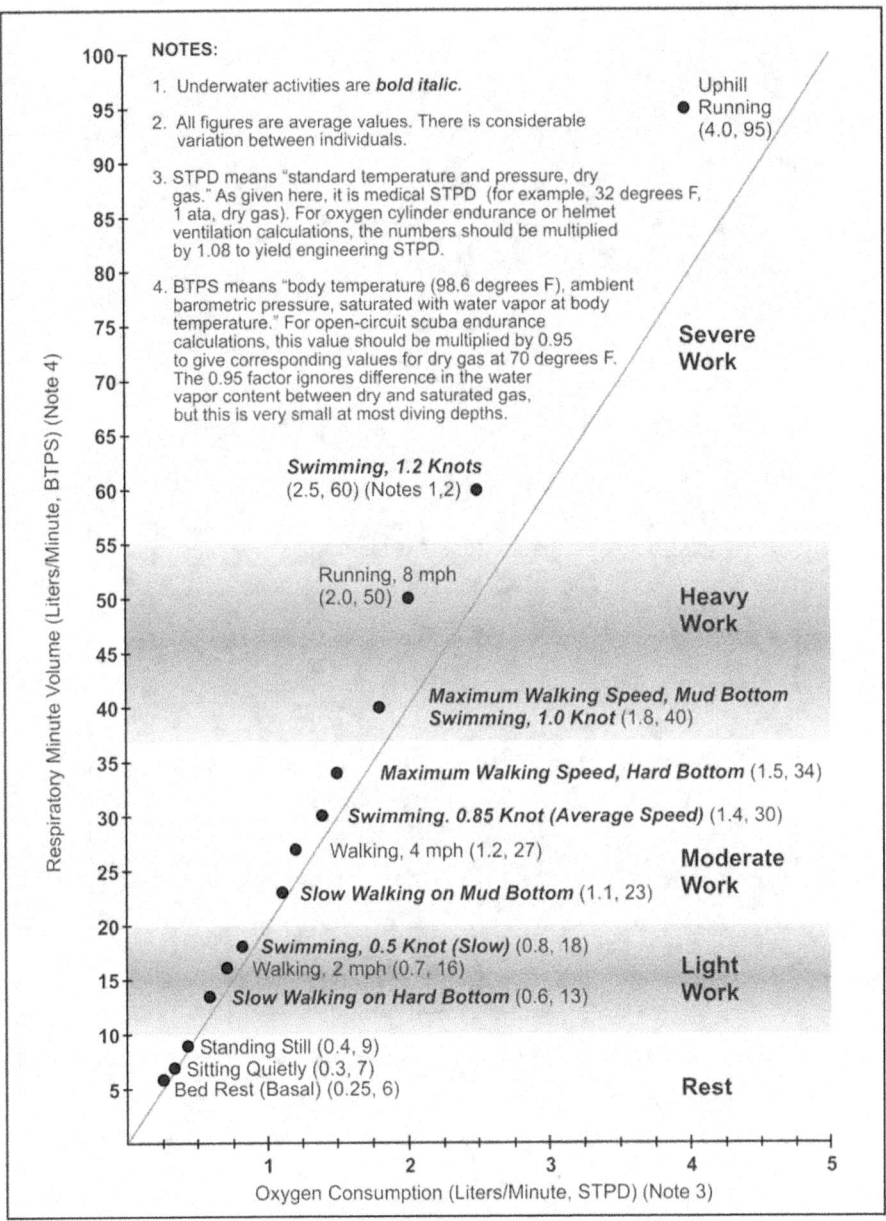

Figure 12-4. Oxygen consumption rates

Chapter 12

Figure 12-5. Filling canister with absorbent

- Inspect and attach the inhalation and exhalation hoses to the mouthpiece (Figure 12-6, page 12-15). Close the dive-surface valve.
- Attach the supply hose to the breathing bag.
- Inspect and attach the breathing bag to the scrubber canister (Figure 12-7, page 12-15). Examine the O-rings on the breathing bag. Lubricate or replace if necessary. Examine the breathing bag and the breathing bag muffs for signs of puncture or wear.

Closed-Circuit Diving

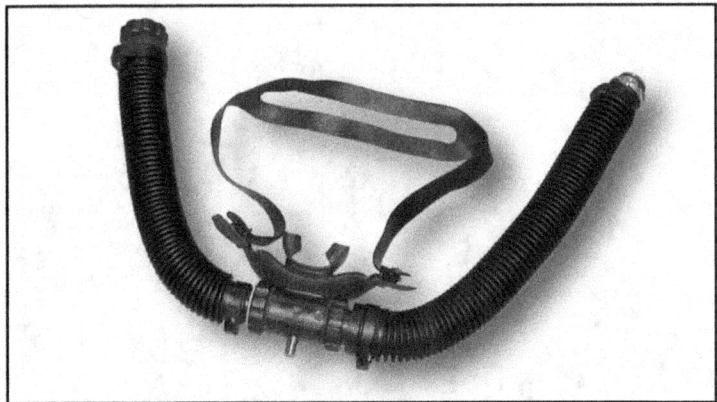

Figure 12-6. Inhalation and exhalation hoses

Figure 12-7. Breathing bag attached to scrubber canister

- Install the scrubber canister in the proper orientation. Secure with the retainer strap. Ensure that the canister lid is to the left and correctly oriented.
- Attach the breathing bag to the demand valve (Figure 12-8, page 12-16). To prevent stripping, use extreme care when threading the plastic breathing bag coupling to the metal demand valve.

Chapter 12

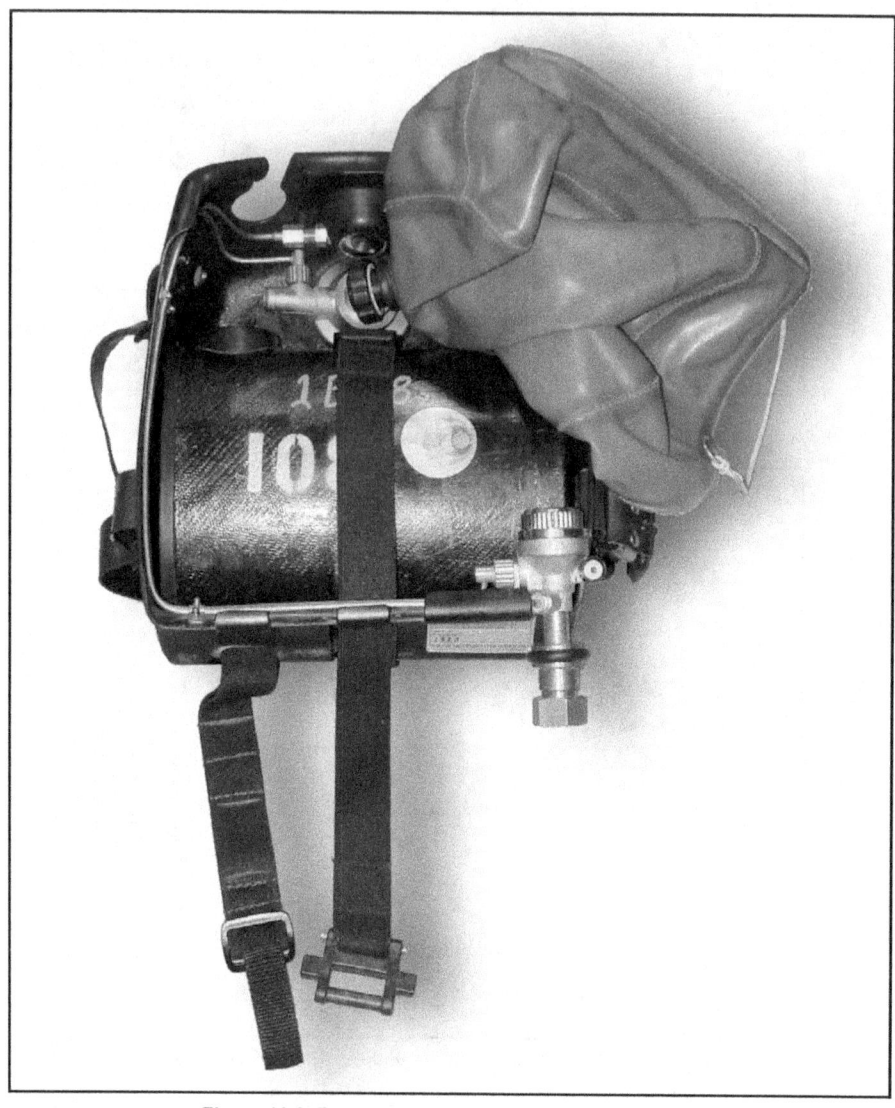

Figure 12-8. Breathing bag attached to demand valve

POSTDIVE PROCEDURES

12-43. The diver checks the MK 25 UBA using DA Form 7533-R (MK 25 MOD 2 UBA Postdive Checklist) (Appendix E).

MALFUNCTION PROCEDURES

12-44. Some of the common equipment problems encountered with the MK 25 UBA and the appropriate actions to be taken by the diver are described below.

CNS OXYGEN TOXICITY

12-45. **Symptoms.** Divers must be aware of the possibilities of the central nervous system (CNS) when breathing pure oxygen under pressure. The most hazardous is a sudden convulsion that can result in drowning or arterial gas embolism. The symptoms of CNS oxygen toxicity may occur suddenly and dramatically, or they may have a gradual, almost imperceptible onset. The letters **V, E, N, T, I, D, C** are a helpful reminder of these common symptoms:

- *Visual*—Tunnel vision, a decrease in the diver's peripheral vision, and other symptoms, such as blurred vision, may occur.
- *Ear*—Tinnitus is any sound perceived by the ears but not resulting from an external stimulus. The sound may resemble bells ringing, roaring, or a machinery-like pulsing sound.
- *Nausea or Spasmodic Vomiting*—These symptoms may be intermittent.
- *Twitching and Tingling*—Any of the small facial muscles, lips, or muscles of the extremities may be affected. These are the most frequent and clearest symptoms.
- *Irritability*—Any change in the diver's mental status, including confusion, agitation, and anxiety.
- *Dizziness*—Symptoms include clumsiness, uncoordinated movements, and unusual fatigue.
- *Convulsions*—The first sign of CNS oxygen toxicity may be a convulsion that occurs with little or no warning. These symptoms may occur singly or together. They occur in no particular order and there is no one symptom which could be considered more serious than another, or which is a better warning of an impending convulsion.

12-46. **Actions.** Diver notifies dive buddy and makes a controlled ascent to the surface while exhaling through the nose to prevent embolism. When on the surface, he inflates the life preserver, shuts the oxygen valve, closes the mouthpiece rotary valve, removes the mouthpiece from his mouth, and signals for assistance.

SYSTEM FLOODING

12-47. **Symptoms.** Symptoms include increased breathing resistance, a gurgling sound, a bitter taste in the mouth (caustic cocktail), and possibly gas escaping from the system.

12-48. **Actions.** The diver immediately attains a vertical position to keep the mouthpiece higher than the scrubber canister, and activates the demand valve bypass button. He notifies his dive buddy and makes a controlled ascent to the surface while breathing in the open-circuit mode (breathe in from mouthpiece, exhale through nose). The diver maintains a vertical position while ascending and activates the bypass button as necessary. When on the surface, he inflates the life preserver, closes the mouthpiece rotary valve, removes the mouthpiece from his mouth, and signals for assistance.

> **CAUTION**
> Do **NOT** shut the oxygen cylinder valve.

Note. If the canister insulator is used, condensation may occur in the demand valve causing a gurgling sound when the demand valve is activated.

Bypass or Demand Valve Stuck Open

12-49. **Symptoms.** Symptoms include increased exhalation resistance and an overabundance of gas accompanied by the sound of oxygen being added to the breathing bag. When the bypass valve is stuck open, the flow rate is approximately one liter/second.

12-50. **Actions.** Diver notifies dive buddy and makes a controlled ascent to the surface while exhaling through the nose to prevent embolism. He must NOT hold his breath. When on the surface, he inflates the life preserver, shuts the oxygen cylinder valve, closes the mouthpiece rotary valve, removes the mouthpiece from his mouth, and signals for assistance.

Breathing Gas Pressure Deficiency

12-51. **Symptoms.** Breathing gas pressure deficiency occurs when the rig is not supplying an adequate amount of pressurized oxygen into the breathing loop to sustain the diver's need. Symptoms include increased breathing resistance and deflation of the breathing bag. Probable causes range from low cylinder pressure to a malfunction in the oxygen cylinder valve, reducer, demand valve, or breathing loop.

12-52. **Actions.** Diver notifies dive buddy and makes a controlled ascent to the surface while exhaling through the nose to prevent embolism. When on the surface, he inflates the life preserver, closes the mouthpiece rotary valve, removes the mouthpiece from his mouth, and signals for assistance.

> **CAUTION**
> Do **NOT** shut the oxygen cylinder valve.

Demand Valve Failure (Automatic or Manual)

12-53. **Symptoms.** Symptoms include demand valve failure to supply oxygen in either mode of operation.

12-54. **Actions.** Diver notifies dive buddy and makes a controlled ascent to the surface while exhaling through the nose to prevent embolism. When on the surface, he inflates the life preserver, closes the mouthpiece rotary valve, removes the mouthpiece from his mouth, and signals for assistance.

> **CAUTION**
> Do **NOT** shut the oxygen cylinder valve.

CLOSED-CIRCUIT OXYGEN EXPOSURE LIMITS

12-55. The USN closed-circuit oxygen exposure limits have been extended and revised to allow greater flexibility in closed-circuit oxygen diving operations. The revised limits are divided into the categories discussed below.

Transit-With-Excursion Limits

12-56. The transit-with-excursion limits (Table 12-2, page 12-19) call for a maximum dive depth of 20 FSW or shallower for the majority of the dive, but allow the diver to make a brief excursion to depths as great as 50 FSW. The transit-with-excursion limits is normally the preferred mode of operation because maintaining a depth of 20 FSW or shallower minimizes the possibility of CNS oxygen toxicity during the majority of the dive yet allows a brief downward excursion if needed. Only a single excursion is allowed.

Closed-Circuit Diving

Table 12-2. Transit-with-excursion limits table

Depth	Maximum Time
21 to 40 FSW	15 minutes
41 to 50 FSW	5 minutes

12-57. A transit with one excursion, if necessary, will be the preferred option in most combat swimmer operations. When operational considerations necessitate a descent deeper than 20 FSW for longer than allowed by the excursion limits, the appropriate single-depth limit should be used. Figure 12-9 depicts the following transit definitions:

- *Transit*—That part of the dive spent at 20 FSW or shallower.
- *Excursion*—The portion of the dive deeper than 20 FSW.
- *Excursion Time*—The time between the diver's initial descent below 20 FSW and his return to 20 FSW or shallower at the end of the excursion.
- *Oxygen Time*—The time interval between when the diver begins breathing from the closed-circuit oxygen UBA (on-oxygen time) to the time when he discontinues breathing from the closed-circuit oxygen UBA (off-oxygen time).

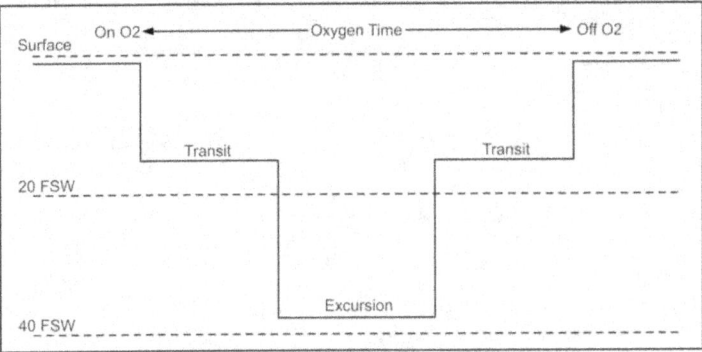

Figure 12-9. Transit-with-excursion limits

12-58. A diver who has maintained a transit depth of 20 FSW or shallower may make one brief excursion as long as he observes the following rules:

- Maximum total time of dive (oxygen time) should not exceed 240 minutes.
- A single excursion may be taken at any time during the dive.
- The diver must have returned to 20 FSW or shallower by the end of the prescribed excursion limit.
- The time limit for the excursion is determined by the maximum depth attained during the excursion (Table 12-2 above). The diver should remember that the excursion limits are different from the single-depth limits.

Example
A dive pair is having difficulty with a malfunctioning compass. They have been on oxygen (oxygen time) for 35 minutes when they notice that their depth gauge reads 55 FSW. Because this exceeds the maximum allowed oxygen exposure depth, the dive must be aborted and the divers must return to the surface.

12-59. If an inadvertent excursion should occur, one of the following situations will apply:
- If the depth or time of the excursion exceeds the limits in the previous paragraph or if an excursion has been taken previously, the dive must be aborted and the diver must return to the surface.
- If the excursion was within the allowed excursion limits, the dive may be continued to the maximum allowed oxygen dive time, but no additional excursions deeper than 20 FSW may be taken.
- The dive may be treated as a single-depth dive applying the maximum depth and the total oxygen time to the single-depth limits shown in Table 12-3 below.

> **Example 1**
> A dive pair is having difficulty with a malfunctioning compass. They have been on oxygen (oxygen time) for 35 minutes when they notice that their depth gauge reads 55 FSW. Because this exceeds the maximum allowed oxygen exposure depth, the dive must be aborted and the divers must return to the surface.
>
> **Example 2**
> A diver on a compass swim notes that his depth gauge reads 32 FSW. He recalls checking his watch 5 minutes earlier and that time his depth gauge read 18 FSW. As his excursion time is less than 15 minutes, he has not exceeded the excursion limit for 40 FSW. He may continue the dive, but he must maintain his depth at 20 FSW or less and make no additional excursions.
>
> *Note.* If the diver is unsure how long he was below 25 FSW, the dive must be aborted.

SINGLE-DEPTH LIMITS

12-60. The single-depth limits (Table 12-3) allow maximum exposure at the greatest depth, but have a shorter overall exposure time. Single-depth limits may, however, be useful when maximum bottom time is needed deeper than 20 FSW. Depths greater than 20 FSW do not allow for an excursion. They may, however, be useful in certain diving situations. Although the limits described in this section have been thoroughly tested and are safe for the vast majority of individuals, occasional episodes of oxygen toxicity up to and including convulsions may occur. This is the basis for requiring buddy lines on closed-circuit oxygen diving operations. These limits have been tested extensively over the entire depth range and are acceptable for routine diving operations. They are not considered exceptional exposure. Specific operational considerations are addressed below.

Table 12-3. Single-depth oxygen exposure limits

Depth (FSW)	Maximum Oxygen Time (Minutes)
25	240
30	80
35	25
40	15
50	10

12-61. The term single-depth limits does not mean that the entire dive must be spent at one depth, but refers to the time limit applied to the dive based on the maximum depth attained during the dive.

12-62. The following definitions apply when using the single-depth limits:
- *Oxygen Time*—The time interval between when the diver begins breathing from the closed-circuit oxygen UBA (on-oxygen time) to the time when he discontinues breathing from the closed-circuit UBA (off-oxygen time). Time of transit-with-excursion limits, above.
- *Depth*—Used to determine the allowable exposure time that is determined by the maximum depth attained during the dive. For intermediate depth, the next deeper depth limit will be used.
- *Depth or Time Limits*—Provided in Table 12-3 above. No excursions are allowed when using these limits.

Closed-Circuit Diving

> **Example**
>
> 22 minutes (oxygen time) into a compass swim, a dive pair descends to 28 FSW to avoid the propeller of a passing boat. They remain at this depth for 8 minutes. They now have the following choices for calculating their allowed oxygen time:
> 1. They may return to 20 FSW or shallower and use the time below 20 FSW as an excursion, allowing them to continue their dive on the transit with excursion limits to a maximum time of 240 minutes.
> 2. They may elect to remain at 28 FSW and use the 30-FSW single-depth limits to a maximum dive time of 80 minutes.

Exposure Limits for Successive Oxygen Dives

12-63. If an oxygen dive is conducted after a previous closed-circuit oxygen exposure, the effect of the previous dive on the exposure limit for the subsequent dive is dependent on the off-oxygen interval.

Definitions for Successive Oxygen Dives

12-64. The following definitions apply when using oxygen exposure limits for successive oxygen dives:
- *Off-Oxygen Interval.* The interval between off-oxygen time and on-oxygen time is defined as the time from when the diver discontinues breathing from his closed-circuit oxygen UBA on one dive until he begins breathing from the UBA on the next dive.
- *Successive Oxygen Dive.* A successive oxygen dive is one that follows a previous oxygen dive after an off-oxygen interval of more than 10 minutes but less than 2 hours.

Off-Oxygen Exposure Limit Adjustments

12-65. If an oxygen dive is a successive oxygen dive, the oxygen exposure limit for the dive must be adjusted as shown in Table 12-4. If the off-oxygen interval is 2 hours or greater, no adjustment is required for the subsequent dive. An oxygen dive undertaken after an off-oxygen interval of more than 2 hours is considered to be the same as an initial oxygen exposure. If a negative number is obtained when adjusting the single-depth exposure limits, a 2-hour off-oxygen interval must be taken before the next oxygen dive. Divers should follow the guidelines below:
- *Maximum Allowable Oxygen Time.* Divers should not accumulate more than 4 hours of oxygen time in a 24-hour period.
- *Exposure Limits for Oxygen Dives Following Mixed-Gas or Air Dives.* When a subsequent dive must be conducted and the previous exposure was an air or MK-16 dive, the exposure limits for the subsequent oxygen dive require no adjustment. If diving mixed gas in conjunction with oxygen, the diver should refer to current U.S. Navy diving regulations for further guidance.

Table 12-4. Adjusted oxygen exposure limits for successive oxygen dives

Type	Adjusted Maximum Oxygen Time	Excursion
Transit-With-Excursion Limits	Subtract oxygen time on previous dives from 240 minutes.	Allowed if none taken on previous dives.
Single-Depth Limits	1. Determine maximum oxygen time for deepest exposure. 2. Subtract oxygen time on previous dives from maximum oxygen time in Step 1.	No excursion allowed when using single-depth limits to compute remaining oxygen time.

OXYGEN DIVING AT ALTITUDE

12-66. The oxygen exposure limits and procedures set forth in the preceding paragraphs may be used without adjustment for closed-circuit oxygen diving at altitudes above sea level (ASL).

FLYING AFTER OXYGEN DIVING

12-67. Flying is permitted immediately after oxygen diving unless the oxygen dive has been part of a multiple-UBA dive profile in which the diver was also breathing another breathing mixture (air or N2O2, HeO2). In this case, the rules found in current U.S. Navy diving regulations apply.

COMBAT OPERATIONS

12-68. The oxygen exposure limits in this chapter are the only limits approved for use by the USN and should not be exceeded in a training or exercise scenario. Should combat operations require a more severe oxygen exposure, an estimate of the increased risk of CNS oxygen toxicity may be obtained from a DMO or the NEDU. The advice of a DMO is essential in such situations and should be obtained whenever possible.

Example 1

90 minutes after completing a previous oxygen dive with an oxygen time of 75 minutes (maximum dive depth 19 FSW), a dive pair will be making a second dive using the transit-with-excursion limits. Calculate the amount of oxygen time for the second dive, and determine whether an excursion is allowed.

Solution

The second dive is considered a successive oxygen dive because the off-oxygen interval was less than 2 hours. The allowed exposure time must be adjusted. The adjusted maximum oxygen time is 165 minutes (240 minutes minus 75 minutes previous oxygen time). A single excursion may be taken because the maximum depth of the previous dive was 19 FSW.

Example 2

70 minutes after completing a previous oxygen dive (maximum depth 28 FSW) with an oxygen time of 60 minutes, a dive pair will be making a second oxygen dive. The maximum depth of the second dive is expected to be 25 FSW. Calculate the amount of oxygen time for the second dive, and determine whether an excursion is allowed.

Solution

The diver first computes the adjusted maximum oxygen time. This is determined by the single-depth limits for the deeper of the two exposures (30 FSW for 80 minutes) minus the oxygen time from the previous dive. The adjusted maximum oxygen time for the second dive is 20 minutes (80 minutes minus 60 minutes previous oxygen time). No excursion is permitted using the single-depth limits.

WATER ENTRY AND DESCENT

12-69. The diver is required to perform a purge procedure before or during any dive in which closed-circuit oxygen UBA is to be used. The purge procedure is designed to eliminate the nitrogen from the UBA and the diver's lungs as soon as he begins breathing from the rig. The procedure prevents the possibility of hypoxia as a result of excessive nitrogen in the breathing loop. The gas volume from which this excess nitrogen must be eliminated is comprised of more than just the UBA breathing bag. The carbon dioxide-absorbent canister, inhalation and exhalation hoses, and diver's lungs must also be purged of nitrogen.

PURGE PROCEDURE

12-70. Immediately before entering the water, the divers should carry out the appropriate purge procedure. It is both difficult and unnecessary to eliminate nitrogen completely from the breathing loop. The purge procedure need only raise the fraction of oxygen in the breathing loop to a level high enough to prevent the diver from becoming hypoxic. For the MK 25 UBA, this value has been determined to be 45 percent. The purge procedure for the MK 25 UBA, (Figure 12-10, page 12-23) is taken from Navy 22600-A3-MMA-010/53833, *Operations and Maintenance Instructions for the MK 25 MOD 2 UBA* dated 16 October 1998. This procedure produces an average oxygen fraction of approximately 75 percent in the breathing loop.

1. Don the apparatus by attaching the neck and waist straps. The upper surface of the UBA should be approximately at the level of the diver's lower chest.

Note. The waist strap should fit loosely to permit complete filling of the breathing bag.

2. Ensure that the oxygen supply valve is closed. Blow all air out of the lungs and insert mouthpiece. Open the dive-surface valve. (The dive-surface valve is left open for the remainder of the procedure.)
3. Empty air out of the breathing bag by inhaling from the mouthpiece and exhaling into the atmosphere (through the nose). Continue until the bag is completely empty.

Note. Be sure not to exhale into the mouthpiece (breathing bag) during the emptying process in Step 3 or 5.

4. Open the oxygen supply valve and fill the breathing bag by depressing the bypass valve completely for approximately 6 seconds. (The oxygen supply valve is left open for the remainder of the procedure.)
5. Empty the breathing bag once more as in Step 3.
6. Fill the breathing bag to a comfortable volume for swimming by depressing the bypass valve completely for approximately 4 seconds. Begin normal breathing.
7. The MK 25 UBA is now ready for diving.
8. If the purge procedure is interrupted at any point, the procedure should be repeated starting with Step 2. It should also be repeated any time the mouthpiece is removed and air is breathed.

Note. Additional purging during the dive is not necessary and should not be performed.

Figure 12-10. Predive purge procedures

12-71. If the dive is part of a tactical scenario that requires a turtleback phase, the purge must be done in the water after the surface swim, before submerging. If the tactical scenario requires an underwater purge procedure, the diver will complete it while submerged after an initial subsurface transit on open-circuit scuba or other UBA. When the purge is done in either manner, the diver must be thoroughly familiar with the purge procedure and execute it carefully with attention to detail so that it may be accomplished correctly in this less-favorable environment.

TURTLEBACK EMERGENCY DESCENT PURGE PROCEDURE

12-72. This procedure is approved for turtleback emergency descents. The diver should—
- Open oxygen supply.
- Exhale completely, clearing the mouthpiece with the dive or surface valve in the surface position.
- Put the dive or surface valve in the dive position and make the emergency descent.
- Immediately upon reaching depth, perform purging under pressure (pressurized phase) IAW the appropriate MK 25 UBA technical manual.

AVOIDING PURGE PROCEDURE ERRORS

12-73. The following errors may result in a dangerously low percentage of oxygen in the UBA and should be avoided. The diver should avoid—
- Exhaling back into the bag with the last breath rather than into the atmosphere while emptying the breathing bag.
- Underinflating the bag during the fill segment of the fill/empty cycle.
- Adjusting the waist strap of the UBA or adjustment straps of the life jacket too tightly. Lack of room for bag expansion may result in underinflation of the bag and inadequate purging.
- Breathing gas volume deficiency caused by failure to turn on the oxygen-supply valve before underwater purge procedures.

Underwater Procedures

12-74. During the dive, personnel should adhere to the following guidelines:
- Know and observe the oxygen exposure limits.
- Check each other carefully for leaks at the onset of the dive. Perform this check in the water after purging, but before descending to a transit depth.
- Swim at a relaxed, comfortable pace as established by the slower swimmer of the pair (100 yards per 3 minutes is the objective swim speed).
- Be alert for any symptoms suggestive of a medical disorder (CNS oxygen toxicity, carbon dioxide buildup).
- Observe the UBA canister limit for the expected water temperature.
- Use minimum surface checks consistent with operational necessity.
- Minimize gas loss from the UBA (keep good depth control; avoid leaks and excessive mask clearing).
- Use proper amount of weights for the thermal protection worn and for equipment carried.
- Use tides and currents to maximum advantage.
- Swim at 20 FSW or shallower unless operational requirements dictate otherwise.
- Maintain frequent visual or touch checks with buddy.
- Wear the appropriate thermal protection.
- Do not use the UBA breathing bag as a buoyancy compensation device.
- Wear a depth gauge to allow precise depth control. The depth for the pair of divers is the greatest depth attained by either diver.
- Do not perform additional purges during the dive unless the mouthpiece is removed and air is breathed.
- Ensure the diver not using the compass carefully notes the starting and ending time if an excursion occurs.
- Provide at least one depth gauge per pair. Because of the importance of maintaining accurate depth control on oxygen swims, both divers in a swim pair should carry a depth gauge whenever possible.

UBA Malfunction Procedures

12-75. The diver shall be thoroughly familiar with the malfunction procedures unique to his UBA. The following procedures are described in the appropriate MK 25 UBA operational and maintenance manual:
- *Ascent Procedures.* The ascent rate shall never exceed 30 fpm.
- *Postdive Procedures and Dive Documentation.* UBA postdive procedures should be accomplished using DA Form 7533-R (Appendix E). Document all dives performed by submitting a combined diving log.

TRANSPORT AND STORAGE OF PREPARED UBA

12-76. Once DA Form 7532-R has been completed, the UBA is ready to dive. At this point, the UBA may be stored for up to 14 days before diving. If stored in excess of 14 days, the carbon dioxide absorbent in the canister must be changed. Before storing, the diver should shut the oxygen cylinder valve and depress the demand valve bypass button to vent the pneumatics subsystem. He should deplete the breathing bag and place the mouthpiece rotary valve in the surface (closed) position. In this configuration, the UBA is airtight and the carbon dioxide absorbent in the canister is protected from moisture that can impair carbon dioxide absorption. He then puts a laminated or hard copy of completed DA Form 7532-R (Appendix E) with the stored UBA.

12-77. High temperatures during transport and storage should not adversely affect carbon dioxide absorbents; however, storage temperatures below freezing may decrease performance and should be avoided. The diver should check the manufacturer's recommendations regarding storage temperatures.

12-78. If an operation calls for an oxygen dive followed by a surface interval and a second oxygen dive, the MK 25 UBA shall be sealed during the surface interval as described above (mouthpiece rotary valve in surface position, oxygen supply valve shut). It is not necessary to change the carbon dioxide absorbent before the second dive as long as the combined canister durations of all subsequent dives do not exceed the canister duration limits specified.

12-79. For certain immediate-use missions, the MK 25 UBA must be shipped with a fully charged oxygen cylinder. The MK 25 UBA oxygen cylinder, composed of aluminum alloy 6061-T6, is manufactured and certified to DOT 3AL specifications. No certificate of equivalency is required when transporting a charged MK 25 UBA oxygen cylinder by any civilian or military means.

This page intentionally left blank.

Chapter 13

Surface Infiltration

A surface vessel can be the primary means of infiltration, an intermediate transport platform, or one of a series of vessels used in a complex infiltration chain. Surface assets include any seagoing vessel from aircraft carriers to Coast Guard cutters, from merchant vessels to charter fishing boats, including CRRCs and kayaks. Selection criteria and planning considerations vary according to the mission requirements and the assets available. This chapter will examine the characteristics of different surface infiltration assets and explain the planning aspects of a CRRC (or other small craft) infiltration and exfiltration.

MOTHER SHIP OPERATIONS

13-1. Missions using large surface vessels are commonly referred to as "mother ship" operations. The term mother ship implies using a large ship as a mission support craft. The mother ship's primary mission is to transport an SFOD from a staging base to a launch point within range of the detachment's organic infiltration assets; for example, CRRCs or kayaks. Vessels suitable for use as a mother ship are generally divided into military and commercial types.

MILITARY SHIPPING

13-2. Military shipping is usually available from U.S. (joint) or allied (combined) navies. Navies have always been the lead elements of "gunboat diplomacy." Because governments have a need to influence situations outside their borders, some assets, often as large as a task force, are often already deployed within range of a potential "hot spot." Regional instabilities mean that friendly naval assets are sometimes the only secure staging base available to a detachment seeking entry into a potentially hostile area. Military shipping has several advantages and disadvantages to other methods of deployment (Figure 13-1, page 13-2).

13-3. Detachments can be assigned or attached to naval task forces as part of a contingency operations package. Special operations command and control elements (SOCCEs) may be required for liaison with the fleet commander and his staff. Higher headquarters may task detachments to use military shipping during contingency operations. Detachments desiring to train with naval assets can take advantage of this by making coordination at the joint headquarters level.

13-4. The U.S. Navy's mobility can be used to extend the range of both fixed- and rotary-wing aircraft into many areas of the world where land-based support is denied. Aircraft carriers normally have at least one cargo aircraft capable of being configured for airdropping a limited number of personnel and a limited amount of equipment. Heliborne operations are not restricted to aircraft carriers; most large naval ships have landing decks that will accommodate one or more helicopters. Available vessels include landing craft, troop transports, aircraft carriers, and a wide variety of other surface craft presently in the USN and Army inventory.

COMMERCIAL SHIPPING

13-5. Commercial or civilian shipping includes any vessel that is not a "flagged" military vessel. These may include civilian pleasure craft, charter or commercial fishing vessels, or merchant shipping. Detachments select commercial shipping as an alternative to military shipping based on mission requirements. Factors may include a lack of available military shipping or a desire to reduce or conceal the military signature associated with naval vessels. Civilian shipping also has several advantages and disadvantages to other methods (Figure 13-2).

Advantages	Disadvantages
Coordination can be made through joint headquarters. Sophisticated communications and navigation support is available to the SFOD throughout the embarked phase. Precise navigation can be maintained to the debarkation point. Weather, tide, and current updates are available. Additional planning and training can be conducted en route. Large quantities of supplies can be transported. Relatively unaffected by weather. Provides long-range delivery and simplicity during debarkation. Exfiltration can be planned in conjunction with infiltration using the same assets. Close air support and naval gunfire can be coordinated and requested.	Sponsor identification cannot be concealed. Training with Navy shipping assets is more difficult. Army personnel do not conduct enough joint operations with the Navy. Surface craft require more transit time to make long-distance transits. This factor may not meet the mission's time constraints. Hydrographic characteristics of the area and enemy shore defenses may force larger vessels to remain further offshore. An intermediate transport vessel such as a CRRC or kayak is usually required to transport the team to the BLS or a final debarkation point.

Figure 13-1. Military shipping considerations

Advantages	Disadvantages
Sponsor ID is concealed. Civilian shipping blends in with local boat traffic. The detachment has more control of the route. Civilian crew members may be able to provide the detachment with invaluable "local knowledge." Commercial shipping may be more responsive to the detachment's requirements.	Lack of fire support. Sophisticated communications and navigation support may not be available. Operational security may be difficult to achieve and maintain. Exfiltration support may be limited or nonexistent. Suitable vessels/crew may be difficult or impossible to obtain.

Figure 13-2. Civilian shipping considerations

13-6. If the mission requires concealment of the sponsor's identity, commercial shipping becomes a very viable means of infiltration. It requires less training and coordination than submarine infiltration and can be used when an OTH airdrop is not possible. If a detachment has the requisite ship-handling skills, or if it can be augmented with suitable crew, it is possible to conduct a "stand-alone" infiltration. If not, external assets are required to support the detachment's infiltration plan.

13-7. Small craft can be purchased outright, rented, leased, or otherwise acquired with or without a crew. Larger vessels will probably require a professional crew and are therefore beyond the capability of most detachments. Detachments must select an appropriate vessel, ensuring that it is seaworthy, has adequate facilities to support the detachment while embarked, and will not draw undue attention to itself due to being "out-of-place." Figure 13-3, page 13-3, lists the characteristics that a detachment should consider when choosing a craft.

Surface Infiltration

- Vessel type common in the area of operations, low visual signature.
- Viable reason for being in the area.
- Reasonable cruising speed.
- Adequate fuel supply and sufficient range, including contingencies.
- Sufficient space for equipment, personnel, and mission preparation.
- Adequate navigation and communications equipment (augment with detachment assets as backup).
- Redundancy and reliability (twin engines required, preferably diesel inboards).

Figure 13-3. Desirable characteristics for choosing a craft

ACTIONS EN ROUTE TO THE DEBARKATION POINT

13-8. Upon assignment of a transport vessel, detachment personnel should become familiar with the ship's characteristics, including exact troop locations and dimensions of storage areas. Learning the ship is best done by coordinating with the vessel commander. Personnel then prepare and package the equipment according to the dimensions and weights compatible with the assigned craft. All equipment must be waterproofed and marked for easy identification.

13-9. Planning for the debarkation of swimmers is the most important consideration when using surface craft. Some large vessels cannot deliver the operational elements within swimming range of the BLS because of hydrographic characteristics of the area or danger of enemy detection. This situation requires additional planning, preparation, and coordination for intermediate transport from the point of debarkation to the launch point in the vicinity of the BLS.

13-10. Once the SFOD selects the debarkation method, it designates specific duties and stations and conducts rehearsals. The detachment should continue debarkation rehearsals en route with all personnel involved in the operation.

CRRC MISSION PLANNING FACTORS

13-11. Once the detachment is safely embarked on its mother ship, it can finalize mission planning and conduct rehearsals to the extent allowed by the support vessel. At this time, the detachment will formalize the infiltration plan from the debarkation point to the swimmer launch point using internal assets. These assets are usually CRRCs or kayaks. Because many of the planning considerations for kayaks are similar or less complicated, this portion of the chapter will concentrate on CRRCs.

TACTICAL LOADS

13-12. While the CRRC is capable of transporting up to ten personnel, it is highly recommended that no more than six personnel be embarked. The weight ceiling for the CRRC is 2,000 pounds. Any weight above this amount significantly reduces the CRRC's efficiency. Experience has shown that six personnel with mission-essential equipment, on average, are closer to this weight ceiling. (This data pertains to the F-470/55 hp engine combination.) For extended transits or marginal sea conditions, the mission commander should consider reducing this maximum load to four persons. In a crowded CRRC, those personnel in the forward positions are subjected to a greater degree of physical pounding from the effect of swell and wave activity on the bow. This beating is especially significant during long transit periods; the increased physical stress may diminish operator ability to perform after arrival at the objective. The detachment commander must also take into account the possibility of prisoners, casualties, and the evacuation of friendly forces. It is prudent to allow sufficient boat space to contend with contingencies.

COMMAND RELATIONSHIPS

13-13. SFOD can be called upon to operate CRRCs from a variety of platforms. If CRRCs are embarked on a USN ship, that ship's commanding officer or other officer in tactical command (if so assigned) must ensure CRRC operations are safely conducted. If not embarked, this task rests with the commander of the detachment.

Chapter 13

PLANNING RESPONSIBILITIES

13-14. When planning an OTH CRRC operation, the detachment operations sergeant is normally the focal point because of his knowledge and experience. The planning sequence must consist of continuous parallel, concurrent, and detailed planning. Planning support and information will be required from a variety of sources. Planning must be characterized by teamwork between the supported and supporting elements. As the plan develops and is completed, all aspects should be briefed, reviewed, and understood by planners and decision makers. The CRRC mission planning checklist provides a planning sequence guide that can be modified as required (Figure 13-4).

- Analyze maps/charts of objective and water in the objective area.
- Request reconnaissance of beach landing site.
- Prepare weapons, boats, and equipment.
- Review training area regulations/manuals dealing with navigation.
- Review after-action reports from liaison visits to the local area.
- Request advice of allied naval liaison officers that are familiar with the local area.
- Request a local and experienced guide for transit.
- Request air support for possible alternate means of withdrawal and MEDEVAC operations.
- Review meteorological and tidal data.
- Request and review GPS, way points, and additional local navigation charts.
- Request any other manuals helpful to navigation in the local area in addition to Sailing Directions and Summary of Corrections.
- Review reconnaissance reports on file of the BLS, estuaries, and harbor entrances to be transited.
- Study photographs, satellite imagery of the BLS, estuaries, and harbor entrances to be transited.
- Request remote piloted vehicle support or fixed and rotary-wing visual reconnaissance, if available.
- Obtain any navigation advice the ship's navigator may be able to lend and develop a plan (NAV track) with him.
- If available, have oceanographer/meteorologist provide detailed data, discuss intended navigation track to BLS and return, and determine potential obstacles along route.
- Select primary, alternate, and contingency BLSs and consider emergency plans (abort criteria).

Figure 13-4. CRRC mission planning checklist

CRRC LAUNCH AND RECOVERY OPERATIONS

13-15. Personnel must perform launch and recovery operations in a specific manner. The following general guidelines apply.

13-16. **Ship Handling.** The ship's stability is of prime importance for launch and recovery evolutions. An appropriate heading must be found to minimize pitch and roll to achieve the best launch and recovery condition. In addition, winds may cause a cross swell; consequently, close coordination between the bridge and the launch and recovery control station is required. Minimizing roll of the ship is the major concern. Handling, launching, and recovering CRRCs can become extremely difficult as seas increase to Sea State 3. If possible, ship course and speed changes should be avoided during CRRC launch and recovery operations. This avoidance reduces the possibility of cross-wave action across the stern gate or sill.

13-17. **Communications Systems.** Hand-held portable radio and phone communications should be established between the bridge and launch or recovery control stations. When the CRRC is ready to launch, the control station will request the bridge to maneuver the ship into the launch heading. Also, the control station establishes communications with the CRRC through hand signals or voice via portable bullhorn or ship's well-deck public address (PA) system. Before the scheduled launch time, the CRRC officer should make a GO or NO-GO recommendation based on the prevailing sea state to the ship's commanding officer through the control station. The ship's commanding officer has the ultimate responsibility and authority to launch and recover CRRCs.

13-18. The mobility and versatility of the CRRC allows it to be launched and recovered from a variety of platforms. They are—
- Fixed-wing aircraft (launch only).
- Rotary-wing aircraft.
- Surface ship (either wet-well or over-the-side).
- Combatant craft.
- Submarine (submerged or surfaced).

13-19. Specific techniques for CRRC launch and recovery operations are addressed in the following:
- Commander, Naval Special Warfare guidance.
- NWP 79-0-4, *Submarine Special Operations Manual–Unconventional Warfare*.
- NSW Combatant Craft TACMEMO (XL-2080-2-89).
- Dry-Deck Shelter (DDS) TACMEMO (XL-2080-4-89).
- Mass Swimmer LO/LI SOP (DDS Technical Manual, Volume III).

SEA STATES

13-20. The reference for understanding and judging sea conditions is the American Practical Navigator Bowditch, Volume I. The sea-keeping characteristics of the CRRC permits operations up to Sea State 3. It is important to understand that rough weather greatly increases the risks associated with launch and recovery of CRRCs, as well as increases fuel consumption and transit times. Another important consideration is that high winds associated with higher sea states adversely affect the maneuverability of the CRRC. Higher sea states and corresponding winds will result in slower speeds and longer transit times. CRRC launch, recovery, and transits will not be routinely conducted in Sea State 3 or greater.

13-21. The mission commander must carefully weigh the tradeoffs when he decides whether operating in marginal sea conditions is feasible or advisable. The advantage offered by operating in higher sea states is a reduced vulnerability to detection by visual means and enemy electronic sensors. Higher sea states will degrade the ranges at which electronic sensors are effective, enhancing the clandestine nature of the operation and contributing to tactical surprise.

13-22. It is also important to understand that sea conditions can fluctuate rapidly. During the planning phase of an OTH operation, the commander must consider the forecasted meteorological factors that could affect the sea state during the operation. If it is likely that sea conditions at launch time will be Sea State 3, the commander should ask, "Are there any forecasted meteorological factors that may cause the sea state to go to 4 or 5 during the operation?"

13-23. Another important consideration is that sea states and surf conditions are not necessarily related to one another. A low sea state does not mean a benign surf zone; sea state conditions and surf zone conditions must be considered independent of one another.

THE SURF ZONE

13-24. A key planning consideration for a CRRC insertion will be the characteristics of the surf zone at the BLS. CRRC will not be operated in surf zones with significant breaker heights greater than 8 feet. If significant breaker heights develop above 8 feet, the judgment of the designated CRRC officer will dictate alternate options for transiting the surf zone. Significant breaker heights are determined per reference. Detailed information on surf zone characteristics can be found in Chapters 4 and 5. Surf zone conditions are reported by a SUROBS as discussed earlier.

SUROBS KEY ELEMENTS

13-25. The critical considerations for CRRC surf zone operations are the **significant wave height** (height of the highest one-third of the breakers observed), the **period** (the time interval between waves measured to the nearest half-second), and the **breaker type** (spilling, plunging, or surging). The higher the significant wave height, the shorter the period, the greater the percentage of plunging or surging waves, and the greater the danger to CRRC surf zone operations. These critical factors must be considered both individually and in combination. The commander should also carefully evaluate the effect on CRRC operations, especially in light of boat or engine maintenance conditions and coxswain or boat-crew experience level. Trained and experienced CRRC personnel should conduct this evaluation. It should also be understood that, although these three elements are the most critical considerations, the other information contained in the SUROBS is important for navigation and surf zone negotiation planning.

PLANNING AND EXECUTION CONSIDERATIONS

13-26. Surf conditions are critical in making the final determination to launch or not launch the CRRC. Once a CRRC is launched and arrives outside the surf zone at the BLS, the mission or boat commander must make the final beaching decision. If prelaunch SUROBS, visual sightings, and mission necessity warrant further surf evaluation, the mission commander may use his own scout swimmers for this purpose. The commander must consider the surf conditions as they exist at the time, his mission and command guidance, boat or engine maintenance conditions, and coxswain or boat crew training and experience level. As the CRRC officer of the boat crew and the on-scene commander trained in this capability, he is best qualified to make the final beaching decision. It is also important to understand that surf conditions, like sea states, can change rapidly. During the planning phase, the commander should routinely evaluate the forecasted meteorological factors that could adversely affect surf conditions.

SURF LIMITS

13-27. Figure 6-1, page 6-9, sets forth recommended operating limits for CRRC surf zone operations. The table plots breaker height versus breaker period, and is applicable to both spilling and plunging waves, or combinations of both. As already noted, plunging breakers are more dangerous to CRRC operations than spilling breakers, and greater care and judgment must be exercised as the percentage of plunging breakers increases. Surging breakers are not included in the table. These recommended surf limits are set forth as a guide; they are not intended to usurp the judgment of officers exercising command.

DISTANCE

13-28. The point at which the CRRC is launched for OTH operations is required to be at least 20 nm from shore, but usually no more than 60 nm (normally, 35 miles is sufficient). In determining the actual distance to execute a launch, the sea state, weather, transit times, and enemy electronic detection capabilities should be considered. The commander should keep the launch platform protected and far enough away to prevent operational compromise, while minimizing the distance for the raid force.

CONTINGENCY PLANNING

13-29. A commander must include provisions for the unexpected when planning any operation. Plans must address the following contingencies and how to effectively resolve them:
- *OBM Breakdown.* Bring tools and spare parts or a spare motor, if space and time permit. Make sure a trained mechanic is available. Ensure spare parts include extra plugs and fuel filters.
- *Navigational Error.* Study permanent geographical features and known tides, currents, and winds; take into account sea state; and establish en route checkpoints. Intentionally steer to the left or right of the target so, once landfall is reached, a direction to the target is already established.
- *Low on Fuel or Out of Fuel.* Run trials with a fully loaded boat in various sea states to get exact fuel consumption figures. Take enough fuel for a worst-case scenario. Carry one extra full fuel bladder.
- *Emergency MEDEVAC Procedures.* Know location of nearest medical facility. Develop primary and alternate plans for evacuating casualties.

Surface Infiltration

PLANNING FOR A RENDEZVOUS AT SEA

13-30. The most critical aspect of exfiltration operations is the rendezvous at sea. An OTH rendezvous is the most difficult because of the precision navigation required to effect linkup. These difficulties are compounded when coupled with the limited navigational equipment available in small craft. More than one GPS and redundant and robust communications can make an at-sea rendezvous relatively easy; however, these items are electronic. They are subject to shock and water damage, power failure, propagation problems, and enemy interference. These are serious problems and are frequently encountered during MAROPS. Chances are one or all of the systems could fail at the crucial moment. At a minimum, the waterborne detachment must have access to a GPS, a compass to shoot bearings with, and a compass to steer the CRRC. To avoid a long swim home, it is a good idea to make the designated rendezvous.

13-31. The most crucial portion of any rendezvous procedure is **prior planning**, which begins with a NAVPLAN. The commander should start with the worst-case scenario and try to cover all contingencies. The NAVPLAN must include primary and alternate plans that are well developed, realistic, and achievable. Every attempt should be made to adhere to it as closely as possible. Both primary and contingency plans must be complete with built-in fudge factors to allow for variances in time of transit and actual fuel consumption. All plans must include primary and alternate rendezvous locations, with each rendezvous location having its own signal plan. The signal plan should encompass audible and visual recognitions—challenge and reply, time windows, and radio frequencies with a method of changing from primary to alternate. Plans must be thoroughly briefed for all possible contingencies and backup plans. Figure 13-5, page 13-8, provides a sample rendezvous plan.

13-32. On occasion, the CRRC formation will still be out of position and the mother ship will have to (if possible) vector the detachment to the rendezvous, or move from the designated RP to the detachment to effect the linkup (depending on the capabilities of the mother ship).

RAIDING CRAFT DETECTABILITY AND CLASSIFICATION COUNTERMEASURES

13-33. Ideally, SFODs conducting a waterborne infiltration will have done a complete area assessment and selected a minimum-risk route (MRR) for their infiltration. MRRs will seek to avoid known or suspected enemy detection and surveillance facilities or capabilities. Information on enemy capabilities can be obtained from the N-2, S-2, or the Joint Intelligence Center (JIC). All of these sources can provide an analysis of enemy detection facilities in the vicinity of the target and their respective capabilities. If mission requirements dictate an infiltration or exfiltration route that will pass within detection range of an enemy installation, consideration must be given to detection countermeasures.

13-34. Detection threats are determined by the ranges of the detection systems, the expertise of the operators, and the amount of time that the detachment spends inside the threat system's detection range. The most common enemy detection systems include active and passive sonar arrays and shore-based shipboard and airborne radar systems. Other technical assets include forward-looking infrared radar (FLIR) or thermal imaging systems and NVGs or optics. These are most likely to be encountered in conjunction with coastal patrol craft, coast watchers, and informants such as "loyal citizens" reporting unusual activities to the authorities. Any one of these can be defeated by deception, distraction, or bypassing. The detachment's goal is to prevent the CRRC formation from being detected and identified or classified as a threat. Unfortunately, the better integrated and more sophisticated the enemy's "defense in-depth" is, the more difficult it will be to defeat.

13-35. The easiest way to defeat a detection system is to avoid it. Detachment personnel should locate the enemy radar positions and establish fans of coverage. They should take these fans and the level of operator proficiency into consideration during mission planning. They should make sure to select routes and destinations outside the system's detection range. If the system cannot be avoided, personnel should select a route that will take advantage of terrain masking. The detachment may be able to find a corridor or gap in the coverage that will allow it to bypass the threat, or a natural obstruction that will shield the detachment from detection. This method is particularly useful with shore-based and shipboard (low-altitude) radar systems, as well as active visual observation methods such as coast watchers. Some high-altitude radars may have a blind spot immediately under them that can be exploited by a detachment posing as commercial shipping that would be able to launch from a mother ship close enough to get under the radar's minimum detection range. Unfortunately, the greatly increased detection ranges of airborne radars may cancel any air operation even if the infiltration aircraft is able to imitate civilian traffic.

---

1. Pickup sites:
    a. Primary Rendezvous: 3245 34n 11734 23w 0100-0130 13 Jan 08.
    b. Alternate Rendezvous: 3237 30n 11729 30w 0230-0300 13 Jan 08.
2. VHF primary: 32.90 MHz; HF primary ___; UHF primary_____
    alternate: 36.50 MHz;   alternate __;   alternate _____
3. Call signs:
    a. Detachment—Frogman.
    b. USS Shoe Factory—Paint Chipper.
4. Signal:
    a. Visual: **Signal**   **Reply**
        Primary: IR strobe   three red
        Alternate: directional white strobe   three red
    NOTE: Think of direction flashed or shone and the possibility of compromise by the enemy.
    b. Audible:
        Primary: pinger   vector by ship/visual signal.
        Alternate: M 80s   vector by ship/visual signal.
    c. Challenge and Reply Signal:
        Primary time and place: approach primary visual signal.
        Direction: then primary signal.
        Alternate communications: code words.
5. Procedure:
    a. Navigate to rendezvous site and check navigation by all means available.
    b. Initiate primary signal at beginning of window.
    c. If ship is visible, wait for proper visual reply.
    d. Proceed to mother ship and begin recovery procedures.
6. Contingency plans to cover at least the following cases:
    a. Unable to make primary rendezvous due to time; move directly to alternate rendezvous.
    b. Threat at or near primary rendezvous; move to alternate.
    c. Ship is not visible at rendezvous point:
        (1) Recheck navigation and confirm position.
        (2) Carry out rendezvous procedure.
        (3) At end of rendezvous time window, move to alternate rendezvous and carry out rendezvous procedure.
    d. Mother ship not at alternate rendezvous or threat at alternate rendezvous:
        (1) Move back to shore.
        (2) Cache.
        (3) Carry out secondary night rendezvous procedure.
    NOTE: Think of emergency pickup or hot extraction plan and fire support plan. Plan and brief thoroughly and when conducting operation, never rely fully on GPS. Always cross-check navigation with speed, time, and distance.

Figure 13-5. Sample rendezvous plan

13-36. Detachments facing a surface radar threat have to consider the strength of the radar signal versus the quantity and quality of local traffic. The greater the signal strength, the greater the range or the better the target discrimination. Because CRRCs have such a small signature, it is possible to lose them in the "clutter" of false returns in congested areas. Specific techniques for confusing enemy radars to avoid detection or classification include the following:

- Coordinate with naval assets for EA-6B air support for the mission. These aircraft may be able to jam enemy radar in or near the detachment's vicinity, the vicinity of the target, or the BLS.
- Launch the CRRCs from the support vessel at greater intervals with each craft traveling in pairs to an RV point off of the BLS (or at the BLS) where the entire force converges. With greater separation, the craft are less vulnerable to detection. GPS navigation and a foolproof linkup plan will simplify the operation.
- Move at slower speeds to reduce the wake size and thus the radar signature.
- If the detachment is infiltrating in the vicinity of commercial shipping lanes, it can take advantage of merchant traffic to mask the movement of the CRRC formation. A good technique is to place a larger vessel (commercial or fishing) between the detachment and any threat radars. Traveling in the wake or to the flank of the larger vessel will keep the detachment masked by the ship's structure, wake, or waves. This technique carries its own risks of compromise by the same merchant traffic that the detachment is using to screen its movement.
- Group the CRRCs together to provide a radar return similar to a commercial or fishing vessel. Move at the same speeds as indigenous vessels and other harbor or local traffic simulating their radar return.

13-37. If traffic in the area is tightly controlled by the enemy, sonar arrays, when present, may prevent the use of CRRCs. The detachment should seek S-2, N-2, or JIC assistance to determine if the enemy possesses this capability and the locations where it is used. It should avoid these locations wherever possible. Methods to reduce the risks presented by sonar include the following:

- Take advantage of the presence of large commercial vessels that tend to wipe out sonar's ability to detect OBMs due to the increased background noises.
- Paddle the CRRCs into the BLS from a designated point outside of sonar range.
- Coordinate with naval assets to increase underwater noise by whatever means available; for example, high-speed coastal ships and hydrophones.
- Determine the types, times, and locations where small fishing vessels are active and, if possible, plan the operation to coincide with their activity.

13-38. Detachment personnel should seek S-2, N-2, or JIC assistance to determine if the enemy possesses a FLIR or other thermal-imaging capability and the locations where it is used. Again, these locations should be avoided wherever possible. Methods to reduce the risks presented by FLIR include the following:

- Ensure that the CRRCs are kept as close to the operating water temperature as possible.
- Use covers made of insulating material to place over the OBM cover when pursued, when within range of a possible FLIR or thermal imager, or when within range of thermal devices on the BLS.

13-39. To reduce the possibility of compromise by other visual imaging techniques such as NVGs or optics, the detachment should plan operations during rain, fog, or other inclement weather. Smoke pots or smoke grenades can be used to conceal movement by floating or dragging them behind the CRRC during withdrawal, contact, or pursuit. Wherever possible, the detachment should minimize the risks by avoiding known patrolled areas.

This page intentionally left blank.

Chapter 14

Submarine Operations

Submariners' ability to pass undetected through the oceans anywhere in the world makes them an ideal infiltration asset. Using submarines in conjunction with SF units has proven to be an effective method of conducting clandestine operations in the past, and this effectiveness will continue into the future.

WW II saw the first extensive use of submersibles as an SO infiltration platform. The Italians, British, and Germans all made extensive use of submarines to infiltrate operators. The United States, especially in the Pacific theater, pioneered missions and techniques that are still valid and occasionally used today.

Today submarines are used to clandestinely transport infiltration teams to debarkation points in the vicinity of their AO. The teams then traverse from the submarine to their BLS and into their AO. The submarine's ability to surface and quickly disembark full teams with extensive equipment before again submerging and disappearing make the submarine ideal for the landing and withdrawal of teams for clandestine operations along coastal areas. The submarine's ability to operate while submerged and the team members' ability to lock out (debark) and lock in (embark) while the submarine is submerged offer SF dive teams the utmost in undetectability when infiltrating dive teams.

Commanders can use SF teams from either nuclear-powered (U.S.) or diesel (HN) submarines. Depending upon which is used, certain special considerations must be given to the technique of debarking and embarking. This chapter discusses the planning required for operations from a conventional or nuclear submarine, and the procedures used by swimmers and boat teams. Submarine swimmer and boat team launch and recovery operations are highly complex and require extensive rehearsal and training. Specific procedures for the conduct of these operations are detailed in NWP 79-0-4.

SUBMARINES

14-1. The U.S. Navy maintains a fleet of submarines, some of which are suitable for use to infiltrate or exfiltrate SFODs. Currently, the U.S. Navy is converting some ex-fleet ballistic missile submarines (SSBNs) to guided-missile submarines (SSGNs) for use as delivery platforms for SOF missions. Some allied navies are also using submarines that can be used by detachments. These are mainly diesel-electric boats.

14-2. Submarine training is arranged at "presail" conferences in a process similar to the joint airborne/air transportability training (JA/ATT) conferences used to obtain training opportunities with USAF assets. Detachments desiring to conduct submarine training should plan well in advance and follow up all coordination. As the time approaches for the SUBOPS, teams must schedule and complete requisite training, such as tower training.

Chapter 14

14-3. When the time comes to embark on the operation, teams must finish preliminary planning, conduct dockside training, stow cargo and equipment, and prepare for life while embarked. Dockside training is a critical event because it allows the detachment to achieve familiarity with the submarine and determine where gear will be stored during transit and operations. It also allows the detachment to demonstrate its proficiency to the submarine's commander and crew.

PRELIMINARY PLANNING

14-4. Before embarkation planning begins, the detachment leader should make a liaison visit to the submarine to ascertain the features of the particular submarine. This liaison visit is needed because differences exist in both internal and external configuration among the various classes of submarines as well as among submarines of the same class. The detachment should coordinate the following items of information with the submarine commander during the liaison visit:

- Time and place of embarkation.
- Number of troops to be embarked.
- Location and capacity of cargo storage areas and amount of cargo to embark.
- Accommodations for embarked troops.
- Communications support available from the submarine and compatibility of communications equipment (radios and frequency range).
- Mission and concept of the operation to include enemy and friendly situations.
- Special equipment needed by the embarked troops to carry out debarkation and recovery operations. These may include towing bridles, towing lines, scuba and swimming equipment, DPDs (Appendix F), signal lights, and submarine rigging equipment.
- Time and place of debarkation and recovery of exfiltration teams, as well as type of debarkation and recovery procedures to be used.
- Miscellaneous information such as general capabilities and limitations of the submarine, fire support available aboard the ship, and medical support available.

STOWAGE OF TROOP CARGO AND EQUIPMENT

14-5. A submarine has spaces to store troop cargo and equipment both inside and outside the pressure hull. Hatches are usually 25 inches in diameter. Another hatch is located in the conning tower but is used primarily by the ship's crew. Equipment should be rigged or packaged to fit through the smallest possible hatch that could be used.

14-6. **Cargo Spaces Within the Pressure Hull.** Cargo and equipment should be packaged with handles for lowering down through the hatch. Personnel should waterproof all packages and mark them for identification. They should also pack individual clothing and equipment in waterproof bags. Outboard motors must be purged of fuel and no fuel may be stored inside the pressure hull of a submarine. Inflation systems for the CRRCs should use air and not CO_2. (CO_2 is a contaminant risk in a submarine's sealed atmosphere.) Small amounts of pyrotechnics may be stowed with the ship's flares in a fireproof locker.

14-7. Personnel should select equipment that will not be needed during the voyage; for example, engines and team gear. Finally, all equipment loaded on the submarine must be clearly labeled as to team, boat, and order of disembarkation.

14-8. **Cargo Spaces Outside the Pressure Hull.** Free-flooding cargo spaces located outside the pressure hull are available to store detachment equipment. Because these spaces are exposed to the sea, only items that are not susceptible to pressure damage may be stored in them. These spaces include line lockers and internal spaces in the sail. Personnel should remove or pad sharp objects or sea growths in these spaces before storing equipment inside them.

14-9. Personnel must store fuel bladders externally. When preparing fuel bladders, personnel should only fill them to three-fourths capacity and purge them of all air. Personnel should then Z-fold the end of the fuel line and secure it with zip ties. This measure prevents water from infiltrating the line under pressure and contaminating the fuel. Personnel should stow two fuel bladders per kit bag and secure the kit bag inside the locker. They may also store fuel in the free-flooding areas of the submarine's sail. Some submarines are equipped with pressure-proof lockers outside the pressure hull. Personnel should use these lockers to stow flammables, such as outboard motor fuel. Some items such as deflated boats that are not subject to pressure damage can be stored externally. The boat's inflation valves must be closed to prevent seawater from entering the airtight compartments of the inflatable boat. For short submerged trips, personnel can stow deflated boats on deck and lash them securely with steel cables.

14-10. When stowing equipment in the external spaces, everything must be secured in place. Personnel should use a "donut" as a single point-release system. They should run all lines from the line's attachment points through a donut and secure them back to themselves. To release the equipment, personnel first attach a safety line and clip it in; then they cut the donut with a hook knife. The securing lines will fall free, releasing the load.

TROOP LIFE WHILE EMBARKED

14-11. The detachment should embark troops aboard submarines for the shortest time consistent with operational needs. The troops should embark at forward bases or at sea near the objective area. Daily troop routine must be compatible with the daily operating routine of the submarine. Conditions aboard the submarine are such that the following restrictions usually apply:

- Movement between compartments is held to a minimum.
- Physical exercise except for isometrics is not practical.
- Water discipline is strict. No laundry or shower facilities for troops should be anticipated.
- Troops are allowed topside only occasionally and then in limited numbers.
- Meals are planned ahead with the submarine's mess officer.
- All troop activities are coordinated between the troop commander and the submarine commander.
- Submarine operating procedures are precise and strictly observed. Embarked troops are not normally allowed to take part in the operation of the submarine.
- On the conventional submarine, the detachment personnel will normally be billeted in the forward torpedo room, taking over the bunks of the submariners who are normally assigned that space. They might even have to "hot rack" in shifts, just as the sailors are doing elsewhere in the submarine.
- The sailors that comprise the submarine's crew were selected to meet rigid and demanding requirements for submarine duty. Care should be taken by embarked troops to avoid creating ill will between the submarine crew and themselves.

DEBARKATION

14-12. There are several ways of debarking swimmers or boat teams from submarines. Tactical debarkation may be conducted from a surfaced or submerged submarine. Regardless of the method used, the following general procedures should be observed:

- Crew members and troops are briefed on the debarkation plan and the specific duties and stations of individuals.
- Equipment to be debarked or used during debarkation is inspected and prepared for debarkation before manning debarkation stations.
- Equipment should be staged in the order that it will be debarked.
- The crew and the operators should man debarkation stations on order.
- Debarking troops are oriented in relation to the landing area and briefed on sea and surf conditions.
- A rehearsal must be conducted before any tactical debarkation.

Chapter 14

SURFACE DEBARKATION OF SWIMMERS

14-13. The coastal defense capabilities and shallow near-shore depths of most threat nations would normally prevent a submarine from getting close enough to the shore to disembark swim teams. The most probable reason for a detachment to leave a surfaced submarine as swimmers would be to transfer to another vessel or helicopter. Swimmers may debark from the conning tower of a broached (partially surfaced) submarine. Debarkation of more than one pair of swimmers usually requires that the upper and lower conning tower hatches be opened simultaneously, a practice that is not always safe when broached. The submarine usually surfaces with decks awash in such cases. Submarine crewmen man and open all hatches and assist each pair of swimmers in handling equipment. Pairs of swimmers are called topside by number from the control room. If space permits, the first pair of swimmers may take station in the conning tower before the surfacing. Swimmers debark in pairs with their equipment on the leeward side of the submarine. The submarine commander usually orders the debarkation of each pair.

SUBMERGED DEBARKATION OF SWIMMERS AND BOAT TEAMS

14-14. Swimmers may debark through the escape trunk of a submerged submarine. The successful use of submarines requires that operators be proficient in basic military scuba and small-boat operations. Among the advantages of lockin and lockout is the fact that the submarine stays completely submerged, thus contributing to stealth and complete deniability. If the water depths allow, the submarine can also be used to get the detachment closer to shore. Its disadvantages include—
- Amount of mission support personnel who do not go on the operation.
- Amount of time needed to prepare for (train-up) and execute (hours at periscope depth).
- Level of difficulty involved in the actual lockin and lockout procedure.

14-15. The three phases of training required to prepare a detachment to conduct submarine lockout operations are tower training, dockside training, and an actual operation at sea (underway). Tower training is used for initial or refresher training of trunk operators and to teach the entire detachment lockin and lockout procedures. It also allows the detachment to conduct free-swimming ascents and to review emergency procedures. Selected personnel will qualify as trunk operators and supervisors. Dockside training allows the detachment to achieve familiarity with the submarine and determine where gear will be stored during transit and operations. It also gives the detachment a chance to demonstrate their proficiency to the submarine's commander and crew.

14-16. Submarine lockin and lockout procedures are complicated and require a well-trained team to execute safely. The foremost members of the team are the submarine's commander and his navigator. Detachments should be aware that a submarine's commander exercises a degree of authority not normally encountered elsewhere and behave accordingly toward him.

14-17. The detachment's liaison to the commander and his staff is the conn liaison officer. This person will be an officer or senior NCO who is not otherwise involved in the operation. The submarine commander will give the order to lock out the divers through the conn liaison officer.

14-18. The trunk supervisor should be a senior dive supervisor. He will be the diving supervisor for the lockout. He provides the trunk supervisor's dialog and can override the trunk operator and run the trunk himself from outside.

14-19. The trunk operator oversees all conduct inside the trunk. He is the only one who talks. He must memorize the trunk operators dialog, the emergency tap codes, and know the appropriate emergency procedures. Remaining team members will make up the boat teams and specialty teams, for example, rigging, derigging, and recovery support.

14-20. Detachments provide the trunk operators, conn liaison officer, trunk supervisor, and rigging team. Submarine crew member duties may include supervising trunk operations and manning and opening the lower hatch of the escape trunk. Swimmers form in the forward torpedo room with their equipment.

14-21. The submarine's escape trunk is a watertight extension of the submarine's pressure hull. It has three hatches. The lower hatch is used for entry into the trunk, the top hatch is for emergency egress or access while in port, and the side door is for locking divers in and out. The side door can be cycled from inside the submarine. The escape trunk has two depth-pressure gauges to monitor pressure in the trunk. When "blowing" (pressurizing) the trunk, operators must take care not to exceed (NTE) 15 psi above bottom pressure. Primary communications are provided by the 31MC, a sound-powered intercom. Emergency communications are conducted using a tap code posted on a placard inside the trunk. The operator should use a hammer or dive knife to tap on the trunk.

14-22. In the event of an air-related emergency, the trunk has a built in breathing system (BIBS) with enough second stages for the trunk's occupants. The detachment will also provide a dive tank with additional second stages attached to the regulator. There are three types of lighting in the trunk—AC lighting, battle lanterns, and chem lights. The number of operators able to lock out is dependent on the size of the escape trunk and the equipment they are locking out. It will vary with the different classes of submarines.

14-23. Submarine lockout procedures require careful coordination between all elements. Critical to the success of the operation are the sequencing of the lockout actions (Figure 14-1) and the rigging of the submarine.

1. The submarine rises to a depth of 25 to 30 FSW (or conning tower depth).
2. Trunk operator readies the trunk.
3. Rigging team enters the trunk.
4. Trunk hatch is shut.
5. Trunk is flooded to the bubble line.
6. Trunk pressure is equalized with seawater pressure.
7. Side door cracks open.
8. Divers exit trunk.
9. Divers rig deck.
10. Divers reenter trunk.
11. Side door is shut.
12. Trunk is drained and pressure remains constant.
13. Trunk pressure is equalized with pressure inside the submarine.
14. Divers exit the trunk.
15. Boat team enters the trunk and continues the cycle.
16. Boat team prepares the boat on the surface, tethered to the salvage buoy.
17. Trunk is shut, drained, and pressure equalized.
18. Divers enter trunk with OBM, paddles, gear, and weapons.
19. Divers with motor assist boat team on surface with preparing the boat.
20. Trunk is shut, drained, and pressure equalized.
21. Conn liaison officer determines that operators are away.
22. Derigging team enters the trunk.
23. Trunk is flooded and pressurized, and derigging team exits trunk and derigs.
24. Divers reenter trunk and continue lockin.
25. Submarine proceeds to rendezvous location.

Figure 14-1. Sequence of lockout procedures

EMERGENCY PROCEDURES

14-24. Submarine lockouts are a complicated, difficult procedure where many things can go wrong. To reduce the risks, divers should observe the following measures:
- Except in a depth excursion, divers must never let go of a rigging line. For the submarine to maintain depth control, it must have water flowing over its control surfaces. During a lockout,

Chapter 14

there is a constant 1- to 3-kt current caused by the submarine's motion. Anyone or anything not secured to a line will be swept away.

- If the submarine does take a depth excursion greater than 50 FSW, divers must kick off the submarine and swim away from it at 90 degrees. If the submarine is turning, they should swim into the direction of the turn. This method gives the divers a reasonable chance to avoid the submarine's screw as it will form a wider arc to the outside of the turn than the bow where the divers were.

- All divers should never release a line to assist another diver. It is easier to lose or recover one diver than it is two.

14-25. Individual ship's emergency bills contain instructions on escape trunk operations. There are rigid protocols to control all steps in the lockout process. This procedure includes a dialog for the trunk operations that must be memorized and followed strictly. Personnel should obtain the current dialogs from naval liaisons before starting the detachment's train-up. The teams conduct dry and wet rehearsals before debarkation.

Note. Currently, only DDS operations are authorized for the underwater lockout of combat divers and their equipment. The use of foreign submarines, the planned conversion of USN ballistic submarines, or the launching of submarines purpose-built for SOF operations may require detachments to be familiar with the rigging of a submarine (Figure 14-2) and the preceding paragraphs. The described lockin and lockout procedures using the forward-escape trunk generally remain the same regardless of differences in submarines, weather conditions, and combat operations. Teams conducting submarine training or actual operations (both referred to as SUBOPS) must use the appropriate references and unit SOPs, and perform liaison with the submarine crew to finalize mission parameters.

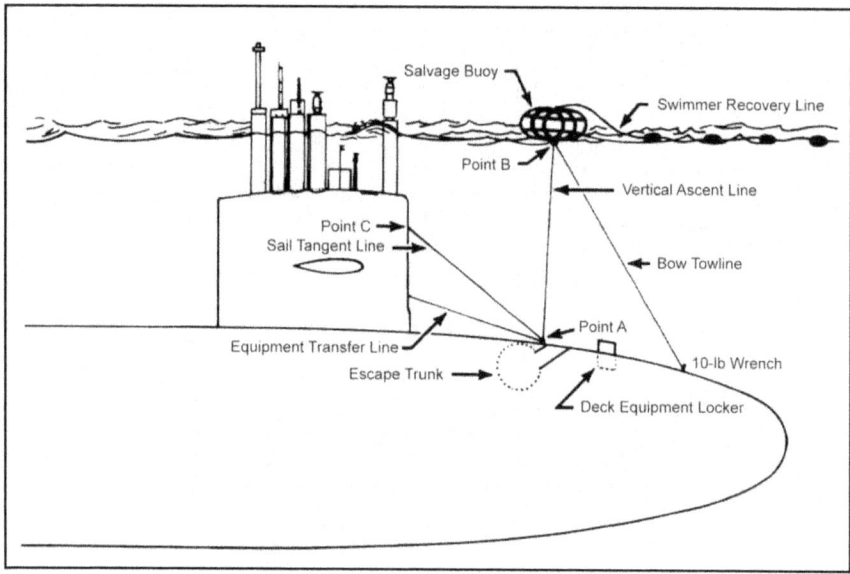

Figure 14-2. Submarine lockout rigging

Surface Debarkation of Boat Teams

14-26. Since locking out a 12-man team with its equipment, boats, and engines is a long process (up to 4 hours or more), the trend in SUBOPS has moved toward the submarine surface launch. A well-rehearsed team can be launched in less than 15 minutes. Inflatable boat teams may debark from a fully surfaced submarine by placing the boats over the side (dry deck) or by loading the boats on the deck and launching as the submarine dives from under the boats (wet deck).

14-27. Before debarkation, personnel will assume stations in the passageway leading to the hatch. All personnel and equipment must be lined up in the order that they will debark before the submarine surfaces. Personnel should request time warnings from the navigator and use them to control preparations. When the submarine surfaces, the first person through the hatch will be a crew member. Crew members man and open all deck hatches. Personnel should remember to stay away from hatches until the crew reports them open and secured. Crew members also rig deck lines and provide a wave lookout and lifeguard with a throw buoy. Once the hatch is secured and the order given by the crew, the detachment will begin debarkation. Boat teams are called topside, by number, through their assigned hatches.

14-28. The team assigned to inflate and rig the boat leads. The boat is passed through the hatch, carried to its station, and inflated. Additional boat teams follow on order and prep their boats. To reduce confusion and improve safety, the team should only bring equipment and personnel on deck as required. A lifeguard should be posted and all equipment secured into the safety lines to prevent its loss overboard in case a wave washes over the deck. The team should snap the boats into the safety line with a pelican hook or similar quick-release so that they can be freed if the submarine does an excursion.

14-29. Team members should inflate the boats with compressed air from the submarine or from 100 free triiodothyronine (ft3) bottles attached to the CRRC's pneumatic inflation system. When using air (80 or 100 ft3 scuba cylinder) to inflate the CRRC, the team should ensure that the proper air fittings, not the standard CRRC CO2 manifold, are installed in the CRRC. Only rolled floors can be used for submerged lockout operations, to include DDS wet operations. If the boats will be inflated from CO2 bottles, the team should retrieve them from the external (free flooding) storage compartments. Once the boats are inflated, the coxswain changes out the used bottle for a fresh one before departing the submarine. While the boat is inflating, a team member brings the OBM on deck. Another team member retrieves the fuel bladders from stowage. The coxswain attaches the OBM, secures it to the transom, puts it on tilt, and prepares to start it. When the fuel is connected and the Z-fold in the fuel line is released, the coxswain starts the OBM and runs it for a few seconds. Once the OBM checks out, the team loads the rucksacks and other operational equipment. Once these steps are completed, the options are to do a dry- or wet-deck launch.

14-30. When the boats are launched over the side of a fully surfaced submarine (dry-deck launching), the team loads and lashes all equipment (waterproofed) before launching. Boats should be launched in the lee of the submarine and may be launched from either forward or aft of the conning tower. Troop debarkation is expedited if the team uses caving or pilot ladders, or if the boats are positioned to take full advantage of hull appendages that offer handholds and footholds. The team then launches the boat stern first and loads it by the long count from the bow.

14-31. The other method of surface debarkation of boat teams is by the submarine submerging from under the fully loaded and manned CRRCs (wet-deck launching). The team loads and lashes all equipment in the boats and then takes their places in the boats. The submarine can be under way or stopped in the water during these preparations. When all boat teams are ready for launching, all submarine personnel go below, secure the hatches, and prepare to submerge. The boat teams will signal the crew that they are ready to launch by striking the hull three times with a paddle. The submarine comes to a speed of 1 to 3 kts and slowly submerges. As the submarine's deck sinks below the surface, the coxswain starts the OBM and stands by to drive off the submarine as soon as the boat becomes waterborne. The boat teams stand by to paddle at a fast stroke to clear the submerging submarine quickly in case the OBM stops (or fails to start). Wet-deck launchings are preferred to dry-deck launchings because the boats are fully loaded and manned before being placed in the water, and the submarine can be under way while all preparations for launching are made. Wet-deck launchings do require somewhat more training and experience than dry-deck launchings.

Kayaks

14-32. Kayaks are launched similarly to CRRCs except that they can be partially preassembled below decks and then handed out the weapons loading hatch for final assembly on deck. When the kayak is assembled in the passageway, personnel lock the keel board in position, mesh but do not lock the gunwales, and leave out ribs 3, 4, and 5 and the Number 1 man's seatback. They attach the coaming and pre-position the spray cover. Personnel bundle the three ribs and secure them inside the kayak. Some mission equipment can now be stowed in the ends of the kayak. Kayakers should take care not to overload the kayak; in its partially assembled state, it is highly susceptible to damage from rough handling. They should cinch the center section of the kayak to keep it from opening up as it is passed through the weapons loading hatch.

14-33. Once the kayak is on deck, personnel insert the three remaining ribs and the Number 1 man's seatback, inflate the sponsons, load the remainder of the equipment, and secure the spray cover. As with CRRCs, the option exists to do a dry or wet launch. If doing a dry launch, personnel should lower the loaded kayaks over the side with help from the crew. Personnel should always maintain control of the kayak with bow and stern painters. The paddlers then embark using a pilot's ladder (caving ladder). With a wet-deck launch, the paddlers get in the kayaks and secure their spray skirts while the crew goes below and prepares to dive. When the team is ready, they signal the crew by banging on the submarine's hull. The submarine then submerges from beneath the kayaks. As the kayaks become waterborne, the team sets a fast paddling cadence to clear the submarine.

Safety

14-34. All personnel engaged in deck launches must be equipped with flotation and signaling devices. For detachment personnel, this means a UDT vest with whistle and strobe or flare. All ship-handling tasks are performed by the crew members. This guidance especially includes the hatches. Specific safety measures that all personnel should follow are listed below:

- Install (crew) and use (everybody) lifelines.
- Post a lifeguard or lookout.
- Have the lookout sound-off with WAVE if a wave is about to wash over the deck.
- Use safety boats during training.
- Always work on the windward side of gear.
- Avoid becoming trapped between equipment and the submarine by a wave (or knocked overboard).
- Never store fuel or CO_2 inside the pressure hull of a submarine.

Dry-Deck Shelter and Mass Swimmer Lockout

14-35. Any U.S. submarine can carry SFODs; however, the complicated procedures associated with the conventional lockin or lockout methods using escape trunks restrict their use. This problem prompted the Navy to develop alternative methods of transporting and launching or recovering divers. The Navy's solution was to modify several submarines specifically to carry swimmers and their equipment more effectively. This change included the installation of chambers called dry-deck shelters. DDSs provide specially configured nuclear-powered submarines with a greater capability of deploying SOF. They can transport, deploy, and recover SOF teams from CRRCs, DPDs (Appendix F), or SEAL delivery vehicles (SDVs), all while remaining submerged. In an era of littoral warfare, this capability substantially enhances the combat flexibility of both the submarine and SOF commander.

14-36. DDSs are specifically designed to transport swimmer delivery vehicles; however, they are also used to deliver boat teams. Navy fleet divers and an SDV detachment operate the DDS. It is comprised of three compartments—the trunk, which is similar to the submarine's escape trunk; the hangar; and a recompression chamber. The operators enter the trunk; it is then flooded and they lock onto the flooded hangar where their boats and equipment are staged. Inside the hangar, there are Navy divers who will assist the operators and their equipment to the surface. The Navy divers are equipped with scuba and have demand regulators with multiple second stages for use by the operators. There are also several demand regulators on the wall of the hangar for the operators to use while they are waiting their turn to ascend with the divers.

DDS Description

14-37. Overall, the DDS is 9 feet wide, 9 feet high, 38 feet long, and displaces 30 tons. It consists of three interconnected compartments made of HY-80 steel within a fiberglass fairing, each capable of independent pressurization to a depth of at least 130 feet. The forward-most compartment—a sphere—is the hyperbaric chamber that is used for treatment of injured divers. In the middle compartment or transfer trunk, operators enter and exit the submarine and either of the other compartments. The third compartment—the hangar—is a cylinder with elliptical ends that houses either the SDV or up to twenty SOF personnel with CRRCs (Figure 14-3).

Figure 14-3. Dry-deck shelter (without fairing)

14-38. The portable DDS is designed for temporary installation on modified host submarines. It can be installed in about 12 hours and is air transportable, further increasing SO flexibility. The DDS is fitted aft of the submarine's sail structure (Figure 14-4). It is connected to the submarine's upper weapons shipping hatch to permit free passage between the submarine and the DDS while the submarine is underway and approaching the objective area. Modifications to a submarine allow it to serve as a DDS host ship. These include mating hatch modifications; addition of electrical penetrations, valves, and piping for ventilation; divers' air; and draining water.

Figure 14-4. Dry-deck shelter (underway)

Chapter 14

14-39. The DDS may be transported to its host ship by barge, trucked over land, or flown via C5A aircraft. Each DDS has a specially designed truck called a "transporter" for this purpose. Complete on-loading and testing to make the DDS ready for manned operations at sea takes from 1 to 3 days.

14-40. To date, DDSs have been operated from four former ballistic missile submarines and six submarines of the Sturgeon Class. Installations on additional Los Angeles and Seawolf class submarines, as well as Virginia class attack submarines, are planned in the future. Two shelters can be installed aboard Benjamin Franklin class submarines; other submarine classes are single-shelter ships.

14-41. The two major mission areas for the DDS are SDV launch and recovery and mass swimmer lockout (MSLO). For SDV missions, Mark 8 Mod 0 or Mod 1 SDVs are generally used, although the DDS can also accommodate the Mark 9 SDV. The miniature wet submersible sits on a cradle within the DDS hangar until ready for use. After flooding the hangar and equalizing it with outside pressure, the DDS operators open the hangar's large outer door and wheel out a track onto the topside surface of the submarine. The cradle and SDV roll out on the track, and the SDV departs. DDS operators then return to the submarine or may remain outside. After conducting their mission, SDV operators locate the submarine by means of an active pinging sonar. When the SDV returns, divers secure it to its cradle, winch it back into the DDS, and shut the hangar outer door.

14-42. The advantages of the DDS are that its large size allows an entire team to lock out in one cycle and that the operators are supplied with divers to support the lockout procedure. These operations may be performed with the submarine surfaced, submerged, or awash. Generally, the SOF team uses CRRCs with outboard motors. The rafts are stored rolled up within the hangar until ready for use, and then are inflated on deck. Though less clandestine than SDVs, CRRCs can insert more personnel into an area more quickly, an advantage for some missions against particular threats.

14-43. DDSs are operated and maintained by members of SDV Team One in Pearl Harbor, Hawaii, and SDV Team Two in Little Creek, Virginia. The SDV Teams report to Navy Special Warfare Groups One and Two, which report to NAVSPECWARCOM in Coronado, California. USSOCOM in Tampa, Florida, oversees all of these organizations. Memorandums of agreement (MOAs) provide the proper interface between the SOF community and the submarine crews, squadrons, and submarine-type commanders.

14-44. SOF may train aboard a DDS at one of two specially outfitted training facilities at the SDV Team locations. Each training facility has the capacity for conducting wet, pressurized training of DDS operators.

Future of the DDS Program

14-45. Four Los Angeles (SSN 688) class submarines, in addition to USS Dallas (SSN 700), are slated for conversion to be DDS host ships. To date, modifications are nearly complete on USS Los Angeles (SSN 688) and USS Buffalo (SSN 715). Conversion of USS Philadelphia (SSN 690) has begun, and work on USS La Jolla (SSN 701) commenced in Fiscal Year 2000.

14-46. Following these conversions, USS Jimmy Carter (SSN 23) will undergo similar modification. Several hulls of the Virginia class attack submarine will serve as DDS hosts as well. All Virginia class submarines will also contain an integral lockout trunk capable of deploying nine divers and their equipment. It is possible that some of the early USS Ohio (SSBN 726) class submarines may also become host platforms.

14-47. With an expected service life of at least 40 years, DDSs will likely continue to support the missions of SDV deployment and MSLO, serving SOF and submarine warfare specialties for many years to come.

ACTIONS OF BOAT TEAMS AFTER DEBARKATION

14-48. After the boat teams have debarked from the submarine, each boat moves a short distance from the submarine and "lays" awaiting the other boats (if any) to join. When all the boats are together, they adopt the prescribed formation and begin the movement to the BLS.

14-49. When the distance involved is great and the tactical situation allows, OBMs may be used to ease this movement. Pounding surf noise will partially cover OBM noise to enemy personnel located on the beach. However, the chance of the motor being heard by enemy patrol boats should be considered.

14-50. Another method of easing the boat teams' movement is to have the submerged submarine tow the manned boats by a line attached to the submarine's periscope. This procedure can be done when the water is deep enough.

WITHDRAWAL BY SUBMARINE

14-51. Tactical recovery of swimmers or boat teams can be made by a surfaced or submerged submarine. Selecting a RP will require careful coordination with the submarine's commander. The selected point must meet the combined requirements of the submarine and the exfiltrating detachment. A recovery area is specified by chart coordinates and located offshore in water deep enough for safe operation of the submerged submarine, yet close enough for the detachment to safely reach. A detailed rendezvous plan must be prepared and coordinated with the submarine's crew before the detachment debarks. It must include primary and alternate recovery times and places as well as a detailed communications plan.

14-52. Swimmers or boat teams move to the recovery area upon completion of the mission. The detachment should attempt to contact the submarine before arriving at the RP. It should arrive early enough to prepare for the recovery. Upon arrival, the detachment initiates a prearranged signal to enable the submarine to locate its precise position. Upon rendezvous, the detachment is normally towed out of the recovery area by the submerged submarine to a location where reembarkation can occur safely. This procedure is known as a Fulton recovery (Figure 14-5).

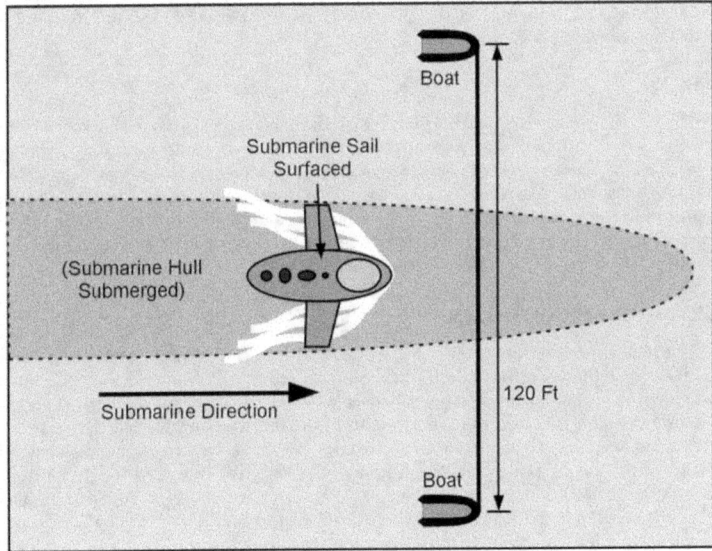

Figure 14-5. Fulton recovery

VECTORING OR HOMING

14-53. Effecting linkup with a submerged submarine requires the detachment to be very precise in its navigation. The detachment must obtain a good fix while navaids are still visible (before crossing the horizon line). Personnel must perform actual current calculations and maintain an accurate DR track. It is unlikely that physical contact between troops to be recovered and a submerged submarine can be made at

Chapter 14

night without using a vectoring or homing device. Some techniques used for vectoring the submarine to the troops to be recovered are discussed below.

Underwater Sound

14-54. The submarine's passive listening devices may be used to give accurate bearings to any source of underwater sound. This sound must be sufficiently well-defined to be distinguished above the background noise. The sounds made by striking two lengths of pipe together or by rattling a canteen containing several nuts and bolts underwater can be detected by the submarine up to 2,500 meters from the source. The swimmers can start the signals at a prescribed time, maintain them for approximately 10 seconds, and repeat periodically (about once per minute) until physical or visual contact is made with the submarine. On all but the darkest of nights, the submarine should be able to make visual contact with the swimmers or boat teams long before they are aware of the submarine's presence. This method is preferred.

Radar

14-55. Surface search radar may be used to give accurate bearings to an inflatable boat. The effects of sea return at short ranges can be countered by hoisting a portable radar reflector in the boat. However, submarine commanders are very reluctant to activate surface search radar in hostile waters because it may reveal the submarine's presence to the enemy.

Infrared Light

14-56. The detachment can use an IR strobe light hoisted on a whip antenna for the submarine to home in on. IR light may be used as a homing beacon for a boat team equipped with night vision equipment. The submarine must be equipped with an IR light source that is activated periodically for the boat team to observe and home in on.

Visible Light

14-57. The detachment may also use a visible light (flashlights) as a beacon for the submarine or the boat team to home in on. Visible light of low intensity may be used with underwater sound to effect physical contact between the submarine and the boat teams. The submarine locates the boat teams from a distance using its underwater sound receivers. As it approaches the teams, a low-intensity light source can be placed against the eyepieces of the periscope inside the submarine and be seen by the teams on the surface. The submarine commander may activate the anchor light on the bow of the submerged submarine in some situations to aid the teams on the surface to locate the submarine.

SURFACE RECOVERY OF SWIMMERS

14-58. The submarine may surface after the swimmers have made physical contact with it and recover them via the conning tower or main deck hatch. Surface recovery of swimmers may be mandatory when some men are wounded or otherwise physically unable to make a submerged recovery. A surface recovery can be done in minimum time if the submarine surfaces with decks awash and recovers the swimmers through the conning tower hatch. Surface recoveries can be expedited if the detachment gives the submarine's crew a caving ladder that can then be deployed by the crew for the detachment to climb. If time and the tactical situation allow, the submarine can surface and recover the team and all of its equipment below decks in the same manner as a debarkation. If time does not allow, the detachment may have to sink the mission equipment before reembarking on the submarine.

SUBMERGED RECOVERY OF DIVERS

14-59. Divers may be recovered aboard a submerged submarine that is under way. Underway recovery of swimmers is made easy by the swimmers stretching a line across the path of the approaching periscope. During this pickup and lockin, the submarine proceeds at a minimum speed of 1 to 3 kts. Recovery of more than four swimmers during a single pass by the submarine is impractical. After the periscope makes contact with the taut rope, the swimmers on both ends of the rope haul themselves hand-over-hand up to the periscope. The crew can provide a descending line rigged from the periscope to the escape trunk to aid the swimmers to descend and lock in. All lockin and lockout procedures must be done with scuba equipment to provide the swimmers with an air source during descent to the escape trunk. Swimmers move along the descending line to the escape trunk door. The trunk is flooded, lighted, and opened to receive swimmers before starting recovery operations. Swimmers "lock in" in groups of four per cycle as in debarkation. The last pair of swimmers to descend retrieves the buoyed surface line and air sources.

RECOVERY OF BOAT TEAMS

14-60. A submarine can also recover boat teams through the conning tower or a main deck hatch. Upon arrival in the recovery area, two boats connect a 75- to 100-foot hauling line between them, using towing bridles and quick-release (Pelican) towing hooks. Additional boats are secured to these boats with towing bridles attached to the aft towing rings of the lead boat and to the forward towing rings of the rear boat. After all the boats have been connected together, the line is stretched taut along an azimuth perpendicular to the azimuth on which the submarine will approach. As the periscope makes contact with the rope, the boats will be pulled in line behind the periscope. The boat teams may have to maneuver their boats so they will all be together and alongside each other. The submarine will continue towing the boat teams out of the recovery area to a place where the submarine can surface and the troops can reembark safely.

Wet-Deck Recovery

14-61. When the submarine has reached a safe area for surfacing, it signals by such means as a bobbing periscope or blinking light for one of the boats to release its towing hook. The submarine then surfaces to a point that its decks are awash. The boat teams maneuver their boats onto the water-covered deck (fore or aft of the conning tower), unload, and begin deflating the boats and preparing to stow them. The broached submarine completes "blowing the tanks" and surfaces completely. This method lets the boat teams safely stow their boats away and reembark via the conning tower or main deck hatch.

Dry-Deck Recovery

14-62. A dry-deck recovery of boat teams is executed the way as a wet-deck recovery except that the submarine surfaces completely. The boat teams come alongside, unload the boats, pull the boats up onto the deck, and prepare to stow the boats. Wet-deck recoveries are faster and easier than dry-deck ones but require more training and experience to execute.

THE FUTURE

14-63. The changing socio-politico-economic situation in the rest of the world has given military and political decision makers new perspectives on the requirement to project force into the littoral (near-shore) environment.

14-64. The newest class of submarine (NSSN), the Virginia class, is designed to meet this need. This new attack submarine is the first submarine designed to satisfy the requirements of regional and near-land missions in the post-Cold War era. This submarine will support the full spectrum of special operations—search and rescue, intelligence collection, sabotage, diversionary attacks, and special missions. It will have the additional space to embark SO personnel and their equipment. The NSSN will have a nine-man lockout and lockin chamber and be compatible with the DDS so that it can disembark those personnel clandestinely.

14-65. The submarine's proven ability to go anywhere it is required and return undetected makes it an ideal support vessel for the SFOD looking to arrive "unannounced."

This page intentionally left blank.

Chapter 15

Riverine Operations

In areas of the world having limited land transportation infrastructure and abundant water surfaces, inland waterways provide natural transportation routes. The peripheries of waterways are also logical places for the development and growth of population centers. In some developing countries, inland waterways are the only major arteries for economic circulation.

Riverine transport of local products may need military operations to keep waterways open and, in some instances, to transport area produce to maintain the local economy. Water routes are also operationally important to an insurgent or enemy force, particularly in situations where an external aggressor supports and directs an internal insurgency. Such a situation dictates that forces engaged in counterinsurgency operations or foreign internal development programs adopt a doctrine and concept of operations that includes interdiction and control of waterways.

SFODs operating in underdeveloped regions of the world may find themselves using riverine techniques for their own transport and protection or for training indigenous forces as part of a FID or counterinsurgency aid program. This chapter covers aspects of the doctrine governing riverine operations and those techniques that a detachment might find useful in planning and executing smaller-scale operations.

ENVIRONMENT

15-1. SFODs conduct riverine operations to cope with and exploit the unique characteristics of a riverine area, to locate and destroy hostile forces, or to achieve or maintain control of a riverine area. Joint riverine operations combine land, riverine, naval, and AIROPS, as appropriate, and are suited to the nature of the specific riverine area in which operations are to be conducted.

15-2. Riverine operations are not amphibious operations, although many of the techniques applicable to amphibious operations are useful in, and applicable to, the riverine environment. Riverine operations are distinct in that they use specialized craft, equipment, and techniques.

15-3. The SFODs conduct riverine operations in an area to achieve or maintain control of a waterway system and its adjoining land areas, or to deny their use to the enemy. These operations can be conducted on, across, along, or emanating from a waterway system. They combine the characteristics of ground, naval, and AIROPS that require the use of tactics, techniques, organization, and equipment particularly adapted to the nature of the riverine area.

15-4. SFODs should thoroughly understand the riverine environment when planning a riverine operation. In a riverine area, water craft are the principal means of transport. In such areas, indigenous personnel often settle along the waterways because they are the only usable means of travel between villages. Civilian traffic and settlements conceal the enemy's movements and mining and ambush operations. Control of waterways is necessary in riverine areas.

15-5. Water lines of communications (LOCs) dominate a riverine environment. It consists of several major rivers and tributaries or an extensive network of minor waterways, canals, and irrigation ditches. Military movements use air and water transportation extensively due to the lack of a suitable road net. Suitable land for bases, airfields, and artillery firing positions may not always be available. The topography of the land,

Chapter 15

the location of the civilian population, the restrictions on using agricultural land, or a combination of these factors make land unavailable when needed.

APPROACH TO PLANNING

15-6. Planning for riverine operations is a continuous process from receipt of the initiating directive to termination of the operation. It necessitates concurrent, parallel, and detailed planning by all participating forces. Plans must be detailed enough to give all participants complete information, yet at the same time be simple and flexible enough so they can be modified as the tactical situation changes.

15-7. Although, in actual combat, safety may not be paramount, it still is a vital consideration. Safety is essentially the preservation of resources. Accidents occurring in the course of combat operations reduce the effectiveness of the overall effort. Therefore, as much as possible, safety shall be a prime consideration in the planning and execution of riverine operations. During exercise or rehearsals, safety shall be paramount, and appropriate consideration shall be given to the accomplishment of the mission without incurring unnecessary losses of men or equipment.

15-8. Weather, terrain, and hydrography take on more importance in riverine operations than in conventional land and sea operations and, under some circumstances, may be controlling factors in any concept of operations. Consequently, thorough knowledge and consideration of the environment assume great importance in planning riverine operations. Because of the problems associated with position location and orientation in various environments, consideration should be given to the use of a gridded aerial mosaic to supplement topographic maps. Obstruction removal and mine countermeasures are critical points when planning for riverine operations.

15-9. Plans for countering an ambush will depend upon whether the unit is to force passage through the ambush, or to land and destroy the ambushing force. If the mission of the unit specifies the destruction of ambushing forces encountered en route, plans for landing elements will be required.

15-10. Rules of engagement should include comprehensive regulations on search, seizure, or destruction of indigenous property, and the conditions under which fires may be directed against hostile forces or inhabited areas. Procedures for the handling of detainees or suspects and PWs should be delineated.

15-11. If forces assigned have not had previous experience or training in riverine operations, plans should provide for training of commanders and staffs in the peculiarities of riverine operations. Such training should include joint training, if feasible, and should include training in control and coordination of assault craft gunfire.

ORGANIZATION AND COMMAND

15-12. Riverine operations can be joint operations undertaken primarily by Army and Navy forces. Participating forces must coordinate and integrate efforts to achieve a common objective. DOD and Joint Chiefs of Staff (JCS) directives prescribe joint forces command arrangements to ensure coordination and integration. Joint command organizations centrally direct the detailed action of a large number of commands or individuals and common doctrine among the involved forces. Flexibility in the organization ensures control and coordination of these forces in varying operational environments.

15-13. METT-TC is the basis for the task organization. Considering the total forces available, riverine operations require a balance between types of forces. A special consideration in task organization for riverine operations is the amount of troop lift and fire support available from Navy, Army aviation, and Air Force units. The major factors determining naval support requirements are—

- The extent to which navigable waters permit moving naval support to, within, and around the AO.
- The size of Army forces needed in the objective area.
- The availability of other means of transportation.
- The desirability of using other means to deliver forces.

SECURITY RESPONSIBILITIES

15-14. The relationship between Army and naval elements stationed on a land or afloat base is one of coordination and mutual support. The Army and naval elements assign their appropriate share of forces for local base defense as the base commander directs. The main mission of the naval force in base defense is to provide gunfire support and protection against any threat from the water.

15-15. During tactical operations, the Army commander provides, plans, and coordinates security elements (ground or air) along the route of the movement. The naval element commander tactically controls the movement and maneuver of watercraft under the operational control of the Army commander being supported.

15-16. The senior naval commander embarked is in tactical control while the afloat base is en route from one anchorage to another. Higher headquarters normally directs or approves the relocation of the afloat base. The naval commander of the riverine force oversees moving naval vessels and watercraft between riverine bases and support facilities outside the riverine area. The Army commander in the riverine area provides the security of movement for vessels within the area.

CONCEPT OF RIVERINE OPERATIONS

15-17. Units conducting riverine operations use water transport extensively to move troops and equipment throughout the area. Riverine operations normally start from areas where ground forces and watercraft marshal and load and where forces can effect coordination. The start point may be at a land base next to a navigable waterway, at an afloat base on a navigable waterway, or in an existing AO. Once troops are aboard, the watercraft proceed to designated landing areas within an assigned AO for offensive operations.

15-18. Unit plans include control measures, such as phase-line checkpoints, for the entire operation. The commander controls the unit's movement either from a command and control (C2) boat located within the movement formation or from an airborne command post. Maneuver unit commanders, embarked in C2 craft, leave these craft to take charge of their units.

15-19. The withdrawal of troops from the AO is a tactical movement back to the watercraft loading areas. Units are loaded in reverse sequence to that used in the riverine assault landing. The maneuver unit using a perimeter security provides the necessary loading area security throughout the withdrawal operation. A tactical water movement back to base areas or to another AO is performed after loading.

PLANNING FOR RIVERINE OPERATIONS

15-20. Riverine operations require detailed planning at all levels and close coordination with a supporting naval river assault squadron. Units conducting riverine operations must be ready to begin ASAP after receiving orders. Boat operators require training in operation, maintenance, and navigation. As a minimum, training consists of briefings in the marshaling or staging area to acquaint Army personnel with embarkation and loading procedures, required action during the water movement, security at the landing area, and landing procedures.

15-21. Plans for riverine operations must be detailed enough to give all participating units complete information. Yet, they must be simple and flexible enough to be modified as the tactical situation changes.

15-22. Plans for a riverine operation are usually developed in the following sequence. The commander develops the—
- Scheme of maneuver based on METT-TC.
- Assault plan based on the scheme of maneuver.
- Water movement plan based on the assault plan and the scheme of maneuver. (The water movement plan includes composition of the riverine force, organization of movement serials, formation to be used, movement routes, C2 measures, mine countermeasures, plans for fire support, and immediate reaction to ambush.)
- Loading plan based on the previous plans and the scheme of maneuver.

Chapter 15

- Marshaling plan, when required, based on the previous plans and the scheme of maneuver.
- Deception plan, when required, based on the mission.

CONDUCT OF RIVERINE OPERATIONS

15-23. Units are trained and prepared to conduct riverine operations on short notice. Applying lessons learned in previous riverine operations keeps SOPs current. Training and adequate unit SOPs allow marshaling activities to focus on the pending tactical operation.

15-24. Units prepare for the tactical operation, move to their loading areas, and load onto assigned watercraft according to the water movement table and information in the watercraft loading table. Bulk supplies and ammunition are transported to the loading site where they are loaded and lashed in designated watercraft. Several units may use the same loading site. Therefore, loading must be completed and watercraft moved to their assigned rendezvous area according to the time schedule in the water movement table.

15-25. All water movements outside of the base areas are tactical moves. They are similar to the approach march of a movement to contact in ground operations; speed of movement and security of the formation are essential. The intent of the operation is to move directly to the objective. However, the unit is prepared for combat at any point along the movement route. The terrain and the enemy situation normally require advance, flank, and rear guards to protect the main body during the move.

WITHDRAWAL FROM RIVERINE OPERATIONS

15-26. While preparing for riverine operations, planners determine the availability of waterways in the AO, the tide and current for the scheduled period of the operation, and suitable loading sites. This information, updated during the operation, is the basis for planning the riverine withdrawal.

15-27. Active use of watercraft during an offensive maneuver simplifies deception in the initial stages of a riverine withdrawal. The quantity of available hydrographic information increases as a result of this employment.

15-28. When possible, riverine withdrawal is timed so watercraft can approach loading areas with the current on the rising tide, load during slack high water, and depart with the current on the falling tide.

15-29. Due to the security problems that accompany large riverine movements and using predictable routes, loading during the last hours of daylight and moving during darkness should be considered. Moving reconnaissance forward along possible withdrawal routes several hours ahead of the movement group is a useful deception measure.

15-30. Loading, normally the most critical phase of the withdrawal, requires detailed planning when selecting troop assembly areas, loading areas, loading control measures, and watercraft rendezvous areas.

15-31. There are many varieties of riverine operations. In general, riverine operations are conducted to—
- Establish and maintain control of riverine LOCs.
- Deny, by interdiction, barrier, or surveillance operations, use of riverine LOCs by hostile forces.
- Locate and destroy hostile forces, bases, and supplies contained within a riverine area.

15-32. The entire riverine maneuver may include the following factors:
- Intelligence collection.
- Planning.
- Embarkation of troops and equipment.
- Patrol, barrier, interdiction, and surveillance operations.
- Riverine assault operations.
- Naval riverine close fire support.
- Close air support.
- Naval gunfire or firebase support.
- Repositioning of forces.

- Resupply of the riverine force until termination of the mission.
- PSYOP and civic action programs.
- Reembarkation and withdrawal.

INTELLIGENCE REQUIREMENTS

15-33. Certain information is necessary to enable the force commander to direct operations, detect and prevent enemy movement by waterways, and to reduce the threat of mines and ambushes to friendly forces. Intelligence requirements should include, but are not limited to, the following:

- Hydrographic information, including waterway depth, width, bottom composition, currents, tidal ranges and currents, and bank characteristics.
- Navigational hazards, including natural and man-made waterway obstacles such as vegetation, debris, fish traps, and barricades.
- Location of bridges and under-bridge clearances.
- Location, strength, and activities of enemy units in the objective area at the beginning of and during operations.
- Loading points and departure points for hostile watercraft.
- Routes followed by the enemy on inland waterways, including staging areas.
- Evasion tactics used by the enemy, including camouflage and deception.
- Delivery points for material being carried on inland waterways.
- The enemy supply system, with emphasis on riverine transport routes.
- Location of arms and supply caches.
- Identification of warning systems used by the enemy to protect against patrol craft.
- Identification of points where the enemy usually crosses rivers and canals.
- Identification of enemy watercraft. Emphasis should be placed on determining whether they are owned by the enemy or are impressed from the local populace.
- Enemy swimmer capabilities, equipment, and methods of operation.
- Enemy mining and ambush operations (with particular emphasis on early warning of ambush sites) and tactics.
- Enemy tactics concerning antipersonnel devices such as traps, camouflage pits, or stakes and spikes driven into the ground.
- Location, capabilities, and tactics employed by enemy antiaircraft elements.
- Identification of enemy intelligence and counterintelligence elements in the objective area.
- Susceptibility of the populace to enemy pressures to provide information about friendly forces and operations.
- Identification of guerrilla, paramilitary, or similar groups in the objective area.
- Identification of individuals, groups, or organizations in the objective area that may be exploited by the enemy for espionage, sabotage, or subversive activities, or by friendly forces for intelligence.
- Weather, to include temperature, precipitation, humidity, visibility, winds, fog, cloud cover, ice incidence, and effect of weather at various seasons on river characteristics.
- Astronomical conditions, to include sunrise, sunset, moonset, and moon phase.
- Identification of civilian uses of waterways, to include types of watercraft, traffic pattern and density, civil registration, and licensing system.
- Determination of medical characteristics of the AO, to include plant and animal ecology, terrain, climatological and disease incidence data, and sanitary conditions ashore.

There are also elements of information that must be analyzed and determined whether they should be designated as specific intelligence requirements. Figure 15-1, page 15-6, states the questions that help determine these requirements.

Chapter 15

> ☐ Which waterways in the area are suitable for military operations, including depths, widths, bottom composition, and shore characteristics?
> ☐ What are the capabilities of weapons available to the enemy to attack craft at anchor and underway, in both major and minor waterways?
> ☐ What are the enemy's water mine and attack swimmer capabilities, to include available ordnance, related equipment, and tactics?
> ☐ What navigational hazards (such as natural and man-made waterway obstacles, vegetation, debris, bridge wreckage, fish traps, and barricades) exist in the area, and what are their locations?
> ☐ What are the tidal ranges and daily tidal predictions at representative locations in the AO throughout the year?
> ☐ What are the parameters of tidal and riverine currents in the area at different tidal stages and at different times of the year?
> ☐ What are the precise locations of all bridges in the area spanning navigable waterways, what types of bridges are they, and what are their under-bridge clearances at high and low tide?
> ☐ What are the locations of enemy crossing points for major and minor waterways, and what are his tactics when he crosses?
> ☐ What are the enemy's tactics, including the use of mines and booby traps in opposing friendly troops and land, water, and air vehicles?

Figure 15-1. Determining intelligence requirements

RIVERINE MOVEMENT

15-34. Riverine movement is the use of craft on riverine waterways to provide tactical mobility. Riverine movement may simply be a means of transporting land forces in the AO or may be the means by which patrolling and area control functions are accomplished on waterway systems.

15-35. Waterways navigable by patrol craft are limited by both natural elements and man-made obstacles. Natural elements include—

- *Currents.* The river current along a route will have a significant effect on fuel consumption and speed capability.
- *Depth of Waterways.* The depth of the various waterways in an AO will affect route selection as well as anticipated SOA. The lower the depth, the greater care needed to avoid grounding.
- *Tides.* In many riverine areas, tides will be an important factor in riverine movement planning and execution. Delta tides generally are semidiurnal, with two highs and two lows daily. The daily tidal variations in a delta region can be significant (as much as 12 to 14 feet or greater in some areas). Many waterways will be navigable by larger craft (those with drafts over 4.5 feet) only at high tide. Conversely, high-tide water levels can restrict under-bridge clearances. During dry seasons, some waterways will only be navigable during high-tide periods. Tidal changes will also have an effect on currents.
- *Natural Obstacles.* Natural obstacles in the riverine environment include floating debris, rapids or whirlpools, low-hanging branches and limbs, underwater stumps and roots, and sandbars.

FORMATIONS

15-36. The detachment must also determine the type of formation to use in riverine operations. Factors that will affect formation selection are as follows:

- The physical characteristics of the waterway itself (width, depth, current, and shore terrain).
- Perceived enemy capabilities and current area threat.
- Size of the force.

- The SOA, mutual support, and maneuverability requirements.
- Patrol firepower requirements.
- Patrol detection avoidance requirements.

15-37. After considering the above factors, the SFOD can decide the type of formation that will best suit the operation. The basic formations are—
- Column.
- Staggered column.
- Online.
- Wedge.
- Diamond.

Types of Movement

15-38. Riverine forces may use several different types of movement to maximize security and firepower. A typical movement will entail a combination of techniques and formations over its course. They are as follows:
- *High Speed*—used in AOs with low to medium threat level, good visibility, and in an area that the force is familiar with the waterway.
- *Low Speed*—used in all threat levels, poor visibility, unfamiliar waters, and waters that provide restricted maneuverability.
- *Bounding Overwatch*—used in AOs with a high threat level where maneuverability, channel width, visibility, and fields of fire are limited.
- *Traveling Overwatch*—same as bounding overwatch.

15-39. Local area knowledge, customs, and practices are imperative to the successful completion of the assigned mission. The immediate identification of the abnormal is essential to maintaining an effective security posture and to gaining the tactical advantage over a potential threat. To identify the abnormal, the force must understand what is normal for the AO.

RIVER PILOTING

15-40. This section provides general river piloting information that is appropriate for safe and effective river operations and areas of the riverine environment. The commander should consider the following factors when planning troop movement using the rivers and estuaries.

15-41. **Currents.** The strength of river currents varies widely from river to river, and from season to season for a particular river. Speeds on some sections of river can range 5 or 6 kts under average conditions. At extreme high-water stages, current strengths may be much greater—9 kts or more in narrow and constricted areas.

15-42. **Speed.** River currents sometimes attain such speeds that navigation upstream is not feasible, although capably handled boats can be taken down safely. Some vessels have power enough to ascend certain rapids, but as a general rule, rapids should be avoided in favor of canals and locks that bypass them, unless the coxswain has local knowledge or engages the service of a local guide or pilot.

15-43. **Characteristics.** River current characteristics are of the utmost importance to larger deep-draft vessels. The surface current acting on a small boat may actually be contrary to that which grips a deep-draft vessel's keel near the river bottom. Even surface currents vary from bank to midstream. Friction of the bank and bottom slows the water.

15-44. **Techniques.** If the detachment uses small craft, it must remember the following guidelines:
- Use the strength of midstream current for downstream run.
- When headed upstream, run as close to the bank as safely possible, even turning into small coves, to take advantage of the countercurrent.

Chapter 15

- Cut running time and conserve fuel by running courses that make the river's current work for the detachment, or minimize its adverse effect.
- At each bend, follow the curved line roughly running the trend of the river as a whole. Keep about a quarter of the river's width off the outside bank.
- On some river charts, follow the contour markings showing the topography ashore because it can give clues to the river itself. Contour lines crowded close together indicate a cliff rising steeply from the bank. This feature will usually mean there is a good chance of deep water close under the bluff. The proper study and use of available charts and maps can also be used to—
   - Determine landmarks to steer by.
   - Provide course and fire support checkpoints.
   - Help identify potential danger areas.

RUNNING AN INLET

15-45. One of the worst places to be in violent weather is an inlet or narrow harbor entrance, where shoal water builds up treacherous surf that often cannot be seen from seaward. When offshore swells run into shallower water along the beach, they build up steep waves because of resistance from the bottom. Natural inlets on sandy beaches, unprotected by breakwaters, usually build up a bar across the mouth. When the swells reach the bar, their form changes rapidly—they become short, steep-sided waves that tend to break where the water is shallowest. The SFOD should consider this when approaching from offshore. A few miles off, the sea may be relatively smooth but the inlet from seaward may not look as bad as it actually is. Breakers may run clear across the mouth, even in a buoyed channel. Shoals shift so fast that buoys do not always indicate the best water. If the SFOD must run the inlet, it should focus on the guidance below:

- Do not run directly in.
- Wait outside the bar and watch the action of the waves as they pile up at the most critical spot in the channel, which will be the shallowest. Usually waves come in sets (groups) of three, sometimes more. The last sea will be bigger than the rest and by watching closely it is possible to pick it out in the successive sets.
- Make sure the boat is ready.
- When ready to enter, stand off until a big wave has broken or spent itself on the bar, and then run through behind it.
- Watch the water both ahead and behind the boat; control the boat's speed and match it to that of the waves.
- Watch for an ebbing current; it builds up a worse sea on the bars than the flood does because the rush of water out works against and under the incoming swells. Try to wait until the flood has had a chance to begin.
- Watch for deeper water in the inlets; the best running time is just before the tidal current turns to ebb.

PILOTING GUIDELINES

15-46. The SFOD must follow specific guidelines for river piloting. These principles are as follows:

- A compass is of little use on most rivers. Generally, the most useful items are a good chart or map and the proper use of binoculars to sight from one navigational aid to the next. Do not dispense with compasses on all inland rivers; some are wide enough that the unit can continue in a fog using a compass, speed curve, watch or clock, and due caution. In addition, some rivers feed into large lakes, bays, or the sea.
- Not even the best charts and publications can "tell all" about a particular body of water. Rivers are particularly prone to seasonal or irregular changes; if the unit is new to a specific stretch of river, it should take every chance to ask experienced local people about hazards or recent changes.
- Perhaps the greatest asset of a boat coxswain in the riverine environment is the power of observation. Much of river piloting depends on the acquired skill of reading the river, interpreting what is seen or observed. Figure 15-2, page 15-9, lists some indicators that can be useful in helping to read the river.

Riverine Operations

Man-Made Indicators	Natural Indicators
Buoy System	Birds Standing in the Water
Fishing Nets	Ripples/Choppy Water
Pilings	Waves or Wakes
Stakes	"Vees"
Boat Hulks	Undulations in Flat Water
Road or Ramps to Waterway	Animal Tracks to Water's Edge
Ferry Landings	Trash Piles
Mooring Stations	Steep Banks
Piers (New and Old)	Swampy or Shallow Banks
	Swirling Water
	Fast-Flowing Water

Figure 15-2. Indicators for river piloting

- In general, lightly rippled water, where no wind is blowing, usually indicates shallow water. A long undulating wave indicates deep water and fast current. A smooth surface usually indicates deep water that is slower moving. However, no flat statement can be made about what certain surface conditions reveal about relative water depths.
- In unfamiliar waters, the wake of the vessel can give a clue to the safety of the course. As it rolls off into shallow water, its smooth undulations give way to sharper formations, even cresting on the flats in miniature breakers. When the waves reach a shoal or a flooded area where submerged stumps are close to the surface, the difference will show. If the wake closes up to the stern, and appears short and peaked, the coxswain should make a definite course change away from the side of the channel where this telltale sign appears, and make it as quickly and as safely as practical.
- When entering sloughs between islands or between an island and the bank, the coxswain should beware of submerged wing dams at the upstream end. To be safe, he should enter and leave from the downstream end.
- Islands occurring in mid-river often leave a secondary channel for small boats on the side opposite of the deep channel.
- When a river cuts a channel behind a section of bank, a "towhead" is formed. Sometimes these are filled in or dammed across at the upper end by river deposits, forming a natural protected harbor that can be entered from the lower end.
- Channels normally run deeper near steep banks and shallower near swamps or banks with shallow gradients.
- When moving from one channel into another that is perpendicular to the first, the coxswain should navigate at right angles to the head-on current. He should pass on the downstream side of the perpendicular current, proceed upstream a short distance, and then turn back into the flow of the new oncoming current. A powerboat should never move into fast water at full throttle, but enter the current at about half throttle until the coxswain is sure of what lies ahead.
- When anchoring or beaching on or near an island or sandbar, the crew should use the downstream rather than the upstream end. If the anchor drags or the boat somehow goes aground on the upstream end, the current will push the vessel harder ashore. Water at the downstream end is likely to be quieter, and the eddies that normally exist there may help to free the vessel.
- The characteristics of the bottom vary widely on inland waterways. Particularly in their lower reaches, river bottoms are often mud. When anchoring in mud, the crew should use a broad fluke anchor of a design that will dig down until it reaches a good holding such as a Danforth or Plow anchor. An anchor with spidery arms and flukes like the kedge anchor would pull through

Chapter 15

the mud and provide no holding power at all. The kedge anchor is best used on rocky bottoms and is acceptable in areas where grass and weeds cover the bottom. Anchoring over rocky bottoms or in areas full of roots and snags, it is best to rig a "trip line," a light line from the anchor's crown to a small buoy at the surface. If the anchor snags and will not come free in the normal fashion, a crew member should pick up the buoy and raise the anchor with the trip line, crown first.

- When making fast to a bank, the coxswain should first check that the depth is adequate and the area is free from underwater obstacles. He should approach the shore slowly. He should also avoid vertical banks that may be in a stage of active caving; exposed tree roots in the bank may be evidence of recent erosion.
- On inland rivers it is generally not necessary to allow for tidal changes common on the coast when docking, tying up, or making fast to fixed positions, except of course, on tidal rivers. On nontidal waterways, however, there is always the chance of change in level with hard thunderstorms or other heavy rainfalls. These may occur many miles away or in a mountainous region or range that receives a downpour that is not readily seen from the river vantage point. Tying up to a barge or float that will itself rise or fall with the change in level is advantageous.
- Navigating a river in early spring, one may encounter high water, flood conditions, racing currents, and floating debris. During late summer and early fall, with low water levels and slower currents, there are normally few, if any, obstacles, except perhaps more numerous shoals. At extreme flood conditions, river navigation is not recommended without the services of an experienced pilot. The navigator should avoid ripples, boils, or other indications of disturbed water. Disturbed water can cause loss of control and can force the boat onto obstructions that could tear the bottom out of the boat or injure personnel.
- "Sand boils" may be caused by sand piling up on the riverbed. During flood stages, these whirlpool-like disturbances can be so violent that they can throw a boat out of control. In more favorable months, they may be no worse than surface eddies—felt, but no danger to a boat.
- A V-shape eddy in the surface of the water generally indicates an obstruction lying parallel to the current. The surface eddy of a submerged obstacle, at or very near the surface of the water, points upstream as its wake divides downstream around it. Do not confuse this eddy with the condition of two currents converging at the downstream end of a middle bar. That condition may also show as a V-shaped eddy, but pointing downstream. Current speed and the size of the obstruction together determine the size of the V. It is only an indication of the size of that portion of the obstruction lying at or very near the surface; it does not indicate the total size of the obstruction. Give it wide berth.
- A roiled surface that is localized at one point usually indicates an obstruction, such as a log or a tree, lying perpendicular to the current. Do not confuse the roiled wave with a long undulating wave, which indicates deep water and fast currents discussed earlier.
- Sweepers are trees rooted in the river bottom and low-hanging branches of trees that have been pulled into the river or stream by collapsing banks. Sweepers are extremely dangerous. Collision with sweepers can overturn a boat or tear out the bottom. Personnel can be impaled on tree branches or trapped underwater and held by the roots and branches.
- In all waters heavily laden with silt, it is wise to carry protection against it such as raw-water strainers, freshwater cooling systems, cutless-type underwater bearings, and pump impellers that will handle mud and sand better than bronze gears.
- Floating and partially submerged debris such as tree trunks and branches are a hazard for small boats; keep a sharp lookout in waters where they have been reported. Floating debris is usually at its worst in the spring months, when floodwater levels have swept away downed trees and other materials from above the normal water line.

EMERGENCIES

15-47. **Overboard.** All craft will be prepared for man overboard. Frequent drills to aid rapid identification of a man overboard are indicated. Rapid small craft action is mandatory in riverine currents in order to be effective.

15-48. **Fire.** Depending on the severity of the fire, it may or may not be necessary to debark troops. If it is necessary, rehearsed emergency debarking procedures will be followed, and designated craft will assist with debarkation and firefighting. The possible necessity for grounding the craft that is on fire should be considered.

15-49. **Breakdown.** All riverine craft and ships should be prepared to tow other craft and ships in case of a breakdown. If a breakdown that would require slowing the entire formation does occur, a decision will be made whether to declare the disabled ship a straggler or slow the formation. The detachment of escorts for stragglers may be necessary. Preselected temporary anchorages may be used if a slower speed detains the whole formation and prevents it from reaching its destination on schedule.

This page intentionally left blank.

Chapter 16
Aircraft in Support of Maritime Operations

Air-water missions encompass all operations where an SFOD uses an aircraft as an intermediate delivery system to travel the majority of the distance to its objective or BLS. The detachment completes its infiltration by means of a boat, surface swim, or subsurface operations. These operations include water landing, water jumps, helocasting, external raft delivery system (ERDS), rolled or tethered duck, hard duck, and recovery operations. Although aircraft provide the most practical and rapid means of transporting infiltration swimmers to the vicinity of the BLS, air transportation can be more complicated than other means, such as surface craft or submarines. There is a wide variety of assault-type aircraft, as well as tactical and utility types that can be used to infiltrate a team with or without a CRRC or Zodiac.

Once the infiltration method has been identified, detachments should make reference to the governing manuals or regulations. USSOCOM Manual (M) 350-6, *Special Operations Forces Infiltration/Exfiltration Techniques*, is the governing regulation for SOF air operations. Detachments with specific questions about the requirements (safety/operational) for conducting any type of air operation must refer to that manual. Additional supporting or amplifying information is contained in FM 3-05.210, *Special Forces Air Operations*, Chapter 14, Air-Water Operations. USASOC Reg 350-2, *Training Airborne Operations*, is currently the guiding regulation for water jumps. Personnel should refer to it for detailed train-up and rigging instructions and requirements.

DELIVERY METHODS

16-1. Aircraft provide the most practical and rapid means of transporting infiltrating detachments from their home station to a debarkation point in the vicinity of the BLS. The type of aircraft and the delivery method selected determine the planning and preparations required during Phase I. Delivery methods include—

- Conventional static-line parachute techniques.
- Military free-fall parachute techniques.
- Water landing by amphibious aircraft.
- Helocast or free-drop from helicopter.

16-2. There are advantages and disadvantages of using air assets to deploy a team conducting a waterborne infiltration. Figure 16-1, page 16-2, lists these factors.

16-3. In all infiltrations by aircraft, the pilot and flight crew must have the appropriate qualifications for the intended delivery method and be current in those qualifications. The aircrew must coordinate all aspects of the flight with the operational element. The pilot briefs on in-flight communications, navaids, abort plans, and other related general flight procedures. The detachment informs the pilot and flight crew of the number of personnel to be infiltrated, the type and quantity of accompanying supplies, the DZ or landing zone (LZ) markings, and other mission-related information. Rehearsals for each phase of the infiltration must be conducted and should include the actual pilot and flight crew for the mission.

Chapter 16

Advantages	Disadvantages
Air assets provide a rapid means of delivery throughout the world.	Airdropping of personnel and equipment makes linkup at sea difficult at night.
A wide variety of aircraft, including Air Force and Army helicopter assets, can be used for waterborne infiltrations.	The enemy's radar capability may require the aircraft to release the team OTH to avoid detection.
Precise navigation can be maintained to the debarkation point.	If something goes wrong once the team has left the aircraft, there are no readily available means to extract the team. This fact only applies to USAF aircraft, not to Army helicopters, which can remain on station to extract the team in case of mission abort.
Training is more easily conducted because of collocation of Air Force bases and Army posts.	

Figure 16-1. Team deployment using air assets

16-4. During flight to the debarkation point, the pilot keeps the troop commander informed of the aircraft's location and any changes in the infiltration plan. All personnel in the infiltrating detachment must know their relative position along the flight route in case of an emergency abort or enemy action. As the aircraft nears the debarkation point, the pilot provides advance warning for final personnel and equipment preparations. Infiltrating detachments are delivered at the debarkation point by water landing, helocasting, or parachute operations.

FIXED-WING AIRCRAFT

16-5. Fixed-wing aircraft are used for long-range (strategic) infiltration of MAROPS detachments. They give the commander the ability to rapidly deploy a team from home station to a crisis point virtually anywhere in the world. They do not require forward staging, nor do they require intermediate transportation assets or transshipment points. No other infiltration method can provide the speed, simplicity, or range of USAF assets.

16-6. In most situations, the detachment will prefer an OTH parachute-delivery method of infiltration. Airdrops will be conducted during periods of limited or reduced visibility onto unmarked water DZs using the CARP. Parachutists exit the aircraft on the pilot's command and attempt to group in the air as closely as possible. Once in the water, they sink the air items, sterilize the DZ, and begin movement to the BLS or launch point.

16-7. When conducting training to gain proficiency, detachment personnel should follow all applicable field and technical manuals that pertain to the conduct of parachute operations involving water DZs. They should also adhere to local safety range SOPs pertaining to safety boats and DZ operations while conducting waterborne operations.

INFLATABLE BOATS

16-8. A variety of inflatable boats can be airdropped from USAF high-performance aircraft. This operation is commonly referred to as a "hard duck." The usual aircraft for hard ducks is the C-130. Other aircraft, for example, the C-17, can also be used; however, practical considerations and USAF regulations governing the conduct of static-line ramp jumps mixing bundles and troops effectively restrict tactical hard duck operations to C-130s.

16-9. The Zodiac F-470 is the boat most likely to be used by SFODs for paradrop and long-range infiltrations. The Zodiac F-470 is rigged on a SOCEP designed to mate with the boat's contour. Details of platform construction, dimensions, and rigging are contained in FM 4-20.142, *Airdrop of Supplies and Equipment: Rigging Loads for Special Operations*. In general, a SOCEP is built with a 3/4-inch marine or exterior-grade plywood floor, 4- by 4-inch stringers, and 2- by 6-inch wooden support members. It is reinforced with 1/2-inch steel plates and stiffeners. When assembling the load, the area between the boat and the bottom of the platform is filled with sandbags to raise the total airdrop weight of the package to that specified in FM 4-20.142 (the accompanying load must weigh at least 650 lb, but not more than 1,170 lb). Any remaining space is filled with "honeycomb" dunnage to match the bottom profile of the boat. This material cushions and supports the boat and its load to prevent damage caused by load shifts following opening shock or impact on the water.

16-10. Fuel and equipment are secured to the inside of the boat. The motor can either be secured inside the boat or on the transom. If the motor is secured inside the boat, it must be well padded and lashed securely to prevent it from shifting with opening shock or impact on the water. Mounting the motor on the transom requires adding a special dovetail to the back of the platform and requesting a special waiver through the USAF. Both the motor and the transom must be supported to prevent damage to the boat from opening shock. If two boats are being stacked, one on top of the other (double duck), the motors should be stowed internally. A single duck uses one G-12D 64-foot cargo parachute; a double duck uses two. The parachutes are attached to the platform by a sling assembly and equipped with a quick-release device that detaches the parachute from the sling upon contact with the water. When the fully prepared boat is delivered to the departure airfield, ground support equipment is required to load the platform onto the aircraft. This equipment can be a flatbed truck with rollers, a K-loader, or a special long-tine forklift (or a standard forklift with tine extensions).

16-11. Rigging the drop platform IAW FM 4-20.142 is performed by the rigger section. It is obviously in the best interests of the operational team to have knowledgeable personnel supervise the loading and securing of individual and team equipment in the CRRC. This equipment should be secure enough, in case the boat should tumble or invert during the free-fall period before parachute deployment, so that no equipment will fall out.

16-12. When rigging for a deliberate water jump, jumpers will suit up as for a combat swim operation, to include exposure suits where required. The jumper has the option of wearing his swim fins or securing them to his lower outside leg above the ankle. If the fins are worn, consideration should be given to tying up the tip of the fin to the jumper's lower leg (elf shoes) with 80-pound test line to reduce the risk of tripping and falling. Fins worn tied to the lower leg must be secured with a retainer cord and duct tape to prevent them from being lost on opening shock. Jumpers wearing fins must exercise caution when approaching and exiting jump doors or the ramp. Aircraft configuration and section SOPs will determine which option a detachment uses. Masks are worn around the jumper's neck with the faceplate rotated behind the head. All equipment will be secured to the jumper with breakaway tape (80-pound test). Jumpers will have their swim equipment inspected by a dive supervisor separately from the jumpmaster personnel inspection.

16-13. Because rucksacks are secured inside the boat during the airdrop, personnel should waterproof and place LCE, two to three rations, extra water, a pair of boots, a set of ACUs and socks, and other equipment needed for minimal mission accomplishment (survival) in a kit bag. The kit bag is then rigged with a single point release harness (no lowering line). This kit bag replaces the rucksack. The rationale for using the kit bag is that the cue for jumper exit is pilot chute inflation only. The chance still exists that the G-12D may not open or may malfunction. If the jumpers wait for the G-12D (because of its slow-opening characteristics) to open fully, they will be so far from the boat that assembly on it will be extremely difficult if not impossible. If this happens and the boat sinks or is rendered inoperable, the team still has the option to attempt to swim or drift with tides and currents to the nearest landmass and attempt mission accomplishment or escape and evade.

16-14. Exact procedures for jumpers exiting an aircraft during deliberate waterborne operations can be found in USASOC Reg 350-2. Individual jumpers release both canopy release assemblies as soon as they feel their feet touch the water. They immediately swim away from the canopy to preclude entanglement with the sinking parachute. This movement is especially critical because the exposed weapon is prone to entanglement with suspension lines. Jumpers should then attempt to assemble on the boat before taking off the rest of the harness. It is not difficult to swim in the harness and the reserve can be dropped if it is in the way. Once at the boat, the harness can be sunk before or after climbing in.

Chapter 16

16-15. There are two boat-loading methods that may be used in waterborne operations. Each method works as follows:
- Depending on the size of the detachment, one (single duck) or two (double duck) CRRCs can be prepared on a single platform and loaded with all the team's equipment (minus LCE and weapons). The obvious disadvantage is that should the chute not deploy, the team will have no backup with which to accomplish the mission, as its entire operational equipment is stowed in the duck.
- A second option is to use two separately rigged inflatable boats. The primary disadvantage of this method is that current USAF regulations require that two separate passes must be made (boat–personnel–boat–personnel). In addition to disclosing the location of the drop, multiple passes make it extremely difficult for the two boat teams to link up in the water at night.

16-16. Whichever option the detachment selects, some consideration must be given to preparing a survival bundle that can be carried on the infiltration aircraft and dropped to the team in case of catastrophic failure of the primary platform. This bundle will allow the team to survive until other U.S. assets can recover them.

16-17. The boat is hooked to the left-side anchor line cable, and the jumpers use the right one. The first jumper (jumpmaster) positions himself at the tailgate hinge just behind the boat platform, with the remainder of the jumpers behind him. When the platform starts moving, the jumpers follow it to the edge of the tailgate and hold until the lead jumper sees the pilot chute of the G-12D deploy. The team then exits the aircraft and attempts to assemble on the boat in the air or water.

16-18. The most difficult phase of a water jump is assembly on the boat in the water. The detachment must be proficient at grouping in the air and landing downwind of the boat. Consideration should be given to using static-line ram-air parachutes (MC-5) to facilitate grouping the detachment. This practice is critical because the boat, with its higher windage, can be blown downwind faster than detachment members can swim to catch it. Both the boat and the detachment members must be marked to facilitate locating and recovering them. Detachment members should also consider using strobe lights, chemlights, flashlights, and flares (marking and pen). Chemlights should be disposed of carefully because they float and are easily seen from the air. Marking techniques must include a method of elevating the signal so that it is visible above the waves. Available technology includes the "Steiner" assembly aid for the boat and inflatable marker buoys (civilian diver "safety sausages") for the individual detachment members. Either option allows the attachment of a light that can then be raised above the level of the surrounding waves so that it is visible to personnel in the water (or conversely on the boat) to facilitate assembly or recovery. One of the initial tasks of the first person to reach the boat is to assemble and deploy the boat's marking system to aid the remainder of the team to assemble on the boat. The primary concern when selecting a marking system must be its effectiveness and the ease of employment. The marking methods have minimal tactical significance due to the OTH nature of the operation.

16-19. The first man to the boat moves around to the rear and smells for gas (in case the fuel bladders have burst or leaked). If there is no gas smell, he then feels in the area around the motor and gas cans, again looking for spilled gas. Once he has determined that there is no spilled gas, he enters the boat. If there is evidence of spilled gas, a decision must be made on the extent of the damage and an estimate as to the relative danger of attempting to start the motor. He then disconnects the risers by taking the pins out of the clevises and cuts the center ring. This technique allows the platform to separate from the boat and sink under the weight of the sandbags. If the motor is not already in place, the first man mounts it and picks up personnel before heading for the BLS.

HELICOPTERS

16-20. Helicopters are an ideal intermediate transportation platform. They are a readily available asset that does not require interservice coordination for use, are an integral asset for Phase I training, and can be used as an on-site exfiltration or MEDEVAC craft in the early stages of the operation. Certain U.S. Army helicopters possess in-flight refueling and precision navigation capabilities that enables their operational utility and range to be greatly extended. These extended ranges can provide a more timely response for mission requirements.

Aircraft in Support of Maritime Operations

16-21. If a suitable transfer point or staging area can be found within the helicopter's planning range, it has a number of significant advantages as an infiltration platform. Helicopters, although they lack the speed and endurance for long-range or strategic infiltration, make an excellent intermediate delivery platform for a detachment that is already forward-deployed. This forward posture can be achieved with land bases or naval assets. Land bases such as SF operational bases or special operations task force sites are normally selected because of their ability to support AIROPS. They should include a suitable airstrip and associated support structures for an attached helicopter contingent. Aircraft carriers, amphibious assault ships, and similar naval assets are especially configured for supporting helicopter operations and will frequently have additional space available to the detachment for mission planning and train-up. Heliborne operations are not restricted to aircraft carriers; most large naval ships have landing decks that will accommodate one or more helicopters.

16-22. Many U.S. Army helicopters are ideal for helocast operations. Inflatable assault boats or CRRCs can be loaded inside as cargo or sling-loaded beneath the helicopter. The drop point can be precisely plotted, giving the detachment an exact start point for their transit navigation. A major advantage is the ability to drop the detachment and its equipment intact within a reasonable transit range of the BLS. This capability greatly simplifies the detachment's linkup and getting underway. Unlike fixed-wing assets, the helicopter has a limited loiter and recovery capability. In case of catastrophic mission failure during the infiltration, the helicopter can provide direct assistance to the detachment.

16-23. Because a helocast does not require the altitude of a paradrop, the aircraft can fly closer to the target area and still be OTH. Depending on the enemy's detection capabilities, helicopters can drop detachments 12 to 15 miles offshore as opposed to the minimum 35 to 40 miles required for a fixed-wing paradrop. The shorter transit distance is a significant advantage of the helicopter-assisted infiltration and greatly reduces the time and fuel required for the detachment to reach the BLS. Helicopters can also loiter at the drop point to ensure the detachment has assembled on the boats and is underway to the BLS.

16-24. There are four different options available to detachments when using helicopters to infiltrate detachments. They are explained below.

Limp Duck

16-25. This method involves one or two CRRCs, with mission equipment, loaded internally in a CH/MH-47. CRRCs are launched out the ramp while the aircraft maintains an altitude NTE 5 feet and a forward speed NTE 5 kts. To facilitate launching the CRRCs, the team can place rollers underneath the Zodiacs. The simplest are 3- or 4-inch-diameter polyvinyl chloride (PVC) tubes that extend the width of the Zodiac. Four of them evenly spaced under the CRRC will support the boat's weight and allow it to be pushed off the ramp. Either technique will make the boat easier to handle and expedite launching. This technique also reduces the risk of injury to the team from entanglement or falling. Exact procedures for helocasting Zodiacs are outlined in USSOCOM M350-6.

Kangaroo Duck or K-Duck

16-26. The K-duck method consists of one CRRC with a reduced mission load and a rigid floorboard, rigged underneath a UH/MH-60. There are two acceptable rigging methods. The first is a harness developed by Natick Labs that envelops and supports the CRRC below the helicopter. The harness is secured to the fast-rope attachment points. Rigging instructions are included with the harness. The second method is an accessory developed by Zodiac that is assembled with the boat when it is initially inflated. This hanger is captured by the sling-load hook under the helicopter. In both cases, the duck is dropped by the crew chief or pilot and the team casts after it on order. Because it is an external load, the duck will affect the flight characteristics of the helicopter. The team should always coordinate with the aircrew to determine any limitations imposed by the load.

Rolled Duck

16-27. This option is made up of a single CRRC rigged with a roll-up floorboard, a pneumatic inflation system with either a CO2 bottle or an 80/100-ft3 air cylinder, paddles, a motor, and fuel bladder. The team ensures that the boat is properly prepared for pneumatic inflation. It pads the motor and fuel bladder to protect the CRRC, then secures them in the stern of the boat. The entire assembly is rolled from the bow to the stern, trapping the motor, fuel, and paddles inside the bundle. The team secures the bundle with a quick-release assembly that can be removed rapidly once the boat is in the water. The bundle is loaded inside the helicopter with the boat's crew. To deploy the team, the helicopter hovers and the bundle is pushed out. The team follows it, helocasting with all of their individual and mission equipment. Although the bundle is fairly heavy (approximately 400 lb), its displacement and the air trapped inside the CRRC ensures that it will float. The first person to reach the boat ensures that the bundle is upright; the transom is up so that when the boat unrolls it is right-side up. If this is not done, the boat will have to be righted (capsize drill) and the team runs the risk of losing or damaging the motor and other equipment stored inside the rolled bundle. The team then releases the straps around the boat, pulls the transom out, and slowly opens the valve on the pneumatic inflation system. The team must take care during the initial inflation to ensure that the boat does not fill too rapidly, which can cause damage to the internal baffles. Team members must also be prepared to support the boat so that the motor does not roll off of it during inflation. After the boat is inflated, they prepare it for navigation by closing the valves and inflating the keel. The team then embarks on the mission.

Tethered Duck

16-28. The last method is a variation of the rolled duck, with the bundle secured to the end of a fast rope. The team pushes the bundle and follows by sliding down the fast rope. This technique has the advantage of grouping the team on the boat immediately. This benefit is particularly useful at night, in marginal seas, or during riverine operations where there is a current that might disperse the team.

16-29. All of these techniques require training and rehearsals. Details for qualifying personnel as castmasters or other designated personnel are contained in USSOCOM M350-6.

RECOVERY OPERATIONS

16-30. Another major advantage of helicopters is their ability to exfiltrate a detachment. If stealth is no longer a mission requirement, detachments should plan on a "dry" extraction. If the enemy situation allows it, helicopters can pick a detachment up directly from the operational or target area. This capability greatly reduces the detachment's exposure and eliminates the risks and planning or coordination associated with a maritime withdrawal. It also allows the detachment to exfiltrate its maritime equipment such as zodiacs, motors, and other items.

16-31. If the mission requires a clandestine withdrawal that can only be achieved by a maritime extraction, the detachment must first exfiltrate itself to a helicopter landing zone (HLZ) or helicopter recovery zone (HRZ). The operational parameters and mission planning steps are identical to those required to plan an infiltration, and they must be completed (with appropriate contingencies) during the initial mission planning to ensure that resources (fuel) and conditions (sea state, tides, currents, weather, illumination) are supportive of the operation. The proposed recovery zone must be OTH from the enemy's viewpoint.

16-32. There are three recovery options when using helicopters to pull a detachment from the water—a Delta Queen, a caving ladder, and the fast rope insertion and extraction system (FRIES).

DELTA QUEEN

16-33. The Delta Queen is used with CH-47 helicopters. The aircraft lands in the water and the CRRC is pulled inside the aircraft. Helicopters should not land in saltwater during training operations. If using a CRRC with motor, passengers should lower themselves within the boat so the CRRC operator has the best view possible of the aircraft, and to protect themselves from colliding with the aircraft ceiling. One person on each side of the CRRC is designated to guide the boat into the aircraft. As the aircraft passes overhead to the pickup point, the CRRC will move in trail toward the ramp. When the ramp is lowered into the water and the crew chief signals the CRRC, the coxswain will increase speed to penetrate the rotor wash and move up onto the ramp. Upon contact with the ramp, the coxswain cuts the engine power and raises the motor out of the water. As the boat comes in contact with the ramp, the two designated personnel jump out, pull the boat into the aircraft, and tie the bow line of the CRRC to an interior hard point on the aircraft as quickly as possible. When not using a motor, the crew chief lowers a rope hooked to the aircraft's winch that has a 10-pound padded weight attached. The line is dropped behind the boat and dragged forward across it. The detachment secures it to the CRRC, and the aircraft's winch pulls the boat aboard. The Delta Queen is normally restricted to calm waters (Sea State 0) because of the risks to the aircraft, its crew, and the detachment caused by wave action.

CAVING (WIRE) LADDER

16-34. The detachment uses caving ladders to recover personnel into the aircraft when it is unable to land. These ladders can be deployed off the ramp, the center hole, and the cabin door of an MH/CH-47 or out the doors of an MH/UH-60. Procedures for preparing the aircraft and prerequisite training for the detachment are outlined in USSOCOM M350-6.

FAST-ROPE INSERTION AND EXTRACTION SYSTEM

16-35. Detachments can also be recovered by the FRIES. The detachment must don the extraction harnesses before the helicopter's arrival. The helicopter simply hovers over the grouped swimmers as they attach their harnesses to the FRIES rope. This extraction technique is only viable for a short distance. The extended ranges normally involved in MAROPS will limit its usefulness. It should only be used when the aircraft cannot land or cannot hover low enough to use ladders, and the tactical situation requires moving the detachment to a more secure location for exfiltration. Procedures for preparing the aircraft and prerequisite training for the detachment are outlined in USSOCOM M350-6.

16-36. These techniques (ladder and FRIES) cannot be used to recover the MAROPS equipment, such as the boats, motors, or other items. Once the detachment has positive confirmation of the inbound aircraft, designated personnel must sterilize the recovery zone by ensuring that all equipment that will not be recovered is sunk. They secure all the equipment that will be abandoned inside the boat. Personnel open all the valves and then slit the main buoyancy tube all the way around the boat. They must not forget the speed skegs under the boat. Personnel should be aggressive; a simple puncture may leave air trapped inside the CRRC that would prevent it from sinking. If the plan is to abandon or sink the equipment, personnel must consider weighting the boat with sandbags or some other heavy, dense item to ensure negative buoyancy.

OVER-THE-HORIZON COMPUTATIONS

16-37. Waterborne operations will include the airdrop of one or more inflatable rubber boats used to reach the BLS. This airdrop must be completed beyond the visible horizon so as to prevent detection by coastal radar or visual observation. The following information explains how to calculate the distance beyond the horizon.

16-38. It is a fairly simple matter, using geometry and algebra, to calculate the critical distances. Fortunately, these calculations have already been done. The results are tabulated in Tables 12 (Distance of the Horizon) and 13 (Geographic Range) of Pub No. 9, *The American Practical Navigator*. Using these tables, the radar-horizon distance for any radar located any known altitude above sea level (ASL) can be calculated. These tables are simplified and presented here as Tables 16-1 and 16-2, page 16-8. Table 16-1 lists various horizon distances for various radar altitudes.

16-39. Table 16-2 lists aircraft-to-point-of-radar tangency distances for various aircraft altitudes. By using Tables 16-1 and 16-2, an individual can find that distance at which an aircraft flying at a given altitude ASL can just be detected by a coastal radar located a given height ASL. The data from Table 16-1 is simply added to the data from Table 16-2, giving the total distance at which the air or surface craft just becomes visible to the radar. To remain safely OTH and reduce the risk of compromise by radar detection, a safety factor equaling 5 to 10 percent of the total distance must be added to determine the final safe OTH distance.

Table 16-1. Distance to horizon

Radar Altitude ASL in Feet	Distance to Horizon (nm)	Radar Altitude ASL in Feet	Distance to Horizon (nm)
1	1.064	50	7.524
2	1.505	60	8.242
3	1.843	70	8.902
4	2.128	80	9.517
5	2.379	90	10.094
6	2.606	100	10.640
7	2.815	500	23.792
8	3.010	1,000	33.647
9	3.192	1,500	41.209
10	3.365	2,000	47.584
20	4.758	2,500	53.201
30	5.829	3,000	58.297
40	6.729		

Table 16-2. Distance to point of tangency

Aircraft Altitude in Feet	Distance to Point of Tangency (nm)	Aircraft Altitude in Feet	Distance to Point of Tangency (nm)
10	3.365	600	26.063
20	4.758	700	28.063
100	10.640	800	30.095
200	15.048	900	31.921
300	18.429	1000	33.647
400	21.280	1250	37.169
500	23.792		

16-40. The airdrop of an inflatable rubber boat (rubber duck operation) normally takes place OTH to preclude detection from coastal radar or visual observation. Although OTH is a commonly used term, few actually understand its implications in planning for waterborne operations.

16-41. Figure 16-2, page 16-9, indicates a typical coastal radar on the surface of the earth. If a straight line (tangent to the earth) is drawn from the radar until it touches the earth's surface, it can be seen that everything beyond is OTH.

16-42. Referring to Figure 16-2, the detachment member should consider that an aircraft is preparing to drop a rubber duck. If the aircraft is flying higher than the radar-horizon tangent, it can be detected (assuming a radar of sufficient power).

16-43. However, if the aircraft is flying below the radar-horizon tangent as in Figure 16-2, it is below the horizon and cannot be detected by radar due to its line-of-sight limitation. The distance of the radar-horizon tangent will vary depending on the height of the radar ASL. The enemy can be expected to locate his radar as high as practical ASL to provide maximum range. In the same way, the point at which the aircraft just becomes visible to the radar also depends on the altitude of the aircraft ASL. Figure 16-2 also shows that, as the aircraft gets closer to the radar, it must fly lower and lower to stay below the horizon.

16-44. Based on the previous information, the following is a discussion of the impact on operational planning considerations when an airdrop of an inflatable boat is required:
- In all likelihood, the most feasible method of infiltration for use with a parachute jump is a rubber duck. Even an extremely low jump would put an infiltration team well beyond combat swim range.
- A CRRC infiltrating from beyond the horizon will require a large amount of fuel to make it to the BLS or even to within reasonable paddling range of the BLS. The heavier the load in the boat, the more fuel it will require to reach the BLS. Fuel consumption and transit time with a realistic load must be carefully determined through extensive rehearsals.

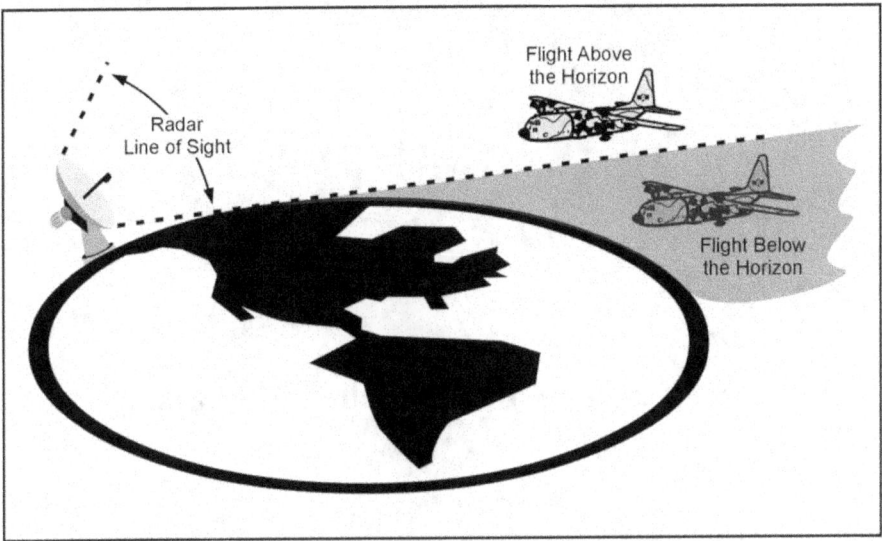

Figure 16-2. Flight in relation to radar line of sight

16-45. Due to the extended distances involved, the team must be highly confident with GPS and radio navigation operations. All team members need to be highly skilled in DR techniques, nautical charts, tidal and current computations, and offset navigation. Training, rehearsals, and detailed planning ultimately determine the team's ability to conduct OTH rubber boat infiltrations. This sustainment training (Appendix B) must be conducted on a frequent basis with all team members involved.

This page intentionally left blank.

Chapter 17
Search Dives

Search dives are one of the most important and challenging services that an SF dive detachment performs while operating in a mission support role. The steps to plan, prepare for, and execute search, general reconnaissance, or mission support dives are similar. For the purposes of this manual, they will be addressed together. Recovery operations will be addressed separately within this chapter.

The objective of this chapter is to enable the combat dive supervisor or dive detachment, given the appropriate equipment, to plan, brief, and conduct a safe and effective search dive. When selecting the search technique, the dive supervisor must consider water conditions, personnel capabilities, time available for the search, and the object of the search.

SEARCHES

17-1. Searches are often required on short notice with little opportunity for detailed planning. They never happen in good conditions. To expedite planning and get an effective force on site as rapidly as possible, the detachment will use the standard troop-leading procedure (Figure 17-1). This well-known process is designed to facilitate small-unit operations. It allows the detachment to start movement with an effective force while mission planning is still underway. This factor is especially important because the longer a search is delayed, the more difficult it becomes and the less chance it has of being successful.

1. Receive the mission.	5. Reconnoiter.
2. Issue a warning order.	6. Complete the plan.
3. Make a tentative plan.	7. Issue the complete order.
4. Start necessary movement.	8. Supervise.

Figure 17-1. Standard troop-leading procedure

RECEIVE THE MISSION

17-2. SF dive teams are frequently called upon to conduct search dives. These missions range from simple survey dives to body recoveries. They may also include the recovery of objects ranging in size from weapons to automobiles. Depending on command policies, divers may also be asked to assist local law enforcement agencies with humanitarian assistance (HA) or evidence collection.

17-3. An important part of receiving the mission is to objectively analyze the mission requirements versus the detachment's capabilities. Because searches are frequently conducted in marginal conditions where there may be significant risk to the involved divers, a complete risk analysis is imperative. Limitations imposed by training levels, available equipment, personnel, and environmental hazards must be considered before committing to the mission. In some cases, objectively weighing risk versus benefit will determine that divers are not the appropriate solution or that additional resources or training are required to safely and effectively accomplish the mission.

17-4. Divers should determine the degree of urgency required, especially when dealing with HA calls. These are operations where life and limb of potential victims are at significant risk, and responding detachments may be under considerable pressure to respond rapidly, accepting an increased risk level and limited preparation time. HA responses are usually divided into rescue or recovery operations.

17-5. Rescue operations are conducted when there is a reasonable chance to save the life of a person involved in a water-related accident. Rescue mode is only adopted when the detachment can be "on scene" and operational within 1 hour of the occurrence of the incident. Well-developed SOPs are critical for a timely response to a rescue operation. While operating in a rescue mode, detachments may reasonably adopt a reduced planning or execution cycle and accept an increased level of risk to get divers in the water if there is a possibility of saving life and limb. While it is generally accepted that drowning victims begin to suffer irreversible brain damage after approximately 6 minutes without oxygen, mitigating circumstances such as extremely cold water or air pockets in submerged vehicles have contributed to successful rescues after extended periods of submersion. The 1-hour window is intended to afford the victim every benefit of the doubt while clearly defining and limiting the detachment's response options.

17-6. If the detachment cannot be on-scene within 1 hour or the rescue operation has lasted more than an hour without success, then the operation should be considered a recovery. At that point the detachment should terminate operations, recover all divers from the water, and regroup for recovery operations. Recovery operations do not have the urgency of saving life; therefore, they must be considered routine. The shortcuts and increased risk acceptance reserved for rescues are no longer appropriate. All mission planning steps should be followed, a detailed mission analysis completed, and a complete dive plan prepared and briefed.

Issue a Warning Order

17-7. The commander must give the dive detachment and the DLSMF enough information to begin mission preparation. The detachment and facility will have to assign individual and subunit tasks, designate special equipment required, and assemble the equipment and support required for the mission.

17-8. The unit must identify and assemble the equipment necessary to conduct a search dive. Examples of the types of equipment needed include the following:
- Ascent or descent, search and marking lines (floating or nonfloating).
- Tether lines (with quick releases) for the divers and safety divers.
- Safety and surface marker buoys to mark divers and search area.
- Snap links, carabiners, or clevises.
- Lead weights or other suitable anchor systems, as required for bottom conditions and current.
- Lift bags, inflation source, and rigging tackle (lashings and clevises) as required.
- Body bags or stokes litter as required for body recovery.
- Underwater lights.
- Depth or fish finders, metal detectors, magnetometer, and GPS or other survey equipment.
- Compasses (surface and subsurface).
- Reels.
- Gloves.
- Boats (safety and support).

Make a Tentative Plan

17-9. Detachment personnel identify the critical mission components and address them. They identify the divers for the mission and conduct a predive medical records screening to ensure all divers meet the medical qualifications to dive. Once all divers and support personnel have been identified, the detachment leader establishes a timetable and posts it where all personnel involved in the dive operation can see it. He ensures that the divers know they are responsible for the predive checks and the packing of their individual dive gear, to include tanks, regulators, depth gauges, and special equipment as assigned. He assigns areas of responsibility to the support personnel and has them prepare general and special equipment. All personnel should use SOPs wherever applicable to expedite planning and preparation.

17-10. The detachment leader verifies the medical support and evacuation plan. He identifies the chamber or treatment facility and ensures the personnel there are aware of the dive operation and will be available to provide treatment should it become necessary. He also conducts any other predive coordination as required and prepares a dive plan for the conduct of a search dive.

Search Dives

START NECESSARY MOVEMENT

17-11. The commander quickly moves an advanced echelon (ADVON) or liaison team on site. This team must make initial coordination with the requestor as quickly as possible. Personnel on the ADVON must have the training and experience to conduct a reconnaissance and finalize the plan. They must also be prepared to interact with the personnel already on site. The requestor should beware of the potential conflict of interest between his desires for requesting the search and the dive team's safety and control requirements. The remainder of the team and its equipment should follow as quickly as it is assembled.

RECONNOITER

17-12. The ADVON determines and defines the search area. It acquires as much information about the situation as possible. The key to a successful search is locating the missing object as precisely as possible before divers enter the water. Information will come from two sources—witnesses and the incident scene itself.

17-13. Witnesses are often the best source of information. Interviews should be conducted. The ADVON should attempt to speak directly with any witnesses and use standard interview techniques. The leader should segregate them, interview them separately, and try not to let them talk among themselves. Doing so risks creating a group consensus that may obscure key elements of information. The team should use elicitation to extract additional details that the witnesses may have overlooked in their account. If possible, the interviews should occur at the incident scene. Personnel should use incident reports, any maps, photographs, or sketches of the incident site to focus the interviews.

17-14. Who, what, where, when, and how are all critical elements of information that can only be provided by witnesses. The first question to ask witnesses is how the object was lost. Was it lost from a vessel, from the shore, or during a water-crossing exercise? If it was lost from a vessel, what was the vessel's speed and course or bearing? The ADVON should obtain a physical description of the object to be recovered. This description should include size, weight, shape, color, and material or composition. Cost and the sensitivity or classification of the lost equipment may also be useful when determining the urgency of the search operation.

17-15. The incident scene will provide additional, valuable clues to the potential location of the search object. Questions that must be answered include the following:
- What is the depth in and around the potential search area?
- What is the visibility?
- What type of bottom (sand, mud, silt, rock, coral, vegetation) will divers encounter?
- What were the tides or currents when the object was lost? What are they now?
- Are there any prominent landmarks or reference points?

17-16. The physical characteristics of the search object must also be considered when determining the size and shape of the area to be searched. Buoyancy becomes a critical factor. Strongly negative items will most likely be located near the last observed position. Weak negatives and neutrally buoyant objects are most likely to be widely dispersed. Positively buoyant objects can usually be located and recovered with a surface search. Modifiers include strong currents or large surface areas that are more readily affected by the current. Confined bodies of water with little or no current that might cause displacement of the search object will obviously be easier to search than rivers and open bodies of water affected by tides and currents. Obviously, the longer an object is exposed to current, the farther it will be dispersed from the last known point. Greater depths also cause more dispersion. The further an object sinks before it strikes the bottom, the larger the potential search area.

17-17. The ADVON should use the known (observed) characteristics of the object and the incident scene to determine the probable search zone. The zone starts at the last known position and extends in the direction of flow for a distance determined by time-distance-speed computations given the measured current. The ADVON can modify this linear model by creating a circle of probable error (CPE) with its center point on the current line. At selected points downstream from the last known position, personnel should construct additional CPEs. The plotted size of the CPE will increase based on the amount of time that has passed since the incident, the distance the object may have drifted, and a subjective evaluation based on the experience of the individual conducting the survey. The area to be searched is then defined by drawing lines from the start point to the outsides of the succeeding CPEs and adding an "insurance factor." The resulting search zone is cone-shaped (Figure 17-2).

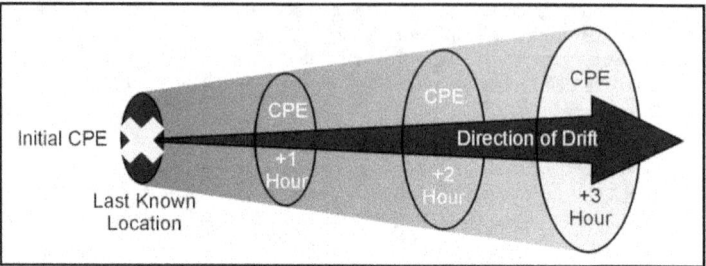

Figure 17-2. Estimating circle of probable error

17-18. All of these factors become significant problems when detachments are called upon for HA (body recovery) missions. Bodies, unless heavily laden with equipment, are usually "weak negative" or neutral, at least initially. Given their relatively large surface area, bodies are subject to considerable dispersion from their last known location, especially in rivers or channels with a current. In rivers, the search area is constrained by the banks. Obstacles such as snags, brush piles, undercuts along the bank, "hydraulics," and eddies act as "strainers" that capture and hold bodies (or any other search object). A significant element of on-site reconnaissance includes identifying these strainers and prioritizing them for search (Figure 17-3).

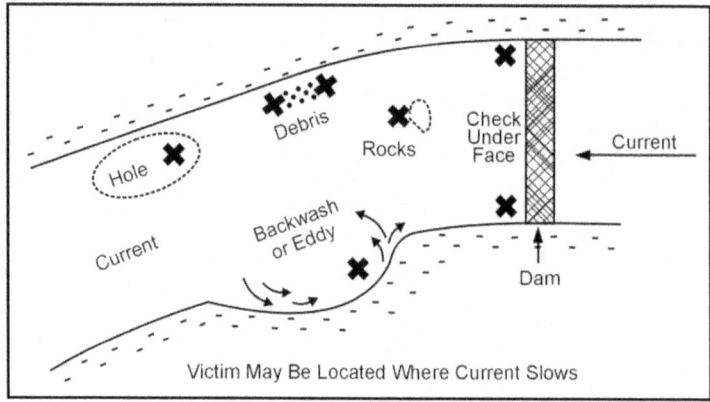

Figure 17-3. Probable river search sites

17-19. Natural decomposition processes further complicate the search. If the water is warm enough, decomposition results in the formation of gasses. The accumulation of these gasses within the body causes it to float. If the water is cold enough, decomposition is retarded and the body takes longer to float. Bodies have also been observed to go through several float-sink-float cycles. Divers should be psychologically prepared for the potential shock of finding the victim, especially in limited-visibility conditions.

Note. If the mission is HA (body recovery), personnel should be prepared to deal with bereaved family members and local officials. Personnel should also remember that these missions are frequently high-profile operations; therefore, they should also have a plan for dealing with media representatives.

COMPLETE THE PLAN

17-20. The detachment begins planning the search operation based on the information compiled. It should follow all safe diving procedures as outlined in current U.S. Navy diving regulations and this manual. The detachment prepares the dive plan in the five-paragraph OPORD format using the following tasks:
- Select a search technique.
- Identify support requirements.
- Conduct a risk analysis.
- Select personnel.
- Assemble the necessary equipment.

17-21. The most critical component of a search plan is selecting an appropriate technique. The selected search method will dictate the personnel and support requirements. Search techniques are selected based on the size of the object to be recovered, the quantity and experience of the available divers, and the physical conditions of the search area. Unfortunately, limited resources or an objective risk assessment may force the detachment to adopt a less-than-ideal technique. Planners must take into consideration the water conditions, personnel capabilities, the time and equipment available, and the object of the search. For planning purposes, the most common techniques are as follows:
- Tended line search.
- Circle line search.
- Running or modified running jackstay.
- Checkerboard jackstay.
- In-line search.
- Free-swimming searches:
    - Expanding box.
    - Reciprocal pattern.
- Towed search.

17-22. These techniques are listed, in order, from simplest (fewest resources) and most precise (greatest control) to more complicated (greater resource requirements) and less precise (less control).

ISSUE THE COMPLETE ORDER

17-23. The detachment conducts a predive briefing to ensure that all personnel are familiar with the plan. The briefing starts with a roll call to ensure all personnel concerned with the dive are present at the dive briefing, to include support personnel. After the briefing, the detachment leader should answer all questions to ensure that involved personnel are aware of their individual responsibilities. To ensure the dive is conducted smoothly, the detachment should always conduct rehearsals. All personnel must be thoroughly familiar with the part they are to play in the search. If ideal water conditions (good visibility, clear weather, and current under 1 knot) do not exist, rehearsals are essential.

SUPERVISE

17-24. Designated personnel conduct a predive inspection. The divers then enter the water using normal scuba procedures. When all divers are in the water, the dive group leader will give an OK when the group is ready to descend. The dive time begins as soon as the first diver's head leaves the surface. The group leader records the dive time and maintains an account of the elapsed dive time. He ensures that normal scuba descent procedures are followed and that the entire group maintains contact during the dive. In conducting any search, personnel should remember the following:
- All equipment must be totally prepared before entering the water. Tenders must be on standby; safety divers must be rigged and ready to enter the water immediately.
- Divers should always be assigned a particular area to search. After it has been searched, the area must be marked with buoys and on a chart to avoid needlessly searching it again.

Chapter 17

- If the area has a muddy or loose sandy bottom, divers should take care to avoid stirring up the silt. If possible, they should remain more than 3 feet above the bottom so that fin movements will not roil up bottom mud.
- Planners should ensure that a surface craft is available and capable of supporting the planned dive and that the coordination has been conducted.

17-25. The dive supervisor must monitor the conduct of the dive throughout. He should rotate divers as required, paying particular attention to nitrogen absorption times and general exposure stress. Comfortable, rested divers are more effective and less likely to suffer adverse complications like DCS or hypothermia. The supervisor ensures the dive time does not exceed the planned limit, and that normal scuba ascent procedures are followed. He should maintain a continuous log of all aspects of the dive. This log will prove invaluable during postoperational analysis. If the search is in support of a law enforcement operation, the log will be invaluable if the detachment is required to testify. He should conduct continual risk analysis to ensure that changing conditions do not negate the risk abatement measures adopted at the start of the operation. The dive supervisor also—

- Ensures all personnel are accounted for before leaving the dive site.
- Conducts postdive procedures.
- Ensures all postdive cleaning and maintenance procedures are followed.
- Conducts a postdive debriefing of the divers.
- Ensures all records and reports are submitted.

SEARCH METHODS

17-26. Once the detachment has received the mission, completed an initial assessment, and begun mission planning, the CDS must select an appropriate search technique. Outlined below are the characteristics, requirements, and some illustrations for the different types of searches (Figures 17-4 through 17-20, pages 17-6 through 17-15). The CDS should use these considerations as a guide when selecting a search method.

Operating Principles	Used at night or in murky water conditions. Can be performed by one diver and one other person. Is very quick. Used to locate small objects. Can be performed with inexperienced divers.
Equipment	A single line for attaching to the diver. At least one underwater compass is suggested.
Procedure	The surface tender pays out a specific length of line to the diver. A single tended diver may be used to conduct the search. The diver sweeps from one side of the search zone to the other. At the completion of the sweep, the tender feeds additional line (based on limits of visibility) to the diver, who tensions the line (moves away from the tender until the line is taut), reverses direction, and repeats the process in the opposite direction until the search is completed. The diver and his surface tender can communicate with a series of tugs (line pull signals). Assistance (safety diver) can be deployed directly to the diver down the tether. The tended line search can be conducted from shore or from a boat (Figures 17-5 and 17-6, page 17-7).

Figure 17-4. Characteristics of tended-line search

Search Dives

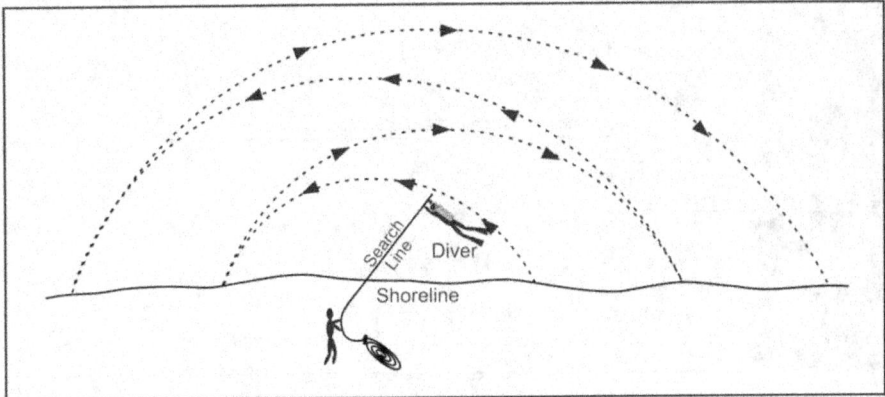

Figure 17-5. Tended-line search

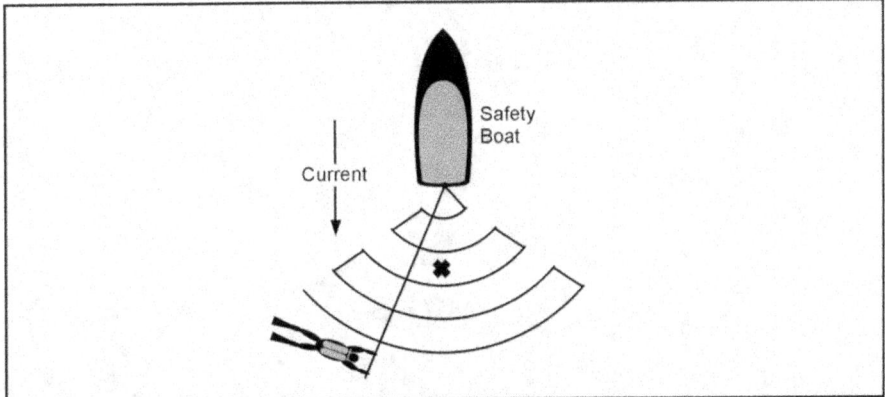

Figure 17-6. Open water tended-line search

Chapter 17

Operating Principles	Used at night, in murky water, or limited-visibility situations. May be used with limited or inexperienced personnel. May be used to locate small objects. Is fairly quick, depending on the number of divers used. Works well in open water with minimal current. Is the easiest method to support from a boat.
Equipment	Anchor. Buoy (attached to the anchor by a line). Search line of the desired length.
Procedure	A line as long as the water is deep is attached to a buoy; the anchor is secured to this line. A search line is attached to the anchored line. The divers should be spaced along the search line. The divers swim in a circle around the anchor. Once a circle is completed, the divers move out along the search line and swim in another circle (Figure 17-8).

Figure 17-7. Characteristics of circle line search

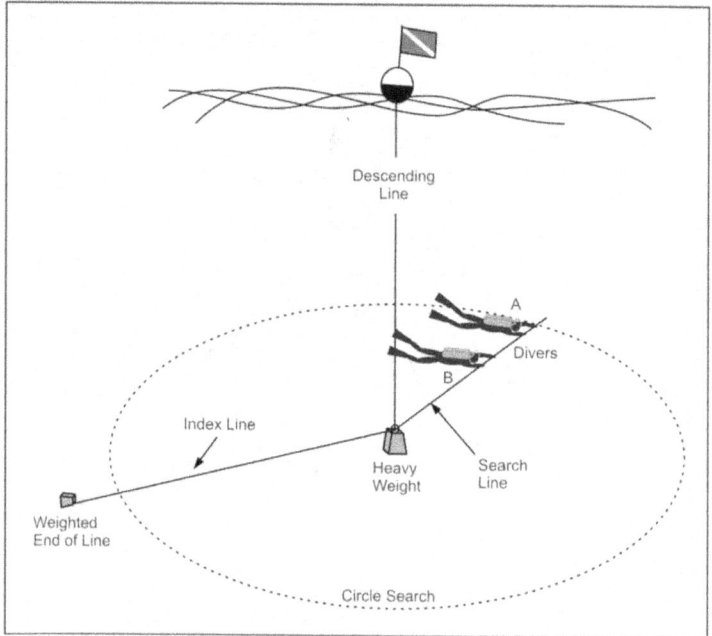

Figure 17-8. Circle line search

Note. An alternate method is for one buddy team to anchor the line to a pole and "wind" themselves up to it. Normally, at least two divers are required for the circle line search; however, any number of divers can be placed on the search line, depending upon the objective of the search. Two divers are adequate to search for objects as large as a helicopter, since the search line would snag the object.

Search Dives

Operating Principles	• Used in clear or murky water or limited-visibility situations. • Requires experienced personnel. • Requires time-consuming setup and execution procedures. • Used to locate small or medium-sized objects. • Requires one diver for every 5 meters of line in clear water.
Equipment	• Two 50-meter grid lines. • One 25-meter search line with a snap link at each end. • Four buoys attached to a length of line equal to the depth of the water. • Diver compass. • Necessary equipment for establishing control ranges. • Four anchors and a delivery craft.
Procedure	• Four buoys with lines (length must equal water depth) and anchors attached are placed in a 50- by 25-meter rectangle. (The rectangle size can vary according to the area to be covered.) • The two grid lines are attached between the buoys on the long sides of the rectangle. • The divers space themselves about 5 meters apart on the search line (lanes searched by each diver should overlap). • On command from the OIC, the divers proceed to the bottom, attach the search line to the grid lines with snap links, and start the first sweep. • At the end of the sweep, the divers ascend, shift to an overlapping lane, and repeat the process (Figure 17-10 and Figure 17-11, page 17-10).

Figure 17-9. Characteristics of running jackstay search

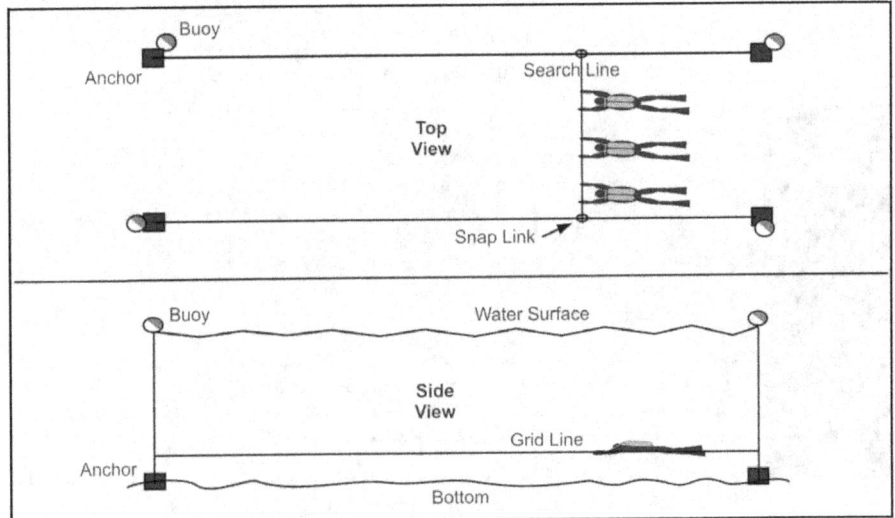

Figure 17-10. Running jackstay search

Note. The prescribed search area (50 by 25 meters) is used as a guide only. Visibility, current, size of the object, and experience of the dive team will determine exactly how large the search pattern should be.

Chapter 17

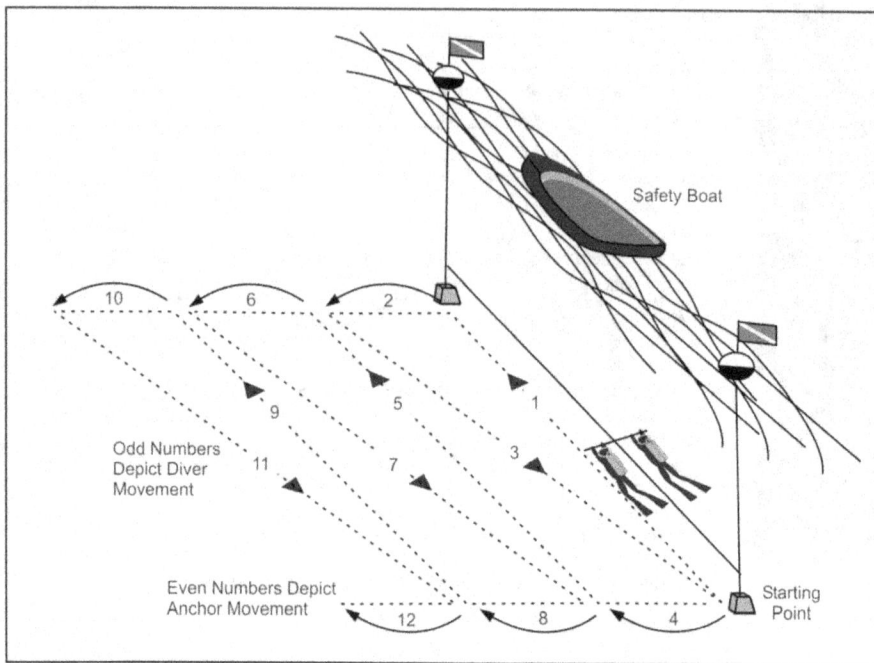

Figure 17-11. Modified running jackstay (open water) search

Operating Principles	Conducted like the jackstay. Same requirements and conditions as the jackstay. Two or more grids must be in place before beginning. Used to find objects that are less than 1 cubic foot in size or hard to see because the color matches the bottom. Used in conjunction with other searches to cover a large area. May be used with good or bad visibility. Used when the current is less than 1 kt. Is the slowest but most thorough of the searches.
Equipment	The checkerboard search requires more equipment than any other search. A minimum of eight buoys with enough weight and line to anchor them is required. An equal amount of bottom markers are required. Bottom markers must be highly visible. A minimum of one underwater compass is required.

Figure 17-12. Characteristics of checkerboard jackstay search

Search Dives

Procedure

If sufficient line, buoys, and weights are available, use the following procedure (Figure 17-13):

- Mark off the corners of the search area.
- Emplace the intermediate buoys and bottom markers between the corner buoys.
- Search and record each area searched. Any search technique may be used. After completing one search grid, move as directed to the next search grid.
- After the object has been found or the entire area has been searched, recover the bottom markers and intermediate buoys. If another area is going to be searched, leave the corner buoys in place to show the area has been searched.

If sufficient line, buoys, weights, and bottom markers are not available, use the following modified procedure (Figure 17-14, page 17-12).

- Mark the first area to be searched with buoys.
- Search the marked area and record the marked area.
- Buoys C1 and C2 are emplaced and area B1 is searched and recorded. Buoys B1 and B2 are removed and used to mark search area C1.
- After each grid has been searched, leave the outer markers and move to the next search grid.
- Leave the buoys in place and emplace additional buoys to mark the next area to be searched.
- Search the area, then remove the intermediate buoys (leaving the outer buoys) and mark the next area.
- Continue this pattern until the item is found or the search is called off.

Figure 17-12. Characteristics of checkerboard jackstay search (continued)

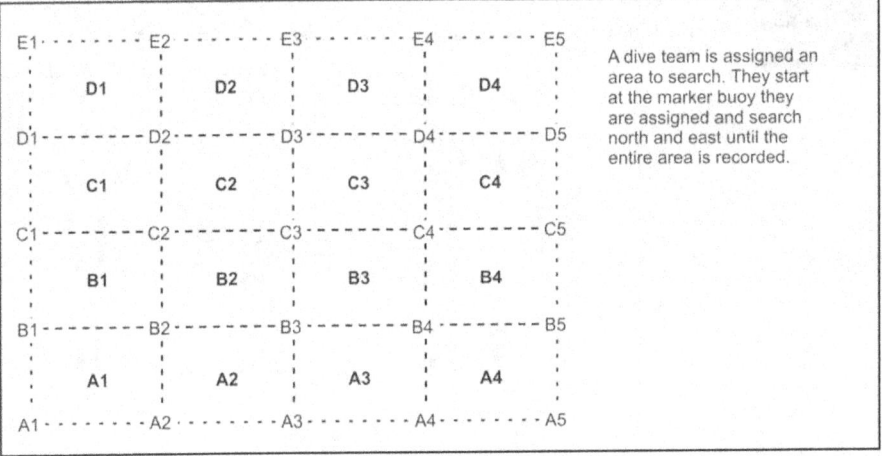

A dive team is assigned an area to search. They start at the marker buoy they are assigned and search north and east until the entire area is recorded.

Figure 17-13. Checkerboard search

Chapter 17

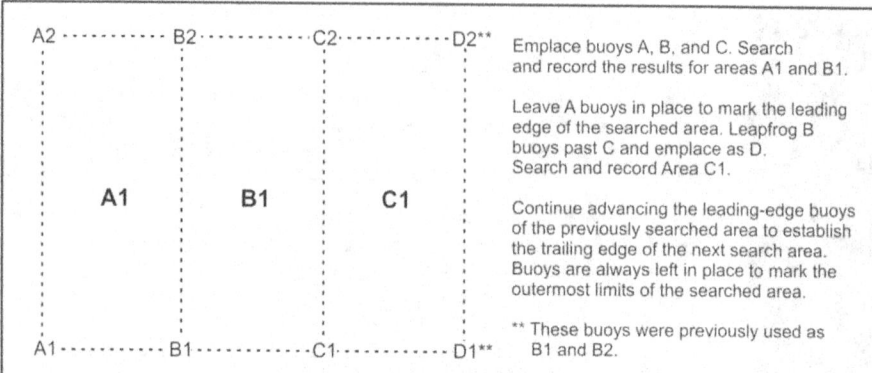

Figure 17-14. Modified checkerboard search

Operating Principles	Used in clear water. Requires experienced personnel. Requires, as a minimum, one diver for every 5 meters of line. Is fairly quick. Used to locate medium- to large-sized objects.
Equipment	Compasses and a line for the divers to hold onto to maintain alignment.
Procedure	Divers line up on a line. Two flank men and the center man swim an azimuth a set distance, then pivot (Figure 17-16).

Figure 17-15. Characteristics of in-line search

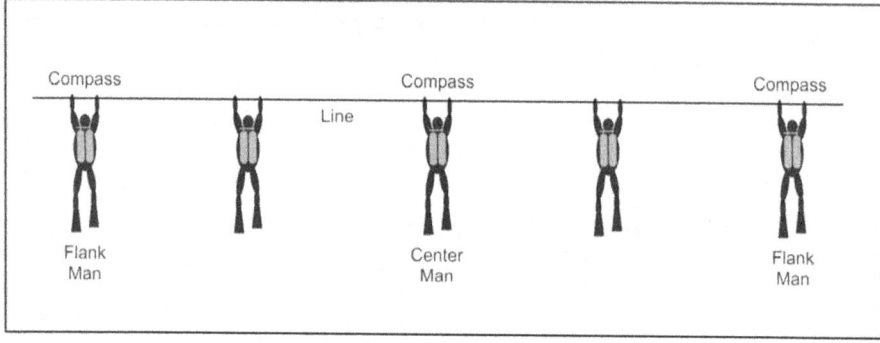

Figure 17-16. In-line search

Search Dives

Operating Principles	☐ Used in clear water. ☐ Requires experienced personnel (free-swimming search patterns). ☐ Can be conducted with minimum personnel. ☐ Is fairly quick. ☐ Used to locate medium-sized to large objects (lacks precision control). ☐ Minimum support and planning requirements.
Equipment	Compasses.
Procedure	☐ Divers enter the water at the last known location (expanding-box search) or up-current of the last known location (reciprocal-pattern search). ☐ *Expanding-Box Search.* The dive team swims an azimuth a set distance, then pivots 90 degrees clockwise or counterclockwise and continues swimming, extending each leg of the pattern by a fraction (50 to 80 percent) of the available visibility to ensure overlap of the search pattern (Figure 17-18). ☐ *Reciprocal-Pattern Search.* The dive team swims an azimuth a set distance (first leg), then pivots 90 degrees and displaces left or right by a fraction (50 to 80 percent) of the available visibility to ensure overlap of the search pattern. Divers then take a reciprocal (to the original) azimuth and continue swimming "legs" until the entire search zone is covered (Figure 17-19, page 17-14).

Figure 17-17. Characteristics of expanding-box and reciprocal-pattern searches

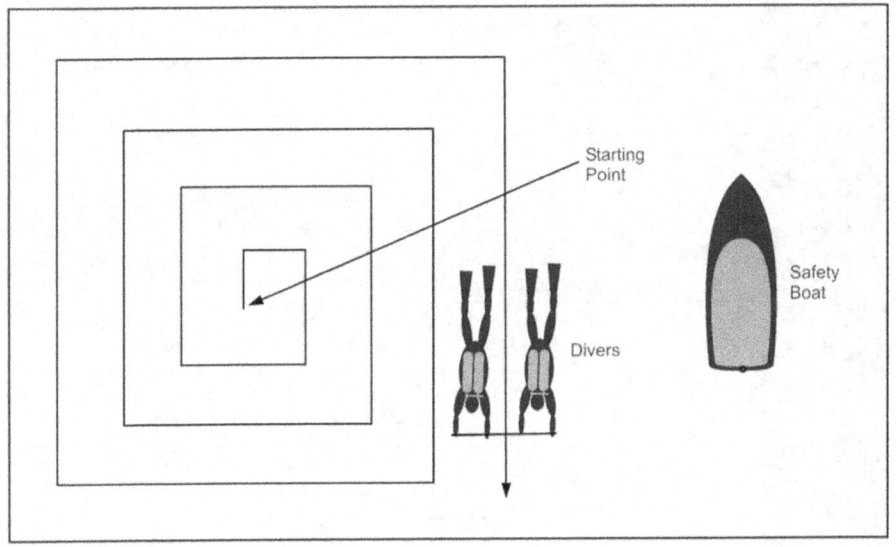

Figure 17-18. Expanding-box search

Chapter 17

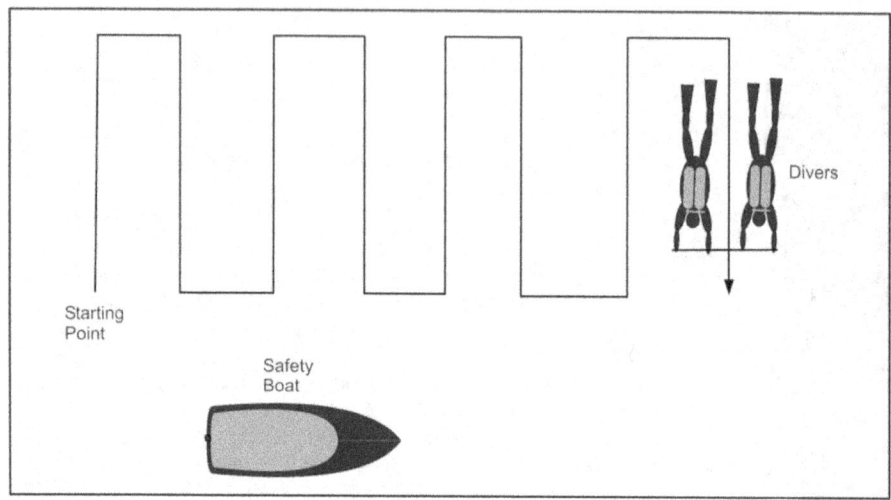

Figure 17-19. Reciprocal-pattern search

Operating Principles	☐ Used to cover large areas. ☐ Used to find large or bright objects. ☐ Used during daylight in clear water. ☐ Used by experienced personnel only. A towed search is quite dangerous and must be strictly supervised. ☐ Is very quick.
Equipment	☐ Towing boat. ☐ Safety boat. ☐ Towline. ☐ Plane board. ☐ Full-face mask with communications (recommended).
Procedure	☐ One or two divers are towed on a plane board behind a boat. The boat must be kept going slowly. The towrope must have adequate scope (be long enough) to allow the divers to maneuver the full depth of the water column and ensure diver safety from the boat propellers. Communications between the divers and the support craft are critical. Special "speed up" and "slow down" signals must be coordinated. Where possible, wireless underwater communications with full-face masks are used so that divers can communicate directly with the dive supervisor, who will be controlling the divers and the boat or boat driver. Full face masks are recommended because they are less likely to free-flow and flood or be lost.

Figure 17-20. Characteristics of towed search

Search Dives

Procedure (continued)	☐ Use caution during course changes, especially when running consecutive sweeps, so that the towboat does not run back over the divers or foul the towrope. The safest method is to stop the boat, allow divers to surface, assume the new heading, reposition the divers, and resume the search. If an item of interest is observed during one of the passes divers notify the towboat using underwater communications (usually some form of through-water single side band) so that the general location can be marked for a follow-up inspection by another standby dive team. Alternately, if the search object is found, divers can simply release the tow, surface, and signal the tow or safety boat so that the area can be marked for a detailed search and recovery.

Figure 17-20. Characteristics of towed search (continued)

SHIP BOTTOM SEARCH

17-27. The ship bottom search is a specialized type of search used to emplace or remove simple underwater mines. The diver must be familiar with the general characteristics of ship bottom structures and design (Figure 17-21, page 17-16). The following characteristics of the ship must be considered when planning the search:

- Electrolysis plates.
- Keel.
- Hull.
- Propeller screws.
- Propeller shafts.
- Rudder.
- Saltwater intakes.
- Sonar dome.

17-28. Divers use the following methods when conducting ship bottom searches:

- *Tended-Line Search.* This method requires one or two divers and a surface tender that leads the divers back and forth, regulating their movements toward the keel by lengthening the line they hold.
- *In-Line Search.* This preferred method of search requires three or more divers. The divers hold the rope, form a line from the keel to the waterline, and sweep slowly from bow to stern. The keel man establishes the direction while the waterline diver keeps the line taut.

17-29. The mines most commonly emplaced on ship hulls are the—

- *Magnetic-Type Limpet.* This mine is timed to explode at a time selected by the emplacer.
- *Propeller-Driven Nautical-Mile Limpet.* This mine detonates after the ship has traveled a distance selected by the emplacer.

RIVER SEARCHES

17-30. Many water-related training incidents occur during the conduct of small-unit river-crossing operations. These incidents range from lost weapons and equipment to the occasional drowning. The resulting search and recovery operations are often complicated by the tactical nature of the river-crossing operation. The terrain or meteorological conditions that forced a deliberate river crossing may also impede search and recovery operations. Rivers usually have limited visibility and may have high current. If the river is at or near flood stage, floating debris ranging in size from branches to entire trees compounds these problems. The risks to the diver of being struck and entangled or impaled are significant. Because river searches pose significant hazards to the divers, additional safety precautions are required. Only additional training and specialized support equipment can mitigate the risks to an acceptable level.

Operating Principles	Divers— • Use tended line or in-line search techniques. • Require coordination with the ship's commander (and harbormaster if in port) before divers enter the water. • Require that all sonar outlets/intakes and screw actuators are off and red-tagged. • Begin at bow and finish at stern. • Search one-half of the vessel at a time. • Do not surface on a side between the vessel and another vessel or between the vessel and a pier or dock. • Will not handle or remove suspicious devices below. They will mark the devices and notify explosive ordnance disposal (EOD) teams for disposition or disposal. Special teams conduct search of rudder, screw, and screw shaft.
Equipment	• Open-circuit gear. • 1/2-inch nylon line, 100 feet long. • Depth gauges. • Watch. • Underwater lights. • Snap links. • Gloves.
Procedure	• Get a diagram of the ship's hull from the ship's engineer. • Ensure each diver knows the search procedure and search objective before the dive. • Notify the harbormaster and the ship's captain **before** the search begins. • Ensure the ship's captain notifies the chief to shut down and secure all machinery on or near the ship that might affect the dive (vents, exhaust ports, and engines) IAW current U.S. Navy diving regulations. • Ensure the sonar is shut down completely. • Once the vessel is secure, announce DIVERS UNDERWATER and repeat at 10-minute intervals. • Start at the bow and work toward the stern. Pay close attention to all hatches and vents. Take special precautions in the vicinity of the screws and saltwater intakes. • If ordnance is discovered, inform EOD unit. Search personnel should not try to disarm ordnance. • Sweep vessels three or four times if there is sufficient time. • Use specially designated teams to search sensitive areas; for example, props, rudders, and through-hull fittings. • Initiate a diver head count at the completion of any ship bottom search.

Figure 17-21. Characteristics of ship bottom search

17-31. All other aspects being equal, current is the most significant complication to conducting river searches. Detachments must establish safety limits based on maximum flow rates and levels of training and experience. Practical upper limits for conducting searches in current are 0.75 kts in training and 1.5 kts operationally. If the current exceeds the maximum safe limit, then the detachment must determine if the flow rate can be expected to diminish (recent periods of heavy rain, hydroelectric facilities upstream) so that searches can be conducted within the recommended safe limits.

Search Dives

17-32. Detachment personnel can determine the flow rate (current speed) by placing a semisubmerged object in the main current and measuring the time it takes to cover a known distance. The recommended distance is 100 feet (or multiples thereof for more accurate averaging) because a current speed of 1 kt per hour equals 100 fpm. This recommendation is based on a trained and conditioned combat diver's sustained swimming speed of 1 kt per hour (about 2,000 yards per hour or 100 fpm). The force of a 1-kt current acting on the diver's body will cause similar fatigue levels. Any additional current causes dramatically increased drag or fatigue and may be enough to strip or flood the diver's mask and cause the diver's regulator to free-flow. If an object in the current should strike the diver, or the diver should break loose from the search line and strike a fixed object, the risk of severe injury climbs rapidly with increased current.

17-33. The size of the river will determine the quantities of specialized equipment required to safely conduct search operations. Small rivers can be searched effectively using a shore-based plan. Medium-sized rivers (for the purposes of this manual) are ones that can be spanned with a 120-foot rope bridge. Larger rivers will require some degree of boat support to search successfully. This discussion will focus on medium-sized rivers. The techniques explained for searching medium-sized rivers are readily adapted to other circumstances. We use medium-sized rivers as the norm because they are the ones most likely to be encountered by small units in the course of deliberate river-crossing operations.

17-34. The detachment should base the search plan and equipment requirements on upstream, search zone, and downstream areas. The upstream area provides a safety or early warning zone to protect the diver from debris floating downstream in the current. It also serves as the anchor spot for the diver's tended line and a staging area for the safety diver. This point can be a shore position (small), a rope bridge (medium), or a boat (large). The search zone is the area where the diver is operating and the adjacent shores (Figure 17-22, page 17-18). The diver's equipment, tended line, tender, and a spotter are located within this zone. It also includes the C2 site and a staging area for any outside support. The downstream area is a safety zone in case the diver loses his tether or is in some way injured and requires assistance. As a minimum, the safety zone should have a means of recovering a loose diver, usually a low rope bridge (small or medium) or a safety boat (large). There should always be a safety swimmer stationed downstream of the search zone.

17-35. When setting up a search zone, the detachment must identify the last known location of the search object. Personnel should construct a low-water one-rope bridge. This bridge should be located at a point at least three times the depth of the water upstream from the last known location of the search object. This rope marks the upstream end of the search zone. It is used to anchor the diver's (and safety diver's) tended line. The area upstream from the rope bridge is a safety zone that should be monitored by a spotter. The spotter must be located far enough upstream, based on current velocity, to provide adequate advance warning of debris floating in the current that might endanger the diver or operation.

17-36. Detachment personnel should calculate or estimate the most probable length of the search zone and construct another low-water, one-rope bridge below (distance is a judgment call) the downstream end of the search zone. This rope is rigged just above water level so that a diver who releases (or breaks loose from) his tether line can surface, inflate his BC, extend his arm over his head, and catch the rope in the crook of the elbow to keep from being swept further downstream. The diver should then secure himself to the rope with a prepared safety line (with snap link) and move hand-over-hand to the near shore (same side of the river as the support element). If the diver is in distress, he can maintain his position on the rope by bending the elbow and clasping the wrist with his other hand (similar to cast and recovery procedures) until the safety swimmer can reach him and render assistance.

17-37. The detachment prepares the diver for the search by rigging him with a safety harness or a suitably rigged heavy-duty buoyancy compensator with an integral harness. Harnesses must be strong enough to support the diver's full weight including the anticipated additional drag of the current on the diver's body. The harness must include appropriate hard points (usually D-rings) for attaching additional equipment and securing the diver's tether. Divers operating in current will require additional weight to ensure negative buoyancy and provide stability on the bottom against the force of the current. The traditional weight belt may prove inadequate to hold or support the potentially large quantities of weight needed. It may be necessary to use integrated weight systems or harnesses similar to those used by commercial divers. The BC must be adequate to lift the additional weight, and divers must receive additional training to ensure they can ditch the BC in an emergency. CDS inspections prior to entering the water must be especially thorough; several dive rescue fatalities have occurred as a direct result of the increased weight and inadequate training or preparation.

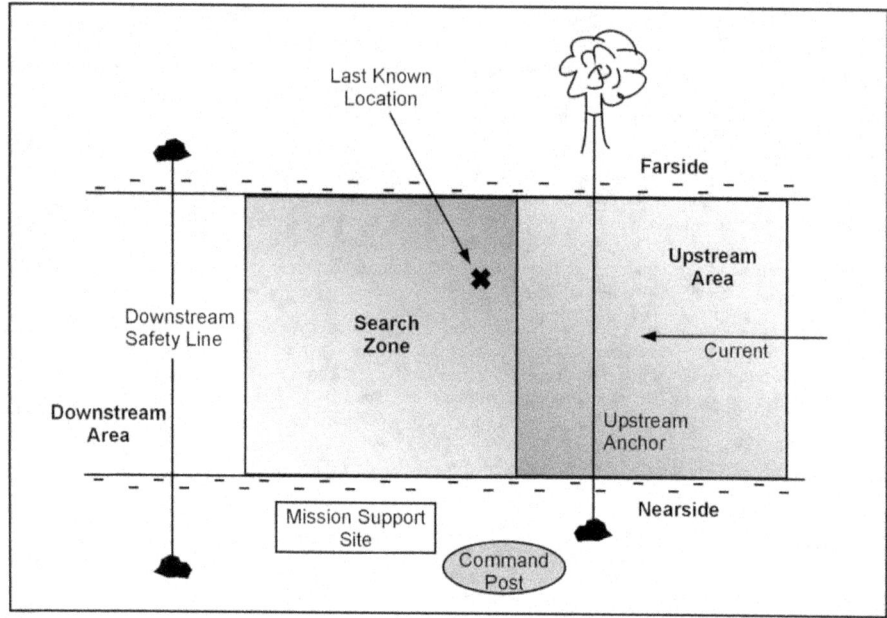

Figure 17-22. Search zone setup

17-38. The diver's tether is his lifeline. It is used to control the search pattern, communicate with the diver (line-pull signals), and locate or send assistance to a distressed diver. The diver's tether (safety line) must be secured to the diver's harness with a quick-release shackle. This method frees up both hands for searching, or allows the diver one free hand to navigate in limited visibility or to fend off obstacles in the water while searching with the other hand. This quick-release must be strong enough to support the diver's full weight against the flow of the current and capable of one-handed release while under tension. Snap hooks and carabiners are unacceptable because they cannot reliably be released with one hand while under tension. They also present a risk of "line trapping" (catching another line inside the gate and further entangling the diver). All divers conducting tended line searches should be fastened to the line. This binding ensures positive control, secure communications, exact location, and a ready guideline for the safety diver to render assistance. If the tether becomes irretrievably tangled upstream from the diver, he can unfasten himself, surface (immediately), and catch the downstream safety line. If the diver should become entangled or otherwise require outside intervention, his signals (line pulls) or lack thereof to the tender initiates the emergency response drill.

17-39. The dive is conducted as an overlapping series of tended-line searches. To ensure complete coverage of the search area, the upstream rope bridge will have evenly spaced anchor points stretching the width of the search zone. The tender (the safety boat or swimmer) will pass the diver's tether through the anchor point (usually a snap link), the diver will enter the water, and the tender will belay the diver against the force of the current. The diver sweeps from left to right in a pendulum fashion. At the extreme end of each sweep, the tender feeds out additional tether line, and the diver allows the force of the current to tension it before continuing his sweep in the opposite direction. After the diver has searched to the extent of his tether, he is recovered to the surface and moved back upstream while the tether is repositioned to the subsequent anchor point. The diver then submerges and repeats the search in the new zone. Once the width of the river has been searched, the diver is recovered and rests or changes out while the rope bridge with its anchor points is repositioned downstream for the next phase of the search (Figure 17-23).

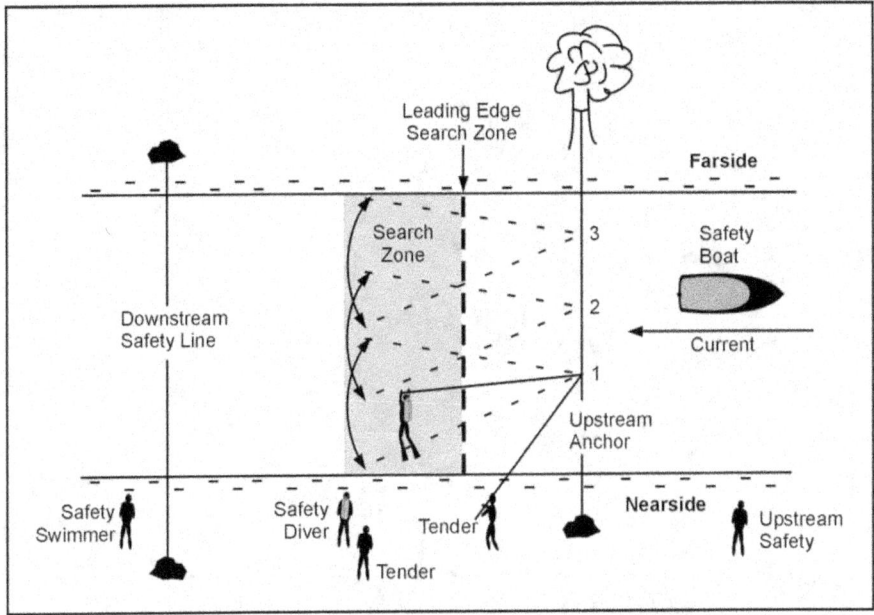

Figure 17-23. River search procedures using a rope bridge

17-40. If the river is too large to use a rope bridge for the upstream control measure, the tether or diver must be anchored or managed from a boat (Figure 17-24, page 17-20). The boat must use two anchor systems of sufficient weight and scope to prevent dragging. These anchors are deployed upstream to left or port and right or starboard of the bow so that when sufficient line (scope) is paid out, the boat is stable and centered between and downstream of them. As each search zone or sweep is completed, the boat leapfrogs the anchors traversing across the river until the entire search zone is covered. Boat personnel should always keep the lead anchor set and bound the trail anchor forward to shift the search zone.

17-41. The safety diver and his tender are on continuous standby at the edge of the search area. The safety diver should be rigged similarly to the search diver with the addition of an extra air source (extra air cylinder with attached regulator) secured to the safety diver's harness, turned on, and pressurized before entering the water. If an intervention is required, the search diver's tender notifies the CDS, who commits the safety diver. Before the safety diver enters the water, he clips onto the search diver's tether with a safety line. The safety diver then follows the search diver's tether until he makes contact with the search diver. The safety diver can then provide an additional air source to the stricken diver while working to ascertain the problem and resolve it. If the safety diver must leave the stricken diver and return to the surface, he leaves the additional air source with the diver. Detachments must practice interventions, including training scenarios for out-of-air emergencies, entanglements, and injuries.

Chapter 17

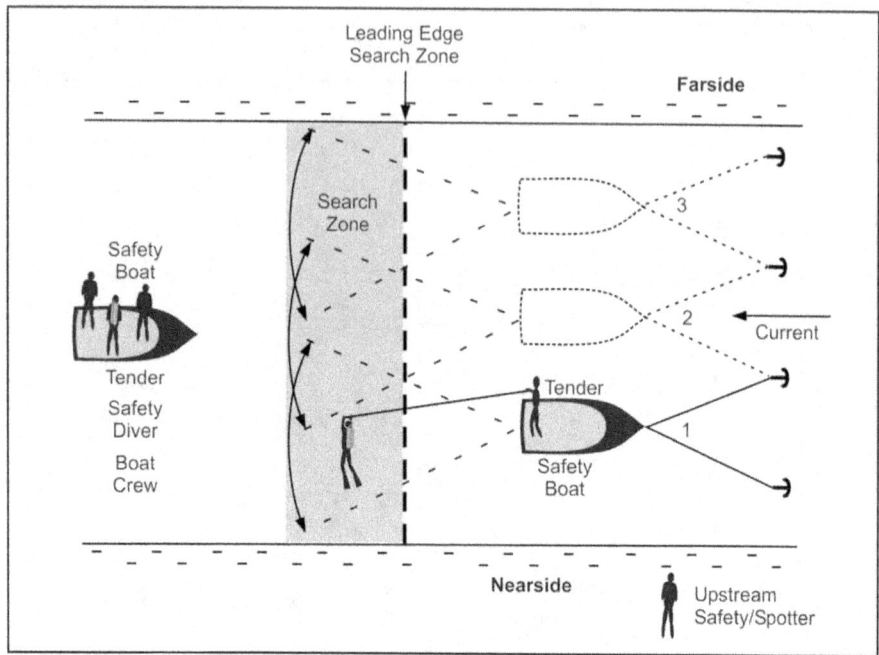

Figure 17-24. River search procedures using a boat

17-42. If divers are operating as buddy teams, the tethered search may not be practical. Alternate methods of controlling the divers must be considered, depending on the current, visibility, and experience levels of the involved divers. One alternative is to modify the circle-line search procedures using similar buoy, anchor, and search line setups. The divers enter the water, assemble at the buoy, descend to the anchor (clump), and start sweeping back and forth in an arc on the downstream side of the anchor. At each extreme of the arc, the divers move the limit of visibility further out (down) the search line and repeat the sweep. Divers control their own movements and turning points. This technique is not as precise as a surface-controlled tended-line search, nor does it provide the direct contact with the surface and the tender or CDS. If the divers should suffer a mishap and release the line, it may be difficult to locate them underwater.

Appendix A
Weights, Measures, and Conversion Tables

Tables A-1 through A-5, pages A-1 and A-2, show metric units and their U.S. equivalents. Tables A-6 through A-15, pages A-2 through A-5, are conversion tables.

Table A-1. Linear measure

Unit	Other Metric Equivalent	U.S. Equivalent
1 centimeter	10 millimeters	0.39 inch
1 decimeter	10 centimeters	3.94 inches
1 meter	10 decimeters	39.37 inches
1 decameter	10 meters	32.81 feet
1 hectometer	10 decameters	328.08 feet
1 kilometer	10 hectometers	3,280.84 feet

Table A-2. Liquid measure

Unit	Other Metric Equivalent	U.S. Equivalent
1 centiliter	10 milliliters	0.34 fluid ounce
1 deciliter	10 centiliters	3.38 fluid ounces
1 liter	10 deciliters	33.81 fluid ounces
1 decaliter	10 liters	2.64 gallons
1 hectoliter	10 decaliters	26.42 gallons
1 kiloliter	10 hectoliters	264.17 gallons

Table A-3. Weight

Unit	Other Metric Equivalent	U.S. Equivalent
1 centigram	10 milligrams	0.15 grain
1 decigram	10 centigrams	1.54 grains
1 gram	10 decigrams	0.04 ounce
1 decagram	10 grams	0.35 ounce
1 hectogram	10 decagrams	3.53 ounces
1 kilogram	10 hectograms	2.20 pounds
1 quintal	100 kilograms	220.46 pounds
1 metric ton	10 quintals	1.10 short tons

Table A-4. Square measure

Unit	Other Metric Equivalent	U.S. Equivalent
1 square centimeter	100 square millimeters	0.16 square inch
1 square decimeter	100 square centimeters	15.50 square inches
1 square meter (centaur)	100 square decimeters	10.76 square feet
1 square decameter (are)	100 square meters	1,076.39 square feet
1 square hectometer (hectare)	100 square decameters	2.47 acres
1 square kilometer	100 square hectometers	0.39 square mile

Table A-5. Cubic measure

Unit	Other Metric Equivalent	U.S. Equivalent
1 cubic centimeter	1,000 cubic millimeters	0.06 cubic inch
1 cubic decimeter	1,000 cubic centimeters	61.02 cubic inches
1 cubic meter	1,000 cubic decimeters	35.31 cubic feet

Table A-6. Temperature

Conversion	Formula
Fahrenheit to Celsius	Subtract 32, multiply by 5, and divide by 9
Celsius to Fahrenheit	Multiply by 9, divide by 5, and add 32

Table A-7. Approximate conversion factors

To Change	To	Multiply By	To Change	To	Multiply By
Inches	Centimeters	2.540	Ounce-inches	Newton-meters	0.007
Feet	Meters	0.305	Centimeters	Inches	.394
Yards	Meters	0.914	Meters	Feet	3.280
Miles	Kilometers	1.609	Meters	Yards	1.094
Square inches	Square centimeters	6.451	Kilometers	Miles	0.621
Square feet	Square meters	0.093	Square centimeters	Square inches	0.155
Square yards	Square meters	0.836	Square meters	Square feet	10.764
Square miles	Square kilometers	2.590	Square meters	Square yards	1.196
Acres	Square hectometers	0.405	Square kilometers	Square miles	0.386
Cubic feet	Cubic meters	0.028	Square hectometers	Acres	2.471
Cubic yards	Cubic meters	0.765	Cubic meters	Cubic feet	35.315
Fluid ounces	Millimeters	29.573	Cubic meters	Cubic yards	1.308
Pints	Liters	0.473	Millimeters	Fluid ounces	0.034
Quarts	Liters	0.946	Liters	Pints	2.113
Gallons	Liters	3.785	Liters	Quarts	1.057
Ounces	Grams	28.349	Liters	Gallons	0.264
Pounds	Kilograms	0.454	Grams	Ounces	0.035
Short tons	Metric tons	0.907	Kilograms	Pounds	2.205
Pounds-feet	Newton-meters	1.356	Metric tons	Short tons	1.102
Pounds-inches	Newton-meters	0.113	Nautical miles	Kilometers	1.852

Table A-8. Area

To Change	To	Multiply By	To Change	To	Multiply By
Square millimeters	Square inches	0.002	Square inches	Square millimeters	645.160
Square centimeters	Square inches	9.155	Square inches	Square centimeters	6.451
Square meters	Square inches	1,550.000	Square inches	Square meters	0.001
Square meters	Square feet	10.764	Square feet	Square meters	0.093
Square meters	Square yards	1.196	Square yards	Square meters	0.836
Square kilometers	Square miles	0.386	Square miles	Square kilometers	2.590

Table A-9. Volume

To Change	To	Multiply By	To Change	To	Multiply By
Cubic centimeters	Cubic inches	0.061	Cubic inches	Cubic centimeters	16.390
Cubic meters	Cubic feet	35.310	Cubic feet	Cubic meters	0.028
Cubic meters	Cubic yards	1.308	Cubic yards	Cubic meters	0.765
Liters	Cubic inches	61.020	Cubic inches	Liters	0.016
Liters	Cubic feet	0.035	Cubic feet	Liters	28.320

Table A-10. Capacity

To Change	To	Multiply By	To Change	To	Multiply By
Milliliters	Fluid drams	0.271	Fluid drams	Milliliters	3.697
Milliliters	Fluid ounces	0.034	Fluid ounces	Milliliters	29.573
Liters	Fluid ounces	33.814	Fluid ounces	Liters	0.030
Liters	Pints	2.113	Pints	Liters	0.473
Liters	Quarts	1.057	Quarts	Liters	0.946
Liters	Gallons	0.264	Liters	Gallons	3.785

Table A-11. Statute miles to kilometers and nautical miles

Statute Miles	Kilometers	Nautical Miles	Statute Miles	Kilometers	Nautical Miles
1	1.61	0.87	60	96.60	52.16
2	3.22	1.74	70	112.70	60.85
3	4.83	2.61	80	128.80	69.55
4	6.44	3.48	90	144.90	78.24
5	8.05	4.35	100	161.00	86.93
6	9.66	5.22	200	322.00	173.90
7	11.27	6.08	300	483.00	260.80
8	12.88	6.95	400	644.00	347.70
9	14.49	7.82	500	805.00	434.70
10	16.10	8.69	600	966.00	521.60
20	32.20	17.39	700	1127.00	608.50
30	48.30	26.08	800	1288.00	695.50
40	64.40	34.77	900	1449.00	782.40
50	80.50	43.47	1000	1610.00	869.30

Appendix A

Table A-12. Nautical miles to kilometers and statute miles

Nautical Miles	Kilometers	Statute Miles	Nautical Miles	Kilometers	Statute Miles
1	1.85	1.15	60	111.00	69.00
2	3.70	2.30	70	129.50	80.50
3	5.55	3.45	80	148.00	92.00
4	7.40	4.60	90	166.50	103.50
5	9.25	5.75	100	185.00	115.00
6	11.10	6.90	200	370.00	230.00
7	12.95	8.05	300	555.00	345.00
8	14.80	9.20	400	740.00	460.00
9	16.65	10.35	500	925.00	575.00
10	18.50	11.50	600	1110.00	690.00
20	37.00	23.00	700	1295.00	805.00
30	55.50	34.50	800	1480.00	920.00
40	74.00	46.00	900	1665.00	1035.00
50	92.50	57.50	1000	1850.00	1150.00

Table A-13. Kilometers to statute and nautical miles

Kilometers	Statute Miles	Nautical Miles	Kilometers	Statute Miles	Nautical Miles
1	0.62	0.54	60	37.28	32.40
2	1.24	1.08	70	43.50	37.80
3	1.86	1.62	80	49.71	43.20
4	2.49	2.16	90	55.93	48.60
5	3.11	2.70	100	62.14	54.00
6	3.73	3.24	200	124.28	108.00
7	4.35	3.78	300	186.42	162.00
8	4.97	4.32	400	248.56	216.00
9	5.59	4.86	500	310.70	270.00
10	6.21	5.40	600	372.84	324.00
20	12.43	10.80	700	435.00	378.00
30	18.64	16.20	800	497.12	432.00
40	24.85	21.60	900	559.26	486.00
50	31.07	27.00	1000	621.40	540.00

Table A-14. Yards to meters

Yards	Meters	Yards	Meters	Yards	Meters
100	91	1000	914	1900	1737
200	183	1100	1006	2000	1829
300	274	1200	1097	3000	2743
400	366	1300	1189	4000	3658
500	457	1400	1280	5000	4572
600	549	1500	1372	6000	5486
700	640	1600	1463	7000	6401
800	732	1700	1554	8000	7315
900	823	1800	1646	9000	8230

Table A-15. Meters to yards

Meters	Yards	Meters	Yards	Meters	Yards
100	109	1000	1094	1900	2078
200	219	1100	1203	2000	2187
300	328	1200	1312	3000	3281
400	437	1300	1422	4000	4374
500	547	1400	1531	5000	5468
600	656	1500	1640	6000	6562
700	766	1600	1750	7000	7655
800	875	1700	1859	8000	8749
900	984	1800	1969	9000	9843

This page intentionally left blank.

Appendix B
Description of Subject Area and Critical Task List

Detachment members must know the following critical tasks. Each subject area listed in the Training Frequency Matrix contained in Chapter 1 is outlined below. Each section is keyed to the matrix and lists the critical tasks that must be accomplished during training. The detachment can use these tasks to help design a MAROPS-specific training plan or the commander may use them to evaluate the detachment's training status. All of these critical tasks are initially trained as a part of the program of instructions of various courses of instruction conducted at the Special Forces Underwater Operations School, Naval Air Station, Key West, Florida.

COMBAT DIVER (51 Tasks, 1 Skill, and 5 Knowledges)

B-1. Module A: General Subjects (1 Task, 1 Skill)
- Objective: Use diving forms and regulations.
- Individual Task: 331-CDS-3402: Employ Diving Forms and Regulations.
- Individual Skill: Red Cross BLS: Perform Cardiopulmonary Resuscitation (CPR).

B-2. Module B: Waterborne Operations (10 Tasks)
- Objective: Complete waterborne operations and a buoyant ascent.
- Individual Tasks:
  - 331-CDQ-3241: Complete a Free-Swimming Ascent From Depth Without an Air Source.
  - 331-CDQ-3219: Complete an Underwater Search.
  - 331-CDQ-3220: Complete a Ship Bottom Search.
  - 331-CDQ-3233: Determine an Individual Pace Count for Underwater Operations.
  - 331-CDQ-3236: Complete a Submarine Lockout/Lockin Drill.
  - 331-CDQ-3237: Calculate the Effects of Ocean Tides, Currents, and Waves on Waterborne Operations.
  - 331-CDQ-3238: Complete a Closed-Circuit Swim as a Member of a Swim Team.
  - 331-CDQ-3245: Maintain Open-Circuit Diving Equipment.
  - 331-CDQ-3246: Waterproof Bundles and Combat Gear for Waterborne Operations.
  - 331-WIC-0004: Complete Waterborne Premission Planning.

B-3. Module C: Pool Training (8 Tasks)
- Objective: Demonstrate confidence in an underwater environment, identify the proper diving procedures and safety requirements, and use and maintain individual diving equipment.
- Individual Tasks:
  - 331-CDQ-3206: Complete Selected Stress Exercises (SFUWO School ONLY).
  - 331-CDQ-3207: Tie Selected Knots Underwater.
  - 331-CDQ-3208: Employ Water Entry Techniques.
  - 331-CDQ-3210: Employ Buddy Breathing Techniques Using Open-Circuit Scuba.
  - 331-CDQ-3215: Employ Underwater Hand Signals.
  - 331-CDQ-3239: Employ Individual Swim Gear.

Appendix B

- 331-CDQ-3241: Complete a Free-Swimming Ascent From Depth Without an Air Source.
- 331-CDQ-3244: React to Emergencies During Open- and Closed-Circuit Diving Operations.

B-4. Module D: Open-Circuit Diving (11 Tasks)
- Objective: Complete open-circuit diving operations and underwater navigation.
- Individual Tasks:
  - 331-CDQ-3206: Complete Stress Exercises.
  - 331-CDQ-3208: Employ Water Entry Techniques.
  - 331-CDQ-3209: Clear a Regulator.
  - 331-CDQ-3213: Don Open-Circuit Scuba Equipment for Diving Operations.
  - 331-CDQ-3215: Employ Underwater Hand Signals.
  - 331-CDQ-3221: Navigate Underwater to a Designated Point.
  - 331-CDQ-3240: Ditch and Don Open-Circuit Equipment Underwater.
  - 331-CDQ-3242: Complete a Deep Dive.
  - 331-CDQ-3244: React to Emergencies During Open- and Closed-Circuit Diving Operations.
  - 331-CDQ-3245: Maintain Open-Circuit Diving Equipment.
  - 331-CDS-3415: Employ High-Pressure Compressed Gas Systems to Charge Divers' Supply Cylinders.

B-5. Module E: Closed-Circuit Diving (6 Tasks)
- Objective: Complete closed-circuit diving operations.
- Individual Tasks:
  - 331-CDQ-0001: Buddy Breathe with Closed-Circuit Equipment.
  - 331-CDQ-3208: Employ Water Entry Techniques.
  - 331-CDQ-3215: Employ Underwater Hand Signals.
  - 331-CDQ-3221: Navigate Underwater to a Designated Point.
  - 331-CDQ-3238: Complete a Closed-Circuit Swim as a Member of a Swim Team.
  - 331-CDQ-3243: Perform Pre and Postdive Procedures on a Closed-Circuit Oxygen Rebreather.

B-6. Module F: Diving Physics, Physiology, Injuries, and Dangerous Marine Life (1 Task, 5 Knowledges)
- Objective: Integrate the principles of physics into diving operations and identify and classify dangerous marine life, the physiology of diving, and the treatment of respiratory- and pressure-related diving injuries.
- Individual Task: 331-CDQ-3234: Administer First Aid to Diving Injuries.
- Individual Knowledges:
  - 331-2K1-4022: Identify Anatomy and Physiology of the Respiratory System.
  - 331-2K1-4023: Identify Anatomy and Physiology of the Cardiovascular System.
  - 331-2K2-3226: Integrate the Principles of Physics Into Diving Operations.
  - 331-2K2-3231: Identify Dangerous Marine Life.
  - 331-2K2-3430: Determine the Effects of Increased Pressure on the Human Body.

B-7. Module G: U.S. Navy Diving Tables (2 Tasks)
- Objective: Apply U.S. Navy no-decompression diving tables to combat diving operations. Apply altitude diving tables to diving operations above 1000 feet above sea level, and recognize the maladies associated with altitude diving.
- Individual Tasks:
  - 331-CDQ-3227: Employ U.S. Navy Decompression Tables.
  - 331-CDQ-3235: Employ Altitude Dive Tables.

B-8. Module H: Small Boat Operations (5 Tasks)
- Objective: Employ small boat operations.
- Individual Tasks:
  - 331-CDQ-0003: Complete Small Boat Operations.
  - 331-CDQ-0005: Complete a Special Operations Combat Expendable Platform (SOCEP) for Airborne Infiltration.
  - 331-CDQ-0006: Complete Waterborne Operations Using Helicopters.
  - 331-CDQ-0007: Complete a Tactical Water Airborne Operation.
  - 331-CDQ-0008: Navigate a Small Boat.

B-9. Module I: Surface Infiltration (6 tasks).
- Objective: Conduct long-distance surface infiltration, waterproof bundles and combat equipment, perform reconnaissance and security of a beach landing site, and conduct cache activities.
- Individual Tasks:
  - 331-CDQ-3208: Employ Water Entry Techniques.
  - 331-CDQ-3218: Complete a Long-Distance Surface Infiltration as a Member of a Swim Team.
  - 331-CDQ-3221: Navigate Underwater to a Designated Point.
  - 331-CDQ-3239: Employ Individual Swim Gear.
  - 331-CDQ-3246: Waterproof Bundles and Combat Gear for Waterborne Operations.
  - 331-CDQ-0010: Complete Beach Landing Site (BLS) Operations.

COMBAT DIVE SUPERVISOR (27 Tasks, 5 Knowledges)

B-10. Module B: Combat Diving Operations (9 Tasks)
- Objective: Plan and supervise vertical and horizontal dives, apply appropriate diving regulations, brief a dive plan, and conduct a boat inspection. Conduct predive and postdive personnel inspection of closed-circuit and open-circuit equipment.
- Individual Tasks:
  - 331-CDS-3400: Complete a Combat Diving Supervisor's Personnel and Equipment Inspection.
  - 331-CDS-3401: Supervise Combat Diving Operations.
  - 331-CDS-3402: Employ Diving Forms and Regulations.
  - 331-CDS-3403: Complete a Boat Inspection.
  - 331-CDS-3404: Plan Diving Operations.
  - 331-CDS-3405: Brief a Dive Plan.
  - 331-CDS-3407: Complete a Tabletop Inspection of a Closed-Circuit Oxygen Rebreather.
  - 331-CDS-3414: Supervise Operator-Level Maintenance on Selected Individual Diving Equipment.
  - 331-DMT-3603: Compute the Minimum Safe Distance for an Underwater Blast.

B-11. Module C: Tides and Currents (3 Tasks)
- Objective: Analyze tide and current data, identify wind drift currents, determine water depth, and determine tidal current set and drift.
- Individual Tasks:
  - 31-CDQ-3237: Calculate the Effects of Ocean Tides, Currents, and Waves on Waterborne Operations.
  - 331-CDS-3408: Employ Nautical Charts and Navigational Aids.
  - 331-CDQ-0008: Navigate a Small Boat.

Appendix B

B-12. Module D: Submarine Operations (1 Knowledge)
- Objective: Plan and supervise submarine operations.
- Individual Knowledge: 331-2K2-3409: Determine Diving Supervisor Responsibilities During Submarine Operations.

B-13. Module E: Diving Operations (5 Tasks)
- Objective: Conduct diving supervisor personnel inspection; conduct predive and postdive personnel inspection of closed- and open-circuit; employ altitude dive tables in planning an altitude dive.
- Individual Tasks:
    - 331-CDQ-3235: Employ Altitude Dive Tables.
    - 331-CDS-3400: Complete a Combat Diving Supervisor's Personnel and Equipment Inspection.
    - 331-CDS-3402: Employ Diving Forms and Regulations.
    - 331-CDS-3407: Complete a Tabletop Inspection of a Closed-Circuit Oxygen Rebreather.
    - 331-CDS-3414: Supervise Operator-Level Maintenance of Selected Individual Diving Equipment.

B-14. Module F: Diving Equipment (4 Tasks)
- Objective: Inspect, maintain, and field-repair open and closed-circuit diving equipment and inspect a closed-circuit rebreather for functionality; Drive and maintain a Diver Propulsion Device (DPD).
- Individual Tasks:
    - 331-CDS-3400: Complete a Combat Diving Supervisor's Personnel and Equipment Inspection.
    - 331-CDS-3406: Operate a Diver Propulsion Device (DPD).
    - 331-CDS-3407: Complete a Tabletop Inspection of a Closed-Circuit Oxygen Rebreather.
    - 331-CDS-3414: Supervise Operator-Level Maintenance of Selected Individual Diving Equipment.

B-15. Module G: Medical Aspect of Diving (1 Task, 2 Knowledges)
- Objective: Identify the effects of increased pressure on the human body, manage stress as it relates to diving operations, recognize the signs and symptoms of oxygen toxicity and take appropriate action during diving emergencies.
- Individual Task: 331-CDQ-3234: Administer First Aid to Diving Injuries.
- Individual Knowledges:
    - 331-2K2-3430: Determine the Effects of Increased Pressure on the Human Body.
    - 331-2K2-3232: Identify the Hazards of Diving in Polluted Waters.

B-16. Module H: Recompression Chamber Operations (5 Tasks, 1 Knowledge)
- Objective: Use U.S. Navy treatment tables during hyperbaric chamber operation and supervise chamber operation, use an air compressor and air bank, fill high pressure air and oxygen cylinders, calculate useable gas volume, and use a mobile compressed air and oxygen cascade system.
- Individual Tasks:
    - 331-CDQ-3227: Employ U.S. Navy Decompression Tables.
    - 331-CDS-3415: Employ High-Pressure Compressed Gas Systems to Charge Divers' Supply Cylinders.
    - 331-CDS-3417: Supervise Hyperbaric Chamber Operations.
    - 331-DMT-3602: Complete a 165-Foot Hyperbaric Chamber Dive.

Description of Subject Area and Critical Task List

- 331-DMT-3616: Employ the U.S. Navy Dive Treatment Tables During Treatment of a Combat Diving Casualty.
- Individual Knowledge: 331-2K2-3634: Determine Hyperbaric Chamber Operational Procedures.

B-17. Module I: Diving Physics (1 Knowledge)
- Objective: Apply physics to diving operations, employ U.S. Navy decompression tables, and conduct altitude dive planning.
- Individual Knowledge: 331-2K2-3226: Integrate the Principles of Physics Into Diving Operations.

DIVING MEDICAL TECHNICIAN (39 Tasks, 1 Skill, and 11 Knowledges)

B-18. Module B: Medical Planning for Diving Operations (2 Tasks)
- Objective: Plan required medical support for combat diving operations.
- Individual Tasks:
  - 331-DMT-3600: Determine a Combat Diver's Fitness for Diving Duty.
  - 331-DMT-3601: Plan Medical Support of Diving Operations.

B-19. Module C: Diving Physiology (0 Tasks, 3 Knowledges)
- Objective: Identify the anatomy and physiology of the cardiopulmonary and neurological systems, and complete a physical and neurological examination of a diving casualty.
- Individual Knowledges:
  - 331-2K1-4022: Identify Basic Anatomy and Physiology of the Cardiopulmonary System.
  - 331-2K1-4023: Identify Basic Anatomy and Physiology of the Neurological System.
  - 331-2K1-4027: Identify Anatomy and Physiology of the Nervous System.

B-20. Module D: Altitude Diving (3 Tasks)
- Objective: Identify and treat altitude diving maladies.
- Individual Tasks:
  - 331-CDQ-3235: Employ Altitude Dive Tables.
  - 331-DMT-3601: Plan Medical Support for Altitude Diving.
  - 331-DMT-3619: Treat Maladies Associated With Altitude Diving.

B-21. Module E: Diving Physics (1 Knowledge)
- Objective: Apply physics to diving operations, and employ U.S. Navy decompression tables.
- Individual Knowledge: 331-2K2-3226: Integrate the Principles of Physics Into Diving Operations.

B-22. Module F: Diagnosis and Treatment of Diving Injuries (25 Tasks, 5 Knowledges, 1 Skill)
- Objective: Diagnose and treat diving-related pressure injuries, and perform scuba life-saving and accident management.
- Individual Tasks:
  - 331-CDQ-3227: Employ U. S. Navy Decompression Tables.
  - 331-CDQ-3244: React to Emergencies During Open- and Closed-Circuit Diving Operations.
  - 331-DMT-3603: Compute the Minimum Safe Distance for an Underwater Blast.
  - 331-DMT-3600: Determine a Combat Diver's Fitness for Diving Duty.
  - 331-DMT-3604: Complete a Differential Diagnosis of a Diving Casualty.
  - 331-DMT-3605: Treat a Diver With Barotrauma.
  - 331-DMT-3607: Treat a Diver for Pulmonary Overinflation Injuries.
  - 331-DMT-3608: Treat a Combat Diver for Decompression Sickness.
  - 331-DMT-3610: Complete a Neurological Examination of a Combat Diving Casualty.
  - 331-DMT-3612: Treat a Diver With Immersion Hypothermia.
  - 331-DMT-3614: Administer First Aid for Aquatic Skin Diseases.

Appendix B

- 331-DMT-3615: Treat Underwater Blast and Sonar Injuries.
- 331-DMT-3616: Employ the U. S. Navy Treatment Tables During Treatment of a Combat Diving Casualty.
- 331-DMT-3619: Treat Maladies Associated With Altitude Diving.
- 331-DMT-3620: Treat Injuries Inflicted by Dangerous Marine Life.
- 331-DMT-3624: Complete a Physical Examination of a Combat Diver.
- 331-DMT-3625: Treat a Diver With Carbon Monoxide Poisoning.
- 331-DMT-3626: Treat a Diving Accident Victim.
- 331-DMT-3627: Insert an Endotracheal Tube Airway in a Diving Casualty.
- 331-DMT-3628: Operate as the Inside Tender During Hyperbaric Chamber Operations.
- 331-DMT-3629: Complete Invasive Procedures in a Hyperbaric Environment.
- 331-DMT-3630: Treat a Combat Diver for Oxygen Toxicity and Hypoxia.
- 331-DMT-3631: Treat a Drowning or Near-Drowning Victim.
- 331-DMT-3632: Implement Adjunctive (Pharmacological) Therapy on the Treatment of Diving Injuries.
- 331-DMT-3633: Employ a Portable Hyperbaric Chamber to Transport a Diving Casualty.
- Individual Knowledges:
  - 332-2K1-4022: Anatomy and Physiology of the Respiratory System.
  - 332-2K1-4023: Anatomy and Physiology of the Cardiovascular System.
  - 331-2K2-3430: Determine the Effects of Increased Pressure on the Human Body.
  - 331-2K2-3613: Identify the Effects of Inert Gas (Nitrogen) and High Pressure Nervous Syndrome (HPNS).
  - 331-2K2-3634: Determine Hyperbaric Chamber Operational Procedures.
- Individual Skill: 331-2S1-4016: Conduct Water Rescues.

B-23. Module G: Dangerous Marine Life (1 Task, 1 Knowledge)
- Objective: Identify and classify dangerous marine life, and apply first aid for injuries inflicted by dangerous marine life.
- Individual Task: 331-DMT-3620: Treat Injuries Inflicted by Dangerous Marine Life.
- Individual Knowledge: 331-2K2-3231: Identify Dangerous Marine Life.

B-24. Module H: Recompression Chamber Operations (8 Tasks, 1 Knowledge)
- Objective: Use U.S. Navy treatment tables during hyperbaric chamber operations and perform duties of inside tender.
- Individual Tasks:
  - 331-CDQ-3227: Employ U. S. Navy Decompression Tables.
  - 331-DMT-3602: Complete a 165-Foot Hyperbaric Chamber Dive.
  - 331-DMT-3610: Complete a Neurological Examination of a Combat Diving Casualty.
  - 331-DMT-3616: Employ the U.S. Navy Dive Treatment Tables During Treatment of a Combat Diving Casualty.
  - 331-DMT-3627: Insert an Endotracheal Tube Airway in a Diving Casualty.
  - 331-DMT-3628: Operate as the Inside Tender During Hyperbaric Chamber Operations.
  - 331-DMT-3629: Complete Invasive Procedures in a Hyperbaric Environment.
  - 331-DMT-3632: Implement Adjunctive (Pharmacological) Therapy in the Treatment of Diving Injuries.
- Individual Knowledge: 331-2K2-3634: Determine Hyperbaric Chamber Operational Procedures.

Description of Subject Area and Critical Task List

WATERBORNE INFILTRATION (20 Tasks, 7 Knowledges)

B-25. Module A: General Subjects (1 Task)
- Objective: Use the swim fins and face masks.
- Individual Task: 331-CDQ-3239: Employ Individual Swim Gear.

B-26. Module B: Mission Planning (1 Task)
- Objective: Conduct waterborne infiltration mission planning; select intermediate and terminal delivery systems; select infiltration and exfiltration sites; select debarkation point, launch point, and beach landing sites; prepare a navigation plan; and issue a waterborne mission operations order.
- Individual Task: 331-WIC-3801: Complete Waterborne Premission Planning.

B-27. Module C: Small Boat Operations (5 Tasks, 1 Knowledge)
- Objective: Prepare small craft (CRRCs) for operations; rig, load equipment, and perform maintenance; inspect, maintain, calculate fuel consumption and operate outboard motors; conduct mother ship operations; complete cast and recovery of surface swimmers from an intermediate delivery system; conduct a tactical movement using small boats; and complete a small boat waterborne infiltration.
- Individual Tasks:
  - 331-WIC-3800: Complete Small Boat Operations.
  - 331-WIC-3804: Complete a Special Operations Combat Expendable Platform (SOCEP) Airborne Infiltration.
  - 331-WIC-3806: Complete Mother Ship Operations.
  - 331-WIC-3807: Complete a Tactical Movement With a Small Boat.
  - 331-WIC-3818: Navigate a Small Boat.
- Individual Knowledge: 331-2K2-3819: Determine Weather Conditions in Waterborne Operations.

B-28. Module D: Surface Infiltration (3 Tasks)
- Objective: Conduct long-distance surface infiltration, waterproof bundles and combat equipment, perform reconnaissance and security of a beach landing site, and conduct cache activities.
- Individual Tasks:
  - 331-CDQ-3218: Complete a Long-Distance Surface Infiltration as a Member of a Swim Team.
  - 331-CDQ-3239: Employ Individual Swim Gear.
  - 331-CDQ-3246: Waterproof Bundles and Combat Gear for Waterborne Operations.

B-29. Module E: Air Operations (3 Tasks)
- Objective: Prepare/rig personnel, their equipment, and CRRCs for helicopter operations, including the launch and recovery of surface swimmers, limp duck, and K-duck; rig surface swimmers for an airborne infiltration; and prepare/rig small boats for SOCEP delivery.
- Individual Tasks:
  - 331-WIC-3804: Complete a Special Operations Combat Expendable Platform (SOCEP) Airborne Infiltration.
  - 331-WIC-3811: Complete Waterborne Operations Using Helicopters.
  - 331-WIC-3816: Complete a Tactical Water Airborne Operation.

B-30. Module F: Nautical Charts and Navigation (2 Tasks, 1 Knowledge)
- Objective: Navigate a small craft between two known points using a navigation plan prepared during mission planning; locate a position on a nautical chart, describe lines of longitude and latitude, and determine direction and distance between two points; employ nautical chart

Appendix B

information and piloting reference materials; identify wave types and weather formations; and use electronic navigation aids.
- Individual Tasks:
  - 331-CDQ-3237: Calculate the Effects of Ocean Tides, Currents, and Waves on Waterborne Operations.
  - 331-WIC-3818: Navigate Small Boat.
- Individual Knowledge: 331-2K2-3819: Determine the Weather Considerations for Waterborne Operations.

B-31. Module G: Kayak Operations (3 Tasks, 2 Knowledges)
- Objective: Prepare, assemble, and disassemble a kayak; perform PMCS on the kayak; employ a kayak on ocean and inland waterways; portage and cache a kayak; perform capsize drill and storm procedures; and complete tactical movement and beach landing procedures.
- Individual Tasks:
  - 331-CDQ-3237: Calculate the Effects of Ocean Tides, Currents, and Waves on Waterborne Operations.
  - 331-WIC-3818: Navigate a Small Boat.
  - 331-WIC-3820: Complete Kayak Operations.
- Individual Knowledges:
  - 331-2K2-3819: Determine Weather Considerations for Waterborne Operations.
  - 331-2K2-3824: Identify the Characteristics of a Kayak.

B-32. Module H: Medical Aspects of Waterborne Infiltration (3 Knowledges)
- Objective: Identify the various forms of marine life that can present a hazard to personnel and other common water injuries, to include heat and cold injuries, underwater blast injuries, drowning, and ear problems. Determine and apply appropriate first aid for the injuries.
- Individual Knowledges:
  - 331-2K2-3826: Determine the Medical and Survival Considerations Applicable in Waterborne Operations.
  - 331-2K2-3231: Identify Dangerous Marine Life.
  - 331-2K2-3232: Identify the Hazards of Diving in Contaminated Waters.

B-33. Module I: Submarine Operations (2 Tasks)
- Objective: Use a submarine as a launch platform for surface swimming, small boat, or kayak waterborne infiltration including wet-deck and dry-deck operations.
- Individual Tasks:
  - 331-WIC-3800: Complete Small Boat Operations.
  - 331-WIC-3827: Employ a Submarine as an Infiltration Platform.

CRITICAL TASK LIST

Cache Closed-Circuit Equipment Underwater 331-CDQ-0002

Condition: As a member of a waterborne infiltration swim/dive team, given—
  1. An open water training area NTE 15 feet in depth.
  2. A beach landing site (BLS).
  3. Closed-circuit diving equipment.
  4. A hypothetical combat situation.
  5. The requirement to cache closed-circuit equipment underwater.

Standard: Cache closed-circuit equipment in the selected underwater site.

Description of Subject Area and Critical Task List

Complete Selected Stress Exercises
331-CDQ-3206

Condition: As a combat diver undergoing stress exercises, given—
1. All open-circuit scuba gear with face mask, swim fins, and 16-pound weight belt.
2. An outdoor or indoor Olympic-size swimming pool.
3. A blacked-out face mask.
4. Specially fabricated Velcro hand and leg restraints.
5. The requirement to successfully complete selected stress exercises.

Standard: To increase or maintain confidence in diving/operating in an underwater environment, successfully complete selected stress exercises consisting of a—
1. Scuba gear exchange.
2. One-man confidence swim.
3. Two-man confidence swim.
4. Weight-belt swim.
5. Drown proofing proficiency examination.

Buddy Breathe With Closed-Circuit Equipment
331-CDQ-0001

Condition: As an SF combat diver buddy team member—
1. In an outdoor swimming pool or ocean training area.
2. Given an MK 25 MOD 2 Draeger LAR V UBA.
3. And one buddy member who has run out of oxygen/air.

Standard: Complete buddy breathing procedures to safely ascend to the surface using the—
1. MK25 MOD 2 Draeger LAR V UBA.
2. Face-to-face method.
3. Swimming method.
4. Cadillac method.

Tie Selected Knots Underwater
331-CDQ-3207

Condition: As a combat diver, given—
1. Surface swim gear.
2. A requirement to use ropes and knots to rig, recover personnel or equipment from underwater, or construct underwater demolition systems.
3. An underwater object or rope on which to secure the knot.
4. Two sections of rope adequate to tie the selected knot.

Standard: Tie the following knots underwater at a depth NTE 12 feet. Each knot or combination of knots will be tied while making a single free dive from the surface in the following sequence: (Only one breath of air may be used for each sequence.)
1. Square knot.
2. Girth hitch with extra turn.
3. Bowline.
4. Girth hitch with extra turn and bowline.
5. Girth hitch with extra turn, bowline, and square knot.

Appendix B

Employ Water Entry Techniques
331-CDQ-3208

Condition: As a combat diver, given the requirement to enter the water wearing a complete set of open- or closed-circuit diving equipment.

Standard: IAW U.S. Navy diving regulations, enter the water using one of the following methods, as appropriate:
1. Front jump or step-in.
2. Rear roll.
3. Side roll.
4. Front roll.
5. Rear step-in.
6. Beach entry.

Clear a Regulator
331-CDQ-3209

Condition: As a combat diver, given—
1. A complete set of open-circuit diving equipment.
2. A requirement to clear a single-hose regulator of excess water (for example, flooded regulator).

Standard: IAW U.S. Navy diving regulations, perform the following:
1. Calmly locate and gain control of the regulator (if it has become dislodged from your mouth).
2. Replace the regulator into his mouth.
3. Purge the regulator.
4. Resume breathing normally.

Employ Buddy Breathing Techniques Using Open-Circuit Scuba
331-CDQ-3210

Condition: As a combat diver in a field environment, given—
1. Open-circuit scuba equipment.
2. The requirement to employ buddy breathing techniques using open-circuit scuba.

Standard: IAW U.S. Navy diving regulations, in a controlled environment or a diving emergency, perform the following buddy breathing techniques with the assistance of your buddy diver:
1. Face-to-face method.
2. Side-by-side swimming method.
3. The ascent method.

Don Open-Circuit Scuba Equipment for Diving Operations
331-CDQ-3213

Condition: As an SF combat diver in a field environment, given—
1. A diver's weight belt and wrist compass.
2. Swim fins, open-circuit scuba diving equipment set, face mask, diver's depth gauge, buoyancy compensator military (BCM), adjusting tool dive, and chronograph (diver's wristwatch).
3. Current U.S. Navy diving regulations.

Standard: IAW U.S. Navy diving regulations and with the aid of a diving buddy, inspect, prepare, and don the open-circuit equipment in proper sequence to prepare for a diving supervisor's inspection and diving operation.

Employ Underwater Hand Signals
331-CDQ-3215

Condition: As a combat diver, given—
1. The current U.S. Navy diving regulations.
2. A complete set of swim gear.
3. An underwater breathing apparatus (open-circuit or closed-circuit).
4. A requirement to conduct an underwater infiltration.
5. A requirement to communicate underwater with the combat diving supervisor or diving buddy.

Standard: Correctly give and respond to underwater hand signals IAW current U.S. Navy diving regulations.

Complete a Long-Distance Surface Infiltration as a Member of a Swim Team
331-CDQ-3218

Condition: As a combat swimmer during day or night, given—
1. A complete set of surface infiltration swim gear.
2. Environmental protection suit if required by the water conditions.
3. Mission-essential equipment and equipment lines with snap links, team swim rope, individual buddy lines, and snap links.
4. A water operational area with a beach landing site (BLS).
5. A requirement to conduct a surface infiltration swim.

Standard: Complete a team surface infiltration swim, with equipment, by—
1. Maintaining noise and light discipline, team integrity, and personnel and equipment accountability.
2. Using the line, column, or wedge formation.
3. Arriving at the BLS in condition to execute a follow-on mission.

Complete an Underwater Search
331-CDQ-3219

Condition: As a member of a waterborne search team, given—
1. Current U.S. Navy diving regulations.
2. FM 3-05.212, *Special Forces Waterborne Operation*, Chapter 17.
3. USASOC Reg 350-20, *USASOC Dive Program*, Chapter 4.
4. AR 611-75, *Management of Army Divers*, paragraph 2-20.
5. The necessary personnel and equipment.
6. An open-water operational area.
7. A requirement to conduct an underwater search.

Standard: As a member of a waterborne search team, complete an effective search to locate an object IAW the cited references.

Complete a Ship Bottom Search
331-CDQ-3220

Condition: As a member of a waterborne search team, given—
1. Current U.S. Navy diving regulations.
2. FM 3-05.212, *Special Forces Waterborne Operations*, paragraphs 17-3 through 17-30.
3. USASOC Reg 350-20, *USASOC Dive Program*, Chapter 4.
4. AR 611-75, *Management of Army Divers*, paragraph 2-20.
5. The necessary personnel and equipment.
6. An appropriate marine vessel.
7. A requirement to conduct a ship bottom search.

Standard: As a member of a waterborne search team, complete an effective ship bottom search IAW the cited references.

Appendix B

Navigate Underwater to a Designated Point
331-CDQ-3221

Condition: As a combat diver during day or night conditions, given—
1. Current U.S. Navy diving regulations.
2. FM 3-05.212, *Special Forces Waterborne Operations*, paragraphs 11-55 through 11-62.
3. USASOC Reg 350-20, *USASOC Dive Program*, Annex D.
4. AR 611-75, *Management of Army Divers*, paragraph 2-19.
5. A complete set of open- or closed-circuit diving equipment.
6. A wrist compass or swimmer attack board.
7. A body of open water with a minimum depth of 10 feet.
8. A requirement to navigate underwater to a designated point.

Standard: Navigate underwater to a designated point on land, or at sea, with a degree of accuracy NTE 125 meters left or right of the designated point, without exceeding the capabilities of your equipment.

Employ U.S. Navy Decompression Tables
331-CDQ-3227

Condition: As a combat diver, given—
1. Current U.S. Navy diving regulations.
2. FM 3-05.212, *Special Forces Waterborne Operations*.
3. Pencil and paper.

Standard: IAW the cited references, calculate the—
1. Maximum bottom time without decompression.
2. Minimum surface interval.
3. Residual nitrogen time.
4. Decompression stops.
5. Repetitive dives.
6. Minimum surface interval before flying for various dives.

Determine an Individual Pace Count for Underwater Operations
331-CDQ-3233

Condition: As a combat diver, given—
1. A complete set of open- or closed-circuit diving equipment.
2. All required safety and support personnel and equipment.
3. A 100-meter rope suspended horizontally in the water at a depth not to exceed 33 feet.
4. A stopwatch or a watch with a sweep second hand.
5. A diver's slate to record results.
6. An indoor or outdoor swimming pool, or an open-water operational area.

Standard: During daylight, swim a designated 100-meter course a minimum of three repetitions, recording the resulting times and kick counts, and averaging them to determine your underwater pace count. Swim at a normal pace for all iterations.

Administer First Aid to Diving Injuries
331-CDQ-3234

Condition: As an SF combat diver, diving supervisor, diving medical technician, or a waterborne infiltration operator, given—
1. Current U.S. Navy diving regulations.
2. A real or simulated patient exhibiting symptoms of a diving-related injury.

Standard: IAW the cited references, administer first aid to personnel with diving-related injuries.

Employ Altitude Dive Tables
331-CDQ-3235

Condition: As an SF waterborne team member, given—
1. Current U.S. Navy diving regulations.
2. A requirement to calculate the data for an altitude dive.

Standard: As an SF waterborne team member, complete the following tasks:
1. Discuss the effects of altitude on the human body.
2. Discuss the effects on the human body of the decreased partial pressure of oxygen at altitude.
3. Discuss the problems that may be encountered when going from sea level to altitude to conduct a dive.
4. Discuss the maladies peculiar to diving operations at higher altitudes.
5. Discuss the altitude diving rules.
6. Compute/determine the required data for an altitude dive (greater than 1,000 feet mean sea level [msl]) using the appropriate tables and compensation IAW the cited references.

Complete a Submarine Lockout/Lockin Drill
331-CDQ-3236

Condition: As an SF waterborne team member, given—
1. FM 3-05.212, *Special Forces Waterborne Operations*.
2. NWP 79-0-4, *Submarine Special Operations Manual–Unconventional Warfare*.
3. A complete set of surface swim or combat diver equipment.
4. A submarine or a shore-based submarine escape trunk training facility.
5. All safety and recovery equipment.
6. The requirement to complete a submarine lockin/lockout drill.

Standard: IAW the cited references, complete a swimmer or diver deployment and recovery evolution (submarine lockout/lockin) from an underway submarine or a shore-based submarine escape trunk training facility under the diving supervisor's supervision.

Calculate the Effects of Ocean Tides, Currents, and Waves on Waterborne Operations
331-CDQ-3237

Condition: As a combat diver, given—
1. Tide Tables and Tidal Current Tables.
2. A nautical chart for the operational area.
3. FM 3-05.212, *Special Forces Waterborne Operations*, Chapter 3.

Standard: IAW the cited reference, calculate the effects that waves, currents, and tides can have on waterborne operations.

Complete a Closed-Circuit Swim as a Member of a Swim Team
331-CDQ-3238

Condition: As a combat diver during day or night, given—
1. FM 3-05.212, Special Forces Waterborne Operations, Chapter 12, and current U.S. Navy diving regulations.
2. A complete set of closed-circuit swim gear.
3. An environmental protection suit, if required by the water conditions.
4. Mission-essential equipment, and equipment lines with snap links, team swim rope, individual buddy lines, and snap links.
5. An open-water operational area and a BLS.
6. A requirement to conduct a closed-circuit team infiltration swim.

Standard: Complete a team closed-circuit infiltration swim, with equipment, IAW the cited references by—
1. Maintaining noise and light discipline, team integrity, and personnel and equipment accountability.
2. Using the line, column, or wedge formation.
3. Arriving at the BLS in condition to execute a follow-on mission.

Appendix B

Employ Individual Swim Gear   **331-CDQ-3239**

Condition: As a dive team member, given—
   1. Individual swim gear.
   2. Current U.S. Navy diving regulations.
   3. FM 3-05.212, *Special Forces Waterborne Operations,* Chapter 10.

Standard: Properly don, adjust, and use the diver's face mask, swim fins, and buoyancy device IAW cited references.

Ditch and Don Open-Circuit Equipment Underwater   **331-CDQ-3240**

Condition: As a combat diver, given—
   1. Current U.S. Navy diving regulations.
   2. FM 3-05.212, *Special Forces Waterborne Operations,* Chapter 11.
   3. A complete open-circuit diving equipment set.
   4. A swimming pool or open water training site NTE 15 feet in depth.
   5. A requirement to ditch and don the open-circuit scuba equipment underwater.

Standard: IAW the cited references, perform the following:
   1. Ditch and don the open-circuit equipment underwater.
   2. Remove the open-circuit scuba equipment underwater and execute a free-swimming ascent to the surface.
   3. Execute a free dive, recover, and don the open-circuit scuba equipment underwater.

Complete a Free-Swimming Ascent From Depth Without an Air Source   **331-CDQ-3241**

Condition: As a combat diver, given—
   1. Current U.S. Navy diving regulations.
   2. FM 3-05.212, *Special Forces Waterborne Operations*, paragraphs 11-43 through 11-50.
   3. NAVSPECWARCEN Instruction 1540.5C.
   4. A compressed air source at depth.
   5. Swim gear.
   6. A buoyant ascent tower or an open-water operational area.
   7. A requirement to complete a free-swimming ascent from depth without an air source.

Standard: Complete a free-swimming ascent from an operational depth, without an air source and injury, directly to the surface IAW the cited references.

Complete a Deep Dive   **331-CDQ-3242**

Condition: As a combat diver, given—
   1. Current U.S. Navy diving regulations.
   2. AR 611-75, *Management of Army Divers,* paragraphs 2-19 and 2-20.
   3. USASOC Regulation 350-20, *USASOC Dive Program,* Annex D.
   4. A complete set of open-circuit scuba swim gear and thermal protection as required.
   5. An open-water operational area.
   6. A requirement to complete a vertical (deep) dive.

Standard: IAW the cited references, complete a deep dive as follows:
   1. Descend normally to depth.
   2. Ascend normally to the surface without exceeding briefed time and depth.
   3. For qualification purposes, reach 130 FSW for initial dives.
   4. For subsequent annual requalification dives, reach 70 FSW but do not exceed 130 FSW.

Perform Pre and Postdive Procedures on a Closed-Circuit Oxygen Rebreather
331-CDQ-3243

Condition: As a combat diver, given—
1. Current U.S. Navy diving regulations.
2. FM 3-05.212, *Special Forces Waterborne Operations*, Chapter 12, and Appendixes D and E.
3. An MK 25 MOD 2 Underwater Breathing Apparatus (UBA) Technical Manual.
4. An MK 25 MOD 2 UBA.
5. An approved carbon dioxide-absorbent compound.
6. A filled oxygen bottle.
7. A requirement to complete a pre and postdive inspection of a closed-circuit oxygen rebreather.

Standard: IAW the cited references, complete a—
1. Predive inspection of an oxygen rebreather before an underwater infiltration swim.
2. Postdive inspection of an oxygen rebreather after an underwater infiltration swim.

React to Emergencies During Open- and Closed-Circuit Diving Operations
331-CDQ-3244

Condition: As a combat diver, given—
1. The appropriate open- or closed-circuit diving equipment.
2. The appropriate technical manuals (TMs).
3. Current U.S. Navy diving regulations.
4. FM 3-05.212, *Special Forces Waterborne Operations*, Chapters 10, 11, and 12.
5. An indoor/outdoor swimming pool or an open-water training/operational area.
6. The necessary safety equipment and personnel.
7. A requirement to perform emergency procedures while conducting open- or closed-circuit diving operations.

Standard: IAW the procedures outlined in cited references, take appropriate action to resolve mechanical or medical emergencies encountered during open- and closed-circuit diving operations

Maintain Open-Circuit Diving Equipment
331-CDQ-3245

Condition: As a combat diver, given—
1. Current U.S. Navy diving regulations.
2. FM 3-05.212, *Special Forces Waterborne Operations*, Chapter 11 and Appendix C.
3. The appropriate manufacturer's technical manuals.
4. The maintenance requirements cards (MRCs) for the diving equipment.
5. A complete set of open-circuit diving equipment.

Standard: IAW the cited references, perform operator maintenance on all equipment contained in a complete set of open-circuit diving equipment.

Waterproof Bundles and Combat Gear for Waterborne Operations
331-CDQ-3246

Condition: As a combat diver or a waterborne infiltration swimmer, given—
1. A standard military rucksack.
2. Various pieces of equipment.
3. Demolitions.
4. Waterproofing materials and/or dry bags.
5. An indoor/outdoor swimming pool and/or an outdoor water training site/operational area.
6. The requirement to waterproof bundles and combat gear for waterborne operations.

Standard: Perform the following:
1. Waterproof selected items of equipment to be carried in the rucksack.
2. Waterproof his weapon, radio, and demolitions and primers.
3. Rig his rucksack for surface or subsurface swim delivery.
4. Adjust his rucksack for neutral buoyancy at various depths (as required).
5. Swim his waterproofed bundle.

Complete a Waterborne Operation Using a Variable-Volume Dry Suit
331-CDQ-3247

Condition: As a combat diver, given—
1. FM 3-05.212, *Special Forces Waterborne Operations,* Chapter 11.
2. Current U.S. Navy diving regulations.
3. A variable-volume dry suit.
4. A requirement to conduct waterborne operations, surface or subsurface, using a variable-volume dry suit.

Standard: IAW the cited references, perform the following:
1. Inspect, size, and don a variable-volume dry suit.
2. Complete a dive using a variable-volume dry suit.
3. Take off a variable-volume dry suit.
4. Complete PMCS on a variable-volume dry suit.

Complete a Combat Diving Supervisor's Personnel and Equipment Inspection
331-CDS-3400

Condition: As a combat diving supervisor, given—
1. FM 3-05.212, *Special Forces Waterborne Operations,* Chapters 10 through 12 and Appendix D.
2. Current U.S. Navy diving regulations.
3. The requirement to complete a combat diving supervisor's personnel and equipment inspection.

Standard: IAW the cited references, complete a combat diving supervisor's personnel and equipment inspection.

Supervise Combat Diving Operations
331-CDS-3401

Condition: As a combat diving supervisor, given—
1. AR 611-75, *Management of Combat Divers,* Chapter 2.
2. Current U.S. Navy diving regulations.
3. FM 3-05.212, *Special Forces Waterborne Operations,* Chapters 10 through 12.
4. USASOC Regulation 350-20, *USASOC Dive Program,* Chapter 1.
5. The requirement to supervise combat diving operations.

Standard: Supervise combat diving operations IAW the cited references.

Employ Diving Forms and Regulations
331-CDS-3402

Condition: As a combat diving supervisor responsible for the completion and submission of diving forms covering all diving operations for the unit, given—
1. Current U.S. Navy diving regulations.
2. USASOC Regulation 350-20, *USASOC Dive Program,* paragraph 1-10j (22) and Annex C.
3. FM 3-05.212, *Special Forces Waterborne Operations,* Chapter 10.

Standard: IAW the cited references and the unit's SOP—
1. Provide the reporting official with the information required to submit a Dive Reporting System (DRS) report to the Navy Safety Center.
2. Complete the following forms as required by local regulations:
   a. DA Form 1262-1 (Diving Site Worksheet).
   b. DA Form 1262 (Command Dive Log).

Complete a Boat Inspection
331-CDS-3403

Condition: As a combat diving supervisor, given—
1. A requirement to inspect a vessel used to support MAROPS.
2. USASOC Regulation 350-20, *USASOC Dive Program*, pages B-6 and B-7.
3. FM 3-05.212, *Special Forces Waterborne Operations*, Chapter 7.
4. The appropriate checklists.

Standard: Inspect the vessel IAW the cited references to determine its seaworthiness and suitability for the intended operation.

Plan Diving Operations
331-CDS-3404

Condition: As a combat diving supervisor, given—
1. A requirement to plan and conduct diving operations.
2. The current U.S. Navy diving regulations and appropriate checklists.
3. AR 611-75, *Management of Army Divers*.
4. USASOC Regulation 350-20, *USASOC Dive Program*, Chapter 1.
5. FM 3-05.212, *Special Forces Waterborne Operations*, Chapters 1, 10, 11, and 12.
6. FM 5-0, *Army Planning and Orders Production*, Appendix G.

Standard: Prepare a five-paragraph dive operations order (OPORD) IAW the cited references.

Brief a Dive Plan
331-CDS-3405

Condition: As a combat diving supervisor, given—
1. A requirement to plan a dive, brief the dive plan, and conduct diving operations.
2. Current U.S. Navy diving regulations and appropriate checklists.
3. AR 611-75, *Management of Army Divers*.
4. USASOC Regulation 350-20, *USASOC Dive Program*, Chapter 1.
5. FM 3-05.212, *Special Forces Waterborne Operations*, Chapters 1, 10, 11, and 12.

Standard: IAW the cited references, brief the dive plan to your detachment members.

Complete a Tabletop Inspection of a Closed-Circuit Oxygen Rebreather
331-CDS-3407

Condition: As a combat diving supervisor, given—
1. FM 3-05.212, Chapter 12 and Appendix D.
2. Current U.S. Navy diving regulations.
3. The requirement to conduct a tabletop inspection of a closed-circuit oxygen rebreather.

Standard: Complete a tabletop inspection of a closed-circuit oxygen rebreather.

Employ Nautical Charts and Navigational Aids
331-CDS-3408

Condition: As a combat diving supervisor, given—
1. A requirement to employ nautical charts and navaids.
2. FM 3-05.212, *Special Forces Waterborne Operations*, Chapters 4 and 5.
3. A nautical chart of the operational area.
4. Chart No. 1, *United States of America Nautical Chart Symbols, Abbreviations, and Terms*.

Standard: IAW the cited references, employ nautical charts and navigation aids to prepare navigation plans and/or to navigate a small craft to a designated point.

Appendix B

Supervise Operator-Level Maintenance of Selected Individual Diving Equipment
331-CDS-3414

Condition: As a combat diving supervisor, given—
1. FM 3-05.212, *Special Forces Waterborne Operations,* Chapters 10 through 12 and Appendixes C and D.
2. OPNAVINSTs.
3. Maintenance Requirements Card (MRC) for the selected equipment.
4. The requirement to supervise operator-level maintenance of selected open- and closed-circuit individual diving equipment.

Standard: Supervise operator-level maintenance of selected open- and closed-circuit individual diving equipment IAW the cited references.

Employ High-Pressure Compressed Gas Systems to Charge Divers' Supply Cylinders
331-CDS-3415

Condition: As a combat diver or a combat diving supervisor, given—
1. Current U.S. Navy diving regulations.
2. FM 3-05.212, *Special Forces Waterborne Operations,* Chapters 10 through 12.
3. A set of open-circuit scuba tanks (steel or aluminum).
4. An oxygen bottle from a closed-circuit diving apparatus.
5. An air compressor and an oxygen booster pump (as required).
6. An air and an oxygen cascade system.
7. All necessary ancillary equipment; for example, pressure gauges, hoses, air purity tester.

Standard: IAW cited references, perform the following:
1. Explain the safety procedures for high-pressure air systems.
2. State the air purity standards for open-circuit diving.
3. Fill a set of scuba cylinders (steel or aluminum), at the proper rate, to the proper capacity, using an air compressor and/or an air cascade system.
4. Fill an oxygen bottle at the proper rate, to the proper capacity, using an oxygen cascade system and/or an oxygen booster pump.

Supervise Hyperbaric Chamber Operations
331-CDS-3417

Condition: As a combat diving supervisor, given—
1. Current U.S. Navy diving regulations.
2. A hyperbaric chamber.
3. The specific chamber's Emergency Procedures/Operating Procedures.
5. A requirement to supervise recompression chamber operations.

Standard: IAW the cited references, perform the following:
1. Identify the parts of a recompression chamber.
2. Discuss the functions of a recompression chamber and its supporting equipment.
3. Discuss the responsibilities of the recompression chamber's support personnel.
4. Discuss the correct use of the equipment and material employed in the treatment of a diving casualty.
5. Supervise the operations of a hyperbaric chamber.

Complete a 165-Foot Hyperbaric Chamber Dive
331-DMT-3602

Condition: As a combat diver, combat diving supervisor, or diving medical technician, given—
1. Current U.S. Navy diving regulations.
2. A hyperbaric chamber.
3. The specific chamber's Emergency Procedures/Operation Procedures.
4. A requirement to complete a 165-foot hyperbaric chamber dive.

Standard: Complete a 165-foot hyperbaric chamber dive IAW the cited references

Compute the Minimum Safe Distance for an Underwater Blast
331-DMT-3603

Condition: As a combat diver, a combat diving supervisor, or a diving medical technician, given—
1. Current U.S. Navy diving regulations.
2. FM 3-05.212, *Special Forces Waterborne Operations,* Chapter 11.
3. A requirement to calculate the minimum safe distance from an underwater explosion.

Standard: IAW the cited references, perform the following:
1. Discuss the mitigating and minimizing factors that further reduce a diver's exposure to an underwater explosion.
2. Compute the minimum safe distance that a submerged diver must be from an underwater blast so that he is not exposed to a blast pressure greater than 50 psi.

Determine a Combat Diver's Fitness for Diving Duty
331-DMT-3600

Condition: As a combat diving supervisor or a diving medical technician, given—
1. AR 611-75, *Management of Army Divers,* paragraph 2-18e.
2. AR 40-501, *Standards of Medical Fitness,* paragraphs 5-9 and 5-10.
3. A diver's physical examination records (DD Form 2808, *Report of Medical Examination* and DD Form 2807-1, *Report of Medical History*).
4. A combat diver.
5. The requirement to determine a combat diver's fitness for diving duty.

Standard: IAW the cited references, perform the following:
1. Discuss the chapters of AR 611-75, *Management of Army Divers,* that identify pertinent information related to military diver rating qualification and disqualification.
2. Discuss the medical fitness standards for combat divers, combat diving supervisors, and diving medical technicians as specified in AR 40-501, *Standards of Medical Fitness.*
3. Evaluate a diver's physical examination documents (DD Form 2807-1, *Report of Medical History* and DD Form 2808, *Report of Medical Examination*) for compliance with AR 40-501.
4. Determine the fitness of a combat diver for diving duty IAW the cited references.

Plan Medical Support of Diving Operations
331-DMT-3601

Condition: As a diving medical technician, given—
1. Current U.S. Navy diving regulations.
2. AR 40-501, *Standards of Medical Fitness.*
3. USASOC Regulation 350-20, *USASOC Dive Program,* Chapter 1.
4. FM 3-05.212, *Special Forces Waterborne Operations,* Chapter 10.
5. Unit SOP.

Standard: IAW the cited references, perform the following:
1. Plan the medical support for a combat diving operation IAW the cited references that includes or discusses the following:
   a. Medical screening requirements.
   b. Requirement for the physical examination and evaluation of detachment divers.
   c. Requirement to use an available hyperbaric/recompression chamber facility.
   d. Emergency medical equipment that is needed for the dive operation.
   e. Detachment's medical training requirement.
   f. MEDEVAC plan of diving casualties.
   g. Environmental considerations in dive planning.
2. Based on the given hypothetical situation scenario, develop a MEDEVAC plan (air and ground) from the dive site to a definitive care facility or hyperbaric/recompression chamber facility.
3. Based on the given hypothetical situation scenario, brief the team's divers (evaluators) on the dive plan and the diving hazards that may be associated with the specific diving site.

Appendix B

Complete a Differential Diagnosis of a Diving Casualty
331-DMT-3604

Condition: As a diving medical technician, given—
1. The appropriate medical examination equipment.
2. *The U.S. Navy Diving Medical Officer Guide* (E.T Flynn, et al).
3. Current U.S. Navy diving regulations.
4. *Bove and Davis' Diving Medicine* (Alfred A. Bove).
5. A real or simulated diving casualty.

Standard: IAW the cited references, perform the following:
1. Identify the four phases of a dive.
2. Identify the medical conditions/injuries that are common to all phases of a dive.
3. Discuss the medical conditions/injuries that are most likely to occur in each phase.
4. Complete a differential diagnosis of a real or simulated diving casualty based on signs and symptoms.

Treat a Diver With Barotrauma
331-DMT-3605

Condition: As a diving medical technician, given—
1. A medical kit.
2. *Bove and Davis' Diving Medicine* (Alfred A. Bove).
3. *The Physiology and Medicine of Diving* (Peter B. Bennett and David H. Elliott).
4. Current U.S. Navy diving regulations.
5. A real or simulated diving casualty suffering from barotrauma.

Standard: IAW cited references, perform the following:
1. Define—
   a. Barotrauma.
   b. Descent barotraumas.
   c. Ascent barotraumas.
2. Discuss the conditions that must be present for barotraumas to occur.
3. Discuss the causes, signs, symptoms, treatment, and prevention of common types of barotraumas.
4. Complete a differential diagnosis of a diver with a real or simulated barotrauma injury, correctly identify the type of barotrauma injury, and recommend an appropriate course of treatment.

Treat a Diver for Pulmonary Overinflation Injuries
331-DMT-3607

Condition: As a diving medical technician, given—
1. A medical kit.
2. *Bove and Davis' Diving Medicine* (Alfred A. Bove).
3. *The Physiology and Medicine of Diving* (Peter B. Bennett and David H. Hill).
4. Current U.S. Navy diving regulations.
5. *The Physician's Guide to Diving Medicine* (Charles W. Shilling, et al).
6. A real or simulated diving casualty suffering from pulmonary overinflation injuries.

Standard: IAW cited references, perform the following:
1. Define pulmonary overinflation syndrome.
2. Discuss the main causes of alveolar rupture.
3. Discuss the types of pulmonary overinflation and associated injuries.
4. Discuss the causes, signs and symptoms, and treatment of pulmonary overinflation and associated injuries.
5. Complete a differential diagnosis of a diver with a real or simulated pulmonary overinflation injury, correctly identify the type of pulmonary overinflation injury, and recommend an appropriate course of treatment.

Description of Subject Area and Critical Task List

Treat a Combat Diver for Decompression Sickness
331-DMT-3608

Condition: As a diving medical technician, given—
1. A medical kit.
2. *Bove and Davis' Diving Medicine* (Alfred A. Bove).
3. *The Physiology and Medicine of Diving* (Peter B. Bennett and David H. Elliott).
4. Current U.S. Navy diving regulations.
5. The *U.S. Navy Diving Medical Officer Guide*.
6. *The Physician's Guide to Diving Medicine* (Charles W. Shilling, et al).
7. A real or simulated diving casualty suffering from decompression sickness injuries.

Standard: IAW cited references, perform the following:
1. Define decompression sickness (DCS).
2. Discuss the direct and indirect bubble effects.
3. Discuss the symptoms of DCS.
4. Discuss Type I DCS:
    a. Musculoskeletal pain-only DCS symptoms.
    b. Cutaneous (skin) DCS symptoms.
5. Discuss Type II DCS:
    a. Neurological DCS symptoms.
    b. Brain DCS symptoms.
    c. Spinal cord DCS symptoms.
    d. Inner ear DCS symptoms.
    e. Pulmonary DCS symptoms.
6. Discuss the treatment and prevention of DCS.
7. Discuss flying and diving after hyperbaric chamber treatment.
8. Discuss the observation of personnel after hyperbaric chamber treatments.
9. Complete a differential diagnosis of a diver with real or simulated DCS injury, identify the type of DCS, and recommend the appropriate course of treatment.

Complete a Neurological Examination of a Combat Diving Casualty
331-DMT-3610

Condition: As a diving medical technician, given—
1. A combat diver requiring a neurological examination.
2. A physical examination medical equipment set (for example, pen light, otoscope/ophthalmoscope, sphygmomanometer, stethoscope, thermometer, and examination gloves).
3. Current U.S. Navy diving regulations.

Standard: Complete a neurological examination of a combat diver IAW the cited reference and identify any neurological deficits with 100 percent accuracy.

Treat a Diver With Immersion Hypothermia
331-DMT-3612

Condition: As a diving medical technician, given—
1. A medical kit.
2. *Bove and Davis' Diving Medicine* (Alfred A. Bove).
3. Current U.S. Navy diving regulations.
4. *The Physician's Guide to Diving Medicine* (Charles W. Shilling).
5. Textbook of Advanced Cardiac Life Support (R. O. Cummins).
6. A real or simulated diving casualty suffering from immersion hypothermia.

Standard: IAW the cited references, perform the following:
1. Discuss the pathophysiology of hypothermia caused by immersion in water.
2. Discuss the signs, symptoms, treatment, and prevention of immersion hypothermia.
3. Treat a real or simulated diving casualty suffering from immersion hypothermia.

Appendix B

Administer First Aid for Aquatic Skin Diseases
331-DMT-3614

Condition: As a diving medical technician, given—
1. An appropriate medical treatment set.
2. *Atlas of Aquatic Dermatology* (Alexander Fisher) and *Diving and Subaquatic Medicine* (Carl Edmonds).
3. *American Family Physician; Marine Envenomations & Aquatic Dermatology* (G. G. Soppe).
4. *New England Journal of Medicine; Marine Envenomations*, (P. S. Auerbach) and *Seabather's Eruption*, (A. R. Freudenthal).
5. WRAIR Communicable Disease Report: *Medical Threats in the Marine Environment* (R. Mahmoud).
6. A real or simulated diving casualty exhibiting an aquatic dermatological condition.

Standard: IAW the cited references, perform the following:
1. Discuss—
   a. The aquatic skin diseases to which combat divers are susceptible.
   b. The causes, signs and symptoms, treatment, and prevention of these diseases.
2. Administer first aid to a diving casualty exhibiting an aquatic dermatological condition.

Treat Underwater Blast and Sonar Injuries
331-DMT-3615

Condition: As a diving medical technician, given—
1. An appropriate medical treatment set including oxygen administration equipment, IV fluids, IV administration equipment, and a nasogastric tube.
2. The Current U.S. Navy diving regulations and the *Medical Officer's Guide* (E. T. Flynn).
3. *The Physiology and Medicine of Diving* (Peter B. Bennett and David H. Elliott).
4. *The Physician's Guide to Diving Medicine* (Charles W. Shilling) and *Diving and Subaquatic Medicine* (Carl Edmonds).
5. U.S. Navy Technical Manual: *Demolition Materials* (C. C. Stevens).
6. *Non-Auditory Effects of Underwater Low Frequency Acoustic Transmissions: A Literature Review* (C. C. Stevens).
7. A real or simulated diving casualty exhibiting signs and symptoms of injury caused by exposure to an underwater explosive blast or a sonar transmission.

Standard: IAW the cited references, perform the following:
1. Discuss—
   a. The physical properties of underwater explosive blasts.
   b. The factors affecting the intensity of an underwater explosion.
   c. Underwater explosive blast injuries.
   d. The calculation of the safe distances from an explosive detonation area.
   e. The injuries caused by sonar transmissions.
   f. The prevention of underwater blast and sonar transmission injuries.
2. Administer emergency medical treatment to a diving casualty exhibiting signs and symptoms of injury caused by exposure to an underwater explosive blast and/or a sonar transmission.

Employ the U.S. Navy Dive Treatment Tables During Treatment of a Combat Diving Casualty
331-DMT-3616

Condition: As a diving medical technician, given—
1. Current U.S. Navy diving regulations.
2. A real or simulated diving casualty exhibiting the signs and symptoms of an injury requiring treatment in a recompression chamber.

Standard: IAW the cited reference, perform the following:
1. Discuss—
   a. The hyperbaric chamber and recompression therapy general guidelines.
   b. The U.S. Navy Decompression Treatment Tables.
   c. Recompression chamber procedures.
2. Given a scenario or a "role playing" diver exhibiting the simulated signs and symptoms of an injury requiring treatment in a recompression chamber, select the appropriate decompression treatment table and implement the rules for recompression of the diving casualty.

Treat Maladies Associated With Altitude Diving
331-DMT-3619

Condition: As a diving medical technician, given—
1. An appropriate medical treatment set including oxygen administration equipment, IV fluids, IV administration equipment, a nasogastric tube, and Acetazolamide.
2. Current U.S. Navy diving regulations and the *Diving Officer's Guide* (E. T. Flynn).
3. *The Physician's Guide to Diving Medicine* (Charles B. Shilling).
4. *Diseases of High Terrestrial Altitude* (P. B. Rock).
5. *Wilderness Medicine: Management of Wilderness and Environmental Emergencies* (Paul S. Auerbach).
6. A real or simulated diving casualty exhibiting signs and symptoms of a malady associated with altitude diving.

Standard: IAW the cited references, perform the following:
1. Discuss—
   a. The physiological effects of diving at altitude.
   b. The medical conditions that may occur during altitude diving.
   c. The altitude dive tables.
2. Solve selected altitude diving problems using the altitude dive tables.
3. Treat a real or simulated diving casualty exhibiting the maladies associated with altitude diving operations.

Treat Injuries Inflicted by Dangerous Marine Life
331-DMT-3620

Condition: As a diving medical technician, given—
1. An appropriate medical treatment set.
2. Current U.S. Navy diving regulations.
3. *Bove and Davis' Diving Medicine* (Alfred A. Bove).
4. *Dangerous Marine Animals* (Bruce W. Halstead).
5. *A Medical Guide to Hazardous Marine Life* (Paul S. Auerbach).
6. *Dangerous Marine Creatures* (Carl Edmonds).
7. *Diving and Subaquatic Medicine* (Carl Edmonds).
8. *A Pictorial History of Sea Monsters and Other Dangerous Marine Life* (J. B. Swenny).
9. A real or simulated diving casualty exhibiting the signs and symptoms of an injury inflicted by marine life.

Standard: IAW the cited references, perform the following:
1. Recognize and identify the dangerous marine life forms that sting or inject venom:
   a. Their characteristics.
   b. Their injury mechanisms.
   c. The signs and symptoms of the injuries caused as a result of their sting/envenomation.
   d. The treatment of the injuries caused by these life forms.
   e. The preventive measures that should be taken to preclude injuries from these life forms.
2. Recognize and identify the dangerous marine life forms that bite:
   a. Their characteristics.
   b. Their injury mechanisms.
   c. Signs and symptoms of the injuries caused as a result of their bites.
   d. Treatment of the injuries caused by these life forms.
   e. Preventive measures that may/should be taken to preclude injuries from these life forms.
3. The preventive measures that may/should be taken to preclude injuries from these life forms.
4. Treat a real or simulated diving casualty exhibiting the signs and symptoms of an injury inflicted by a dangerous marine life form.

Appendix B

Complete a Physical Examination of a Combat Diver
331-DMT-3624

Condition: As a diving medical technician, given—
1. A patient requiring a physical examination.
2. AR 40-501, *Standards of Medical Fitness*, paragraphs 5-9 and 5-10.
3. DD Form 2807-1, *Report of Medical History* and DD Form 2808, *Report of Medical Examination*.
4. A physical examination medical equipment set (for example, pen light, otoscope/ophthalmoscope, sphygmomanometer, stethoscope, tongue depressor, thermometer, reflex hammer, and examination gloves).

Standard: Complete and record a detailed physical examination of a combat diver.

Treat a Diver With Carbon Monoxide Poisoning
331-DMT-3625

Condition: As a diving medical technician, given—
1. An appropriate medical treatment set including oxygen administration equipment, IV fluids, IV administration equipment, and IV corticosteroids.
2. Current U.S. Navy diving regulations and the *Medical Officer's Guide* (E. T. Flynn).
3. *The Physiology and Medicine of Diving* (Peter B. Bennett and David H. Elliott).
4. *The Journal of Neurology, Neurosurgery, and Psychiatry* (BMJ Publishing Group).
5. A real or simulated diving casualty with signs and symptoms of carbon monoxide poisoning.

Standard: IAW the cited references, perform the following:
1. Identify the characteristics, causes, signs, symptoms, diagnosis, treatment, and prevention of carbon monoxide poisoning.
2. Discuss U. S. Navy and civilian source air purity standards.
3. Initiate emergency medical treatment of a real or simulated diving casualty displaying the signs and symptoms of carbon monoxide poisoning.

Treat a Diving Accident Victim
331-DMT-3626

Condition: As a combat diver, combat diving supervisor, or a diving medical technician, given—
1. *Scuba Life Saving and Accident Management Manual of the YMCA* (Excerpt).
2. A simulated/actual diving accident victim in the water, surface or subsurface, conscious or unconscious.
3. A requirement to rescue the victim, administer first aid, and evacuate the casualty to a comprehensive medical care facility.

Standard: IAW cited references, perform the following:
1. Evaluate the situation.
2. Rescue the casualty and remove him from the water.
3. Administer first aid as required.
4. Evacuate the casualty to a comprehensive medical care facility.

Insert an Endotracheal Tube Airway in a Diving Casualty
331-DMT-3627

Condition: As a diving medical technician, given—
1. An unconscious, non-breathing diving casualty with no gag reflex, simulated or real.
2. A qualified assistant performing CPR.
3. Suction equipment available and ready for use.
4. A prepared laryngoscope, endotracheal (ET) tube, and BVM resuscitator or oxygen with a demand valve.
5. Ancillary medical equipment (for example, examination gloves, oral bite block or J tube, suction equipment, adhesive tape, Benzoin, stethoscope, and a 10-cc syringe).
5. *Advanced Trauma Life Support for Doctors* (American College of Surgeons Committee on Trauma) and the *Clinician's Pocket Reference* (L. G. Gomella).

Standard: Insert an endotracheal tube airway in the diving casualty (an intubation mannequin) ensuring that he is not deprived of oxygen for longer than 20 seconds at any time during the procedure.

Operate as the Inside Tender During Hyperbaric Chamber Operations
331-DMT-3628

Condition: As a diving medical technician assigned duties as an inside tender during recompression treatment of a diving casualty, and given—
1. Current U.S. Navy diving regulations.
2. The Local Chamber Emergency Procedures/Operations Procedures.
3. A hyperbaric recompression chamber.
4. A real or simulated diving casualty requiring decompression treatment.

Standard: IAW the cited references, describe the duties of the DMT as an inside tender and operate as the inside tender during hyperbaric chamber operations IAW the cited references.

Complete Invasive Procedures in a Hyperbaric Environment
331-DMT-3629

Condition: As a diving medical technician, given—
1. An appropriate medical treatment set consisting of equipment to perform venipuncture (IV therapy), intubation (respiratory therapy), and urinary catheterization procedures.
2. Current U.S. Navy diving regulations.
3. *Advanced Trauma Life Support (ATLS) for Doctors Manual.*
4. *Clinician's Pocket Reference* (L. G. Gomella).
5. A hyperbaric chamber.
6. A real or simulated diving casualty requiring the diving medical technician to perform invasive procedures in a hyperbaric environment.

Standard: IAW the cited references, perform the following:
1. Discuss the equipment that is needed and the procedures for—
   a. Intravenous (IV) therapy and venipuncture.
   b. Intubation.
   c. Urinary catheterization.
2. Complete the invasive procedures of IV therapy/venipuncture, intubation, and urinary catheterization in a hyperbaric environment as part of a comprehensive treatment protocol of a simulated or real diving casualty.

Treat a Combat Diver for Oxygen Toxicity and Hypoxia
331-DMT-3630

Condition: As a diving medical technician, given—
1. An appropriate medical treatment set including oxygen administration equipment.
2. Current U.S. Navy diving regulations.
3. *Bove and Davis' Diving Medicine* (Alfred A. Bove).
4. *The Physiology and Medicine of Diving* (Peter B. Bennett and David H. Elliott).
5. *Hyperbaric Oxygen Therapy* (Richard A. Neubauer and Morton Walker).
6. *Diving and Subaquatic Medicine* (Carl Edmonds).
7. *The Physician's Guide to Diving Medicine* (Charles W. Shilling).
8. A hyperbaric treatment facility.
9. A real or simulated diving casualty displaying the signs and symptoms of oxygen toxicity or hypoxia.

Standard: IAW the cited references, perform the following:
1. Discuss the causes, signs and symptoms, treatment, and prevention of hypercarbnia, hypoxia, and oxygen toxicity.
2. Treat a diving casualty displaying the signs and symptoms of oxygen toxicity or hypoxia.

Appendix B

Treat a Drowning or Near-Drowning Victim
331-DMT-3631

Condition: As a diving medical technician, given—
1. An appropriate medical treatment set including oxygen administration equipment.
2. Current U.S. Navy diving regulations.
3. *The USN Diving Officer Guide* (E. T. Flynn).
4. *Wilderness Medicine* (Paul S. Auerbach).
5. A real or simulated diving casualty displaying the signs and symptoms of drowning or near-drowning.

Standard: IAW the cited references, perform the following:
1. Restore the pulse and breathing in the casualty.
2. Obtain a patient history.
3. Complete a detailed physical examination of the casualty.
4. Administer supportive care and pharmacological therapy.
5. Evacuate the patient to a comprehensive medical care facility (as required).

Implement Adjunctive (Pharmacological) Therapy in the Treatment of Diving Injuries
331-DMT-3632

Condition: As a diving medical technician, given—
1. An appropriate medical treatment set containing the drugs/medicines needed to implement pharmacological therapy.
2. *Bove and Davis' Diving Medicine* (Alfred A. Bove).
3. *Advanced Cardiac Life Support (ACLS) Manual.*
4. *USN Diving Medical Officer Guide* (E. T. Flynn).
5. A real or simulated diving casualty requiring recompression and possible pharmacological therapy.

Standard: IAW cited references, perform the following:
1. Discuss—
   a. The pharmaceutical therapy used in hyperbaric chamber operations.
   b. The guidelines for the use of pharmaceutical therapy in hyperbaric chamber operations.
2. Identify the pharmacological agents that are hazardous for use in a hyperbaric environment.
3. Based on a given scenario indicating a diving casualty who requires recompression therapy with pharmaceutical therapy, implement the appropriate and correct pharmacological (adjunctive) therapy in support of hyperbaric chamber treatment.

Employ a Portable Hyperbaric Chamber to Transport a Diving Casualty
331-DMT-3633

Condition: As a combat diving supervisor or diving medical technician, given—
1. Current U.S. Navy diving regulations.
2. A portable hyperbaric chamber (for example, hyperlite hyperbaric stretcher).
3. A portable hyperbaric chamber operations manual.
4. A requirement to transport a diving casualty using a portable hyperbaric chamber.

Standard: IAW the cited references, perform the following:
1. Inspect the portable hyperbaric chamber.
2. Place a portable hyperbaric chamber into operation.
3. Using a portable hyperbaric chamber, pressurize a patient to storage depth.
4. Transport the patient to an advanced treatment facility.
5. Transfer the patient from the portable hyperbaric chamber to a fixed hyperbaric chamber.
6. Complete the required maintenance on a portable hyperbaric chamber.

Description of Subject Area and Critical Task List

Complete Small Boat Operations
331-WIC-3800

Condition: As a small boat crew member, given—
1. FM 3-05.212, *Special Forces Waterborne Operations,* Chapters 5, 6, 7, and 13.
2. TM 5-1940-279-13 & P1, *Operation and Maintenance Instructions With Component and Repair Parts Listing for Combat Rubber Raiding Craft.*
3. A combat rubber raiding craft (CRRC) complete with organic equipment and accessories.
4. An outboard motor, fuel line, and fuel tanks/bladders.
5. Other crew members.
6. An outdoor water training area.
7. A requirement to use a CRRC in support of detachment operations.

Standard: IAW the cited references, perform the following duties:
1. Inspect the boat for serviceability prior to assembly.
2. Assemble the boat.
3. Put the boat into service and operate it using both paddles and the OBM.
4. Properly maintain the boat.

Complete Waterborne Premission Planning
331-WIC-3801

Condition: As a detachment member, given—
1. FM 3-05.212, *Special Forces Waterborne Operations,* Chapter 1.
2. A scenario containing the pertinent information dealing with a special operations mission to be conducted in a selected area.
3. A waterborne mission operation order (OPORD) from higher headquarters.
4. A scenario containing all the pertinent data/information about the AO and the target area.
5. Maps, nautical charts, and other reference materials covering the AO.
6. All the required maritime publications that are needed to plan and conduct waterborne operations in the AO.
7. A requirement to conduct premission planning for a waterborne operation.

Standard: IAW the cited reference, complete premission planning to conduct a waterborne mission.

Complete a Special Operations Combat Expendable Platform Airborne Infiltration
331-WIC-3804

Condition: As a detachment member, given—
1. FM 3-05.212, *Special Forces Waterborne Operations,* Chapters 1, 7, and 16.
2. FM 3-21.220, *Static Line Parachuting Techniques and Training.*
3. FM 4-20.142, *Airdrop of Supplies and Equipment: Rigging Loads for Special Operations.*
4. USSOCOM M350-6, *Special Operations Forces Infiltration/Exfiltration Techniques.*
5. TM 5-1940-279-13&P1, *Operation and Maintenance Instructions With Component and Repair Parts Listing for Combat Rubber Raiding Craft.*
6. A combat rubber raiding craft (CRRC) complete with organic equipment and accessories.
7. An outboard motor, fuel line, and fuel tanks/bladders.
8. A Special Operations Combat Expendable Platform (SOCEP).
9. Parachutes and rigging materials as identified in paragraph 2 above.
10. Other boat crew members and supporting rigger personnel.
11. Other crew members and supporting rigger personnel.
12. An appropriate fixed-wing aircraft.
13. A beach landing site (BLS).
14. A requirement to complete a SOCEP airborne infiltration using a fixed-wing aircraft as an intermediate delivery platform.

Standard: IAW the cited references, complete a SOCEP airborne infiltration using a fixed-wing aircraft as an intermediate delivery platform.

Appendix B

Complete Mother Ship Operations
331-WIC-3806

Condition: As a member of a MAROPS team/detachment, given—
1. FM 3-05.212, *Special Forces Waterborne Operations*, Chapters 1 and 13.
2. A combat rubber raiding craft (CRRC) complete with organic equipment and accessories.
3. An outboard motor, fuel line, and fuel tanks/bladders.
4. Other crew members.
5. An outdoor water training area.
6. A requirement to use a mother ship as a launching platform for CRRCs in support of detachment operations.

Standard: IAW the cited references, complete—
1. Launching/casting of a Zodiac F-470 CRRC from an underway mother ship.
2. Recovery of a Zodiac F-470 CRRC by an underway mother ship.
3. Surface swim team casting from an underway mother ship.

Complete a Tactical Movement With a Small Boat
331-WIC-3807

Condition: As a detachment member, given—
1. FM 3-05.212, *Special Forces Waterborne Operations*, Chapters 1, 5, 6, 7, 13, and 16.
2. A combat rubber raiding craft (CRRC) complete with organic equipment and accessories.
3. An outboard motor, fuel line, and fuel tanks/bladders.
4. Other crew members.
5. A requirement to use CRRC tactical formations and tactical movement techniques in support of MAROPS team/detachment mission/operation.

Standard: IAW the cited references, complete—
1. Tactical movements with the CRRC using the overwatch, bounding, and traveling overwatch formations.
2. Basic tactical movements with the CRRC using the line, wedge, "V", and file formations.
3. An infiltration and exfiltration using the CRRC tactical movement techniques.

Complete Waterborne Operations Using Helicopters
331-WIC-3811

Condition: As a member of a MAROPS team/detachment, given—
1. FM 3-05.212, *Special Forces Waterborne Operations*, Chapters 1, 7, and 16.
2. TM 5-1940-279-13 & P1, *Operation and Maintenance Instructions With Component and Repair Parts Listing for Combat Rubber Raiding Craft*.
3. USSOCOM M350-6, *Special Operations Forces Infiltration/Exfiltration Techniques*.
4. A combat rubber raiding craft (CRRC) complete with organic equipment and accessories.
5. An outboard motor, fuel line, and fuel tanks/bladders.
6. Rigging materials sufficient for preparing to tow, capsize equipment, and lifting lines.
7. Other crew members.
8. An appropriate helicopter as available.
9. A requirement to use a helicopter as an intermediate delivery platform during a waterborne mission/operation.

Standard: IAW the cited references, complete waterborne insertions and extractions with the support of rotary-wing aircraft as an intermediate delivery system using limp duck, K-duck, rolled duck, or tethered duck as appropriate to the available airframe.

Description of Subject Area and Critical Task List

Complete a Tactical Water Airborne Operation
331-WIC-3816

Condition: As a member of a MAROPS team/detachment, given—
1. FM 3-05.212, *Special Forces Waterborne Operations,* Chapters 1 and 16.
2. FM 3-21.220, *Static Line Parachuting Techniques and Training.*
3. USSOCOM M350-6, *Special Operations Forces Infiltration/Exfiltration Techniques.*
4. Parachuting and rigging materials as specified in paragraph 2 above.
5. Individual swim gear (face mask, swim fins, buoyancy compensators, B-7 life preserver).
6. Supporting rigger personnel.
7. An appropriate fixed-wing aircraft.
8. An outdoor water training area with a water drop zone (WDZ).
9. A beach landing site.
10. A MACO briefing.
11. A requirement to complete a tactical water airborne operation using a fixed-wing aircraft as an intermediate delivery platform.

Standard: IAW cited references, complete a tactical water airborne operation that includes—
1. Passing a diving supervisor's and jump master's personnel and equipment inspection.
2. Exiting the aircraft using the proper procedures.
3. Implementing the proper procedure when feet enter the water.
4. Implementing the appropriate actions after entering the water.

Navigate a Small Boat
331-WIC-3818

Condition: As a navigator of a small boat, given—
1. FM 3-05.212, *Special Forces Waterborne Operations,* Chapters 1 through 5.
2. A combat rubber raiding craft (CRRC) complete with organic equipment and accessories.
3. An outboard motor, fuel line, and fuel tanks/bladders.
4. CRRC navigation console with compass and a knotmeter.
5. Nautical charts of the operational area.
6. A GPS.
7. Navigational tools (as required) IAW FM 3-05.212, Chapter 5.
8. A scenario containing a hypothetical tactical situation.
9. An operation order (OPORD) from higher headquarters directing the MAROPS team/detachment to proceed from a launch point to a designated area in the area of operations/target area using small boats as the delivery platforms.
10. A requirement to produce a navigation plan prepared during premission planning.
11. A requirement to produce navigational charts based on the navigation plan and the tactical data given in the scenario and the OPORD from higher headquarters.
12. A requirement to navigate a CRRC from a designated start point to a designated ending point in the area of operations/target area using the navigation plan, nautical charts, and the appropriate navigation tools and techniques.

Standard: IAW the cited references, navigate a CRRC from a designated start point to a designated end point using dead-reckoning, piloting, and electronic navigation techniques IAW the navigation plan.

Appendix B

Complete Kayak Operations
331-WIC-3820

Condition: As a member of a MAROPS team/detachment, given—
1. FM 3-05.212, *Special Forces Waterborne Operations*, Chapters 1 and 8.
2. A one- or two-man kayak and paddles.
3. A six-kayak detachment.
4. An outdoor water training area.
5. A beach landing site (BLS).
6. A mother craft.
7. A scenario containing a hypothetical tactical situation.
8. An operation order (OPORD) from higher headquarters directing the MAROPS teams/detachments to conduct waterborne operations kayaks during day or night.

Standard: IAW the cited references, perform the following:
1. Assemble the kayak.
2. Prepare the kayak for waterborne operations.
3. Load the kayak with personal and team equipment.
4. Launch the kayak from a—
   a. Beach.
   b. Beach in disturbed water.
   c. Bank.
   d. Mother craft.
5. Paddle a kayak using the—
   a. Forward paddle stroke.
   b. Backward paddle stroke.
   c. Sweep paddle stroke.
   d. Draw paddle stroke.
   e. Slap support/recovery paddle stroke.
6. Land a kayak on a—
   a. Beach.
   b. Beach with disturbed water.
   c. Mother craft.
7. Operate/maneuver a kayak in the surf.
8. Sail a kayak.
9. Tactically employ a kayak, using the—
   a. Daylight movement formation as follows:
      (1) The file formation.
      (2) The staggered trail (right or left) formation.
   b. Nighttime movement formation as follows:
      (1) Echelon left formation.
      (2) Echelon right formation.
   c. Defensive movement formations as follows:
      (1) The wedge formation.
      (2) The modified wedge formation.
10. Cache—
    a. A disassembled kayak.
    b. An assembled kayak.
11. Portage—
    a. A disassembled kayak.
    b. An assembled kayak.
12. Navigate a kayak to a designated point.
13. React to the following emergency situations:
    a. A storm.
    b. A capsized kayak, using—
       (1) Cockpit method of self-rescue.
       (2) Pump and reentry method.
       (3) Raft method.
14. Perform kayak maintenance and repair:
    a. Short-term and long-term maintenance procedures.
    b. Repair procedures.

Employ a Submarine as an Infiltration Platform
331-WIC-3827

Condition: As a member of a MAROPS team/detachment, given—
1. FM 3-05.212, *Special Forces Waterborne Operations,* Chapter 14.
2. A U.S. Navy submarine or a mobile submarine simulator.
3. An outdoor water training area.
4. A beach landing site (BLS).
5. A combat rubber raiding craft (CRRC) with organic equipment and a 35-hp gasoline engine/outboard motor (OBM).
6. A long-haul Commando kayak.
7. Individual swim gear.
8. Individual scuba gear.
9. A requirement to use a U.S. Navy submarine, equipped with or without a dry-deck shelter, as an intermediate infiltration platform.

Standard: IAW cited references, employ a submarine as an infiltration platform as follows:
1. Complete preliminary mission planning and coordination with the submarine commander and crew.
2. Stow the detachment's equipment and cargo in the appropriate spaces of the submarine.
3. Rig the detachment's inflatable boats for launching from a submarine.
4. Debark the detachment's swimmers/divers/boat teams from the submarine using the appropriate procedures:
    a. Surface debarkation.
    b. Submerged debarkation using the escape trunk.
5. Surface debark the detachment's small boats from the submarine using the appropriate procedures:
    a. Inflatable boats:
        (1) Dry-deck launching.
        (2) Wet-deck launching.
    b. Kayaks:
        (1) Dry-deck launching.
        (2) Wet-deck launching.
6. Employ the proper safety measures during surface launchings.
7. Employ the appropriate actions after debarkation of boat teams.
8. Recover swimmers/divers/boat teams using the appropriate surface procedures:
    a. Dry-deck recovery.
    b. Wet-deck recovery.

Complete Beach Landing Site (BLS) Operations
331-WIC-3831

Condition: As a member of a MAROPS team/detachment, given—
1. FM 3-05.212, *Special Forces Waterborne Operations*, Chapter 9.
2. An outdoor water training area.
3. A beach landing site (BLS).
4. Inflatable small boats with organic equipment and a 35-hp gasoline engine/outboard motor (OBM).
5. Individual swim gear.
6. Individual scuba gear.
7. Individual equipment and weapons.
8. A requirement from higher headquarters to conduct BLS operations in support of a waterborne mission.

Standard: IAW cited references, complete beach landing site (BLS) operations as follows:
1. Ensure boat teams assemble in the designated offshore boat pool.
2. Issue last minute orders, confirm signal to be used during the operation, and review the scout swimmer rendezvous plan.
3. Make sure scout swimmers—
    a. Enter the water.
    b. Negotiate the surf zone.
    c. Swim to the beach.
    d. Conduct the required beach reconnaissance.
    e. Conduct a surf observation and complete a surf observation report (SUROBSREP).
    f. If the beach is secure, signal the boats to come in using the proper procedures.
    g. If the beach or surf conditions are not favorable, signal the boats to initiate the appropriate withdrawal procedures.
    h. If enemy contact is made, or scout swimmers are compromised, signal the boats to initiate withdrawal procedures.
4. Ensure boat team recovers scout swimmers with the appropriate recovery procedures.
5. Use the appropriate signal devices as required by the tactical situation.
6. Establish the appropriate mission signals for—
    a. Safe.
    b. Delay.
    c. Abort.
    d. Absence.
    e. Pickup craft.
7. Establish the appropriate signaling method:
    a. One Lamp System.
    b. Two Lamp System.
8. Employ the appropriate infiltration swim line:
    a. Online.
    b. Column (In-Line).
    c. Column (Offset).
9. Employ the appropriate surface swimming techniques.
10. Employ the appropriate underwater swimming techniques, as required.

Appendix C
Equipment Maintenance

The maintenance material management (3-M) system is the standard Navy system of maintenance and will be used by all SOF diving units for maintaining its diving equipment. The NAVSAFECEN conducts safety surveys of the Divers Life Support Maintenance Facility at least once every 2 years.

DEFINITIONS

C-1. The following definitions are provided to clarify Navy terms referring to the maintenance of dive equipment:
- *3-M System*—provides maintenance and material managers with the means to plan, acquire, organize, direct, control, and evaluate manpower and material resources in support of maintenance.
- *Planned Maintenance System (PMS)*—provides a standard procedure for planning, scheduling, controlling, and performing maintenance on divers life support systems and associated equipment.
- *Maintenance Data System (MDS)*—is the reporting means used by maintenance personnel, maintenance and material managers, and logistic support centers for analyzing maintenance and logistic support problems, correcting or updating maintenance requirements, and providing feedback to the field on maintenance or support problems and solutions.
- *Quality Assurance Program (QAP)*—ensures conformance to technical specifications during maintenance and that certified systems maintain technical specifications at all times.
- *Quality Assurance Officer (QAO)*—plans, monitors, and executes the QAP IAW OPNAVINSTs.
- *Planned Maintenance System Feedback Reports*—provides a timely means for units, logistics centers, and the Naval Sea Systems Command to document the effectiveness of equipment design, reliability, and maintainability. It allows divers to provide input on safety procedures, support problems, and discrepancies in technical manuals or maintenance requirements.
- *Maintenance Requirements Card*—provides detailed procedures for performing maintenance on equipment.
- *Equipment Guide List (EGL)*—is an index card listing the serial number and location of the equipment requiring maintenance.
- *Maintenance Index Page (MIP)*—is a document that lists all maintenance requirements for a specific piece of equipment.
- *List of Effective Pages (LOEP)*—is an index that lists each MIP by line item. The LOEP identifies which MIP to refer to for any equipment in the PMS.

OBJECTIVES

C-2. The following objectives are supported by the 3-M system:
- Achieve uniform maintenance standards and criteria.
- Effectively use available manpower and material resources in maintenance and support efforts.
- Document information relating to maintenance and support activities.
- Improve the maintainability and reliability of systems and equipment by documenting maintenance data for analysis.

Appendix C

- Provide the means for reporting configuration changes.
- Identify and reduce the cost of maintenance and support in terms of manpower and material resources.
- Provide the means to schedule, plan, manage, and track maintenance.
- Provide data on which to base improvements in equipment design and spare parts support.

RESPONSIBILITIES

C-3. The USSOCOM component commanders will promulgate regulations and instructions to delineate the responsibilities at each level of command to facilitate compliance with the 3-M system.

PLANNED MAINTENANCE SYSTEM

C-4. Maintenance will be performed IAW OPNAVINSTs and the appropriate MRC. A record of maintenance actions will be kept in a log for each type of diving equipment with the following minimum entries:

- Start-up and in-service maintenance records.
- Date.
- MRC code.
- Equipment serial number, if applicable.
- REC and Failure Analysis or Inadequacy Report (FAIR) forms.
- Calibration, test, and air sampling results, if applicable.

C-5. EGLs will be attached to the MRCs to assist the maintenance personnel in locating the equipment to be maintained. Each unit will be issued a PMS package that includes an LOEP, maintenance index cards, an MRC deck, and a PMS supplement that includes a lubricants, compounds, and cleaning agents cross guide list. The PMS package will be updated semiannually by the Naval Sea Support Center.

C-6. When diving equipment is inoperative due to planned or corrective maintenance, the item will be tagged-out to ensure that its use is prevented. A tag-out danger ticket will be placed on the item for identification purposes and a tag-out control log will be maintained as a record.

C-7. Reentry control provides a quality assurance process in which system integrity, material condition, cleanliness, and gas purity are maintained, controlled, and documented. REC procedures are required after any time the prescribed boundaries of a certified system are disturbed. All maintenance on the equipment will be documented on REC forms and listed in the REC log.

INACTIVE EQUIPMENT MAINTENANCE

C-8. Inactive equipment maintenance (IEM) is a method of reducing PMS requirements during periods of equipment inactivity. This process includes lay-up, periodic maintenance, start-up, and operation tests. IEM will be used when equipment is not to be used for operational requirements or is in lay-up status for 30 or more days, without degrading material condition or jeopardizing future operational reliability.

C-9. Equipment placed in IEM will be maintained at a readiness level that will require no more than 72 hours of maintenance to place it back in operation for deployment. All IEM actions will be handled IAW the OPNAVINSTs.

CERTIFICATION

C-10. Certain items of maritime equipment require certification. The Naval Sea Systems Command conducts certification of certain scuba systems. Normally, certification tickets are issued for 2 years. Occasionally, provisional tickets are issued for a period of less than 2 years. Hyperbaric chambers are certified by the Navy Facilities Engineering Command.

INSPECTIONS

C-11. Inspections of each DLSMF are conducted at least once each fiscal year by the Naval Safety Center. These are comprehensive inspections of certain scuba systems and the documentation required by the 3-M system.

This page intentionally left blank.

Appendix D
Diving Supervisor Personnel Inspection

OPEN-CIRCUIT DIVER

D-1. The following paragraphs outline the diving supervisor's inspection on an open-circuit diver:
- **Look at Board:** Ask the diver how much air he has. Visually look at the minimum gas value requirement written on the board. Ensure sufficient air for the dive in psig is present.
- **Check the Mask:** Check for tempered glass to ensure it is not broken or cracked and is worn properly; ensure the metal buckles on the sides are stowed in the keepers; make sure the head strap and face seal are serviceable, and the metal seal securing the glass is in place.
- **Check the Gas Supply:** Ensure the air has been turned all the way on and then backed off 1/4 to 1/2 turn.
- **Inspect the Reserve Handle:** Ensure it is in the UP position. Turn it to the DOWN position for 2 seconds and then back up. If a rush of cascading air was heard, the amount of air in the tanks cannot be validated. They must be gauged again.
- **Inspect the Regulator Attachment to the Tanks:** Ensure the second-stage hose is oriented in the 9 o'clock position. Form an OK with the left hand and trace the regulator over the diver's right shoulder to the second stage to ensure it is not misrouted. Inspect the mouthpiece for serviceability. Have the diver take 3 to 5 normal breaths and stow the regulator in the BCM pouch.
- **Inspect the BCM Dump Valve and Lanyard:** Pull down on the dump valve lanyard and visually inspect the activation of the dump valve. Trace the lanyard protector sleeve from top to bottom for serviceability.
- **Inspect the CO2 Firing Mechanism and Lanyard:** Pull the BCM pocket from the Velcro patches on the vest and inspect the CO2 cartridge. Ask the diver, WHAT IS THE STATUS OF YOUR CARTRIDGE? The diver will respond with, ONE 38-GRAM CARTRIDGE, CLEANED, WEIGHED, LUBED, AND NOT YET FIRED. Verify the cartridge is tight, then loosen to make sure it is not frozen in the actuator, then retighten. Inspect the firing lanyard and ball for serviceability and that the firing arm is in the unfired position. Replace the lanyard in the proper position.
- **Inspect the Oral Inflation Hose, Dump Valve, and Power Inflator Mechanism:** Inspect the oral inflation hose from top to bottom for serviceability. Inspect the mouthpiece, depress the dump button at the bottom of the multipurpose valve, and observe the valve in the mouthpiece open and shut. Activate the power inflator button and listen for a rush of air. Verify presence of whistle.
- **Inspect the Low-Pressure Inflation Hose:** Check the low-pressure inflation hose connection to the multipurpose valve; trace the hose up over the diver's left shoulder all the way to the first-stage regulator (verify the hose is at the 6 o'clock position). Ensure the hose is not misrouted under any equipment.
- **Inspect BCM Serviceability:** Check for any significant rips, tears, or sewing that may preclude the BCM from functioning.
- **Inspect the Left-Side Tank Shoulder Strap:** Inspect shoulder strap from the top of the backpack to the bottom. Ensure the strap is not twisted or frayed and the correct quick-release is in place.
- **Inspect the High-Pressure Hose:** Ensure the high-pressure hose is oriented in the 3 o'clock position and is routed under the tank waist strap but over the weight belt on the diver's left side. Verify that the gauge is operational within +/– 100 psig of the calibrated gauge reading.

Appendix D

- **Inspect the Right-Side Tank Shoulder Strap:** Inspect shoulder strap from the top of the backpack to the bottom. Ensure the strap is not twisted or frayed and the correct quick-release is in place.
- **Inspect the Tank Waist Strap:** Inspect the waist strap from the right side of the backpack to the left side. Check for proper routing above the weight belt making sure it is not twisted and that the running end is in the buckle with a right hand quick-release.
- **Inspect the Weight Belt:** Look for a right-hand quick-release and check that the weight belt is not twisted, frayed, or unserviceable and that it is routed over all other straps but underneath the dive tool with scabbard.
- **Inspect the Dive Tool:** Ensure the dive tool is in the scabbard and attached to the diver's right-side BCM strap and when lifted, is unrestricted by the tank waist strap and weight belt. Ensure the MK 124 Day or Night Emergency Signal device is taped to the scabbard, day/smoke side up.
- **Inspect the BCM Waist Strap and Crotch Strap:** Ensure the fastex buckles are fastened. Make sure the straps are routed underneath the diver's weight belt. Maintaining control of the crotch strap, trace the strap underneath the diver and tell the diver to turn and bend forward at the waist. Continue to inspect the straps up through the space between the diver's tanks for any frays or twists. Ensure the straps are routed underneath the diver's weight belt.
- **Inspect the Weight Belt:** Follow the weight belt around to ensure no twists are visible and the weight belt is not misrouted under any straps. Tell the diver to stand erect and turn.
- **Inspect the Fins:** Ensure the fins are not torn or ripped and the boot portion is serviceable. The buckles on both sides are present and the fin strap is stowed in each keeper.
- **Inspect the Watch:** The watch is required by each diver. Ensure it is present, serviceable, suited for the dive, and that it is accessible to the diver.
- **Inspect the Depth Gauge:** The depth gauge is required by each diver. Ensure it is present; the glass is not cracked or broken, the needle is not indicating depth, the straps are serviceable, and that it is accessible to the diver.
- **Inspect the Compass:** One compass is required per dive team. Ensure the compass functions and the lens is not cracked or broken.
- **Inspect the Buoy and Buoy Line:** The dive team is required to have a buoy with buoy line. Ensure the two function and are serviceable.
- **Inspect the Buddy Line:** The dive team is required to have a buddy line. Ensure it is serviceable and is approximately six to eight feet in length. When inspecting the buoy with buoy line and buddy line, sound off appropriately with buoy, buoy line, and buddy line.

CLOSED-CIRCUIT DIVER

D-2. The following paragraphs outline the diving supervisor's inspection on a closed-circuit diver:

- **Look at Board:** Ask the diver how much oxygen in bar he has. Visually look at the minimum gas value requirement written on the board. Ensure sufficient oxygen in bar for the dive is present.
- **Check the Mask:** Check for tempered glass to ensure it is not broken or cracked and is worn properly. Ensure the plastic fasteners are serviceable and the remaining strap is stowed in the keepers, the head strap and face seal are serviceable, and the seal securing the glass is in place.
- **Check the Gas Supply:** Verify oxygen is not in the system by looking at the pressure gauge. It should read 0.
- **Turn on Oxygen Supply:** Check shut, then open the supply valve 1/4 to 1/2 turn to verify that the diver has sufficient oxygen in bar for the dive and that the gauge is operational within +/- 10 bar of the calibrated gauge reading.
- **Fill Breathing Bag:** Depress the lung demand-bypass valve for 4 to 6 seconds, checking for operability and filling of the breathing bag.
- **Turn Off Oxygen Supply:** Turn off the oxygen supply.

Diving Supervisor Personnel Inspection

- **Inspect the Mouthpiece and Dive Surface Valve:** Ensure it is not ripped, torn, or unserviceable. Check to make sure the red dot is up and to your right. Ensure the hoses are not kinked and are even in length to prevent undo fatigue during the dive. Check the neck strap for serviceability; have the diver open his mouth and place the mouthpiece in his mouth.
- **Open the Dive/Surface Valve:** Rotate the dive/surface valve to the up or dive position.
- **Check the Breathing Loop:**
    - Pinch off the inhalation/supply hose. Ask the diver to inhale, CAN YOU INHALE? If he meets resistance, the one-way valve is operational in the exhaust hose. If he does not meet resistance, release the hose, re-pinch off the hose, and repeat the procedure a second time. Release the inhalation/supply hose. Ask the diver, CAN YOU INHALE NOW? He should be able to inhale through the mouthpiece, verifying the operational status of the one-way valve in the supply hose.
    - Pinch off exhalation/exhaust hose. Ask the diver to exhale, CAN YOU EXHALE? If he meets resistance, the one-way valve is operational in the supply hose. If he does not meet resistance, release the hose, repinch off the hose, and repeat the procedure a second time. Release the exhalation/exhaust hose. Ask the diver CAN YOU EXHALE NOW? He should be able to exhale through the mouthpiece, verifying the operational status of the one-way valve in the exhaust hose.
- **Close the Dive/Surface Valve:** Rotate the dive/surface valve to the down or surface position and remove the mouthpiece from the diver's mouth.
- **Inspect Multipurpose Valve Assembly:** Unfasten the assembly from the Velcro strap and inspect the hose for rips, tears, or unserviceability. Ensure the top and bottom portions have tie fasteners. Depress and turn the multipurpose valve to ensure it will open and close. Ensure the whistle is present and serviceable. Check for the C-ring in the bottom and tap the spring tension housing. Secure the assembly with the Velcro strap.
- **Inspect the Left Secumar Cylinder:** Ask the diver, HOW MUCH AIR IS IN THIS CYLINDER? and ensure it is sufficient for the dive. Minimum psig is 2400. Check tightness of the cylinder fitting on the Secumar vest. Crack the valve on the cylinder and listen for the rush of air introduced into the Secumar vest.
- **Inspect the Left-Side Drain Plug:** Ensure the drain plug and lanyard are present, and it is seated all the way into the vest.
- **Check the TSK 2-42 Secumar Life Vest:** Ensure the vest is worn properly and that it is not ripped, torn, or unserviceable.
- **Inspect the Chest Strap:** Ensure the strap is horizontal in front of the LAR V fairing, that it runs through the crotch strap, that it is not twisted, and the buckle is fastened properly.
- **Inspect the Right Secumar Cylinder:** Ask the diver, HOW MUCH AIR IS IN THIS CYLINDER? and ensure it is sufficient for the dive; minimum psig is 2400. Check tightness of the cylinder fitting on the Secumar vest. Crack the valve on the cylinder and listen for the rush of air introduced into the Secumar vest.
- **Inspect the Right-Side Drain Plug:** Ensure the drain plug and lanyard are present, and it is seated all the way into the vest.
- **Check the TSK 2-42 Secumar Live Vest:** Ensure the vest is worn properly and that it is not ripped, torn, or unserviceable.
- **Inspect the Dive Tool:** Ensure the dive tool with scabbard is attached to the LAR V on the right side with the equipment straps provided and no quick-release is used. Ensure the MK 124 Day/Night Emergency Signal device is taped to the scabbard day/smoke side up.
- **Inspect the Crotch Strap:** Ensure it is not twisted in front and check for proper routing under the weight belt.
- **Inspect the Weight Belt:** Check for a right-hand quick-release on the right hip. Ensure the belt is not torn, frayed, twisted, or unserviceable and is routed over the crotch strap. Instruct the diver to turn and trace it around the back checking for the same while maintaining control of the crotch strap.

Appendix D

- **Inspect the Crotch Strap:** Trace it from the front; ensure the strap is not twisted and is routed under the weight belt.
- **Inspect the Horizontal Back Strap:** Ensure it is fastened at the buckle and that the vertical back strap is running through it. Check for twists.
- **Inspect the Vertical Back Strap:** Trace it from the crotch strap area and ensure it is not twisted or misrouted over the weight belt.
- **Inspect the LAR V Back Straps:** Trace the strap from the top left of the diver to the bottom right ensuring no twists exist and that it has a quick-release at the buckle at the right side. Trace the strap from the bottom left of the diver to the top right of the diver and verify the same. Instruct the diver to turn.
- **Inspect the Fins:** Ensure the fins are not torn or ripped and the boot portion is serviceable. The buckles on both sides are present and the fin strap is stowed in each keeper.
- **Inspect the Watch:** The watch is required by each diver. Ensure it is present, serviceable, suited for the dive, and is accessible to the diver.
- **Inspect the Depth Gauge:** The depth gauge is required by each diver. Ensure it is present, the glass is not cracked or broken, the needle is not indicating depth, the straps are serviceable, and it is accessible to the diver.
- **Inspect the Compass:** One compass is required per dive team. Ensure the compass functions, the lens is not cracked or broken, the straps are serviceable, and it is accessible to the diver.
- **Inspect the Buoy and Buoy Line:** The dive team is required to have a buoy with buoy line. Ensure the two function and are serviceable.
- **Inspect the Buddy Line:** The dive team is required to have a buddy line. Ensure it is serviceable and is approximately six to eight feet in length. When inspecting the buoy with buoy line and buddy line, sound off with buoy, buoy line, and buddy line.

D-3. Figure D-1, pages D-4 and D-5, explains how the diving supervisor performs a table-top inspection of each part of the diving equipment and the procedure that must take place.

	DIVING SUPERVISOR TABLE-TOP INSPECTION
Breathing Bag	Verify the breathing bag was observed for 2 minutes, and no leaks were observed by diver. Ask diver, HAVE YOU OBSERVED YOUR BREATHING BAG FOR 2 MINUTES?
Scrubber	Verify the scrubber canister is filled with the approved CO2 scrubber. Ask diver, WHAT IS YOUR SCRUBBER CANISTER FILLED WITH?
Predive Sheet	Verify the predive setup; steps 1 through 17 on DA Form 7532-R have been completed including the heading.
Mouthpiece	Ensure the dive surface valve is in the surface/down position (look inside mouthpiece for screw, and not open). Red dot on mouthpiece is up and to the left of the diver. Check head strap.
Hoses	Check inhalation and exhalation hoses; ensure they are secured to the mouthpiece, and hoses hang free. Ensure collar connects tight to mouthpiece and hoses pose no stress to the diver's jaw (twists).
Exhaust Hose	Check exhaust hose to hose slot (properly installed) to bayonet fitting of canister (turned 90 degrees, secured in fitting, and wave washer present).

Figure D-1. Table-top inspection

Diving Supervisor Personnel Inspection

	DIVING SUPERVISOR TABLE-TOP INSPECTION (continued)
Canister	Ensure canister lid seal is on the left as the diver is wearing the LAR-V. Follow seal down while ensuring a proper seal.
Retaining Strap Rod	Verify C-clips.
Canister Retaining Strap	Verify retaining strap is secured properly. Verify canister and breathing bag alignment and seating.
Breathing Bag	Check breathing bag bayonet fitting is secured to canister.
Inhalation Hose	Check inhalation hose connected to breathing bag, and that muff is stuffed into bag.
Demand Valve	Check tight and not stripped the breathing bag collar connector to demand valve (back off collar connector 1/4 to 1/2 turn to ensure plastic threads did not get stripped during connection, then retighten).
Low- and High-Pressure Lines	Check two-finger, low-pressure line fitting tight to demand valve. Check low- and high-pressure lines to ensure no cuts, kinks, or damage (misalignment), and they follow proper path. Check two-finger, low-pressure line fitting. Tighten to pressure reducer.
Reducer Hand-Wheel	Check reducer hand-wheel is finger tight.
O2 Bottle Retaining Straps	Ensure O2 bottle is secured with twin retaining straps, and verify two C-clamps on secondary (right) retainer strap. Ensure free play in breathing bag safety cord.
Back Straps	Ensure the two diagonal back straps are present and serviceable to include adjustable buckles and two buckle straps present and serviceable (complete through buckle).
Equipment Straps	Ensure two equipment straps are present and serviceable.
Block 18	Initial Block 18 on DA Form 7532-R, verifying the previous steps.

Figure D-1. Table-top inspection (continued)

This page intentionally left blank.

Appendix E

MK 25 MOD 2 Predive and Postdive Checklists

MK 25 MOD 2 PREDIVE CHECKLIST
For use of this form, see FM 3-05.212; the proponent agency is TRADOC.

DIVER NAME			RATE			DATE (YYYYMMDD)	RIG #
O2 BOTTLE: #1	#2	#3	PRESSURE: #1	#2	#3	BAR CO2 ABSORBENT	
S. BOTTLES: 1/1	2/2	3/3	PRESSURE: 1/1	2/2	3/3	PSIG LOCATION	
WATER TEMP			PREDIVE START TIME			PREDIVE COMPLETION	
DIVE #2 TIME					DIVE #3 TIME		

NOTE: Refer to MRC for detailed step-by-step procedures. Inspect all components for dirt, damage, deterioration, or residual water during initial setup.

INITIALS			
DIVE 1	DIVE 2	DIVE 3	
			1. Gauge O2 bottle for minimum operating pressure and record above.
			2. Inspect or charge Secumar bottles, record pressure above.
			3. Inspect and fill scrubber canister with absorbent.
			4. Inspect equipment housing, straps, and hose slots.
			5. Check two-hand tight fittings on low pressure line.
			6. Inspect and check operation of the one-way valves in the supply and exhaust hoses.
			7. Inspect and attach supply and exhaust hoses to mouthpiece, close dive/surface valve.
			8. Inspect and attach the breathing bag to the scrubber canister.
			9. Attach the supply hose to the breathing bag and insert muff into bag.
			10. Attach the exhaust hose to the scrubber canister.
			11. Attach the breathing bag to the demand valve.
			12. Install scrubber canister in proper orientation, ensure canister retainer strap is in proper orientation.
			13. Insert the supply and exhaust hoses into the hose slots.
			14. Set rig on deck and inspect and attach full O2 bottle. Do not open.
			15. Install breathing bag safety cord.
			16. Fasten O2 bottle retaining straps. Check safety cord for loose fit.
			17. Deplete breathing bag. Close dive surface valve. Observe breathing bag for 2 minutes. No leakage allowed.
			18. Diving supervisor checks previous steps.
			19. Diving supervisor opens supply valve, records O2 bottle pressure. (Must be +/- 10 BAR of pressure from Step 1.)
			20. Diving supervisor depresses bypass knob until breathing bag is full. Do not overfill. (Normal fill time 6 to 8 seconds, maximum 12 seconds.)
			21. With the rig inflated, dip complete rig in water. No bubbles from gas leakage are allowed.
			22. Close supply valve, suck down breathing bag, close dive surface valve, and notify dive supervisor.

DIVER SIGNATURE	DIVING SUPERVISOR SIGNATURE
DIVE #2	DIVE #2
DIVE #3	DIVE #3

DA FORM 7532-R, SEP 2004 APD V1.00

Appendix E

MK 25 MOD 2 UBA POSTDIVE CHECKLIST
For use of this form, see FM 3-05.212; the proponent agency is TRADOC.

DIVER NAME	DATE (YYYYMMDD)	RIG #

NOTE 1: Refer to MRC for detailed step-by-step procedures.

NOTE 2: For repetitive dives using the same LAR V rig, a complete postdive need only be performed at the end of the dive day, provided cumulative time of dive does not exceed canister duration limits.

INITIALS	
	1. Ensure dive surface valve is in the surface (closed) position.
	2. Secure the system and remove apparatus from diver.
	3. Rinse the rig in fresh water.
	4. Record oxygen bottle pressure: _____ BAR.
	5. Close supply valve and vent pneumatic system.
	6. Remove oxygen bottle and insert protective plug in regulator. Place protective cap on supply valve and place in rack.
	7. Remove retainer strap from canister.
	8. Disconnect breathing bag from demand valve.
	9. Disconnect mouthpiece and hoses.
	10. Disconnect breathing bag from scrubber canister.
	11. Rinse mouthpiece, hoses, and breathing bag in fresh water; remove excess water; and hang components to dry.
	12. Empty, clean, and dry scrubber canister.
	13. Ensure automatic demand valve is completely dry.
	14. Store equipment in proper condition and place, with dive surface in open position.

REMARKS

POSTDIVE COMPLETION TIME	DIVER SIGNATURE	DIVING SUPERVISOR SIGNATURE

DA FORM 7533-R, SEP 2004

Appendix F
Diver Propulsion Device

This appendix provides information for introductory training and safety on the DPD. This information will familiarize personnel with DPD team composition, organization, and duties. Basic DPD-handling drills and emergency procedures are also discussed. These procedures explain how to safely operate a DPD in all conditions. The DPD (Systems Training Integration and Devices Directorate [STIDD]) is most likely to be used by the SFODs and the USMC reconnaissance units.

Note. When coordinating with sister Services, the SFOD must be aware that the term DPD only applies to the STIDD.

BASIC USE

F-1. The STIDD DPD offers fully equipped divers a fast, rugged, and reliable vehicle for surface and subsurface mobility. Built of marine-grade aluminum, the DPD enables divers to travel farther and faster with more equipment. The STIDD can carry one fully equipped diver with up to 3 ft3 (85,000 cm3) of neutrally buoyant cargo at speeds up to 3.0 kts or to a maximum range up to 6.6 nm. The DPD can also carry two fully equipped divers and their cargo at speeds up to 2.7 kts or to a maximum range up to 5.4 nm. Additional specifications are listed in Figure F-1, pages F-1 and F-2.

F-2. Two personnel can carry and launch the DPD (Figure F-2, page F-2). It is factory trimmed to neutral buoyancy in saltwater. The DPD has a freshwater foam block that can be installed to provide neutral buoyancy in freshwater.

SPECIFICATIONS

Length	Fully deployed—87.8 inches (223cm); stowed configuration for transport—54.3 inches (130cm)
Beam/Depth	Fully deployed—24 inches (61cm); Stowed configuration—24 inches (61cm) in diameter
Weight	With standard lithium ion (Li-Ion) battery—159 lb dry (72 kg)
Cargo Volume	3 ft3 (85,000 cm3) of volume for up to 100 lb (46 kg) of neutrally buoyant cargo
Operating Depth	50 FSW maximum
Speed	1 Diver 2 Divers; Cruise—2.3 kts 2.0 kts 1 Diver 2 Divers; Full—3.0 kts 2.7 kts
Range	@ cruise speed—6.6 nm 5.4 nm (single battery @ full speed—5.2 nm 3.8 nm
Battery	Lithium-ion, 33.7v max–26v nominal
Motor Control	Solid state pulse width modulator (PWM) motor controller

Figure F-1. Diver propulsion device characteristics

Appendix F

Propulsion	Infinitely variable speed 26 VDC electric thruster
Control	Steering and depth control via a single hand-operated control yoke
Navigation	Magnetic compass w/luminous dial and depth gauge
Electrical Gauge	Battery status indicator (BSI) graphically displays battery capacity remaining
Hull	Welded marine-grade aluminum, hard coat anodized
Floatation	Closed-cell PVC foam core composite
Viewport	12-inch (305mm) diameter clear polycarbonate

Figure F-1. Diver propulsion device characteristics (continued)

Figure F-2. DPD collapsed for launch or storage

COMPONENTS

F-3. The DPD hull is divided into two major sections—the fore body hull section and the aft body hull section. Figure F-3 shows the two sections.

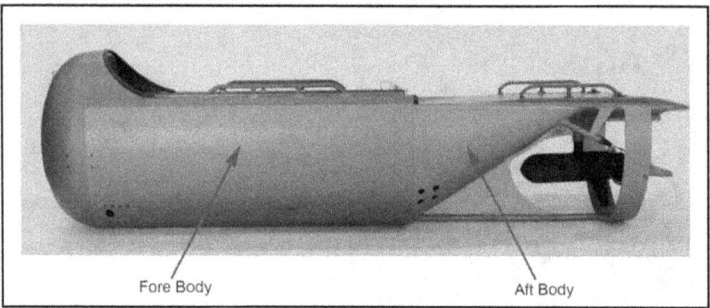

Figure F-3. Hull subsections

FORE BODY HULL SECTION

F-4. The fore body hull section is fabricated from a partially enclosed 24-inch (61 cm) diameter aluminum cylinder, with an ellipsoidal head on the bow. This section provides the swimmer with a comfortable ride, shielding him and the cargo from the drag of flowing water during transits. The edge of this cylinder is shrouded with piping to provide safe ingress and egress for the swimmer, as well as providing lift points during launching and recovery of the DPD.

> **CAUTION**
> 1. Never operate the thruster when it is out of the water. Operating the thruster when dry will destroy the shaft seals and cause the motor to flood.
> 2. Keep appendages clear of propeller path at all times. Failure to do so could cause injury when the propeller is in motion.
> 3. Improper handling or charging of the battery may cause electrical shock and other injuries. Always follow safe battery handling and charging procedures. Use only the DPD "muscles" battery charger to charge batteries or battery function will be damaged.

F-5. A clear 12-inch diameter (305 mm) polycarbonate view port is located at the center of the bow. This area provides the swimmer with a forward view of his surroundings during DPD operations (Figure F-4).

Figure F-4. DPD view port

F-6. DPD navigation controls are located on the center line of the DPD (Figure F-5, page F-4). The navigation instruments are configurable to be either above or below the view port. They include a magnetic compass and a depth gauge. The mounting bracket for the compass and depth gauge includes a fixture for a chemical light stick to be inserted. This light stick will illuminate both instruments during night operations. The throttle (variable speed controller for the thruster) is on the starboard side and the battery status indicator is on the port side. Both are in close reach and can be readily seen by the swimmer for ease of operation. The throttle is designed with a key mechanism attached to a dead-man lanyard. A spare throttle key is permanently attached to the throttle assembly in case the primary throttle key on the lanyard is lost. A foam block installed under the upper bow provides buoyancy and stability.

Appendix F

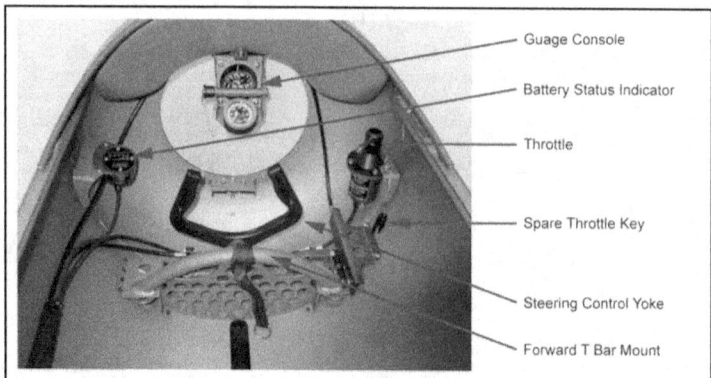

Figure F-5. DPD interior view

F-7. Cargo is stored in the bottom of the fore body section and is secured by a cargo net (Figure F-6). The DPD provides up to 3.0 cubic feet (85,000 cubic cm) of volume.

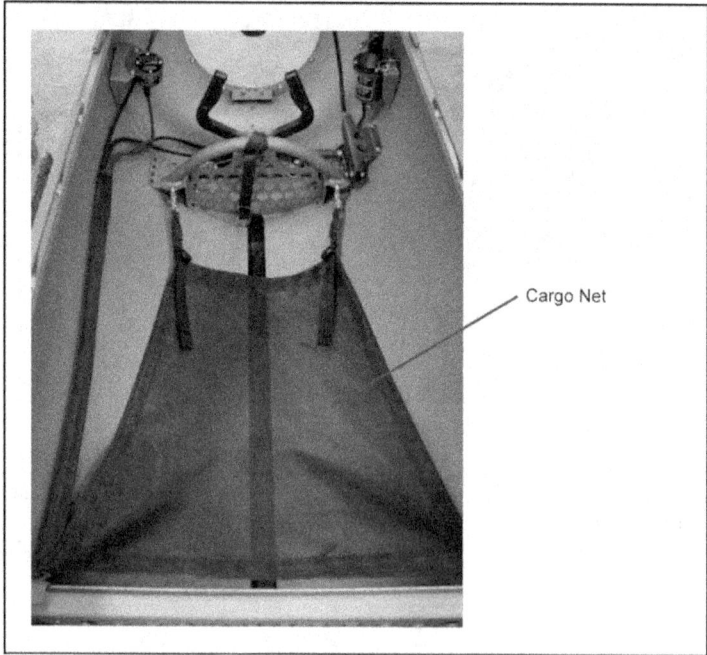

Figure F-6. Cargo area

AFT BODY HULL SECTION

F-8. The aft body section provides four key components necessary for optimum strength and operation:
- *Bulkhead.* The bulkhead strengthens the tail section of the hull and provides support and attachment points for the battery box.
- *Deck.* The deck is constructed of closed cell rigid PVC structural foam and provides the DPD with positive stability and neutral buoyancy in salt water. The deck is assembled with dual handrails to provide additional grab handles for the divers, two rear deck mounting screws to attach to the DPD, and a single deck eye used to attach the passenger's T-bar (Figure F-7).

CAUTION
The deck handrails should not be used to lift the DPD out from the water. Serious damage to the deck may result if these handrails are used for dry weight.

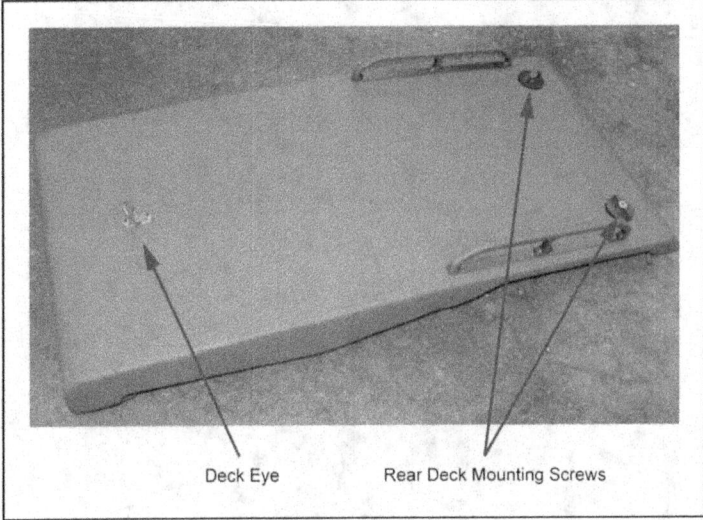

Figure F-7. Deck features

- *Tow Bar.* The T-bar is made up of a variable length web strap with a solid plastic rod at one end (Figure F-8, page F-6). The plastic rod is placed between the diver's legs; the other end of the T-bar is connected to a snap hook. This snap hook can be attached to the deck eye for passenger use or to the forward bulkhead eye for operator use.
- *Thruster Shroud.* The thruster shroud provides directional stability to the DPD when it is underway. It also provides protection for the swimmer, shielding him from accidental contact with the thruster or propeller. Lastly, the shroud protects the prop during operation and transportation (Figure F-9, page F-6). The shroud is fabricated from lightweight, marine-grade aluminum.

Appendix F

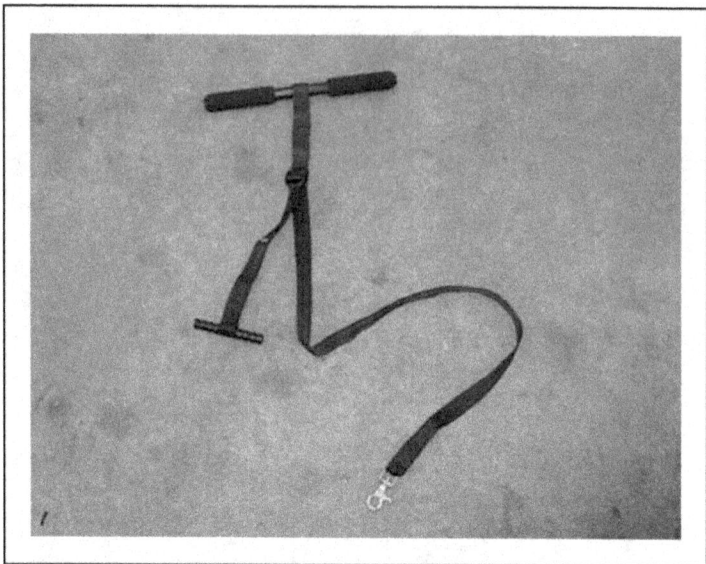

Figure F-8. T-bar

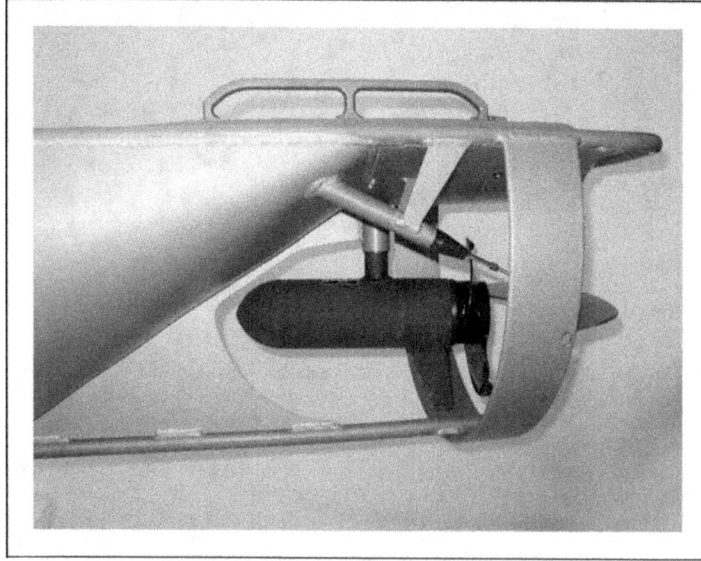

Figure F-9. Thruster shroud

ELECTRIC PROPULSION SUBSYSTEM

F-9. This system consists of an electric thruster driven by a 26VDC battery.

LITHIUM-ION BATTERY

F-10. The Li-Ion battery (Figure F-10) supplies a constant 26 volts to the motor resulting in a constant speed of approximately 2.7 kts for two combat-equipped divers at full speed and 2.0 kts at cruise speed. Speeds of 3.0 kts at full speed and 2.3 kts at cruise speed are attainable with a single combat-equipped diver. Speed values listed are approximate and may vary due to several factors. The battery is housed in a sealed, pressure-proof battery box (Figure F-11, page F-8) that is provided with a pressure relief valve and vacuum test port. The 65-pound box may be removed to facilitate battery charging; however, it is not recommended.

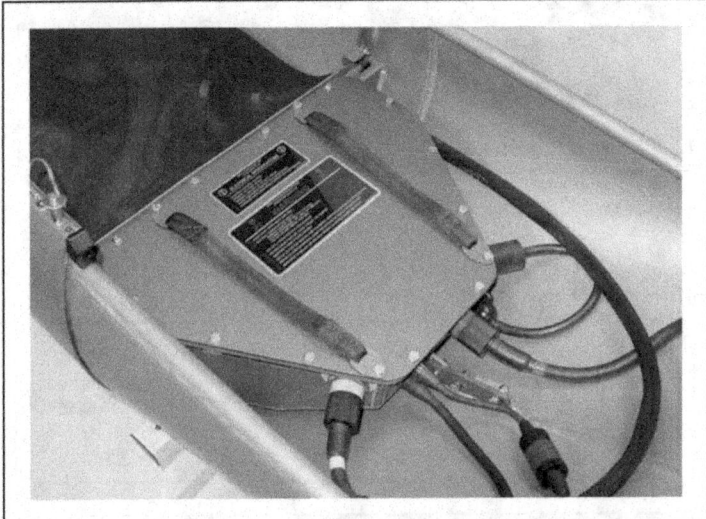

Figure F-10. Lithium-ion battery

EXTENDED RANGE DUAL BATTERY OPTION

F-11. As an option, the capability of dual Li-Ion batteries can be added to the standard DPD's propulsion battery system. This capability doubles the amp hours available for a mission, allowing for longer-range missions over a single battery system. This dual battery configuration becomes available by installing a second identical Li-Ion battery in the cargo area of the DPD. The batteries are designed with a battery link port that, when connected together via the interconnecting cable, allows the second battery to act as a slave power source to the primary battery's power source and control circuitry.

CAUTION

Prior to charging the DPD batteries, the interconnecting cable must be disconnected and all color-coded dummy plugs inserted prior to charging. Failure to do so may result in a battery fire.

Appendix F

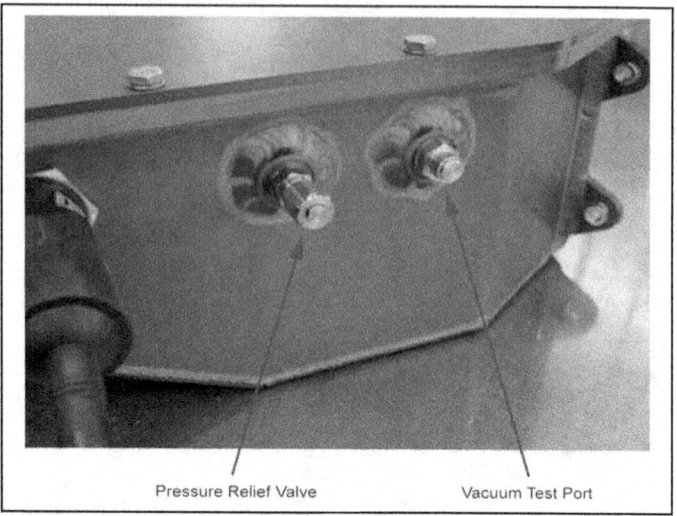

Figure F-11. Battery box

THRUSTER

F-12. A variable speed 26 VDC permanent magnet motor thruster unit powers the DPD. The thruster provides a maximum 74 pounds of thrust (Figure F-12).

CAUTION

Keep appendages clear of propeller path at all times. Failure to do so could cause injury when the propeller is in motion. Never operate the thruster when it is out of the water. Operating the thruster when dry will destroy the shaft seals and cause the motor to flood.

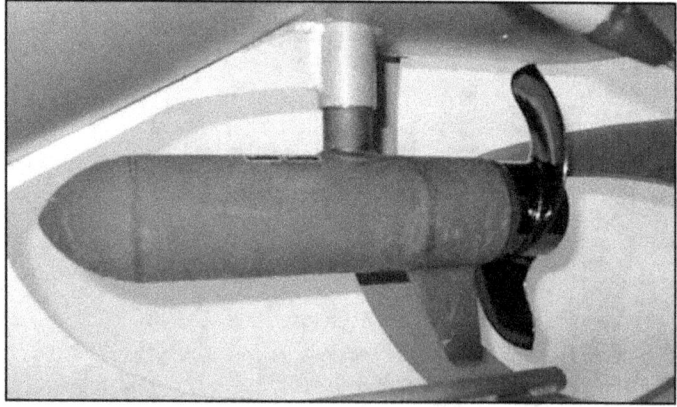

Figure F-12. Thruster

Diver Propulsion Device

THROTTLE

F-13. A throttle controls the speed of the DPD, with detents at the stop, cruise, and full-speed positions (Figure F-13). The throttle operates a pulse width modulator (PWM) that allows for infinitely variable speed.

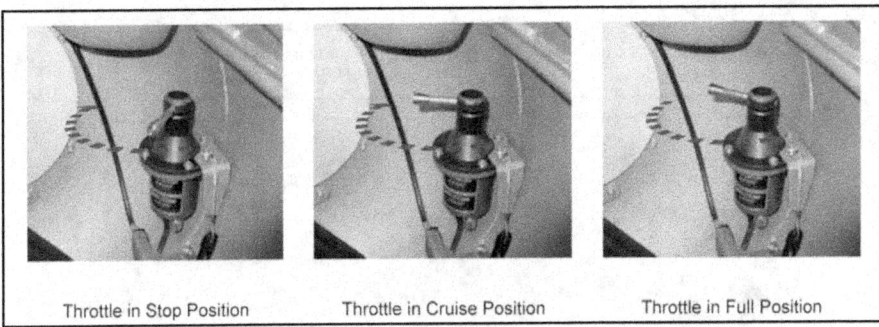

| Throttle in Stop Position | Throttle in Cruise Position | Throttle in Full Position |

Figure F-13. Throttle controls

BATTERY STATUS INDICATOR

F-14. The battery status indicator (Figure F-14) displays the following system information:
- Battery state-of-charge on a multicolor light bar.
- Each LED represents 20 percent of battery charge remaining.
- The left most green LED will turn to amber to warn the driver when only 20 percent battery capacity is remaining.

F-15. For planning purposes, a fully charged battery will operate for approximately 2 hours and 20 minutes at full speed. A fully discharged battery requires approximately 8 hours to recharge.

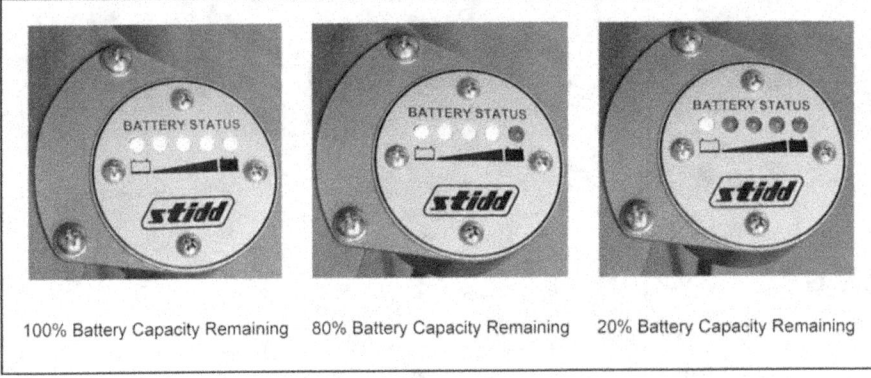

| 100% Battery Capacity Remaining | 80% Battery Capacity Remaining | 20% Battery Capacity Remaining |

Figure F-14. Battery status indicator

DPD BALLASTING

F-16. The DPD is equipped with components to properly ballast the vehicle. These components include nose foam, a rear deck assembly, and trim lead. The trim lead and nose foam are permanently fixed to the vehicle and should not be removed by the user. The deck foam can be removed to access the battery and

Appendix F

thruster steering components by removing the two rear deck mounting screws (Figure F-15). The deck assembly is specific to the vehicle and is marked with the hull number. The user should be sure the correct deck is mounted on the hull. It is critical that the deck and nose foam are properly secured to the vehicle. Should either of these components break loose while diving, the DPD will become negatively buoyant and will go into an uncontrollable dive.

Note. Confirm that the deck and nose foam are securely mounted to the DPD. Loss of either of these items while diving will result in an uncontrollable dive.

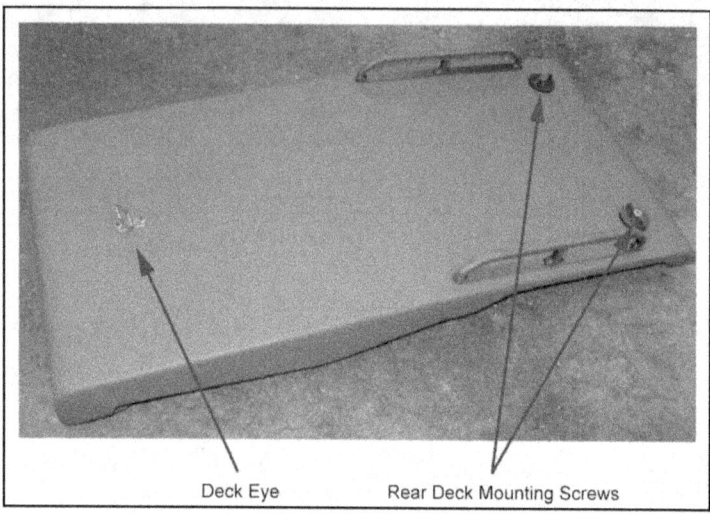

Figure F-15. Rear deck mounting screws

STERN PLANE

F-17. The diver uses the active stern plane (Figure F-16, page F-11) to control the rate of ascent and decent when the DPD is in operation. The active stern plane is mounted immediately aft the thruster directing the thrust in the desired direction. The active stern plane is connected to the control yoke via a push/pull cable and linkage system.

F-18. The DPD control yoke moves the active stern plane through an arc of approximately 60 degrees (+/–30°). The size and location of the active stern plane does not exceed the 24-inch diameter of the craft allowing the DPD to be passed through a 25-inch diameter hatch.

BOW PLANE

F-19. The DPD is equipped with provisions to add bow planes to the depth control yoke. If the DPD's cargo, driver, and passenger are neutrally buoyant, the use of bow planes is not necessary. The bow planes may assist in increasing depth control should the DPD's cargo, driver, and passenger not be correctly ballasted.

F-20. If bow planes are used to compensate for an incorrectly ballasted vehicle (to maintain depth control), the DPD's range and top speed will be reduced. Bow planes are attached by sliding the bow plane into the provided bushing until seated and pinning the bow plane shaft to the torque tube as Figure F-16, page F-11, shows.

> **CAUTION**
> Never ascend faster than 60 feet per minute. Exceeding this rate could potentially result in a gas embolism.

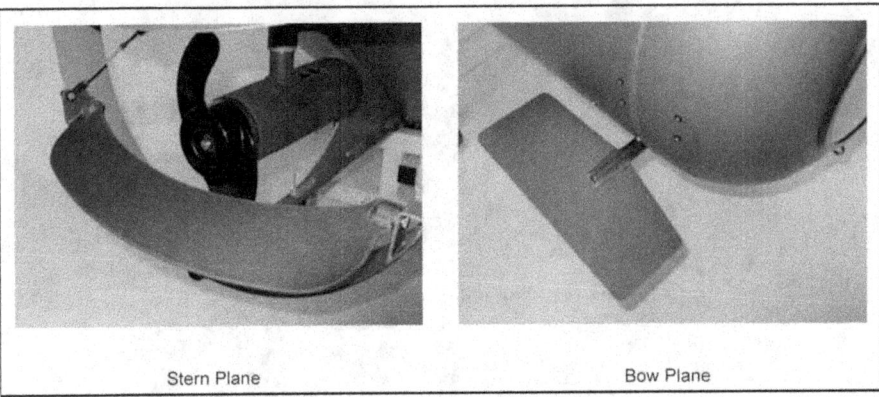

Figure F-16. Stern plane and bow plane

DEPTH AND STEERING CONTROL YOKE

F-21. The plastic control yoke is lightweight and easily used by the swimmer. The control yoke is connected to both the active stern plane (to control depth) and the thruster's tiller arm push pull cable (to control port and starboard steering motion) (Figure F-17, page F-12). This control yoke is securely attached to the transverse torque tube and the swimmer directly controls the rate of diving/surfacing by rotating the yoke forward or aft. The yoke is also attached to one end of a push/pull cable—the other end of the push/pull cable connects to the thruster tiller arm. This aluminum tiller arm is positioned to allow the thruster to turn 35 degrees port and starboard.

GAUGE CONSOLE

F-22. The compass, depth gauge, and light stick holder are part of the gauge console assembly (Figure F-18, page F-12). The diver can mount this assembly to the upper surface of the craft's nose or to the underside of the front view port. The universal mounting brackets (Figure F-19, page F-13) or system provide a means to integrate other equipment. The diver can remove the gauge console from the vehicle for calibration of the depth gauge. The purpose of each console part is as follows:
- A *magnetic compass* is standard equipment on the DPD.
- A *depth gauge* is standard equipment on the DPD. This depth gauge has a maximum depth range of 60 feet and includes an illuminating face when a light stick is used.
- A *light stick holder* provides illumination to both the compass and depth gauge during submerged operations. To insert a new light stick into the holder, the diver removes the end cap and slides it in. Multiple light sticks can be inserted to provide additional illumination. The diver then replaces the end cap. This holder uses the smaller 1/4-inch diameter light sticks.

Appendix F

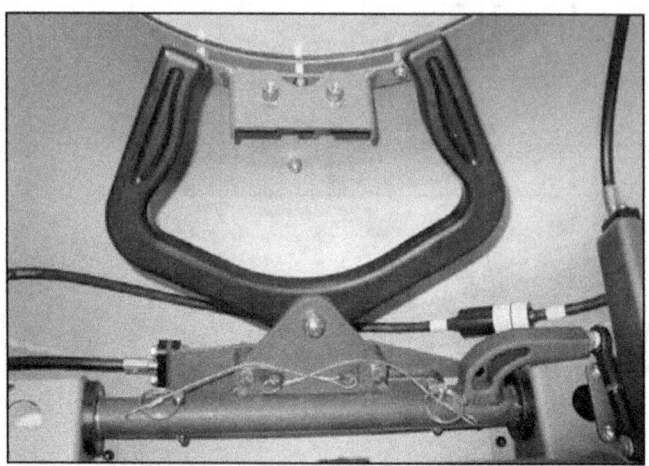

Figure F-17. Depth and steering control yoke

Figure F-18. Gauge console

Diver Propulsion Device

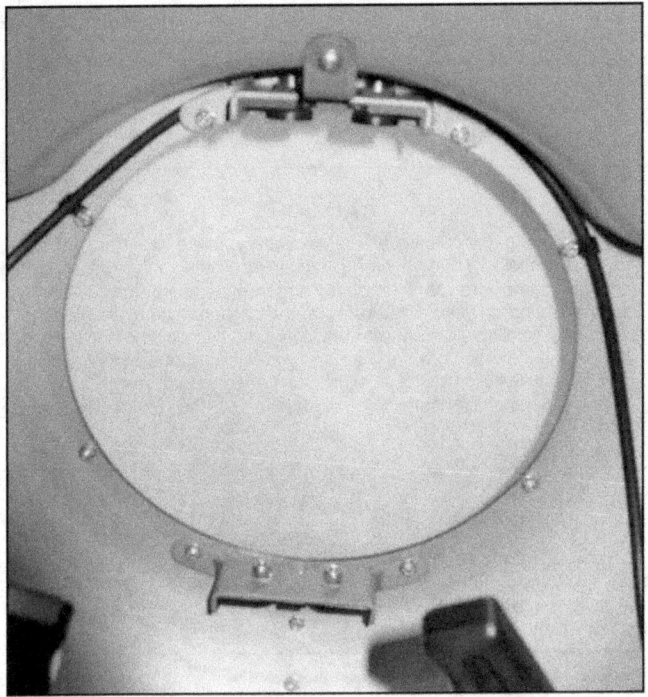

Figure F-19. Universal mounting brackets

PERFORMANCE

F-23. If the vehicle is trimmed to be neutrally buoyant and level, and the operator is closely positioned in the DPD, there will be reduced drag and increased speed/range. Figure F-20 lists the DPD performance specifications.

> - Cruise: Driver Only–2.3 kts; Driver and Passenger–2.0 kts
> - Full Speed: 3.0 kts; 2.7 kts
> - Range @ Cruise Speed: 6.6 nm; 5.4 nm
> - (Single Battery) @ Full Speed: 5.2 nm, 3.8 nm
> - Range @ Cruise Speed: 13.2 nm, 10.8 nm
> - (Dual Battery) @ Full Speed: 10.4 nm, 7.6 nm
>
> *Note.* The speed and range values listed are approximate and may vary due to several factors: currents, cargo loading, neutrality, and operator skill level.

Figure F-20. Performance specifications

F-24. To use the DPD to its full capabilities, it is critical that users of the DPD understand the importance of reducing the hydrodynamic drag generated by the position of both the driver and passenger. To increase the speed and range of the DPD, it is critical that the driver position himself as far down into the fore body as possible, and that the passenger tuck in close to the driver with the objective of reducing the amount of projected surface area beyond the projected cross section area of the DPD's nose cone. Additionally, the

Appendix F

importance of correctly ballasting oneself and all cargo placed in the DPD is critical to obtaining top performance. To keep an incorrectly ballasted DPD on a straight and level course, the control surfaces (bow planes, stern plane, and/or thruster) must be used to compensate for ascending, descending, or turning tendencies of the poorly ballasted vehicle. This constant correcting required to keep the DPD on a straight and level course lowers the top speed and reduces the maximum range. The energy required to keep the DPD straight and level is wasted and cannot be used to contribute to forward speed.

> **CAUTION**
>
> Before making any connections to the battery, make sure the throttle is in the off position. Before diving, confirm that the driver, passenger, and all cargo are neutrally buoyant using appropriate buoyancy compensating devices. Failure to have neutral buoyancy could result in an uncontrollable decent causing injury, death, or loss of the vehicle. All electrical connectors must be sprayed with silicone before making any physical connections. Failure to do so could result in water intrusion into the connection or failure of the connector when unplugged.

PREOPERATIONAL PROCEDURES

F-25. The battery is a critical pressure vessel. Before diving, the battery should be inspected for physical damage. If any physical damage is found on either the enclosure or the electrical connectors, the battery must not be dived. Physical damage could result in a catastrophic implosion of the pressure vessel or the intrusion of water into the battery.

Note. Confirm that the deck and nose foam are securely mounted to the DPD. Loss of either of these items while diving will result in an uncontrollable dive.

F-26. Before operating the DPD, perform the following:
- Install the freshwater foam wedge (Figure F-21, page F-15) before launching, if the DPD is to be used in fresh or brackish water. (Refer to DPD User's Manual.)
- Ensure the throttle key lanyard is installed on the throttle.
- The battery should be inspected for physical damage. If any physical damage is found on either the enclosure or the electrical connectors the battery must not be dived. Physical damage could result in a catastrophic implosion of the pressure vessel or the intrusion of water into battery.
- If not already installed, install battery box in DPD by carefully lowering it by the two straps attached to the cover. Be careful the battery is not placed down upon the loose cables and connector as this could cause damage to the unit. Using the four bolts and flat washers attach the battery to the bulkhead on the front flange and one bolt on the rear tab. Confirm that the battery box is properly secured.
- If not already connected, connect all cables and connector dummy plugs. Apply silicone spray to all connectors. Insert each cable end into its corresponding battery box connector using the color-coded markings. Check that the cables are connected properly and that they are all secure.
- Burp entrapped air from the thruster cable connectors, extended range cable connectors, and E-link/thruster dummy plugs as required. Burp air by sliding back the locking ring and apply pressure around the connector and squeeze toward the battery box forcing trapped air out of the connector. Attach the locking sleeve.
- Confirm the batteries are charged to the appropriate level using the Battery Status Indicator.

Diver Propulsion Device

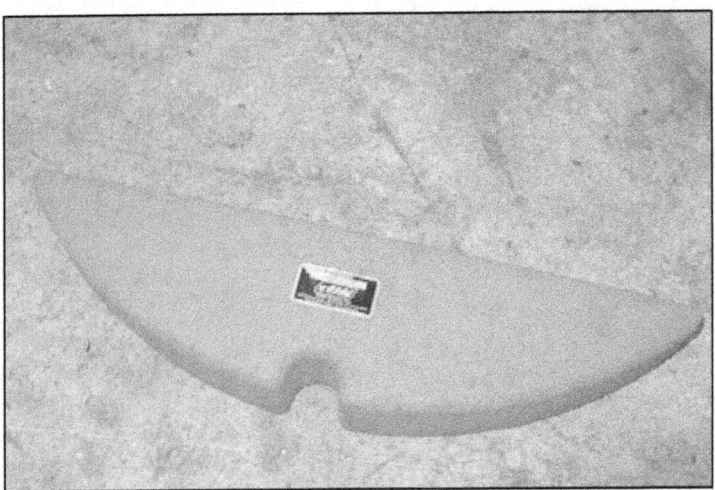

Figure F-21. Freshwater foam wedge

CAUTION

Do not operate the thruster out of the water for more than 3 seconds. Extended operation of the thruster out of the water will damage the shaft seals and cause the motor to flood when submerged.

- Check for proper operation of the throttle. Be sure the propeller path is clear and momentarily turn on the thruster to be sure that it turns.
- Disconnect the throttle key and retest thruster, the thruster should not turn. Reinstall the throttle key.
- Secure the deck with the two deck-mounting screw assemblies. Confirm the deck is properly secured. Failure to properly attach the deck could result in the loss of the vehicle.
- Confirm that the "T" bars are attached to the DPD and are properly adjusted for the divers.
- Check that there is unrestricted movement of the control yoke port or starboard and fore or aft (to check thruster and stern plane motion).
- Visually inspect all mechanical linkages for signs of damage or wear.
- Spray with a water-displacing lubricant, such as silicone spray.
- Ensure the propeller edges are smooth and free of debris. If this edge is rough or nicked, refer to the maintenance section of the propeller.
- Check to see that the prop nut is still secure using a 9/16-inch socket. The nut should be tightened 1/4-turn past snug.

EXTENDED RANGE BATTERY INSTALLATION

F-27. To install the extended range battery, the diver needs the following components (Figure F-22, page F-16):
- Primary Li-Ion Battery with Key Plug (p/n 4510-118).
- Slave Li-Ion Battery with Key Plug (p/n 4510-118).
- Range Kit (p/n 4510-920).
- Range Cable Assembly (p/n 4524-1-606).
- Throttle Shorting Dummy Plug (p/n 4530-9-139).

Appendix F

- Battery Status Indicator Dummy Plug (p/n 4530-9-140).
- Battery Strap (p/n 4528-1-160).
- Battery Foam Cradle (p/n 4527-1-235).
- Freshwater Foam Compensator (if used in freshwater) (p/n 4527-1-40).

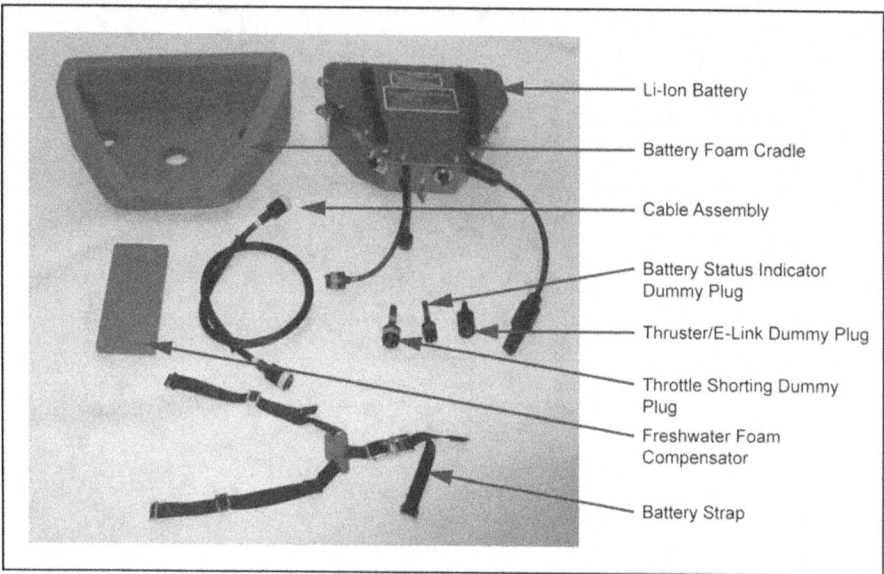

Figure F-22. Extended range battery components

F-28. To install the extended range battery kit, use the following procedures:
- Remove the deck.
- Remove the dummy plug from the primary E-link port and install it onto the extended range battery's thruster port connector (Figure F-23). The primary battery is the battery mounted to the DPD's bulkhead. This plug requires "burping" of air.

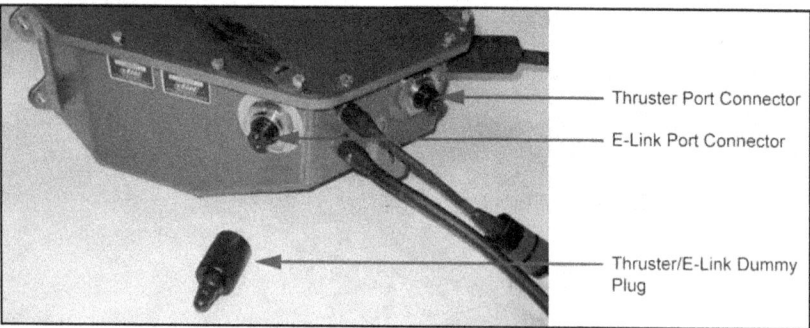

Figure F-23. Installation of dummy plug to thruster port connector

Diver Propulsion Device

- Plug the extended range cable assembly into the E-link port on the primary battery and route the cable through the provided hole in the DPD's bulkhead (Figure F-24). This connector requires "burping" so not to trap air.

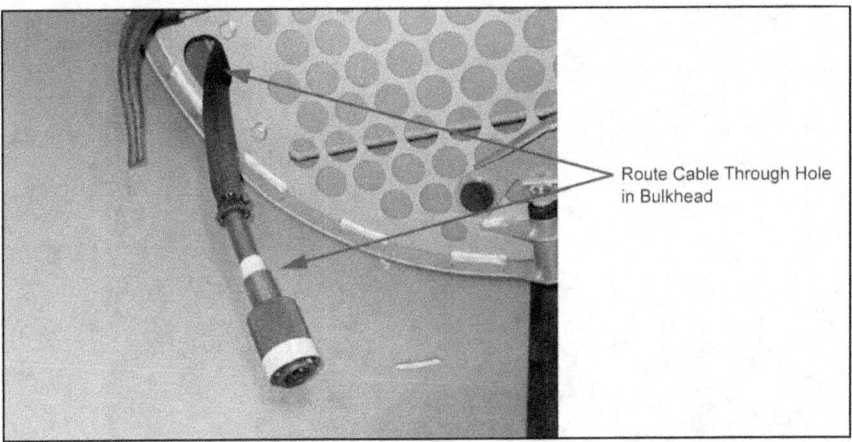

Figure F-24. Cable plugs into the E-link port and routes through bulkhead

- Install the remaining dummy plugs into the slave battery if not already installed using the color-coding.
- Place the foam cradle (and freshwater foam compensator as required) in the hull (Figure F-25). If the freshwater foam compensator is required, place it in the boat first and place the cradle on top of it. Be sure the freshwater foam compensator is located in the cutout on the underside of the cradle.

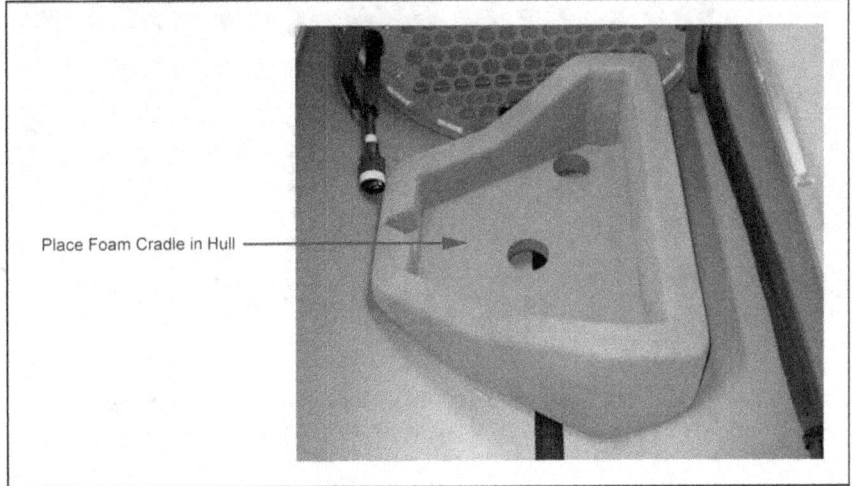

Figure F-25. Foam cradle sits in the hull

Appendix F

- Place the extended range battery into the foam cradle.
- Plug the extended range cable assembly into the series port on the slave battery. This connector requires "burping" of air.
- Install the extended range battery strap. Feed each strap catch through the slots in the DPD fore body and pull to get the catch to lie flush to the hull. The cradle can be shifted around to gain better access to the slots but should be repositioned into its correct position after all the catches are installed. Adjust the strap length as required on all three legs and snap the over-the-center buckle into place (Figure F-26). The battery/foam cradle should be tightly held to the hull. The battery in the cradle should not be able to shift when the buckle is closed. If there is excessive movement, retighten the webbing and lock the buckle down again.
- Reinstall the deck. Tighten the two deck mounting screws and confirm that the deck is securely attached.

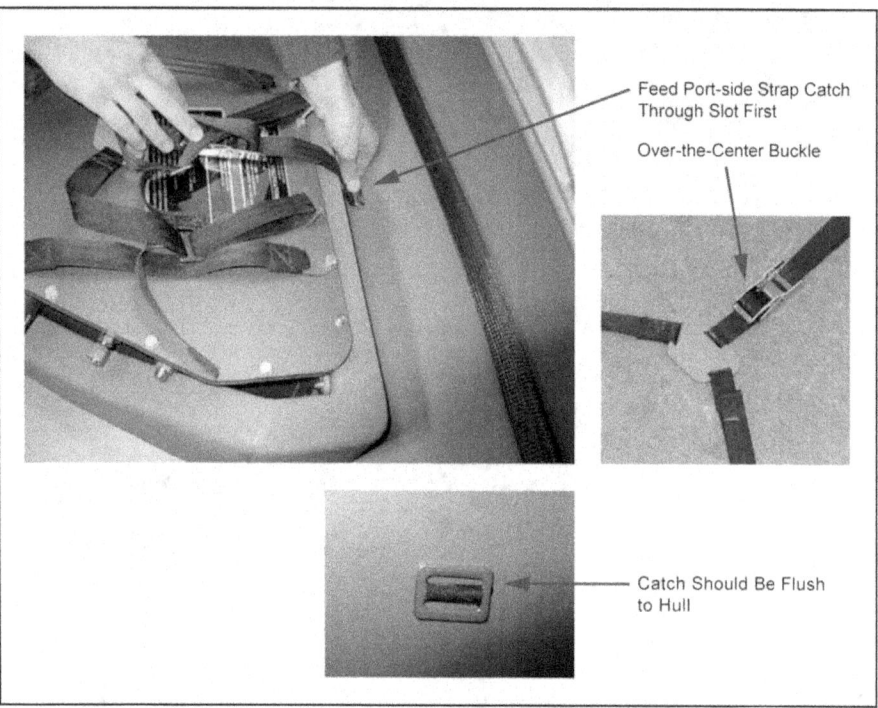

Figure F-26. Battery is secured

PREOPERATIONAL CHECKLIST

F-29. Personnel should complete the following checklist prior to each dive (Figure F-27, page F-19). This checklist should be copied and used each time the DPD is dived, and should also be entered into the DPD's logbook.

Diver Propulsion Device

☐	The shipping container's cover was reinstalled and all latched secured.
☐	The DPD was checked for significant damage to the hull that would impede operation. If damage is found do not dive the DPD.
☐	The fiberglass components were checked for damage exposing the foam.
☐	The hull number on the deck matches the DPD hull number.
☐	The control linkages were sprayed with silicone spray.
☐	Confirm the freshwater wedge is installed if required.
☐	The battery box was inspected for damage. If damage found, replace unit.
☐	The battery is securely installed in the hull.
☐	The battery box has all dummy plugs installed.
☐	The battery box and throttle assembly have vacuum cap installed.
☐	Confirm that the required connectors are burped of entrapped air.
☐	Confirm all connector locking sleeves are properly secured.
☐	The nose foam and deck are securely attached.
☐	The yoke control was inspected for proper operation.
☐	The throttle key lanyard is installed on the throttle.
☐	The gauge console is installed in appropriate location.
☐	The BSI is indicating a full or acceptable battery charge state.
☐	Confirm that the tow bars are at or near the appropriate length and secure to the DPD.
☐	Confirm the two quick-release pins and center detent pin securing the fore and aft bodies are properly engaged.
☐	The thruster and stern plane rotate through the proper range of motion.
☐	The propeller is free of cracks and major nicks.
☐	The propeller is securely fastened.
☐	The thruster was tested momentarily for proper operation.
☐	Confirm that the cargo to be carried is neutrally buoyant using incompressible means of floatation.
☐	Confirm that the cargo is securely stowed under the cargo net.
☐	Confirm bow planes are undamaged and are securely installed and pinned.
☐	The extended range battery strap is secure and the battery is secure.

Performed By: _____ Date: _____

Figure F-27. Preoperational checklist

OPERATIONAL PROCEDURES

F-30. The DPD should only be used by qualified personnel who have been trained in its proper safety, handling, and operational standards. The launching procedure better explains the safety and handling issues associated with the DPD.

LAUNCHING PROCEDURES

F-31. The DPD is extended to its operational length by releasing the two quick-release pins on the top rail and extending the DPD. When the DPD is fully extended, the two release pins should be reinserted and the center pin release handle checked for engagement.

F-32. The DPD is neutrally buoyant and should not be left unattended once launched. For freshwater use, the freshwater foam block should be placed below the battery box to obtain neutral buoyancy.

Appendix F

Two Divers on the DPD—Control

F-33. The DPD is controllable in pitch, yaw, and roll as well as forward speed. Pitch and yaw is controllable through the steering yoke and the rolling of the craft can be achieved by the shifting of the diver or passenger. The forward speed is controlled through using the thruster via the throttle assembly. The types of control are as follows:

- *Throttle Control.* The thruster offers a choice of infinitely variable speeds as provided by the controller using a pulse width modulation system. The speed control should be rotated clockwise to increase forward speed. Detents are provided at the stop (off), cruise speed, and full speed positions. The throttle is equipped with a throttle key lanyard assembly that should be securely attached to the driver as a safeguard against the loss of the vehicle. If the driver departs the DPD accidentally, the lanyard will cut off the DPD throttle.
- *Direction Control.* The DPD control yoke provides the operator with a comfortable, efficient way to control the direction of the craft. The DPD turns to starboard as the yoke is angled to the right and turns to port as the yoke is angled to the left. The DPD's thruster turns at a maximum angle of 35 degrees in either the port or starboard direction allowing rapid turns in either direction.
- *Depth Control.* The DPD control yoke also provides the operator with a comfortable, efficient way to control the depth of the craft. The DPD dives as the yoke is pushed forward and surfaces as the yoke is pulled backward.

Diving the DPD

F-34. Passengers should securely stow all gear in the DPD to maximize speed. In addition, the operator should be sure that his head is down in the hydrodynamic shadow of the bow. The passenger should position himself behind the operator and keep his head in the hydrodynamic shadow of the operator and DPD. The DPD will travel fastest and be the easiest to maneuver if it is kept from listing and has a slight bow-down attitude (5 to 10 degrees), is neutrally buoyant, and the divers stay down in the hydrodynamic shadow of the DPD's nose cone. The following suggestions are offered for optimal DPD operation:

- Equipment should be tested and set to neutral buoyancy prior to being secured in the DPD.
- Divers should be neutrally buoyant at the surface, and when in a horizontal position, the diver should tend to stay horizontal.
- Cargo should be stowed and secured as low as possible so that the diver can get down into the DPD.
- After securing equipment in the DPD, check to ensure that the DPD is not listing and has a slight bow-down attitude (5 to 10 degrees). If the DPD is in a bow-down or bow-up attitude or listing to one side or the other, adjust DPD to level by moving equipment or adding weight. Weight should be added in the smallest possible increments. If the DPD lists while underway, divers can shift slightly from one side to the other when underway to compensate. It is critical to the DPD's performance that the DPD must remain neutrally buoyant after weight has been added.
- Operator should get down as low into the DPD as possible to reduce resistance and therefore increase speed. When operating with two divers, the passenger should be as far aft as possible and should attempt to remain in the diver's shadow. This action will significantly reduce drag.
- Operator should run at cruise speed to maximize range.
- If contact is made with the bottom or any other hard object, immediately check the condition of the bow planes and straighten as necessary.
- Any time the DPD is cached or left on the bottom, it should be safely secured.

Neutral Buoyancy

F-35. It is vital for the safe and effective use of the DPD to confirm that the driver, passenger, and cargo are all neutrally buoyant or correctly ballasted. It is critical to achieve neutral buoyancy using incompressible volumes that will float when placed in water. An item is considered to have an incompressible volume when its outside shape does not change as the water pressure around it increases (as the diver goes deeper). In more detailed wording, the displacement of water by an incompressible volume does not change with increased water pressure or depth. This factor is critical because, if a compressible volume is used to achieve neutral buoyancy, it will only be effective at the depth where it was initially set to neutral. If a compressible volume is dived deeper, the amount of floatation it provides is reduced causing the diver or vehicle to descend. This decrease could lead to uncontrollable descents causing death, injury, or loss of the vehicle. As a minimum, achieving neutral buoyancy using compressible volumes will make the depth control of the DPD difficult. If the DPD's control surfaces (bow and stern planes) are used to overcome the descending or ascending tendencies of a poorly ballasted vehicle, the overall speed and range of the DPD will be reduced. It should be noted that the control surfaces of the DPD may become ineffective when a poorly ballasted DPD is dived. The DPD will become "heavier" as the unit descends into an unrecoverable dive. An example of an incompressible volume for proper ballasting of the DPD's cargo is STIDD's neutral buoyancy unit (NBU). These 6" x 5" x 1" blocks are constructed of a special material that has a low weight per volume but high compression strength (Figure F-28). The NBU is sized to compensate for 1 pound of submerged weight (in SW) of cargo. To determine the number of NBUs required, the operator weighs the cargo with a scale while it is submerged in salt water. The weight equals the number of NBUs required. The NBU is designed with a scoring line so that it can be broken in half to allow for finer buoyancy adjustment. The submerged weight amount should always be rounded to the next highest 1/2-pound increment. This method will prevent the entire DPD and its cargo from being negatively buoyant. Other buoyancy compensating means can be used to ballast the vehicle. It is critical that the components used to ballast the cargo do not compress when dived. Air bladders similar to a diver's buoyancy compensator (BC) is an example of a compressible volume. If an air bladder is used to compensate for the submerged weight of cargo, the bladder will need constant pressure adjustment to maintain neutral buoyancy. As the vehicle is dived deeper, the bladder will require additional pressure to be added to it to maintain its required buoyancy and, when the vehicle ascends toward the surface, air will need to be vented from the bladder for it to maintain its required buoyancy.

Figure F-28. Neutral buoyancy unit

DPD Recovery

F-36. The DPD is equipped with drain holes to facilitate the dewatering process when removing the vehicle from the water. Standard procedure to recover the DPD is to pull the DPD bow first into the support craft allowing the water to completely drain. The DPD should not be hoisted out of the water by mechanical means. The weight of the entrapped water will overload the hoisting handles.

POSTOPERATIONAL PROCEDURES

F-37. After operating the DPD, the operator should perform the following:
- Remove the throttle key from throttle and clip lanyard to DPD.
- Remove bow planes and rinse with freshwater.
- Clean the DPD of weeds and other debris.

Appendix F

- Remove the deck and thoroughly rinse the entire DPD with freshwater.
- Rinse the mechanical linkages with freshwater, dry, and spray with silicone spray water displacing lubricant.
- Remove the gauge console and rinse with freshwater and dry.
- Rinse the gauge console mounting points in the DPD with freshwater, dry, and spray with silicone spray.
- Inspect the DPD for physical damage. Repair/replace components as appropriate.
- Inspect the battery for physical damage. If any physical damage is found, the battery must not be dived. Replace the unit.
- If electrical connector pins were exposed to salt water, rinse with freshwater, dry, and spray with silicone spray.
- Inspect propeller for damage, repair/replace as appropriate.
- Check that the prop nut is still secure using a 9/16-inch socket. The nut should be tightened 1/4-turn past snug.
- Place battery on charge.
- Be sure the DPD is dry and place into shipping container.
- Make sure the shipping container's cover was reinstalled and all latches secured.
- Check the DPD for significant damage to the hull that would impede operation. If damage is found, do not dive the DPD.
- Check the fiberglass components for damage that would expose the foam.
- Make sure the hull number on the deck matches the DPD hull number.
- Spray the control linkages with silicone spray.
- Confirm the freshwater wedge is installed if required.
- Inspect the battery box for any damage. If damage found, replace unit.
- Securely install the battery in the hull.
- Make sure the battery box has all dummy plugs installed.
- Ensure the battery box and throttle assembly have vacuum cap installed.
- Confirm that the required connectors are burped of entrapped air.
- Confirm all connector locking sleeves are properly secured.
- Securely attach the nose foam and deck.
- Inspect the yoke control for proper operation.
- Install the throttle key lanyard on the throttle.
- Install the gauge console in the appropriate location.
- Make sure the BSI indicates a full or acceptable battery charge state.
- Confirm that the tow bars are at or near the appropriate length and secure to the DPD.
- Confirm the two quick-release pins and center detent pin securing the fore and aft bodies are properly engaged.
- Make sure the thruster and stern plane rotate through the proper range of motion.
- Check the propeller for cracks and major nicks.
- Securely fasten the propeller.
- Test the thruster momentarily for proper operation.
- Confirm that the cargo to be carried is neutrally buoyant using incompressible means of floatation.
- Confirm that the cargo is securely stowed under the cargo net.
- Confirm bow planes are undamaged and are securely installed and pinned.
- Ensure the extended range battery strap and battery are secure.

Postoperational Checklist

F-38. Personnel should complete the following checklist after each dive (Figure F-29). This checklist should be copied and used each time the DPD is dived, and should also be entered into the DPD's logbook. Personnel may also have to determine why features are not working and try to repair them at the site. Figure F-30, page F-24, provides a troubleshooting guide that may be used to analyze functioning problems.

- ☐ Confirm that the throttle key is removed from the throttle and lanyard securely clipped to DPD.
- ☐ The bow planes are removed, checked for damage, rinsed with freshwater, dried and placed in the shipping container.
- ☐ The DPD was thoroughly rinsed with freshwater and dried.
- ☐ The gauge console was rinsed with freshwater and dried.
- ☐ The gauge console mounting points were sprayed with silicone spray and gauge console reinstalled.
- ☐ The mechanical linkages were rinsed with freshwater and sprayed with silicone spray.
- ☐ Exposed electrical connectors were rinsed with freshwater, dried, and sprayed with silicone spray.
- ☐ The DPD was checked for significant damage to the hull that would impede operation. If damage is found, repair/replace as required.
- ☐ The fiberglass components were checked for damage exposing the foam core. Repair/replace as required.
- ☐ The hull number on the deck matches the DPD hull number.
- ☐ The battery box was inspected for damage. If damage was found return battery for repair.
- ☐ The battery is securely installed in the hull.
- ☐ The battery box has all dummy plugs installed.
- ☐ The battery box and throttle assembly have vacuum cap installed.
- ☐ The thruster and stern plane rotate through the proper range of motion.
- ☐ The propeller is free of cracks and major nicks. Repair/replace as required.
- ☐ The propeller is securely fastened.
- ☐ The battery was placed on charge.
- ☐ The battery charge/discharge log was completed and filed in the maintenance log.
- ☐ The DPD was placed into the shipping container and all latches secured.

Performed By: _____ Date: _____

Figure F-29. Postoperational checklist

Note. Use only the DPD "muscles" battery charger to charge batteries or battery function will be damaged and warranty voided.

Battery-Charging Procedure

F-39. The operator should charge the Li-Ion battery using the following procedures:
- Remove the DPD deck by unscrewing the deck screws.
- Remove the key plug from the end of the charging cable.
- Confirm that the charger power switch is in the "OFF" position.
- Connect the charger to the AC power source.
- Confirm that the interconnecting cable between the primary and secondary batteries is disconnected and that all color-coded dummy plugs are inserted.
- Plug the charger's cable to the battery-charging whip on the battery.
- Turn the charger power switch to the "ON" position.
- The amber "CHARGE" LED should be illuminated when the charge is in process.

Appendix F

- Check battery status indicator. If any LEDs are lit, the battery has enough power to operate the thruster.
- Check all connections to the battery box.
- Check that the throttle key is installed under the throttle lever.
- Make sure terminals are clean and corrosion free. If corrosion is found, clean, rinse, dry, and apply silicone spray.
- Remove the throttle key and spin the prop by hand to confirm that the prop turns freely.
- Thruster fails to run or lacks power. Reset the battery by removing the throttle key from under the throttle handle. Wait 10 seconds and reinstall.
- Thruster loses power after a short running time. Check battery charge; if low, restore to full charge.
- Thruster runs but no thrust. Check for proper prop installation; replace prop or tighten as appropriate. Minor damage: sand or file to restore a smooth edge. Major damage: replace propeller and drive pin. Remove thruster from DPD. Support prop shaft across a bench vise and tap drive pin through.
- Bent pin; cut pin flush at prop shaft and tap drive pin through.
- Inspect steering cable and linkages for damage. Replace if damaged.
- Inspect steering cable for damage. If no damage is found, apply lubricant to both ends.
- Thruster will not turn or hard to turn. Confirm tiller arm is tight on thruster shaft. Tighten if loose.
- Inspect stern plane cable and linkages for damage. Replace if damaged.
- Stern plane will not move or is hard to move. Inspect stern plane cable for damage. If no damage is found, apply lubricant to both ends. Retest.
- Turn off charger. Unplug charger cable from battery and charger. Reconnect and retry.
- Connect charger to different battery and retry.
- Charger red fault light is illuminated. Connect battery to different charger and retry.
- Momentarily move the throttle position to "wake up" the BSI.
- Confirm key plug is inserted in the battery's charging port.
- Confirm BSI cable is correctly connected to the battery.
- BSI LEDs are not illuminated. Check that the throttle key is installed under the throttle lever. BSI may need recalibrating. Run battery completely down to shut off condition to reset BSI calibration.
- BSI not illuminating all green LEDs when fully charged. If problem persists, replace BSI.
- DPD is not neutrally buoyant. Confirm cargo is neutrally buoyant at all water depths.
- DPD was neutrally buoyant on the surface but sinks when dived deeper. Cargo was incorrectly ballasted and/or a compressible volume was used to ballast the vehicle. Retrim the DPD with an incompressible volume.
- DPD needs significant vertical control input to travel straight and level. Driver/passenger not keeping tight to the vehicle and generating excessive drag. Tuck tighter into the vehicle. Cargo was incorrectly ballasted and/or a compressible causing control correction and induced drag. Retrim the DPD with an incompressible volume.
- DPD does not travel at top speed or range is reduced. Driver/passenger not keeping tight to the vehicle and generating excessive drag. Tuck tighter into the vehicle.

Figure F-30. Troubleshooting guide

- The amber LED will extinguish and the green LED will illuminate when the charge is complete. (Typical recharge time for a fully discharged battery is 8 hours.)
- If the red LED is illuminated a fault is indicated. Turn off the power and ensure that the cable between the charger and the battery is properly connected. Turn the power back on; if the fault condition still exists, consult the troubleshooting section of this manual.
- Disconnect the charging cable from the charging port and reinsert the key plug.
- Spray connectors with silicone before reconnecting.

- Reinstall the deck. Tighten both deck screws.
- Complete the battery charge/discharge log. This log sheet needs to be completed after each charge/discharge.

Note. When batteries are to be stored for a prolonged period of time the battery should be stored fully charged. The vehicle will not function if the charger key dummy plug is not connected to the charging port on the battery box. Ensure the cable bundle does not get pinched between the aft body and the fore body.

This page intentionally left blank.

Glossary

SECTION I – ACRONYMS AND ABBREVIATIONS

3-M	maintenance material management
ACU	Army combat uniform
ADVON	advanced echelon
AGE	air gas embolism
AIROPS	air operations
ALICE	all-purpose, lightweight, individual carrying equipment
AM	amplitude modulation
ANU	Authorized for Navy Use
AO	area of operations
AOIC	assistant officer in charge
AOR	area of responsibility
AR	Army regulation
ARNG	Army National Guard
ASL	above sea level
BA	buoyant ascent
BC	buoyancy compensator
BCM	buoyancy compensator military
BDP	beach defense posture
BIBS	built-in breathing system
BLS	beach landing site
BOI	basic ocean items
BUD/S	Basic Underwater Demolition/SEAL
C	Centigrade or Celsius
C2	command and control
CARP	computed air release point
C/C	closed circuit
cc	cubic centimeters
CDS	combat dive supervisor
CNS	central nervous system
COA	course of action
COG	course over ground
COMM	communications
COMNAVSURFLANT	Commander, Naval Surface Force, Atlantic
COMNAVSURFPAC	Commander, Naval Surface Force, Pacific
conn	conning
COPPs	Combined Operation Assault Pilotage Parties

CPE	circle of probable error
CRRC	combat rubber raiding craft
CSAR	combat search and rescue
CTS	course to steer
CWO	chief warrant officer
DA	direct action
DCS	decompression sickness
DDS	dry-deck shelter
DGPS	Differential Global Positioning System
DLSMF	Divers Life Support Maintenance Facility
DMAHTC	Defense Mapping Agency Hydrographic and Topographic Center
DMO	diving medical officer
DMT	diving medical technician
DOD	Department of Defense
DOE	detachment ocean equipment
DOT	Department of Transportation
DPD	diver propulsion device
DPV	diver propulsion vehicle
DR	dead reckoning
DRS	diver recall system
DSN	Defense Switched Network
DTG	date-time group
DZ	drop zone
E	East
E&E	evasion and escape
EGL	equipment guide list
EMCON	emission control
EOD	explosive ordnance disposal
EP	estimated position
ERDS	external raft delivery system
ETA	estimated time of arrival
ETD	estimated time of departure
F	Fahrenheit
FAIR	failure analysis or inadequacy report
FID	foreign internal defense
FLIR	forward-looking infrared
FM	field manual
FMP	full mission profile
fpm	feet per minute
FRIES	fast-rope insertion and extraction system
FSA	free swimming ascent

FSW	feet of seawater
ft	feet
ft3	free triiodothyronine
FTX	field training exercise
GPH	gallon per hour
GPS	global positioning system
GSA	General Services Administration
HA	humanitarian assistance
HALO	high-altitude low-opening
HLZ	helicopter landing zone
HN	host nation
hp	horsepower
HPNS	high pressure nervous syndrome
HRZ	helicopter recovery zone
IALA	International Association of Lighthouses Authorities
IBS	inflatable boat, small
I/C	interconnecting
ICC	Interstate Commerce Committee
IEM	inactive equipment maintenance
IFF	identification friend or foe
IP	insert point
IR	infrared
ITR	intended track
JA/ATT	joint airborne/air transportability training
JCS	Joint Chiefs of Staff
JIC	Joint Intelligence Center
JSOA	joint special operations area
JSOTF	joint special operations task force
kg	kilogram
kt	knot
lb	pounds
LCE	load-carrying equipment
LED	light emitting diode
LOD	line of departure
LOC	line of communications
LOEP	list of effective pages
LO/LI	lockout/lockin
LOP	line of position
LP	launch point
LPU	life preserver unit
LZ	landing zone

MAC	maintenance allocation chart
MAROPS	maritime operations
MARSOF	Marine Corps Special Operations Forces
mb	millibar
MCDC	Marine Combatant Diver Course
MCRP	Marine Corps reference publication
MDMP	military decision-making process
MDS	maintenance data system
MEDEVAC	medical evacuation
MEK	methyl ethyl ketone
METL	mission-essential task list
METT-TC	mission, enemy, terrain and weather, troops and support available—time available and civil considerations
MIP	maintenance index page
MOA	memorandum of agreement
mph	miles per hour
MRC	maintenance requirements card
MRR	minimum-risk route
MSLO	mass swimmer lockout
N-2	Director of Naval Intelligence; Navy component intelligence staff officer
NATO	North Atlantic Treaty Organization
navaids	navigational aids
NAVFAC	Naval Facilities
NAVPLAN	navigation plan
NAVSAFECEN	naval safety center
NAVSEA	Naval Sea Systems command
NAVSEAINST	Naval Sea Instruction
NAVSPECWARCEN	Naval Special Warfare Center
NAVSPECWARCOM	Naval Special Warfare Command
NBU	neutral buoyancy unit
NCOIC	noncommissioned officer in charge
NEDU	Navy experimental diving unit
NGA	National Geospatial-Intelligence Agency
NIMA	National Imagery and Mapping Agency
nm	nautical mile
NOAA	National Oceanic and Atmospheric Administration
NSSN	newest class of submarine
NSW	Naval Special Warfare
NSWC	Naval Special Warfare Center
NTE	not to exceed

NVG	night vision goggle
OBM	outboard motor
O/C	open circuit
OIC	officer in charge
OJT	on-the-job training
OPLAN	operation plan
OPNAVINST	Chief of Naval Operations Instruction
OPORD	operation order
OPSEC	operations security
ORP	objective rallying point
OSS	Office of Strategic Services
OTB	over-the-beach
OTC	officer in tactical command
OTH	over the horizon
PA	public address
PBC	position, bearing, course
PE	practical exercise
PMS	planned maintenance system
ppm	parts per million
psi	pounds per square inch
psig	pounds per square inch gauge
PVC	polyvinyl chloride
PWM	pulse width modulator
QAO	quality assurance officer
QAP	quality assurance program
RDT&E	research, development, test, and evaluation
REC	reentry control
RGD	repetitive group designator
RHIB	rigid hull inflatable boat
RMBPD	Royal Marine Boom Patrol Detachment
RP	rendezvous point
rpm	revolutions per minute
RV	rendezvous (point)
S-2	battalion or brigade intelligence staff officer (Army; Marine Corps battalion or regiment)
SAS	Special Air Service
SBS	special boat section
SCPS	Small Craft Propulsion System
scuba	self-contained underwater breathing apparatus
SDV	SEAL delivery vehicle
SEAL	sea-air-land (team)
SF	Special Forces

SFOD	Special Forces operational detachment
SI	surface interval
SMG	speed made good
SO	special operations
SOA	speed of advance
SOCCE	special operations command and control element
SOCEP	Special Operations Combat Expendable Platform
SOE	special operations executive
SOF	special operations forces
SOG	speed over ground
SOI	signal operating instructions
SOP	standing operating procedure
SOT	special operations technician
SOTF	special operations task force
SR	special reconnaissance
SRP	swimmer release point
SRU	sea reconnaissance unit
SSB	single side band
SSBN	fleet ballistic missile submarine
SSGN	guided-missile submarine
STABO	short tactical airborne operation
STIDD	Systems Training Integration and Devices Directorate
SUBOPS	submarine operations
SUPDIVE	supervisor of diving
SURFREP	surf report
SUROBS	surf observation
SUROBSREP	surf observation report
SWO	staff weather officer
T	time
TAC	tactical
TFC	total fuel consumption
TLD	thermoluminescence dosimeter
TOP	time of passage
UBA	underwater breathing apparatus
UDT	underwater demolition team
U.S.	United States
USAF	United States Air Force
USAJFKSWCS	United States Army John F. Kennedy Special Warfare Center and School
USAR	United States Army Reserve
USASOC	United States Army Special Operations Command
USCG	United States Coast Guard

Glossary

USMC	United States Marine Corps
USSOCOM	United States Special Operations Command
UW	unconventional warfare
VBSS	visit, board, search, and seizure
VHF	very high frequency
VVDS	variable-volume dry suit
W	West
WASS	Wide Area Augmentation System
WOT	wide open throttle

SECTION II – TERMS

abort profile
The decompression schedule used to bring a diver safely to the surface when a dive must be aborted.

absolute temperature
Used in physics formulas relating to gas laws. To convert to absolute temperature; Fahrenheit = reading + 460 degrees; Centigrade = reading + 273 degrees.

ascents
Movement in the direction of reduced pressure (that is, up), whether simulated or due to actual elevation in water or air.

atmospheric pressure
The atmospheric force or weight of air exerted upon an area of matter. One atmosphere = 14.7 pounds per square inch, a constant at sea level, equal to a column of seawater 33 feet by one square inch.

beach landing site line
A line parallel to the shoreline through the beach landing site used in calculating the tidal current offset.

Beaufort Wind Scale
A wind scale correlating wind velocity (in knots) and anticipated wave height (in feet).

bends or caisson disease
An imprecise term denoting any form of decompression sickness.

bottom time
Total elapsed time from when the diver leaves the surface until he begins his ascent. Expressed in whole minutes.

buoyancy
The property of an object to float if it is lighter than the liquid it displaces.

chamber (hyperbaric)
A chamber designed to withstand high internal pressures; used in diving simulations and medical treatment.

closed-circuit scuba
A life support system or breathing apparatus in which the gas is recycled, carbon dioxide removed, and oxygen periodically added.

coxswain
The person responsible for the performance of the crew, boat handling, and equipment distribution. He operates the outboard motor, if used, and issues commands to the crew.

Glossary

decompression schedule
The specific decompression procedure for a given combination of depth and bottom time. It is indicated in feet and minutes.

Draeger LAR V
Closed-circuit breathing rig, UBA-approved for diving use by the U.S. Army and Navy.

lockout/lockin
Entering and exiting of a diver or combat swimmer while submarine is submerged.

offset navigation
The method used to compensate for currents not perpendicular to shore when planning for swimming missions.

piloting
The determination of the position and the direction of the movements of a vessel involving frequent or continuous reference to landmarks, navaids, and depth soundings.

residual nitrogen
Nitrogen gas that is still dissolved in the diver's tissues after surfacing.

variation
The difference between magnetic direction and true direction.

References

REQUIRED REFERENCES
These documents must be available to intended users of this publication.
None

RELATED PUBLICATIONS
These are the sources quoted or paraphrased in this publication.

Army Publications
AR 40-501, *Standards of Medical Fitness*, 14 December 2007
AR 385-10, *The Army Safety Program*, 3 September 2009
AR 611-75, *Management of Army Divers*, 20 August 2007
DA Form 1262 (Command Dive Log)
DA Form 1262-1 (Diving Site Worksheet)
DA Form 2028 (Recommended Changes to Publications and Blank Forms)
DA Form 7532-R (MK 25 MOD 2 Predive Checklist)
DA Form 7533-R (MK 25 MOD 2 UBA Postdive Checklist)
DD Form 2544 (Diving Log)
DD Form 2807-1 (Report of Medical History)
DD Form 2808 (Report of Medical Examination)
FM 3-21.220, *Static Line Parachuting Techniques and Tactics*, 23 September 2003
FM 4-20.142, *Airdrop of Supplies and Equipment: Rigging Loads for Special Operations*, 19 September 2007
FM 21-20, *Physical Fitness Training*, 30 September 1992
FM 55-501, *Marine Crewman's Handbook*, 1 December 1999
USASOC Reg 350-2, *Training Airborne Operations*, 22 January 2009
USASOC Reg 350-20, *USASOC Dive Program*, 12 September 2008
USASOC Reg 385-1, *Accident Prevention and Reporting*, 9 September 2008
USASOC Reg 525-1, *USASOC Reporting Structure (SOCOMREP)*, 27 August 2008
USSOCOM Manual 350-6, *Special Operations Forces Infiltration/Exfiltration Techniques*, 25 August 2004
USSOCOM Reg 350-4, *Maritime Training and Operations*, 30 October 2007

Joint Publications
JP 1-02, *Department of Defense Dictionary of Military and Associated Terms*, 12 April 2001
JP 3-0, *Joint Operations*, 17 September 2006

Navy Publications
COMDTINST 16794.51A, *USCG Auxiliary Boat Crew Manual*, 1 January 2007
COMDTINST M16114.5C, *USCG Boat Crew Seamanship Manual*, May 2009
NEDU Report 5-86, *Central Nervous System Oxygen Toxicity in Closed-Circuit Scuba Divers III*, F.K. Butler, January 1985
NEDU Report 7-85, *Closed-Circuit Oxygen Diving*, F.K. Butler, July 1985
Pub. No. 9, *The American Practical Navigator (Bowditch)*, 1995, Defense Mapping Agency Hydrographic/Topographic Center, Bethesda, Maryland, 2007

References

Report No. 5-86, *Central Nervous System Oxygen Toxicity in Closed-Circuit Scuba Divers III*, Naval Experimental Diving Unit, February 1986

Tidal Current Tables, Atlantic Coast of North America, National Oceanographic and Atmospheric Administration (NOAA), Department of Commerce, Washington, DC, April 2001

Tide Tables, East Coast of North and South America, National Oceanographic and Atmospheric Administration (NOAA), Department of Commerce, Washington, DC, April 2001

Index

A

air assets, 16-1 through 16-3
air decompression tables, 11-7, 11-8
air diving tables, 11-10
altitude diving, 11-22
ascent
 buoyant, 11-16
 free, 10-14
 rates, 11-16, 12-24
assault boats, 1-5, 1-6
atmospheric pressure, 3-2, 3-12, 11-22

B

barometric pressure, 2-3, 2-9
basic kayak (canoe) strokes, 8-4, 8-11
basic ocean items, 8-4, 8-5
beach landing site, 1-6, 1-7, 5-2, 8-16
beaching a boat in surf, 7-10, 7-11
Beaufort wind scale, 2-2
boat team member duties, 7-2, 7-8
bottom characteristics, 4-14, 4-15
bottom time, 11-7, 11-8
buddy lines, 10-13, 12-20
buoys, 4-19 through 4-21

C

caching, 1-12 through 1-14
canoe operational techniques
 capsize procedures, 6-12, 8-12
 launching and landing, 8-9 through 8-11
 paddling, 7-9, 8-11
 surf operations, 8-13, 8-14
charts, 2-10, 4-1 through 4-4, 4-11, 4-16
checkerboard search, 17-11 through 17-13

circle line search, 17-8
closed-circuit diving, 12-1, D-2
cold weather diving, 11-23 through 11-28
combat diving, 10-1
 medical technician, 1-15, 10-5, 10-15, B-5
 supervisor, 1-15, 10-1, 10-4, 17-6, B-3, D-1
compass, magnetic, 4-23 through 4-26, F-11
compass rose, 4-9, 4-10, 4-25, 5-21
conversion factors, 4-9, A-2
currents, 3-6 through 3-15

D

day beacons, 4-13
dead reckoning, 5-8, 5-13
debarkation, 1-4, 1-9, 13-3, 14-3, 14-7
decompression, 11-7 through 11-10, 11-22
depth, 3-3, 4-7, 5-5, 11-7, 12-20
dividers, 4-9, 5-5, 5-6, 6-13
drift, 5-2, 5-19, 5-24

E

eddy current, 3-9
embarkation, 9-3, 14-2
embolism, 11-5, 12-17
environmental factors, 2-1 through 2-10, 8-18
exfiltration, 1-6, 1-11

F

flood current, 3-6
fog, 2-23, 4-8, 8-19

I

infiltration, 1-3 through 1-6, 8-1, 8-2, 9-9, 13-1 through 13-8, 16-5
inflatable boat, 7-1, 7-11, 16-2
injuries, diving, 11-3

in-line search, 17-12, 17-13

K

kayaks, 8-1 through 8-10
 launching and landing, 8-9 through 8-11
 terminology, 8-4 and 8-5

L

landmarks, 4-14
latitude, 4-8, 4-9
launch point, 1-5, 1-9
light lists, 4-17
longitude, 4-7 through 4-9

M

malfunction procedures, 12-17, 12-24
meridians, 4-7 through 4-9
movement to BLS, 1-6, 1-7

N

nautical charts, 4-1 through 4-8
nautical slide rule, 4-6, 5-6
navigation aids, 4-19 through 4-21
nitrogen, 11-7 through 11-11

O

offset navigation, 3-13 through 3-16
over-the-horizon computations, 16-7 through 16-10
oxygen dive, 12-21, 12-22
oxygen exposure limits, 12-18 through 12-22
oxygen toxicity, 12-17 through 12-20

P

piloting, 5-8, 5-17, 5-24, 15-7 through 15-9
plotting symbols, 5-10

Index

R

recovery operations, 13-4, 16-6, 16-7
repetitive dive, 11-8 through 11-12
residual nitrogen, 11-8 through 11-11
running jackstay search, 17-9 through 17-11

S

scout swimmer, 1-6, 7-11, 9-4 through 9-11
scrubber, 12-7, 12-15, D-4
signaling devices, 9-7
signals, 4-23, 9-7, 9-8, 11-15, 17-8
surface swimming, 9-10

T

tended-line search, 17-6, 17-7
tidal bores, 2-15
tidal current, 2-15, 3-11 through 3-15, 4-17
tides, 3-1 through 3-3
towed search, 17-14, 17-15
triangle, 5-19

U

underwater navigation, 11-18
unit diving officer, 10-3

V

vectoring, 14-11

W

war chest, 8-5, 8-6
water entry, 12-22
waterborne operations, phases of, 1-3 through 1-7
waves, 2-14 through 2-17
wrecks, rocks, and reefs, 4-14, 4-15

Z

Zodiac, 7-11
   advantages, of, 7-14
   assembly and maintenance, 7-14, 7-15
   characteristics, 7-13

FM 3-05.212
MCRP 3-11.3A
30 September 2009

By Order of the Secretary of the Army:

GEORGE W. CASEY, JR.
General, United States Army
Chief of Staff

Official:

JOYCE E. MORROW
Administrative Assistant to the
Secretary of the Army
0925810

By Order of the Commandant of the Marine Corps:

GEORGE J. FLYNN
Lieutenant General, U.S. Marine Corps
Commandant
Combat Development and Integration

DISTRIBUTION:

Active Army, Army National Guard, and U. S. Army Reserve: To be distributed in accordance with initial distribution number 111114, requirements for FM 3-05.212.

United States Marine Corps: To be distributed in accordance with PCN 144 000171 00, requirements for MCRP 3-11.3A.

PCN: 144 000171 00   PIN: 081788-000